Global Industrial Impacts of Heavy Metal Pollution in Sub-Saharan Africa

Joan Nyika
University of Johannesburg, South Africa

Megersa Olumana Dinka
University of Johannesburg, South Africa

A volume in the Advances in Environmental Engineering and Green Technologies (AEEGT) Book Series

IGI Global
PUBLISHER of TIMELY KNOWLEDGE

Published in the United States of America by
IGI Global
Engineering Science Reference (an imprint of IGI Global)
701 E. Chocolate Avenue
Hershey PA, USA 17033
Tel: 717-533-8845
Fax: 717-533-8661
E-mail: cust@igi-global.com
Web site: http://www.igi-global.com

Library of Congress Cataloging-in-Publication Data

Names: Nyika, Joan, 1987- author. | Dinka, Megersa, 1976- author.
Title: Global industrial impacts of heavy metal pollution in Sub-Saharan Africa / edited by: Joan Nyika, Megersa Dinka.
Description: Hershey : Engineering Science Reference, 2023. | Includes bibliographical references and index. | Summary: "This book contributes, holistically, to these efforts by presenting the toxic effects of heavy metals to living things from a general viewpoint (Chapter 1) and relating their pollution to industrialization developments in the world (Chapter 2) and in Sub-Saharan Africa region (Chapter 3). The book further discusses the techniques for assaying the heavy metals in environmental samples and particularly, soils (Chapter 4) before discussing how to rate their pollution extent and impacts to the ecology using various indices (Chapter 5). Soil pollution by lead (Pb), chromium (Cr), arsenic (As), mercury (Hg), cadmium (Cd), nickel (Ni), copper (Cu) and zinc (Zn) in soils of SSA region is discussed (Chapter 6 - 13). The approaches to manage (Chapter 14) and measures to control and prevent (Chapter 15) heavy metal pollution in soils of the region are also discussed"-- Provided by publisher.
Identifiers: LCCN 2023011157 (print) | LCCN 2023011158 (ebook) | ISBN 9781668471166 (hardcover) | ISBN 9781668471173 (paperback) | ISBN 9781668471180 (ebook)
Subjects: LCSH: Soil pollution--Africa, Sub-Saharan. | Heavy metals--Environmental aspects--Africa, Sub-Saharan. | Heavy metals--Assaying. | Heavy metals--Toxicology.
Classification: LCC TD879.H4 N95 2023 (print) | LCC TD879.H4 (ebook) | DDC 628.550967--dc23/eng/20230308
LC record available at https://lccn.loc.gov/2023011157
LC ebook record available at https://lccn.loc.gov/2023011158

This book is published in the IGI Global book series Advances in Environmental Engineering and Green Technologies (AEEGT) (ISSN: 2326-9162; eISSN: 2326-9170)

British Cataloguing in Publication Data
A Cataloguing in Publication record for this book is available from the British Library.

All work contributed to this book is new, previously-unpublished material.
The views expressed in this book are those of the authors, but not necessarily of the publisher.

For electronic access to this publication, please contact: eresources@igi-global.com.

Advances in Environmental Engineering and Green Technologies (AEEGT) Book Series

Sang-Bing Tsai
Zhongshan Institute, University of Electronic Science and Technology of China, China & Wuyi University, China
Ming-Lang Tseng
Lunghwa University of Science and Technology, Taiwan
Yuchi Wang
University of Electronic Science and Technology of China Zhongshan Institute, China

ISSN:2326-9162
EISSN:2326-9170

MISSION

Growing awareness and an increased focus on environmental issues such as climate change, energy use, and loss of non-renewable resources have brought about a greater need for research that provides potential solutions to these problems. Research in environmental science and engineering continues to play a vital role in uncovering new opportunities for a "green" future.

The **Advances in Environmental Engineering and Green Technologies (AEEGT)** book series is a mouthpiece for research in all aspects of environmental science, earth science, and green initiatives. This series supports the ongoing research in this field through publishing books that discuss topics within environmental engineering or that deal with the interdisciplinary field of green technologies.

COVERAGE

- Sustainable Communities
- Alternative Power Sources
- Cleantech
- Water Supply and Treatment
- Air Quality
- Contaminated Site Remediation
- Renewable Energy
- Green Technology
- Policies Involving Green Technologies and Environmental Engineering
- Industrial Waste Management and Minimization

IGI Global is currently accepting manuscripts for publication within this series. To submit a proposal for a volume in this series, please contact our Acquisition Editors at Acquisitions@igi-global.com or visit: http://www.igi-global.com/publish/.

The Advances in Environmental Engineering and Green Technologies (AEEGT) Book Series (ISSN 2326-9162) is published by IGI Global, 701 E. Chocolate Avenue, Hershey, PA 17033-1240, USA, www.igi-global.com. This series is composed of titles available for purchase individually; each title is edited to be contextually exclusive from any other title within the series. For pricing and ordering information please visit http://www.igi-global.com/book-series/advances-environmental-engineering-green-technologies/73679. Postmaster: Send all address changes to above address.

Titles in this Series

For a list of additional titles in this series, please visit: http://www.igi-global.com/book-series/

Perspectives on Ecological Degradation and Technological Progress
Veli Yilanci (Faculty of Political Sciences, Canakkale Onsekiz Mart University, Turkey)
Engineering Science Reference • © 2023 • 342pp • H/C (ISBN: 9781668467275) • US $225.00

Nanopriming Approach to Sustainable Agriculture
Abhishek Singh (Sardar Vallabhbhai Patel Agriculture University, India) Vishnu D. Rajput (Academy of Biology and Biotechnology, Southern Federal University, Rostov-on-Don, Russia) Karen Ghazaryan (Biology Yerevan State University, Armenia) Santosh Kumar Gupta (National Institute of Plant Genome Research, India) and Tatiana Minkina (Academy of Biology and Biotechnology, Southern Federal University, Rostov-on-Don, Russia)
Engineering Science Reference • © 2023 • 300pp • H/C (ISBN: 9781668472323) • US $245.00

Contemporary Developments in Agricultural Cyber-Physical Systems
G.S. Karthick (Department of Software Systems, PSG College of Arts and Science, India)
Engineering Science Reference • © 2023 • 320pp • H/C (ISBN: 9781668478790) • US $240.00

Nematode-Plant Interactions and Controlling Infection
Waleed Fouad Abobatta (Horticulture Research Institute, Agriculture Research Center, Egypt) and Rehab Yasin Ghareeb (Plant Protection and Biomolecular diagnosis Dep., Arid Lands Cultivation Research Institute, City of Scientific Research and Technological Applications, Egypt)
Engineering Science Reference • © 2023 • 300pp • H/C (ISBN: 9781668480830) • US $240.00

Perspectives on Global Biodiversity Scenarios and Environmental Services in the 21st Century
Naveen Kumar Chourasia (Government College, Bichhua, India) and Kavita Chahal (Government College, Bichhua, Chhindwara, India)
Engineering Science Reference • © 2023 • 300pp • H/C (ISBN: 9781668490341) • US $250.00

IGI Global
PUBLISHER of TIMELY KNOWLEDGE

701 East Chocolate Avenue, Hershey, PA 17033, USA
Tel: 717-533-8845 x100 • Fax: 717-533-8661
E-Mail: cust@igi-global.com • www.igi-global.com

I dedicate this work to my loving husband, Madaraka and sons, Mwema and Nyika, thank you for the love.

Joan Nyika

Table of Contents

Preface

Industrialization along with its drivers including the need for economic growth, urbanization and population rise around the globe is inevitable, multifaceted, and dynamic. The relationship between industrial growth and environmental quality is indirectly proportional. With rapid growth in industrialization, various environmental issues and a series of pollution kinds emanate and spread at different temporal rates and spatial extents. The environmental issues are as a result of greater carbon emissions and contaminant release during production and manufacturing processes. Additionally, industrialization promotes pollution by releasing solid waste, wastewater (effluents), particulate matter and emissions containing toxic organic and inorganic pollutants such as heavy metals. The pollutants accumulate in various environmental compartments including air, soil and water beyond the natural degradation capacity. This is because most of these pollutants including heavy metals (HM), phenolic compounds, aromatic hydrocarbons and surfactants are recalcitrant and non-biodegradable. As such, they are then transferred to trophic levels and exert their noxious effects on living things and natural resources.

Heavy metals are naturally occurring elements in the earth's crust and have a variety of uses that are increasing in the industrial era. Subsequently, their increased use had resulted to a surge in the aquatic and terrestrial ecosystems as effluents and solid wastes are being produced from manufacturing activities. Anthropogenic activities including foundry, smelting, metal mining, textile making, jewelry production and chemical production are metal-based and result to environmental pollution due to the leaching of such pollutants from open dumpsites, landfill, sewer drains, animal manure, excretion, automobiles, runoffs and roadworks. The use of heavy metals in the manufacture of fertilizers, insecticides and pesticides for various agricultural activities is also a source of pollution. Metal evaporation, corrosion and volcanic activity among other natural processes occurring during weathering, soil pedogenesis and sedimentation also contribute to the environmental heavy metal load though human activities are the largest contributors.

The potential etiology resulting from environmental pollution by heavy metals and the subsequent degradation of land and water resources is one of the most pressing global issues of modern day. Overuse and misuse of natural resources in attempts to meet demands of an unsustainable development pattern worldwide is causing considerable health risks and public concerns to existence of living things. For this reason, it is imperative to understand the pollution potential of various heavy metals, their anthropic sources, ways to assay and rate them and control and preventative measures to take to avoid widespread pollution and its associated effects. Such undertakings are because of existent evidence showing that heavy metals not only cause environmental pollution but they are also atmospheric pollutants that are lethal to humans. Once the metal mix with different elements of the environment such air, soil and water, they become strongly toxic and such effects can spread to ecosystems.

This book contributes, holistically, to these efforts by presenting the toxic effects of heavy metals to living things from a general viewpoint and relating their pollution to industrialization developments in the world and in Sub-Saharan Africa (SSA) region. This is because industrialization is growing and the use of heavy metals now and in the future will become inevitable. Consequently, production of solid waste and effluents containing the toxic metals will also grow. In such a trend, devising solutions geared to eliminating and preventing the toxins from polluting the environment and in particular, soils will be a bold step towards sustainable development. This is because advances to more resilient and pollution-free soils will promote better life on land and water, cleaner water, better consumption, sustainable communities, reduced poverty, food security and proactive climate action, aspects that are engraved in the sustainable development goals (SDGs).

The book further discusses the techniques for assaying the heavy metals in environmental samples and particularly, soils before discussing how to rate heavy metal pollution extent and impacts to the ecology using various indices. The discussion is done with the precognition that techniques of assaying heavy metals in soils are varied, have some merits and demerits and determine accurate and representative rating of pollution in assayed samples. Soil pollution by lead (Pb), chromium (Cr), arsenic (As), mercury (Hg), cadmium (Cd), nickel (Ni), copper (Cu) and zinc (Zn) in soils of SSA region is discussed. The metals are known to be the most lethal in the environment, to natural resources and humans. The approaches to manage and measures to control and prevent heavy metal pollution in soils of the SSA and the globe at large, are also discussed.

Chapter 1 focuses on the chemistry of heavy metals to understand their capacity to induce toxicity in the environment and living things. Some of the metals discussed include As, Cd, Co, Cr, Cu, Hg, Ni, Pb and Zn. The factors that influence the bioavailability of metals for uptake and the mechanisms they use to inflict toxic effects are explored and demonstrated using various equations. The toxicity of heavy metals to living organisms is discussed under several categories including their ability to cause neurotoxicity, nephrotoxicity, carcinogenicity, hepatoxicity, immunological toxicity, cardiotoxicity, genotoxicity and reproductive and developmental toxicity.

In Chapter 2, a general overview of industrial growth globally and its effects on environmental pollution by heavy metals is discussed. Industrialization growth periods are classified into five based on technological advancements. Furthermore, the growth is related to environmental pollution by heavy metals in water, soil and air compartments as well as the trophic transfer of the pollutants to living organisms. In the chapter, the manufacturing, cement, chemical, coal, oil and petroleum, foundry, smelting, milling and mining industries are identified as key sources of heavy metals.

Chapter 3 specifically focuses on heavy metal pollution in sub-Saharan Africa, which just like other parts of the globe is experiencing industrial growth in recent times. The drivers of heavy metal pollution particularly the transportation sector, waste management systems, agricultural activities and the industrial sector are discussed and connected to heavy metal release to the environment. A specific focus on Eastern, Western and Southern Africa in relevance to their heavy metal pollution status is also detailed using specific-country examples.

Chapter 4 discusses the techniques of assaying for heavy metals in soil samples in order to make inferences on the soil status of a given region. The chapter begins with discussing the sampling approaches to use and the precautions to take to preserve and prepare collected samples before analysis and without losing their contents. Direct and indirect methods of assaying for heavy metals are also detailed along with their merits and demerits. Some of the discussed methods apply the principles of spectrometry, spectroscopy, chromatography and electrophoresis.

In Chapter 5, the rating of analyzed soil samples based on their pollution classes is discussed. In this case, the use of indices as tools and indicators of pollution status is explored. Indices used are categorized as single and total complex pollution indices based on their simplicity and applications. The factors to consider when selecting an index for pollution rating are also explored. These include the soil uses, data availability on background values, comparability of the calculated values and the purpose of the study.

Chapter 6 reports of a case study of heavy metal pollution in soils of five suburban regions of Nairobi, Kenya where untreated industrial effluent was used for vegetable irrigation. Soils were assayed for heavy metals using mass spectrometry and both single and total complex indices were used to rate pollution in soils. Using the pollution load index (PLI), soils of the study area had high pollution and ecological risk while Cd and Hg were the most lethal metals to soils.

Pollution by Pb, Cr, As, Hg, Cd, Ni, Cu and Zn were discussed in Chapters 7 to 13 in respective order. In each case, the properties of the metal that make it toxic to soils, the industrial sources of the metal and its physicochemical modifications to soils were discussed. Mining and smelting, unsystematic management of effluent and solid waste and use of agrochemicals were common sources of metal contamination to soils. The mining industry was the dominant cause of heavy metal pollution in soils of SSA region.

In Chapter 14, the management of polluted soils towards their reclamation and rehabilitation was discussed. Techniques used to manage soils were categorized as physical, chemical and biological remediation approaches. They were also compared based on their merits and demerits. Physicochemical techniques were found to be expensive to implement in large-scale polluted areas in addition to causing more pollution through production of sludge while bioremediation was found to be eco-friendlier though its application at field-scale is still at nascent stages. The need to find the best fit technique to remediate a particular heavy metal in a targeted soil was emphasized as the key to optimizing removal efficacy.

Chapter 15 discussed the specific measures that can be taken to prevent and control heavy metal pollution in soils of SSA region. The measures can also be replicated in other areas of the globe. The chapter emphasizes on avoidance of pollutant sources (industrial activities) through uptake of a circular economy, use of advanced technologies to predict, detect, analyze and model heavy metals pollution in soils, sustainable management of waste including effluents, regulated use of agrochemicals to prevent over- and misuse and enactment of effective policies of land management, reclamation and rehabilitation as potential corrective measures.

The book is a useful resource to academicians and researchers in the field of environmental and natural resources management who can use it to gain knowledge on heavy metal pollution extent, sources and effects in SSA. They can also use aspects of the book to teach on sampling of soils suspected to be polluted, analyzing collected samples at field and/or laboratory scale and transforming analyzed data to credible information on the pollution state of a specified area. With the knowledge, they can conduct their teaching with emphasizes on preventing further pollution. Additionally, they can conduct research geared towards taking affirmative action to control and prevent further heavy metal pollution using innovative, sustainable, cost-effective and environmentally friendly technologies. The book is also relevant to regulatory agencies and policymakers charged with sustainable environmental management and planning in SSA and the world. With the concepts outlined in the book, such individuals can prioritize on actions to take towards better management of heavy metal pollution in soils and other natural resources, soil reclamation and land rehabilitation if pollution has already taken place. Furthermore, they can use the information to push for policies and regulatory measures geared to protection of soils and land resources from heavy metal pollution.

The book is therefore one of a kind in the field of environmental science since it shows the influence of the pollution on land resources and particularly soils. It pays attention to SSA region that is vulnerable to soil pollution, although it is least prepared to deal with the resultant effects. The book also cautions on the need to advance industrialization while being environmental sensitive of the impacts of such developments and their capacity to reverse gains towards sustainable development. As such, it is a wakeup call for to rethink industrialization and environmental sustainability as inextricable rather than distinct occurrences.

We hope you shall appreciate it.

Joan Nyika
University of Johannesburg, South Africa & Technical University of Kenya

Megersa Olumana Dinka
University of Johannesburg, South Africa.

Acknowledgment

I would like to thank God, the Almighty for his grace throughout writing this book. It is what gave me resilience to keep going to the end.

I would like to express my gratitude to my postdoctoral research host, Prof. Megersa Olumana Dinka for giving me a platform to be able to write this book and for his invaluable guidance throughout the process. His guidance and review of each chapter improved the quality of the work herein. It has been a privilege working with him and I look forward to many more opportunities to collaborate with him in the near future.

I am grateful to the editors at IGI Global who facilitated smooth processing of this book at every stage including informing of the next steps to take and how to do it. Your great dedication towards completion of this book is highly appreciated.

I am grateful to my mother, Wanjiku and my dad, Nyika for their prayers, love and care and most of all, for the sacrifice to educate me in readiness for the future. I am grateful to my husband, Madaraka, my two sons, Mwema and Nyika for their love and support to keep going as I was writing the book. They stood with me as I burnt the midnight oil to look for material to complete this book and encouraged me not to give up.

Chapter 1
Introduction to Heavy Metals and Their Toxicity

ABSTRACT

Heavy metal pollution is a growing environmental and public health concern globally. The trend is attributable to the rise in concentrations of the contaminants in the environment in modern day propagated by geogenic and mainly anthropic activities. Once in the environment, the metals bioaccumulate and cause toxicity in living systems. In this chapter, heavy metal chemistry was studied and their ability to induce toxicity in the environment and living organisms discussed. Non-essential metals including cadmium, chromium, mercury, lead, and the metalloid arsenic, among others, were discussed. They were found to induce neurotoxicity, nephrotoxicity, cardiotoxicity, hepatoxicity, skin toxicity, genotoxicity, and carcinogenicity in living systems. Specific toxicity was dependent on the type of metal, its bioavailability and exposure time, where toxicity spread via a variety of exposure routes.

INTRODUCTION

Heavy metals (HM) refer to elements of the periodic table with an atomic number of 22 to 34 and 40 to 52 (Kumar et al., 2021). They are also defined as metallic elements whole density is relatively higher compared to that of water (Tchounwou et al., 2012). In another study, HM are defined as elements with an atomic number beyond 20 and are naturally occurring (Ali & Khan, 2018). The term is commonly used to refer to mercury (Hg), lead (Pb), cadmium (Cd), chromium (Cr) and arsenic (As) among other metals whose effects in the environment are lethal (Ali & Khan, 2018). These examples include non-essential metals with no biological function and

DOI: 10.4018/978-1-6684-7116-6.ch001

are toxic to living organisms. The HM metals are also referred to us trace elements because of their minute though toxic concentrations in different environs. The metals have a non-biodegradable, bioaccumulation and persistent nature and some of them have great significance in living things while others are potent and noxious to cause genotypic, mutagenic and enzymatic modifications. For instance, toxic HM are carcinogenic (cancer-causing), damage the nervous system and compromise the immune system of humans (Magalhaes et al., 2015). Depending on the specificity of a given HM, effects to humans and the environment are varied. Using the definition by Tchounwou and others (2012) and based on their toxicity, HM also include metalloids such as arsenic (As), which are elements in the periodic table's stair-step line and distinguish non-metals from the metals. Metalloids sometimes behave as metals or non-metals depending on the reactant. Heavy metals such as Cu and Zn, which are of significance such as serving as protein cofactors in biological functions of living organisms are known as essential elements and are non-toxic in minute quantities but toxic in high levels (Slobodian et al., 2021). Figure 1 shows elements considered as heavy metals and metalloids in the periodic table (Scerri, 2020).

Many tissues and cells of living organisms contain metal elements of the periodic table including heavy metals such as Cu and Zn either in their free state or chemically bound. The metals act as building blocks and are key elements in various biochemical and metabolic processes. According to Djoko et al. (2015), Cu and Zn regulate the innate immune system to fight against foreign organisms invading living things and facilitate normal development and physiology of their cells. They also have a key role in redox and enzymatic reactions. As such, some heavy metals have an essential role in living things whose effectiveness depends on the dose. Even metals considered as essential in livings things such as Ca, Fe, Na, K and Mg in elevated levels have toxic effects while HM such as As, Cd, Hg and Sb among others in low doses have medicinal effects (Skalnaya & Skalny, 2018). The concentration range in which HM levels are considered toxic or beneficial is narrow and hence the need for cautious use of HM.

HM are challenging environmental pollutants with the ability to induce environmental imbalance from their toxic, bioaccumulative and degradation resistant nature (Nyika, 2021). The metals bioaccumulate in soils, water bodies, crops and aquatic life eventually entering food chains due to their nonbiodegradable characteristic (Nyika & Dinka, 2022). Examples of common hazardous HM, their sources and effects in humans are provided in Table 1. Most waste effluents contain HM such as Cd, Cr, Cu, Ni, Pb and Zn, which are responsible for causing environmental and human health risks (Nyika et al., 2019). Their toxicity could occur once HM are ingested through skin absorption, food, water and air. The sources of HM pollutants are either natural through weathering and precipitation but also from human activities including agrochemicals, urban runoff, industrial effluents, smelting, mining, unscientific solid waste disposal and sewage discharge (Tchounwou et al., 2012; Xu et al., 2022). Global industries such as paper processing, wood preservation, microelectronics, textile, plastics, nuclear power stations, petroleum combustion and coal burning power plants and firms contribute to anthropogenic heavy metal pollution in the environment. The elements compete for protein binding sites in cells of living systems with non-toxic metals resulting to mutagenic and cellular malfunctions (Kocadal, Alkas, Battal & Saygi, 2020). This chapter discusses the chemistry of some of the toxic heavy metals and details how they induce toxicity to the environment.

Figure 1. Heavy metals and metalloids as shown in the periodic table (Scerri, 2020)

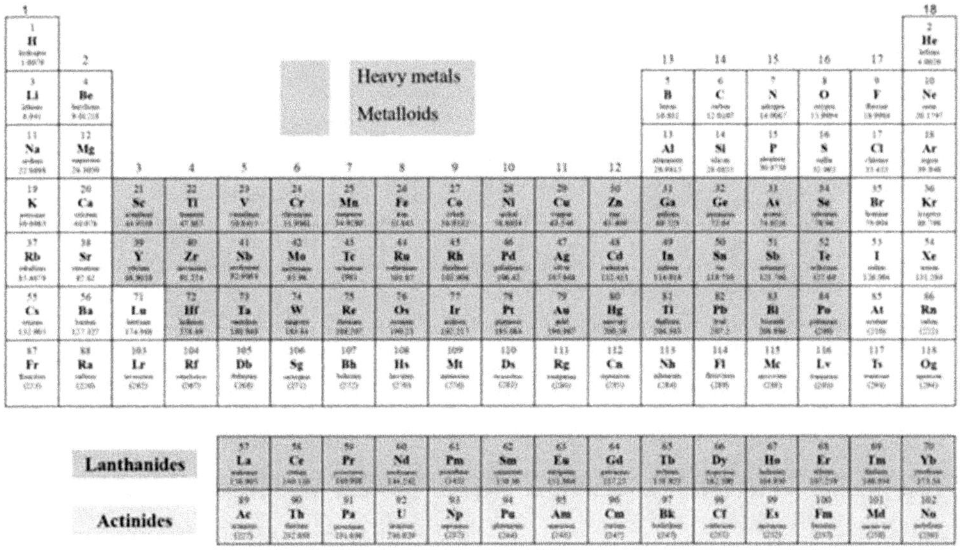

Table 1. Health effects of exposure to common heavy metals and their sources (Kumar et al., 2021; Jyothi, 2020)

Heavy Metal/ Metalloid	Health Hazards	Sources
Antimony (Sb)	Sb containing compounds cause dermatitis, respiratory tract irritation and compromise the immune system. Antimony chloride interferes with the estrogenic system and causes toxicity in many aquatic organisms.	Various industrial processes, vehicle exhaust emissions, incinerators, mining activities and coal-burning power plants can exposure one to Sb.
Arsenic	Exposure and ingestion to high levels of inorganic arsenic results to skin cancer, black foot disease, kurtosis and hyperpigmentation.	Ingested As can be sourced from paints, fungicides, pesticides, herbicides and wood preservatives.

continues on following page

Table 1. Continued

Heavy Metal/ Metalloid	Health Hazards	Sources
Cadmium	Prolonged exposure results to destruction of male reproductive organs, dysfunctions of lungs and kidneys and extreme toxicity in humans.	Ingested Cd can be found in pesticides, fertilizers and plastics.
Cobalt	Co compounds result to endocrine, hepatic, cardiovascular and hematological dysfunctions.	Co exposure mainly occurs during hip alloy replacement modifications.
Copper	Exposure to high levels of Cu results to abnormalities in the metabolic activity and abdominal disorders.	Copper sources include printing, plating and copper polishing activities.
Chromium	Small quantities of Cr (VI) induce allergic reactions while high amounts and prolonged exposure causes cancers of the respiratory tract, skin ulcers and disorders and generally affects normal sensitizing and metabolism activities in living organisms.	During textile manufacturing processes, electroplating and steel fabrication, Cr ingestion is possible.
Iron	High ingestion of Fe results to lethargy, dehydration, abdominal pains, diarrhea and vomiting	Sources of Fe include oral consumption and from intake of iron supplements.
Lead	Elevated levels of Pb result to gastro-intestinal malfunctions, restlessness, insomnia, appetite losses and abnormal functioning of the kidneys, brain, nervous system among other visceral organs.	In paint industries, manufacture of batteries and coal combustion, lead exposure can occur.
Mercury	Once absorbed by the skin, Hg modifies enzymatic and genetic systems of the body, damages the immune and nervous systems and forms organic compounds such as methyl chloride in water resources that are more toxic than elemental Hg forms.	Exposure and ingestion of Hg occurs during volcanic eruptions, coal combustion and in paint and paper processing plants.
Nickel	Prolonged exposure to Ni containing compounds induces allergic reactions and skin irritation.	Activities such as jewelry making, battery and electronic manufacture, plating, steel manufacture and cigarette smoking lead to Ni exposure.
Zinc	Exposure to elevated levels of Zn causes abnormalities in the functioning of the liver, kidney and the gastro-intestinal system.	Oil refining, brass manufacturing and plumbing can lead to Zn exposure.

FACTORS THAT INFLUENCE HEAVY METAL BIOAVAILABILITY

Bioavailability is a term that measures the extent and rate to which HM reach their action site. It is used to approximate ecological toxicity when evaluating HM dosages and exposure rates delivered to various environs such as water and soils (Adams et al., 2020). Bioavailability determines the response of an organism to trace metal toxicity. Different factors influence HM bioavailability as summarized in Figure 2. According to Tchounwou et al. (2012), bioavailability of HM is influenced by physical, chemical and biological factors. Physical factors include sequestration, adsorption and speciation ability as well as temperature; while chemical factors include water partition coefficients, lipid solubility, complexation kinetics and thermodynamic equilibrium speciation (Miranda et al., 2021). In soils, their solid features including soil moisture content, ion exchange capacity, organic matter content and mineralogy influence metal bioavailability since they regulate leaching processes (Nyika, 2021).

Temperature increment enhances solubility of HM by increasing the adsorbate diffusion rate and hence, bioavailability. Biological factors include trophic interactions, physiological and biochemical adaptation of the species affected by HM toxicity and their characteristics. On the other hand, Adams et al. (2020) noted that bioavailability of HM is determined by the availability of chemicals that compete for binding sites with HM, pH of the medium hosting the pollutants, solubility and sorption limits and inherent characteristics of organisms involved. In another study, bioavailability was described to be influenced by the valence state of a given HM, its redox potential, pH of the environment, presence of organic matter and ionic strength of the medium involved and in the case of water, its hardness (Magalhaes et al., 2015).

In rating bioavailability pH is a common factor that influences HM in particular environs since it controls the rate of precipitation, aggregation, polymerization, hydrolysis and the competition of protons for free ligands. A minute pH change results to a great change on HM bioavailability. For instance, a pH change from 6 to 7 in 1.3 Mm phosphate resulted to an 8.8-fold reduction in Cd bioavailability (Olaniran, et al, 2013). Acidic pH favors metal dissociation, increases their solubility and ultimately, toxicity while alkaline pH enhances metal precipitation to hydroxides and oxides hence reduces their toxicity and solubility (Olaniran et al., 2013).

Alkalinity also reduces bioavailability through competition for active binding sites with toxic metals. For example, Nickel is not bioavailable and was reported less toxic at a pH of 8.5 and usually forms complexes with other ligands at such conditions (Olaniran et al., 2013). Addition of phosphate and trisaminomethane buffers enhances precipitation of metals to reduce their bioavailability. Such buffers sequester the pollutants to insoluble complexes at neutral or mildly acidic pH (Nyika, 2021). Zwitterionic buffers on the other hand promote bioavailability.

The redox potential (Eh) that defines oxidation-reduction reactions also influences metal bioavailability so that HM such as Cr, Cu, Fe, Hg and Mn at oxidizing conditions produce hydroxides and oxides whose bioavailability is low while at reducing conditions, the metals have high solubility and hence high bioavailability and toxicity (Olaniran et al., 2013).

The presence of some cations (Ca^{2+} and Mg^{2+}) ions among other divalent metal ions influences bioavailability. The cations complete for binding sites with HM and as such, block the entry of contaminants in cells and hence their bioavailability (Kozvola, Wood & McGeer, 2009). Some of the affected HM whose bioavailability is influenced by Ca^{2+} and Mg^{2+} include Cd, Co, Pb and Zn. Majority of HM have high affinity to sulfur (S^{2-}) and O_2 ions with naturally occurring metals having binary compounds containing such organic elements. For instance, hexavalent chromium {Cr (VI)} found in combination with O_2 produced oxyacids such as dichromate ($Cr_2O_7^{2-}$) and chromate (CrO_4^{2-}) ions that are more bioavailable and toxic compared to trivalent chromium {Cr (III)} (Singh et al., 2013). HM that exists in compounds of soluble nature such as halides, sulfates, acetates and nitrates are more bioavailable compared to insoluble forms (Magalhaes et al., 2015). The presence of natural or anthropogenic-based organic compounds influences bioavailability in that the carbon-containing compounds serve as metal ligands that reduce the quantities of free HM cations and hence, their bioavailability. Organic ions promote complexation and control cation exchange capacity and adsorption reactions with HM to produce coordination compounds whose toxicity is reduced compared to free occurring metals (Bezerra et al., 2009; Miranda et al., 2021).

Figure 2. Factors that influence bioavailability of heavy metals in the environment (Adopted from Magalhaes et al., 2015)

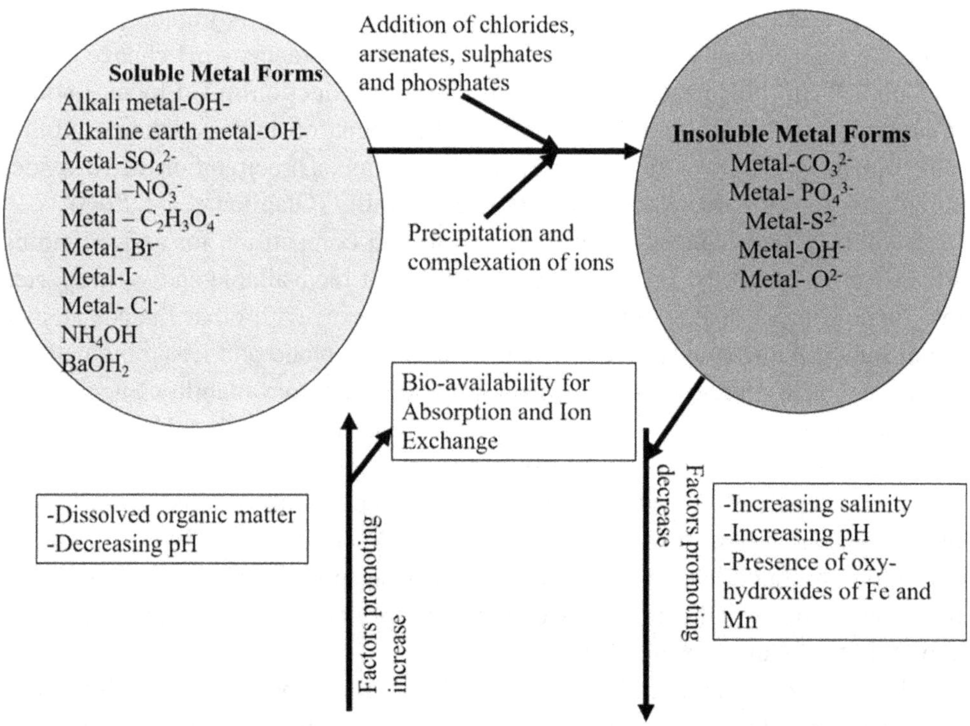

MECHANISMS OF HEAVY METAL TOXICITY

Heavy metals induce cellular damage and oxidative stress in living things through the generation of free radicals. The most common free radicals are reactive oxygen species (ROS) and the reactive nitrogen species (RNS) (Balali-Mood et al., 2021). The mechanism of generating the free radicals depends on the specific heavy metal. Iron for instance, which is a constituent of hemoglobin in the blood of living organisms and is involved in many physiological functions can form free radicals. However, in its free form, the metal generates hydroxyl radicals (OH*) via a combination of reactions including Fenton (Equation 2) and Haber-Weiss (Equation 3) reactions. The mechanism of radical formation is represented in the following Equations 3 to 6.

$$Fe^{3+} + O_2^{-} \rightarrow Fe^{2+} + O_2 \tag{1}$$

$$Fe^{2+} + H_2O_2 \rightarrow Fe^{3+} + OH^{\cdot} + OH^{-} \tag{2}$$

$$O_2^{-} + H_2O_2 \rightarrow OH^{\cdot} + OH^{-} + O_2 \tag{3}$$

$$OH^{\cdot}H_2O_2 \rightarrow H_2O + H^{+} + O^{2\cdot-} \tag{4}$$

$$OH^{\cdot} + Fe^{2+} \rightarrow Fe^{3+} + OH^{-} \tag{5}$$

$$LOOH + Fe^{2+} \rightarrow Fe^{3+} + LO + OH^{\cdot} \tag{6}$$

In the oxidation of Fe, OH· is the commonest free radical that reacts with biomolecules such as DNA, lipids and proteins leading to their damage. The radical also reacts with nitrogenous-based nucleic acid, guanine generating 8-oxo-7,8-dihydro-20-deoxyguanosine (8-oxo-dG) and 2,6-diamino-5-formamido-4-hydroxypyrimidine (FAPy-G) that are both molecular markers of oxidative damage (Valko et al., 2004). In oxidative damage, an imbalance exists between the free radical formation rate and the capacity of cells of living organisms to clear them resulting to an interruption of normal functions of living organisms.

Radicals of all HM including Fe can attack phospholipids among other polyunsaturated fatty acids to generate hydroxyl radicals that oxidize lipid membranes through peroxidation processes (Engwa et al., 2019). The steps involved in lipid peroxidation are as shown in Equations 7-11. Equation 7 represents the initiation stage, Equations 8 – 9 are the propagation stages and 1.10 – 1.11 are the termination stages. In the initial stages, the radicals form a radical lipid after attacking the lipid membrane. The radical lipid them propagates to form a peroxyl lipid radical and consequently damage the lipid molecule. At the end of the peroxidation, two radical lipid molecules and/ or a peroxyl lipid radical stabilize in a reaction to form a lipid molecule such as malondialdehyde that is a molecular marker for the process (Engwe et al., 2019).

$$\text{Lipid} + \text{R} \cdot \text{OH}^{\cdot} \rightarrow \text{Lipid}^{\cdot} \quad (7)$$

$$\text{Lipid}^{\cdot} + \text{O}_2 \rightarrow \text{Lipid} - \text{OO}^{\cdot} \quad (8)$$

$$\text{Lipid} - \text{OO}^{\cdot} + \text{Lipid}^{\cdot} \rightarrow \text{Lipid} - \text{OOH} + \text{Lipid}^{\cdot} \quad (9)$$

$$\text{Lipid}^{\cdot} + \text{Lipid}^{\cdot} \rightarrow \text{Lipid} - \text{Lipid} \quad (10)$$

$$\text{Lipid} - \text{OO}^{\cdot} + \text{Lipid}^{\cdot} \rightarrow \text{Lipid} - \text{OO} - \text{Lipid} \quad (11)$$

Eaton and Qian (2002) also showed that Fe induces damage on proteins in a process where the metal catalyzes the loss of histidine molecules, cross linking of bityrosine, oxidative scission of molecules, carbonyl group introduction and consequently, formation of alkyl-peroxyl (ROO*), alkoxyl (RO*) and alkyl (R*) radicals that are protein-based.

Chromium (IV) ($Cr^{4+)}$ generates free radicals from hydrogen peroxide (H_2O_2) as evidenced in in-vitro experiments (Liu & Shi, 2001). In-vivo studies also showed the generation of the radicals in blood and liver of living organisms as a result of Cr presence. During radical generation, electron reduction resulted to the formation of Cr^{5+} intermediates (Li & Shi, 2001). In another study, Cr (IV) radicals reduced the antioxidant potency inducing oxidative stress and toxicity in proteins, lipids and DNA of living organisms (Aggarwal et al., 2019), and consequently interfere with the normal functioning of the organisms.

Cobalt generated superoxide ($.O_2^{\cdot}$) radicals following a reaction with hydrogen peroxide and as shown in Equation 12 (Engwe et al., 2019).

$$\text{Co}^{2+} + \text{O}_2 \rightarrow \text{Co}^{+} + \text{O}_2^{\cdot -} \rightarrow \text{Co}^{+} - \text{OO}^{\cdot} \quad (12)$$

Copper also forms cuprous (Cu^{1+}) and cupric (Cu^{2+}) ROS that engage in redox reactions. Cupric ions found in ascorbate and glutathione, which are biological reductants can be reduced to cuprous ions that enhance decomposition reactions in the presence of H_2O_2 to form hydroxyl radicals in the Fenton reaction as Lloyd, et al. (1997) documented (Equation 13). The resultant OH* reacts with biomolecules resulting to oxidation of DNA bases and its strand breakage accompanied by low-density lipoprotein oxidation (Burkitt, 2001; Brezova et al., 2003).

$$Cu^{+} + H_2O_2 \rightarrow Cu^{2+} + OH^{\cdot} + OH^{-} \quad (13)$$

Arsenic generates free radicals such as dimethylarsinic $\{(CH_3)_2As^*\}$, dimethylarsinic peroxyl $\{(CH_3)_2AsOO^*\}$, peroxyl (ROO*), H_2O_2, nitric oxide (NO*), singlet oxygen ($^1O_2$) and O_2^{*-} species (Pi et al., 2003). The mechanism in which these free radicals are generated in living things however remains unclear (Engwe et al., 2019).

Exposure to Pb reduces the activity of antioxidants such as glutathione (GSH), catalase (CAT), superoxide dismutase (SOD), glutathione peroxidase (GPx) and glutathione S-transferase (GST) while increasing the activity of H_2O_2 and malonaldehyde (MDA) (Wang et al., 2013).

Cadmium overwhelms the antioxidant defense of cells by generating NO*, OH* and O_2^{-*} radicals (Rani, et al., 2014). The consequence is enhanced damage of the genetic material (DNA) along with increased lipid peroxidation. Presence of the also interferes with the function of the mitochondria, which controls the physiology and energy synthesis of living things (Branca et al., 2018).

Mercury releases ROS that are responsible for its toxic effects on the cardiovascular and central nervous systems of living things. The radicals reduce the activity of SOD and GPx among other antioxidant enzymes leading to interference of the intracellular signaling of receptors in living organisms (Balali -Mood et al., 2021). According to Brown et al. (2017), radicals resulting from methyl-Hg induce the activity of phospholipase D (PLD), which is attributable to may cancers and diseases in humans. In both Hg and Pb toxicity, ROS production results to antioxidant depletion in cells while in some other HM such as Cd, ROS and RNS effects are indirect due to their ability to replace divalent ions such as Cu and Fe in cellular proteins. The replacement of the cations causes oxidative stress and dysfunctions of cellular metabolism (Wu et al., 2016).

Vanadium (V) also generates free radicals in the plasma of living things by reducing nicotinamide adenine dinucleotide phosphate (NADPH) and ascorbate antioxidants that bind on plasma proteins prior to their transportation (Crans et al., 2004). The HM can also generate OH* following a Fenton reaction at neutral pH, which is the physiological condition of many living things. The generation of vanadium radicals is as shown in Equations 14 to 17.

$$V^{5+} + NADPH \rightarrow V^{4+} + NADP^{+} + H^{+} \quad (14)$$

$$V^{4+} + O_2 \rightarrow V^{5+} + O^{2\cdot-} \quad (15)$$

$$V^{5+} + O^{2\cdot-} \rightarrow [V^{5+} - OO^{\cdot} \quad (16)$$

$$V^{4+} + H_2O_2 \rightarrow V^{5+} + OH^{-} + \cdot OH \quad (17)$$

TYPES OF TOXICITY RESULTING FROM HEAVY METALS

Heavy metal toxicity in the environment originates from natural and mainly anthropogenic sources. The rise of industrial activities resulting to release of HM containing solid wastes, wastewater, particulate matter, dust, emissions and gases to the environment has led to the toxicity affiliated with the pollutants. The pollution trend has increased exponentially over the last decade especially to natural (land and water) resources and human beings among other living organisms. The toxicity has escalated due to the persistent, bioaccumulative and non-biodegradable nature of HM (Jan et al, 2015). In this section, the toxicity types that result from HM contamination to living things are discussed under nine categories. The categories are as summarized in Figure 3.

Genotoxicity

Genotoxicity in living things as a result of HM exposure induces genetic material modification in two ways: - 1) carcinogenesis and 2) teratogenesis (Mitra et al., 2022). Carcinogenesis also known as tumorigenesis or oncogenesis refers to the transformation of normal body cells to malignant ones. In this case, the cells reproduce faster and do not reach maturity and as such, cannot function normally. In teratogenesis, chemical compounds (in this case HM) modify genetic material and interfere with its normal physiology. Carcinogenic and teratogenic effects are manifested in the offspring of organisms affected by HM in the form of congenital abnormalities or in the development of malignant tumors (Mitra et al., 2022). HM toxicity was reported to promote ROS production and subsequently induce genotoxicity through false DNA repair and high mutation risks (Kocadal et al., 2020).

Exposure to As for instance is shown to alter DNA through sister chromatid exchanges, deletion and production of micronuclei, mutations and chromosomal abnormalities (Roy et al., 2018). In another study, As toxicity was found to interfere with DNA repair resulting to oxidative stress in cells of living things (Pierce et al., 2012). Mercury toxicity also induces genotoxicity by promoting breakages in the

Figure 3. Categories of toxicity resulting from heavy metals

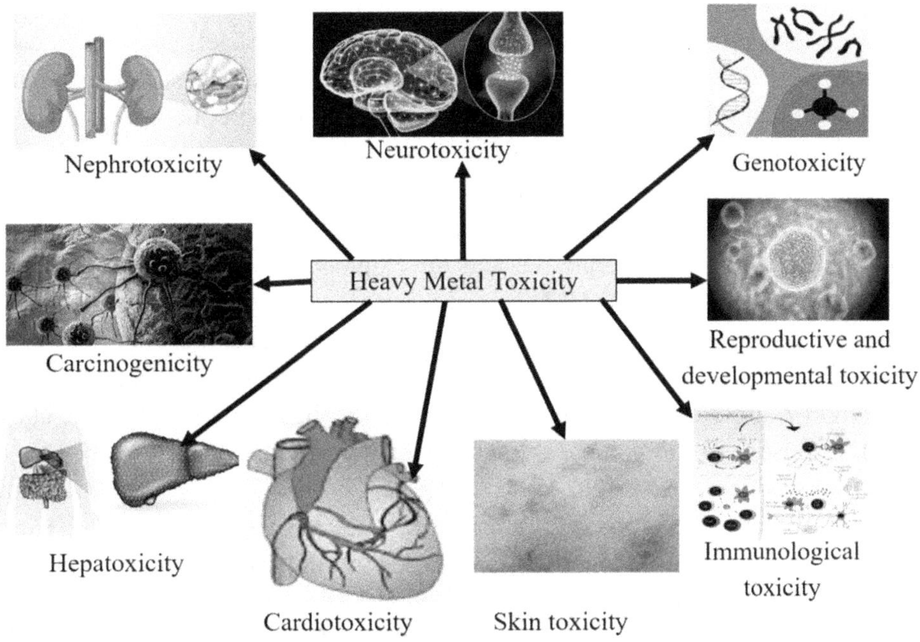

DNA strand and the development of adducts on the genetic material (Mitra et al., 2022). The exposure to Cr (IV) was reported to induce structural modifications on the genetic material of human and animal cells (Fang et al., 2014). These modifications include the breakage of nucleotide strands, cross-linking of inter-DNA strands and errors in chromosomal proteins that constitute the genetic material. Cadmium toxicity activates proto-oncogenes, promotes protein degradation and upregulates cytokines resulting to errors in the genetic material (Kocadal et al., 2020). Lead poisoning was shown to induce DNA breakage, chromosomal aberrations and inhibit DNA synthesis by generating ROS in living things, which caused oxidative stress (Hemmaphan & Bordeerat, 2022).

Reproductive and Developmental Toxicity

Exposure to HM in living organisms has negative effects on their reproductive systems. In a study of a Cu smelting vicinity in northern Sweden (Nordstrom et al., 1978), exposure to a mixture of As, Cu and Pb was associated to depressed birth weights for infants and high frequencies of abortions among women. Similar results were reported among factory and industry workers of Finland exposed to As, Co and Zn, where their rates of abortions were significantly higher compared to populations that were not exposed to the HM (Hemminki et al., 1980). Exposure of As, Hg and Pb to mice during their tenth gestation day resulted to maternal toxicity and negative effects on the fetus development particularly, their cleft palate and weight (Belles et al., 2002). Exposure to a mixture of HM in women induces ovulation disturbances and subfertility (Naz & Batool, 2017). The reproductive disturbances are characterized by hormonal imbalances and a dysfunctional endocrine system, chromosomal abnormality of the oocytes, delayed ovulation, irregular menstrual cycles and a high risk of infertility (Rattan et al., 2017).

The effects on the reproductive and developmental systems differ based on specific exposure to a given HM. In most cases, exposure is among workers in different industries dealing with smelting and processing of the metals. However, the effect of HM exposure is more pronounced in women compared to men according to a WHO (World Health Organization) study published by Apostoli and Catalani (2011). Exposure to inorganic As for instance, reduces the sperm count in male epididymis, reduces the weight of accessory sex organs and testes in addition to causing dysfunctions of gonadotrophins, testosterone and steroidogenesis (Kim & Kim, 2015). In females, As toxicity is associated with endometrial cancer (Salinikow & Zhitkovich, 2008) as well as impaired embryo development commonly known as endometrial angiogenesis (Milton et al., 2017). Cadmium toxicity resulted to hemorrhagic modifications in the uterus and ovaries resulting to persistent ovulation or estrous, which emanate from dysfunctions in the female sex hormones (Sengupta et al., 2015). Exposure to Cr (VI) was associated to spontaneous abortions, pregnancy complications and decreased weights of the uterus and ovaries among women working in industries with such metal exposure (Sengupta et al., 2015). Cobalt poisoning was associated with lengthened estrous periods, malfunctions in the ovaries, infertility, altered sexual behavior and premature menopause among living organisms (Sengupta et al., 2015). Lead toxicity in both male and female mice resulted to infertility, death of fetus and neonates and spontaneous abortions (Dutta et al., 2013). Exposure of females to methyl mercury resulted to altered menstrual cycles, infertility, hormonal imbalances and low survival of offspring (Davis et al., 2001). Exposure to manganese among dried battery factory and mine workers induced birth defects, retarded fetal growth, hormonal imbalances and reduced ovarian follicles that negatively affected reproduction and development (Sengupta et al., 2015). Nickel toxicity was also associated with spontaneous abortion, reduced implantation and inhibited growth of embryos in experiments involving female mice (Obone et al., 1999).

Skin Toxicity

Exposure of HM has negative effects on the skin of living things and in particular, humans. Prolonged exposure to As is associated with skin diseases such as skin cancer, hyperpigmentation and hyperkeratosis (Mitra et al., 2022). Hyperpigmentation, which is a condition that manifests with some skin areas being darker than others is the commonest condition resulting from As toxicity. Hyperkeratosis resulting from As exposure manifests as soles in the hand dorsum, arms, fingers, toes, legs and mainly in the palms (Mitra et al., 2022). Arsenic carcinogenesis usually known as Bowen's disease also results from exposure to the metalloid and its ability to induce chromosomal abnormalities and oxidative stress (Yu et al., 2006). The Bowen's disease and hyperkeratosis induced by As have the capacity to advance to invasive malignancies (Huang et al., 2019).

Exposure and contact with Cr causes dermatitis, which is a condition characterized by delayed hypersensitivity on the skin due to the prolonged presence of allergen (Yoshihisa & Shimizu, 2012). Mercury and mercury-based compounds, once exposed to the skin, induce infections such as the pink disease (acrodynia) where the skin turns pink once it is in contact with the HM (Bar-Zeev et al., 2002). Individuals with tattoos and subsequent exposure to mercury sulfide and cadmium sulfide also develop skin inflammations within the half year after tattooing (Mitra et al., 2022). Boyd and others (2000) also noted that Hg toxicity is attributable to skin problems such as irritation, dermatitis, vesiculation, scaling and swelling. In subjects exposed to Pb poisoning, skin elasticity and hydration was found to be reduced (Rerknimitr et al., 2019). The heavy metal also caused skin lesions and cancers (Jaishankar et al., 2014). In individuals exposed to aluminum, Hg and Cr, the risk to develop skin rashes and ulcers was reported to be high (Jaishankar et al., 2014).

Cardiotoxicity

Exposure to HM causes dysfunctions of the heart to higher vertebrates especially humans. For instance, Cd toxicity induces hypertension, hemorrhage, cerebrovascular and cardiovascular diseases (Lama et al., 2012). The physiological and cellular mechanisms that leads to development of these diseases involve the ability of the HM to release phosphokinase, which distresses the vascular smooth muscle whose function is to regulate homeostasis of blood vessels through their active relaxation and contraction (Messner & Bernhard, 2010). High levels of the metal also inhibit the activity of the K and Na ATPase, an enzyme that regulates the Ca ions in the smooth muscles and ultimately, its functioning. Elevated levels of the ions result to hypertension (Sharifi et al., 2004). Cadmium poisoning is also attributable to strokes, heart failures, myocardial infarctions and is a suspected etiology to deaths due to cardiovascular disease in the United States (Tellez-Plaza et al., 2013).

Exposure to Pb could introduce cardiac diseases, atherosclerosis, thrombosis, hypertension and arteriosclerosis among other heart conditions (Mitra et al., 2022). The mechanisms of activity in this case involves modification of the heart's response to vasoactive agonists, endothelium-based vasorelaxation and its inflammation, abnormal functioning of the vascular smooth muscle, reduced production of vasodilator prostaglandins, increased production of vasoconstrictor prostaglandins, reduced production of nitric oxide (NO) and increased NO synthase (NOS) (Vaziri, 2008).

Toxicity by Hg results to atherosclerosis, acute coronary failure and atherosclerotic lesions that emanate from the presence of oxidized low-density lipoproteins (LDL) (Yoshizawa et al., 2002). The heavy metal also inhibits the activity of paraoxonase, which is an enzyme that regulates the activity of high-density lipoproteins. The inactivity results to carotid artery stenosis, coronary heart disease, acute myocardial infarction among other heart complications (Kulka, 2016). Mercury also forms complexes with thiol (-SH) component from selenium. The complexes inhibit the activity of catalase among other antioxidant enzymes inducing ROS production and a consequent risk to myocardial infarctions (Magos & Clarkson, 2006).

Toxicity due to Co causes temporary systolic cardiac depression and progressive cardiomyopathy that sometimes can be fatal (Packer, 2016). In another study, a mixture of metals containing As, Cd, Cu, Hg and Pb predisposes individuals to cardiovascular diseases by modifying the heart's pathophysiology, increasing inflammatory response and oxidative stress, inducing apoptosis and DNA damage among other peripheral arterial dysfunctions (Sevim et al., 2020). The interaction of a mixture of Cd and Hg trace metals allows their absorption in the cellular and blood systems resulting to modification of protein functional groups and ultimately, cardiovascular diseases (Ali et al., 2020).

Immunological Toxicity

Many immunological defects including immunostimulation and immunosuppression are associated with HM toxicity in living organisms. The HM accumulate in immune cells and modify their functions. They induce immunological reactions that result to health effects in living things. As immunotoxic agents, heavy metals enhance the death of immune cells, alter the production and release of white blood cells and enhance the production of ROS (Wang et al., 2021). The ultimate effects of these modifications are oxidative stress and selective production of antibodies in white blood cells leading to both immunosuppression and immunostimulation.

Laboratory tests involving the use of rodents and mice showed that injection with mercury chloride compromised the immune systems of the animals and made them vulnerable to autoimmune diseases, infections and allergies (Mitra et al., 2022). The exposure to Cd among industrial workers resulted to immunosuppressive responses characterized by increased vulnerability to infections and reduced phagocytosis (natural killing of harmful cells) (Mitra et al., 2022). Cd also induced thyroid toxicity and reduced oxidative stress in experimental studies using mice (Benvenga et al., 2020). Lead exposure at chronic and acute conditions increased human vulnerability to infections, cancer and autoimmune diseases (Hsiao et al., 2011). Mercury also induces hypersensitivity and autoimmunity in living organisms by modifying the activity of the major histocompatibility complexes (MHC), T- and B- cells and subsequently increasing susceptibility to immunological toxicity (Kasten-Jolly et al., 2010). Introduction of Cr in ears of mice skin resulted to immunotoxicology manifested as reduced activity of cytokines and high vulnerability to immunosuppression (Wang et al., 2018). In mice a combination of silver (Ag) and Hg toxicity was found to be immune-compromising. The two HM promote the production of autoantibodies that chemically bind with organic donors leading to immunosuppression (Mishra & Singh, 2020).

Hepatoxicity

The liver breaks down HM and excretes them via the bile to the small intestine. In this case, about 5% of the metals are removed while the rest are retained for enterohepatic circulation (Kim et al., 2021). For this reason, liver cells get exposed to the HM, which can lead to their damage or dysfunction and ultimately, organ failure. In a study by Kim et al. (2021), exposure to Cd, Hg and Pb among adult Korean population was associated to hepatic toxicity. The metals prompted the production of biomarkers of liver injury such as gamma-glutamyl transferase (GGT), alanine aminotransferase (ALT) and aspartate aminotransferase (AST). Accumulation of Mn in the liver as a result of hepatitis also resulted to the dysfunction of the organ as a result of hepatotoxicity (Irina & John, 2020). An animal model also showed hepatoxicity due to Hg by examining the three enzymes in rat liver (Hazelhoff & Torres, 2018). The mechanism of inducing hepatoxicity in the rodents included enzyme inhibition, reduced activity of glutathione peroxidase and production of ROS. The ability of HM such as As, Cd, Cr, Hg, Ni and Pb to produce ROS was associated with hepatoxicity (Renu et al., 2021). The metals cause injury to the liver by inducing epigenetic alterations, cell death (apoptosis) and inflammation.

Experimental examination of the effect of lead acetate (PbA) introduction in liver cells of rats showed lead's their ability to induce hepatoxicity. The metal inhibited the activity of antioxidants such GSH, GST, SOD, CAT and GPx in liver cells while increasing MDA and H_2O_2 concentrations resulting to organ injury (Omobowale et al., 2014). Cadmium was also shown to induce apoptosis and necrosis in rat hepatocytes (Pi et al., 2015). In another study, As resulted to cytochrome C release and ROS generation in rats inducing oxidative stress that caused apoptosis of hepatocytes and damage of the livers lipid bilayer and its subsequent swelling (edema) and injury (Bodaghi-Namileh et al., 2018). Arsenic toxicity has also been closely associated with hepatocellular carcinoma (liver cancer) in children (from Northern Chile) who consumed contaminated water (Liaw et al., 2008).

Carcinogenicity

Majority of HM are toxicants and potential carcinogens. According to the International Agency for Research on Cancer, IARC (2012), As, Cd, Cr and Ni are category 1 HM and their ingestion modifies damage repair processes, enzyme activity and tumor suppression gene expressions. The etiology to HM-induced cancer is higher for industrial workers exposed to the contaminants in their occupation (Kim et al., 2015). Arsenic toxicity disrupts the signaling pathways of cells, induces oxidative stress and leads to genotoxicity in mice and human beings (Kim et al., 2015). Methyl As inhibits DNA repair, enhances free radical generation and induces cell apoptosis through abnormal gene expression increasing cancer etiology (Hartwig & Schwerdtle, 2002). The metal also binds to DNA-associated enzymes such as methyl transferase to inhibit the function of the tumor suppressor genes and create a conducive environment for cancer growth (Martinez et al., 2011). In addition, As binding to DNA proteins results to slow DNA repair, DNA methylation, histone modifications and epigenetic alterations that enable tumor development (Park et al., 2015).

Cadmium causes oxidative stress that enhances the activity of metallothionein (MT), which is a ubiquitous protein in the body involved in transcription. The complex between Cd and MT then accumulates in the kidney causing cellular dysfunction of the glomerular and renal tubules and increases the risk of cancer of the kidney (Kim et al., 2015). The cellular dysfunctions in the kidney further interfere with Ca metabolism, which causes bone damage and disposes living organisms to osteoporosis and osteomalacia (Baba et al., 2013). Cadmium also imitates the divalent chemical state of Zn and therefore interferes with DNA binding sites of Zn, which modifies the production of sex hormones and the ovaria steroidogenic process increasing the etiology to breast and ovarian cancers among women (Yang et al., 2015).

Trivalent and hexavalent forms of Cr have the capacity to cause cancers. Thie was evident among laborers of chromate-processing industries who showed high prevalence to lung cancer (Kim et al., 2015). Carcinogenesis by Cr is as a result of the HM causing DNA damage, oxidative stress and apoptosis. The metal also interacts with MT and nuclear factors among other genomic components promoting tumor development (Garg et al., 2007). Industrial dust from Ni-based refineries has insoluble compounds of the metal such as nickel oxide (NiO) and nickel sulfide (Ni_3S_2) that are cancer-causing (Kim et al., 2015). Inhaling the dust though the nasal cavities and lung over a prolonged period could induce lung and nasal cancers (Kupper et al., 2015). Nickel disrupts free radical generation, management of transcription factors and gene regulation interfering with RNA coding and expression, which promotes carcinogenesis (Zhou et al., 2017). Toxicity due to Pb disrupts DNA transcription by mimicking Zn binding sites, inhibits DNA repair, modifies chromosomal structures and inhibits the function of tumor regulating genes predisposing individuals to cancer (Mitr et al., 2022). By generating free radicals, Hg damages DNA, lipids, and cellular proteins, which promotes cell damage and carcinogenesis (Reczek & Chandel, 2017).

Nephrotoxicity

Nephrotoxicity refers to the disrupted functioning of the kidneys. Occupational, accidental and/or therapeutic exposure to HM may cause renal toxicity. In this case, the metals associated with kidney dysfunctions include As, Ba, Cd, Co, Cu, Hg, Pb, Li, Pt and Tl (Lentini et al., 2017). Kidney toxicity occurs because the HM can be reabsorbed and concentrated in the organ based on the time, dose and nature of exposure. The metals in the diffusible and non-diffusible form are cleared from the blood and sequestered in organs such as the kidney. In the tissue, they are concentrated in the early proximal tubule's luminal fluid where they are conjugated with MT and GSH. The compounds induce renal failure and chronic inflammation of the kidneys after they are reabsorbed through endocytosis. In acute exposure to HM, the contaminants are reabsorbed at the proximal tubule's apical membrane and in the loop of Henle inducing acute kidney injury. By releasing free radicals, HM cause mitochondrial pathway uncoupling, disrupt cellular membrane and induce cellular toxicity in the kidney leading to apoptosis (Lentini et al., 2017). Exposure to HM also alters the functions of nephrons leading to the death of cells. Additionally, it alters glomerulosclerosis affecting the overall kidney function (Orr & Bridges, 2017).

Toxicity by Cd causes conditions such as aminoaciduria (elevated levels of amino acids in urine), phosphaturia (inability to reabsorb phosphate), Fanconi-like syndrome and glucosuria (glucose presence in urine) that indicate nephrotoxicity (Reyes et al., 2013). Excessive and prolonged exposure to Cd causes acidosis of the renal tubules, hypercalciuria (excess calcium in urine) and renal failure (Friberg et al., 2019). Lead also causes nephropathy characterized by tubule atrophy, renal failure, interstitial fibrosis, hyperplasia and glomerulonephritis (Mitra et al., 2022). Many clinical symptoms such as hypotension, chills, vomiting, tremors, profuse salivation, abdominal pain and acute dyspnea that characterize acute tubular necrosis result from acute exposure to Hg in the kidneys (Lentini et al., 2017). In prolonged exposure to Hg, tubular failure characterized by higher urine production enhance Hg-induced chronic kidney injury (Mitra et al., 2022). Thallium poisoning in the kidneys also induces hematuria (blood in urine) and albuminuria (high albumin levels in urine) (Yumoto et al., 2017).

Neurotoxicity

Heavy metals such as As, Cd, Hg and Pb are accumulated in the brain leading to dysfunctions of the central nervous system (CNS). For this reason, HM are associated with neurotoxicological dysfunctions such as cerebral palsy, attention deficit disorder, mental retardation and autism (Singh & Sharma, 2021). The neurotoxic effects result from the ability of the metals to generate free radicals, decrease antioxidant activities and through their interactions with macromolecules such as lipids, proteins and DNA as well as micronutrients (Wang & Du, 2013). Lead toxicity especially among children has a higher capacity to induce neurotoxicity compared to mature brains. The metal ion substitutes Ca ions to cross the blood-brain barrier, which damages the cerebellum, hippocampus and cerebral cortex among other brain parts. The consequence of the damage is development of neurological diseases such as schizophrenia, Parkinson's disease, Alzheimer's disease and mental retardation diseases (Zheng et al., 2003). Exposure to Pb and Mn also results to neurological effects. Using rodents, the introduction of the HM resulted to enhanced neurotransmission levels and motor activity (Wang & Du, 2013). The exposure to methyl-Hg among children resulted to neurological dysfunctions that manifested as immaturity and impairment of the CNS (Hong et al., 2012).

Manganese in moderate levels promotes neuroprotection by inhibiting cell apoptosis but exposure to large quantities induce neurological complications such as Parkinson's and Alzheimer's disease due to altered homeostasis in the brain (Goldhaber, 2003). The trend is attributable to Mn accumulation, which promotes the generation of free radicals and mitochondrial dysfunction leading to brain cell apoptosis (Harischandra et al., 2019). In another study, exposure to Mn was associated with brain and blood-brain barrier damage resulting to psychosis development among humans (Irina & John, 2020). Poisoning by As results to cognitive impairment and a high risk to neurodegenerative diseases due to modifications in the neurotransmitter and synaptic transmission balance (Garza-Lombo et al., 2019). The mechanism of this neurotoxicity is via As and As-methylated metabolites induced cell death and intracellular Ca upturn leading to reduced autophagy.

Cadmium has been associated with neurodegenerative diseases such as amyotrophic lateral sclerosis and multiple sclerosis (Branca et al., 2018). In both children and adults, the HM affects the central and peripheral nervous systems to induce behavioral changes, motor function inadequacies, learning disabilities, mental retardation, olfactory and neurological disturbances and neuropathy (Marchetti, 2014). Usually, Cd neurotoxicity is a function of the metal's ability to disrupt the hormonal system, inhibit expression of neuron genes, induce epigenetic effect and promote cellular apoptosis (Miura et al., 2013). Metals such as Tl promote necrosis of brain cells, capillary alteration engorgement and edema of the cerebral hemispheres as a result of peripheral nerve and brain damage (Mitra et al., 2022). Iron, Cu and Zn prevent neurodevelopment in excess concentrations in the brain (Prohaska, 2000).

CHEMISTRY OF COMMON HEAVY METALS

Nickel (Ni)

Nickel (Ni) is ranked 7th among the most abundant transition metal and the 22nd most abundant substance on earth. It is a silvery white and hard metal. Nickel forms trivalent and divalent ionic forms when reacting with both organic and inorganic substances though the latter is the commonest. The sulfide (NiS and Ni_3S_2) and oxide (NiO, Ni_2O_3) compounds of the metals are insoluble in water though they may dissolve in biological fluids (Cempel & Nikel, 2005). The nitrates, sulfates and chlorides of Ni are soluble in water. In living organisms Ni is part of DNA, amino acids, peptides and adenosine triphosphate complexes. The metal is sourced from its sulfide ores, soils, crude oil and water and is extracted through a combination of magnetic separation, froth flotation and electrolytical casting processes.

The metal shows 0, +2 and +3 oxidation states. It is used in alloys for coin making and stainless steel as well as an industrial catalyst and in unsaturated organic compound hydrogenation. The routes of exposure to Ni include air, food and drinking water. Commercially available forms of Ni include: - 1) pentlandite {$(Ni, Fe)_9S_8$} and laterites such as nickeliferous limonite {(Fe, Ni)O (OH). nH_2O} and garnierite {$(Ni, Mg)_6Si_4O_{10}(OH)_8$} (Buxton et al., 2019). Operations such as refining, alloy manufacturing, plating, smelting, mining, combustion (of wood and fossil fuels) and solid waste disposal all release Ni to the environment.

Mercury (Hg)

Mercury (Hg) is a noxious HM whose extraction is easy since it substitutes gold in ores and forms amalgams with other metals (Kumar et al., 2021). Compared to other HM such as Cd and Zn, Hg has high vapor pressure at room temperature, lower boiling and melting points but larger atomic mass and is oxidized by air slowly to form mercury (II) oxide, which is red. The metal exists in three oxidation states including the mercuric (Hg^{++}), mercurous (Hg^{+}) and metallic (Hg^{0}) (Saturday, 2018). The metal is shiny, silvery white in color and boils at 357 °C. Mercury introduction to the atmosphere is a result of natural degassing of the surface of the earth or evaporation of the metal's vapor found on the surface of the earth. In this case, metallic Hg is released from geogenic origin but it can also emanate from anthropogenic activities such as smelting and mining of cinnabar ore for use in pigmenting and preserving paints, in control and measurement equipment such as thermometers and barometers and in batteries and electrical switching equipment (Kumar et al., 2021). Other uses of the metal include medical applications due to the antiseptic nature of the metal, as a catalyst in manufacture of ethanal, in mercury lamps and in standard cells. All these applications release Hg to the atmosphere causing pollution.

To extract the metal, HgS is heated in a condenser and fuel (such as brushwood, crushed), concentrated via froth flotation and then roasted at 600 °C. In ores that are rich in the metal, scrap iron or quicklime roasting can be done. Equations 18 to 20 represent the reactions during extraction of Hg.

$$HgS + O_2 \rightarrow Hg + SO_2 \tag{18}$$

$$HgS + Fe \rightarrow Hg + FeS \tag{19}$$

$$4HgS + CaO \rightarrow 4Hg + 3CaS + CaSO_4 \tag{20}$$

Mercury exposure can occur through ingestion of foodstuffs, contaminated soils (eg during geophagia among pregnant women) and drinking water containing the metal and/or from the general atmosphere and during dental fillings (Saturday, 2018).

Cadmium (Cd)

Cadmium (Cd) is a silver-white, ductile and soft element in the group IIb metals of the periodic table along with Hg and Zn. It has a high vapor pressure and a boiling and melting point of 765 and 312 °C, respectively. The metal is rare and usually, it is extracted from Zn ores, where it is separated from Zn by distillation or electrolyte deposition (Genchi et al., 2020). The metal is oxidized to cadmium oxide (CdO) on heating and forms a protective coating in the atmosphere. On dissolving Cd in acids, hydrogen gas is produced. Cadmium is also a result of human activities due to its industrial application as a stabilizer in PVC products, in the manufacture of Ni-Cd batteries and in color pigments and its use as a corrosive agent. Use of phosphate fertilizers, combustion of fossil fuels, refining and smelting of Cu and Ni also expose the environment to Cd where it causes soil and water pollution with the metal before its entry in food chains (Rahimzadeh et al., 2017). The metal has an atomic weight of 112.41 and an atomic number of 48. Its atomic weight is a result of mixing eight stable isotopes. Cd has a complete d orbital and two electrons in its s orbital, which makes its oxidation state as +2 in most chemical reactions (Kumar et al., 2021).

Lead (Pb)

Lead has an atomic number of 82 and is a ductile, malleable and soft metal, which is resistant to corrosion and has poor conductivity (Wani et al., 2015). The metal has four naturally occurring isotopes (^{208}Pb, ^{207}Pb, ^{206}Pb, ^{204}Pb) and the first three are stable natural radioactive series end products. The main source of natural lead is lead (II) sulfide (PbS), which is the heavy black mineral galena while mimetesite, $Pb_5(AsO_4)_3Cl$, cerussite, $PbCO_3$, anglesite, $PbSO_4$ and pyromorphite, $Pb_5(PO_4)_3Cl$ are minor ores of the metal (Kumar et al., 2015). Lead's commonest oxidation states are +2 and +4.

The extraction of Pb from galena is through froth flotation where the concentrated ore of the metal is roasted with little oxygen to form PbO, which is then reacted with a flux such as limestone and/or coke and finally, undergoes reduction in a blast furnace. Instead of carbon reduction, fresh PbS can also be used to reduce the ore as shown in Equation 24). Equations 21 - 24 represent the processes involved in Pb extraction.

$$PbS + 1.5O_2 \rightarrow PbO + SO_2 \tag{21}$$

$$PbO + C \rightarrow Pb(Liq) + CO \tag{22}$$

$$PbO + CO \rightarrow Pb(Liq) + CO_2 \tag{23}$$

$$PbS + 2PbO \rightarrow 3Pb(Liq) + SO \uparrow (g) \tag{24}$$

The resultant Pb contains valuable impurities such as Ag, As, Au, Cu, Sb and Sn. Several processes such as oxidation, preferential distillation, electrolysis and zone refining can be applied to purify the metal.

In many alloys, Pb is used as an inert material such as the metal use in batteries as a support for the reducing agent. It is also used as an anti-corrosive priming paint for steel and lead. In this case, calcium orthoplumbate (IV) (Ca_2PbO_4) galvanizes steel while lead tetraoxide (Pb_3O_4) is a primer and lead chromate ($PbCrO_4$) is an ingredient in colored plastics and green paints (Kumar et al., 2021). Exposure to the metal can occur via two main routes: - 1) ingestion of Pb-containing water, dust and food and 2) inhalation of Pb containing particulates generated during combustion of the metal in processes such as the use of lead aviation fuels, striping Pb-containing paints, recycling and smelting (WHO, 2022).

Arsenic (As)

Arsenic (As) is a metalloid of ubiquitous nature, which occurs in sulfide ores as arsenical pyrite (FeAsS), realgar (As_4S_4) and orpiment (As_4S_6) (Kumar et al., 2021). The metalloid is an elusive element due to its ability to change toxicity, reactivity, behavior and color (O'Day, 2006). For instance, realgar is red while orpiment is bright yellow. Its pentavalent and trivalent oxidation states are the commonest. The oxidation states result to three classes of As compounds: - arsine gas, organic and inorganic As compounds. Inorganic compounds such as arsenic trichloride, sodium arsenite and arsenic trioxide are trivalent while arsenates, arsenic acid and arsenic pentoxide are pentavalent in nature.

Arsenic occurs naturally in sulfide ores in association with Ag, Co, Cu, Fe, Ni, Pb and Sb. In large scale, the metal is obtained through smelting the iron arsenide (loellingite, $FeAs_2$) or FeAsS at 650-700 °C anaerobically before condensing the sublimed As. Sulfides of the metalloids are reacted with oxygen to form arsenic trioxide (As_2O_3), which is sublimed. The oxide is used directly as a chemical or is reduced in the presence of carbon to pure As at 700-800 °C. Equations 25 and 26 show the processes involved in extraction of the metalloid from arsenical pyrite.

$$\mathrm{FeAsS} \rightarrow \mathrm{FeS} + \mathrm{As}(\mathrm{g}) \rightarrow \mathrm{As}(\mathrm{s}) \tag{25}$$

$$\mathrm{As_4O_6} + 6\mathrm{C} \rightarrow \mathrm{As_4} + 6\mathrm{CO}\uparrow \tag{26}$$

The resultant metalloid is used in alloys to improve properties of storage batteries. Some compounds of As such as arsenic acid {$AsO(OH)_3$} are used as wood preservatives while gallium arsenide (GaAs) is used as a semiconductor along with Zn. Exposure routes of As include drinking contaminated water, contact with wood preservatives during manufacture or use, cigarette smoking, ingestion of As-contaminated food grown in metal contaminated soils, inhalation of contaminated air and use of cosmetics containing As (Chung et al., 2014).

Copper (Cu)

Copper (Cu) is a group 1b transition metal along with Ag and Au and with an atomic weight of 63.546. The metal has a filled d orbital, which enables it to have three oxidation states, 0 (pure metal), +1 (cuprous) where it loses an electron from 3s energy level and +2 (cupric) like other members of its period (Mustafa & AlSharif, 2018). However, the +1 oxidation state is uncommon. Sulfide ores are the commonest sources of Cu though oxides and carbonates also occur. More than half of all Cu deposits occur as copper pyrite ($CuFeS_2$) while minor deposits of the metal are from copper glance (Cu_2S), azurite ($2\{CuCO_3.Cu(OH)_2\}$) and malachite ($\{CuCO_3.Cu(OH)_2\}$) (Kumar et al., 2021).

Copper pyrite, which contains most of the Cu also contains Fe, which complicates the extraction process. The ore is first crushed and using froth flotation processes, it is concentrated. The concentrate is heated at 1400 °C before adding silica to it. Silica facilitates easy reaction of ferrous sulfide to oxides on the upper layer and a Cu matte in the lower layer with ferrous sulfide and cuprous sulfide. The matte is reacted in excess silica in the presence of oxygen to convert its ferrous sulfide to FeO while Cu_2S is initially converted to Cu_2O and finally, impure metallic Cu (blister Cu). The blister copper is then purified through electrolysis. The extraction process is represented using Equations 27 -29.

$$2FeS + 3O_2 \rightarrow 2FeO + 2SO_2 \tag{27}$$

$$2Cu_2S + 3O_2 \rightarrow 2Cu_2O + 2SO_2 \tag{28}$$

$$2Cu_2O + Cu_2S \rightarrow 6Cu + SO_2 \tag{29}$$

Copper obtained is used in piping, making electricals, roofing and for water tanks. Its alloys including bronze, cupro-nickel and brass are used in making utensils and coins. The metal is also used as a fungicide and in making catalysts. During the various applications of the metal, wastes containing the metal can be released to the environment and end up in soils where they cause pollution.

Zinc (Zn)

Zinc (Zn) is a metal element with stable isotopes of mass 66, 67, 68 and 70 and an atomic number of 30. It has a low boiling point and offers weak metal bonding. The metal has an $1s^2$, $2s^2$, $2p^6$, $3d^{10}$ electronic configuration and does not have an unfilled d subshell. For this reason, the metal manifests few characteristics of transition elements and has only one oxidation state of +2 (Kumar et al., 2021). Numerous ores including willemite ($Zn_4Si_2O_7(OH)_2.H_2O$), silicate ($ZnSiO_4$), hopeite ($Zn_3(PO_4)_2.4H_2O$), smithsonite ($ZnCO_3$), zincite (ZnO), franklinite ($ZnFe_2O_4$), gahnite ($ZnAl_2O_4$), zincosite ($ZnSO_4$), goslarite ($ZnSO_4$-$2H_2O$) and sphalerite (ZnS) are sources of Zn (Robson, 1993). However, ZnS and $ZnCO_3$ are the commonest sources.

The metal is usually sourced from ZnS ore and is extracted by froth flotation or sedimentation processes. Equation 30 shows the reaction involved in extraction of Zn from zincite.

$$ZnO + C \rightarrow Zn + CO \qquad (30)$$

The metal concentrate is reacted in air to produce SO_2 and its oxide (ZnO). The oxide is reacted with coke or smelted via electrolysis while the sulfur dioxide is used to produce sulfuric acid. Reoxidation of the metal quickly occurs as cooling occurs (Equation 31) but can be avoided by spraying Pb on Zn vapor during the process.

$$Zn + CO_2 \rightarrow ZnO + CO \qquad (31)$$

The resultant Zn, which is about 99.99% pure is used as an anticorrosion coating. The metal alloys are also used in manufacture of dry batteries, rood cladding, die-casting and also as a component of various enzymes.

Chromium (Cr)

Chromium is a gray crystalline metal with an atomic weight of 51.996 and a melting and boiling point of 1900 and 2642 °C, respectively (Mondal et al., 2021). The metal has four isotopes Cr-54, Cr-53, Cr-52 and Cr-50. It can lose all its outermost electrons just like V and Ti to acquire a +6-oxidation state. In this case, it acts as a strong oxidizing agent and is found in combination with F and O. Chromium also has +3 and +2 oxidation states with the former being the most stable. Some compounds of the metal also have 0, +4 and +5 oxidation states, which are usually unstable. Cr (III) and Cr (Vi) are the most stable oxidation states of the metal.

Chromium is sourced mainly from iron (II) chromite ($FeCr_2O_4$) and other minor ores such as crocoite ($PbCrO_4$) and chrome ocher (Cr_2O_3) (Kumar et al., 2021). Extraction of the metal involves molten alkali oxidation of chromite to sodium chromate (Na_2CrO_4), which is solubilized in water and then precipitates. By reacting the product with coke, Cr_2O_3 is reduced by Si or Al to Cr metal as shown in Equations 32 -33.

$$Cr_2O_3 + 2Al \rightarrow 2Cr + Al_2O_3 \qquad (32)$$

$$2Cr_2O_3 + 3Si \rightarrow 4Cr + 3SiO_2 \qquad (33)$$

The extracted metal is used in manufacturing nonferrous alloys and in decorative and protective Cr plating. Activities such as spray painting, leather processing and arc welding could expose natural environments and individuals to Cr in addition to ingesting HM-contaminated water and food (Mondal et al., 2021).

Cobalt (Co)

Cobalt (Co) has an atomic weight of 58.933 and is a result of eight isotopes – Co-56, Co-57, Co-58, Co-58m, Co-59, CO-60, Co-60m and Co-61. The metal is gray, ductile, malleable and has natural magnetic properties that are apparent when the metal is combined with other elements. The metal has a high boiling and melting point of 2870 and 1495 °C, respectively. The electron configuration of Co is [Ar] $4s^2 3d^7$ and its oxidation state is +2 and +3 for Co (II) and Co (III), respectively (Yildiz, 2017). Cobalt is found in ores such as cobaltite (CoAsS), smaltite {(CONi) As_2) or $CoAs_2$} or linnaeite (CO_3S_4). The metal is used in making alloys such as magnet steel, vitallium and stellite (Davis, 2009). It is also used as a catalyst in organic synthesis (Hapke & Gerhard, 2020).

CONCLUSION

Heavy metals are important elements that occur naturally and their availability to the environment is as a result of anthropogenic activities in the industrial era. Most of the HM are toxic and their ability to inflict toxicity depends on their bioavailability, level of exposure and time. In enhancing toxicity in living organisms, the metals cause homeostatic imbalances propagated by the production of free radicals such as ROS and RNS. In this chapter, several types of toxicity were associated with HM contamination. They include neurotoxicity, nephrotoxicity, cardiotoxicity, hepatoxicity, skin toxicity, genotoxicity and carcinogenicity. Exposure routes to which the metals cause health harm include inhalation of particulates, ingestion of contaminated food grown on heavy metal polluted soils and water as well as occupational exposure in industries working with HM. This chapter therefore established that HM are ubiquitous in the environment and have the potential to cause harm to the environment and living things. Additionally, their physicochemical characteristics regulate their ability to exert toxic effects in the environment and living organisms.

REFERENCES

Adams, W., Blust, R., Dwyer, R., Mount, D., Nordheim, E., Rodriguez, P., & Spry, D. (2020). Bioavailability assessment of metals in freshwater environments: A historical review. *Environmental Toxicology and Chemistry*, *39*(1), 48–59. doi:10.1002/etc.4558 PMID:31880839

Aggarwal, V., Tuli, H., Varol, A., Thakral, F., Yerer, M., Sak, K., Varol, M., Jain, A., Khan, M., & Sethi, G. (2019). Role of reactive oxygen species in cancer progression: Molecular mechanisms and recent advancements. *Biomolecules*, *9*(11), 735. doi:10.3390/biom9110735 PMID:31766246

Ali, H., & Khan, E. (2018). What are heavy metals? Long standing controversy over scientific use of the term 'heavy metal'- proposal of a comprehensive definition. *Toxicological and Environmental Chemistry*, *100*(1), 619. doi:10.1080/02772248.2017.1413652

Ali, S., Awan, Z., Mumtaz, S., Shakir, H., Ahmad, F., Ulhaq, M., Tahir, H. M., Awan, M. S., Sharif, S., Irfan, M., & Khan, M. A. (2020). Cardiac toxicity of heavy metals (cadmium and mercury) and pharmacological intervention by vitamin C in rabbits. *Environmental Science and Pollution Research International*, *27*(23), 29266–29279. doi:10.100711356-020-09011-9 PMID:32436095

Apostoli, P., & Catalani, S. (2011). *Metal ions in toxicology: effects, interactions, interdependencies*. De Gruyter. doi:10.1515/9783110436624-016

Baba, H., Tsuneyama, K., Yazaki, M., Nagata, K., Minamisaka, T., Tsuda, T., Nomoto, K., Hayashi, S., Miwa, S., Nakajima, T., Nakanishi, Y., Aoshima, K., & Imura, J. (2013). The liver in itai-itai disease (chronic cadmium poisoning): Pathological features and metallothionein expression. *Modern Pathology*, *26*(9), 1228–1234. doi:10.1038/modpathol.2013.62 PMID:23558578

Balali-Mood, M., Naseri, K., Tahergorabi, Z., Khazdair, M., & Sadeghi, M. (2021). Toxic mechanisms of five heavy metals: Mercury, lead, chromium, cadmium, and arsenic. *Frontiers in Pharmacology*, *12*, 643972. doi:10.3389/fphar.2021.643972 PMID:33927623

Bar-Zeev, Y., Greenberg, D., Ling, G., & Lifshitz, M. (2002). Acrodynia: A case report of two siblings. *Archives of Disease in Childhood*, *86*(6), 453. doi:10.1136/adc.86.6.453 PMID:12023189

Belles, M., Albina, M., Sanchez, D., Corbella, J., & Domingo, L. (2002). Interactions in developmental toxicology: Effects of concurrent exposure to lead, organic mercury, and arsenic in pregnant mice. *Archives of Environmental Contamination and Toxicology*, *42*(1), 93–98. doi:10.1007002440010296 PMID:11706373

Benvenga, S., Marini, H., Micali, A., Freni, J., Pallio, G., Irrera, N., Squadrito, F., Altavilla, D., Antonelli, A., Ferrari, S. M., Fallahi, P., Puzzolo, D., & Minutoli, L. (2020). Protective effects of myo-inositol and selenium on cadmium-induced thyroid toxicity in mice. *Nutrients*, *12*(5), 1222. doi:10.3390/nu12051222 PMID:32357526

Bezerra, P., Takiyama, L., & Bezerra, C. (2009). Complexation of metal ions by dissolved organic matter: Modeling and application to real systems. *Acta Amazonica*, *39*, 639–648. doi:10.1590/S0044-59672009000300019

Bodaghi-Namileh, V., Sepand, R., Omidi, A., Aghsami, M., Seyednejad, A., Kasirzadeh, S., et al. (2018). Acetyl- l -carnitine attenuates arsenic-induced liver injury by abrogation of mitochondrial dysfunction, inflammation, and apoptosis in rats. *Environmental Toxicology and Pharmacology, 58*, 11–20. https://doi.org/. 2017.12.005 doi:10.1016/j.etap

Boyd, A., Seger, D., Vannucci, S., Langley, M., Abraham, L., & King, E. Jr. (2000). Mercury exposure and cutaneous disease. *Journal of the American Academy of Dermatology*, *43*(1), 81–90. doi:10.1067/mjd.2000.106360 PMID:10863229

Branca, J., Morucci, G., & Pacini, A. (2018). Cadmium-induced neurotoxicity: Still much ado. *Neural Regeneration Research*, *13*(11), 1879–1882. doi:10.4103/1673-5374.239434 PMID:30233056

Brezova, V., Valko, M., Breza, M., Morris, H., Telser, J., Dvoranova, D., Kaiserova, K., Varecka, L., Mazur, M., & Leibfritz, D. (2003). Role of radicals and singlet oxygen in photoactivated DNA cleavage by the anticancer drug camptothecin: An electron paramagnetic resonance study. *The Journal of Physical Chemistry B*, *107*(10), 2415–2425. doi:10.1021/jp027743m

Brown, A., Thomas, G., & Lindsley, C. W. (2017). Targeting phospholipase D in cancer, infection and neurodegenerative disorders. *Nature Reviews. Drug Discovery*, *16*(5), 351–367. doi:10.1038/nrd.2016.252 PMID:28209987

Burkitt, M. (2001). A critical overview of the chemistry of copper-dependent low density lipoprotein oxidation: Roles of lipid hydroperoxides, α-tocopherol, thiols, and ceruloplasmin. *Archives of Biochemistry and Biophysics*, *394*(1), 117–135. doi:10.1006/abbi.2001.2509 PMID:11566034

Buxton, S., Garman, E., Heim, K., Darden, T., Schlekat, C., & Taylor, M. (2019). Concise review of nickel human health toxicology and ecotoxicology. *Inorganics*, *7*(7), 89. doi:10.3390/inorganics7070089

Cempel, M., & Nikel, G. (2005). Nickel: A review of its sources and environmental toxicology. *Polish Journal of Environmental Studies*, *15*(3), 375–382.

Chung, J., Yu, S., & Hong, Y. (2014). Environmental source of arsenic exposure. *Journal of Preventive Medicine and Public Health*, *47*(5), 253–257. doi:10.3961/jpmph.14.036 PMID:25284196

Crans, C., Smee, J., Gaidamauskas, E., & Yang, L. (2004). The chemistry and biochemistry of vanadium and the biological activities exerted by vanadium compounds. *Chemical Reviews*, *104*(2), 849–902. doi:10.1021/cr020607t PMID:14871144

Davis, B., Price, H., O'Connor, R., Fernando, R., Rowland, A., & Morgan, D. (2001). Mercury vapor and female reproductive toxicity. *Toxicological Sciences*, *59*(2), 291–296. doi:10.1093/toxsci/59.2.291 PMID:11158722

Davis, J. (2009). Nickel, cobalt, and their alloys. Cobalt market review. ASM International.

Djoko, K., Ong, C., Walker, M., & McEwan, A. (2015). The role of copper and zinc toxicity in innate immune defense against bacterial pathogens. *The Journal of Biological Chemistry*, *290*(31), 18954–18961. doi:10.1074/jbc.R115.647099 PMID:26055706

Dutta, S., Joshi, K., Sengupta, P., & Bhattacharya, K. (2013). Unilateral and bilateral cryptorchidism and its effect on the testicular morphology, histology, accessory sex organs and sperm count in laboratory mice. *Journal of Human Reproductive Sciences*, *6*(2), 106–110. doi:10.4103/0974-1208.117172 PMID:24082651

Eaton, J., & Qian, M. (2002). Molecular bases of cellular iron toxicity. *Free Radical Biology & Medicine*, *32*(9), 833–840. doi:10.1016/S0891-5849(02)00772-4 PMID:11978485

Engwa, G., Udoka, P., Nwalo, F., & Unachukwu, M. (2019). *Mechanism and health effects of heavy metal toxicity in humans*. Intech Open. doi:10.5772/intechopen.82511

Fang, Z., Zhao, M., Zhen, H., Chen, L., Shi, P., & Huang, Z. (2014). Genotoxicity of tri- and hexavalent chromium compounds in vivo and their modes of action on DNA damage in vitro. *PLoS One*, *9*(8), e103194. doi:10.1371/journal.pone.0103194 PMID:25111056

Friberg, L., Kjellström, T., Elinder, G., & Nordberg, F. (2019). Cadmium and health: a toxicological and epidemiological appraisal. *Cadmium Health: A Toxicological and Epidemiological Appraisal*. . doi:10.1201/9780429260599

Garg, U., Kaur, M., Garg, V., & Sud, D. (2007). Removal of hexavalent chromium from aqueous solution by agricultural waste biomass. *Journal of Hazardous Materials*, *140*(1-2), 60–68. doi:10.1016/j.jhazmat.2006.06.056 PMID:16879918

Garza-Lombó, C., Pappa, A., Panayiotidis, M., Gonsebatt, E., & Franco, R. (2019). Arsenic-induced neurotoxicity: A mechanistic appraisal. *Journal of Biological Inorganic Chemistry*, *24*(8), 1305–1316. doi:10.100700775-019-01740-8 PMID:31748979

Genchi, G., Sinicropi, M., Lauria, G., Carocci, A., & Catalano, A. (2020). The effects of cadmium toxicity. *International Journal of Environmental Research and Public Health*, *17*(11), 3782. doi:10.3390/ijerph17113782 PMID:32466586

Goldhaber, S. (2003). Trace element risk assessment: Essentiality vs. toxicity. *Regulatory Toxicology and Pharmacology*, *38*(2), 232–242. doi:10.1016/S0273-2300(02)00020-X PMID:14550763

Hapke, M., & Gerhard, H. (2020). Introduction to cobalt chemistry and catalysis. In M. Hapke & H. Gerhard (Eds.), *Cobalt catalysis in organic synthesis: methods and reactions*. Wiley Publishers., doi:10.1002/9783527814855.ch1

Harischandra, S., Ghaisas, S., Zenitsky, G., Jin, H., Kanthasamy, A., Anantharam, V., & Kanthasamy, A. G. (2019). Manganese-induced neurotoxicity: New insights into the triad of protein misfolding, mitochondrial impairment, and neuroinflammation. *Frontiers in Neuroscience*, *13*, 654. doi:10.3389/fnins.2019.00654 PMID:31293375

Hartwig, A., & Schwerdtle, T. (2002). Interactions by carcinogenic metal compounds with DNA repair processes: Toxicological implications. *Toxicology Letters*, *127*(1-3), 47–54. doi:10.1016/S0378-4274(01)00482-9 PMID:12052640

Hazelhoff, H., & Torres, A. (2018). Gender differences in mercury-induced hepatotoxicity: Potential mechanisms. *Chemosphere*, *202*, 330–338. doi:10.1016/j.chemosphere.2018.03.106 PMID:29574386

Hemmaphan, S., & Bordeerat, N. (2022). Genotoxic Effects of Lead and Their Impact on the Expression of DNA Repair Genes. *International Journal of Environmental Research and Public Health*, *19*(7), 4307. doi:10.3390/ijerph19074307 PMID:35409986

Hemminki, K., Niemi, M., Kostinen, K., & Vainio, H. (1980). Spontaneous abortions among women employed in the metal industry in Finland. *International Archives of Occupational and Environmental Health*, *47*(1), 53–60. doi:10.1007/BF00378328 PMID:7429646

Hong, Y., Kim, Y., & Lee, K. (2012). Methylmercury exposure and health effects. *Journal of Preventive Medicine and Public Health*, *45*(6), 353–363. doi:10.3961/jpmph.2012.45.6.353 PMID:23230465

Hsiao, C., Wu, H., & Wan, S. (2011). Effects of environmental lead exposure on T-helper cell-specific cytokines in children. *Journal of Immunotoxicology*, *8*(4), 284–287. doi:10.3109/1547691X.2011.592162 PMID:21726182

Huang, H., Lee, C., & Yu, H. (2019). Arsenic-induced carcinogenesis and immune dysregulation. *International Journal of Environmental Research and Public Health*, *16*(15), 2746. doi:10.3390/ijerph16152746 PMID:31374811

IARC. (2012). Monographs on the evaluation of carcinogenic risk to human. International Agency for Research on Cancer.

Irina, R., & John, R. (2020). Does manganese contribute to methamphetamine-induced psychosis? *Current Emergency and Hospital Medicine Reports*, *8*(4), 133–141. doi:10.100740138-020-00221-6

Jaishankar, M., Tseten, T., Anbalagan, N., Mathew, B., & Beeregowda, N. (2014). Toxicity, mechanism and health effects of some heavy metals. *Interdisciplinary Toxicology*, *7*(2), 60–72. doi:10.2478/intox-2014-0009 PMID:26109881

Jan, A., Azam, M., Siddiqui, K., Ali, A., Choi, I., & Haq, Q. (2015). Heavy metals and human health: Mechanistic insight into toxicity and counter defense system of antioxidants. *International Journal of Molecular Sciences*, *16*(12), 29592–29630. doi:10.3390/ijms161226183 PMID:26690422

Jyothi, N. (2020). *Heavy metal sources and their effects on human health*. Intech Open.

Kasten-Jolly, J., Heo, Y., & Lawrence, D. (2010). Impact of developmental lead exposure on splenic factors. *Toxicology and Applied Pharmacology*, *247*(2), 105–115. doi:10.1016/j.taap.2010.06.003 PMID:20542052

Kim, D., Ock, J., Moon, K., & Park, C. (2021). Associations between Pb, Cd and Hg exposure and liver injury among Korean adults. *International Journal of Environmental Research and Public Health*, *18*(13), 6783. doi:10.3390/ijerph18136783 PMID:34202682

Kim, H., Kim, Y., & Seo, Y. (2015). An overview of carcinogenic heavy metal: Molecular toxicity mechanism and prevention. *Journal of Cancer Prevention, 20*(4), 232–241. doi:10.15430/JCP.2015.20.4.232 PMID:26734585

Kim, Y., & Kim, M. (2015). Arsenic toxicity in male reproduction and development. *Development & Reproduction, 19*(4), 167–180. doi:10.12717/DR.2015.19.4.167 PMID:26973968

Kocadal, K., Alkas, F., Battal, D., & Saygi, S. (2020). Cellular pathologies and genotoxic effects arising secondary to heavy metal exposure: A review. *Human and Experimental Toxicology, 39*(1), 3–13. doi:10.1177/0960327119874439 PMID:31496299

Kozlova, T., Wood, C., & McGeer, J. (2009). The effect of water chemistry on the acute toxicity of nickel to the cladoceran Daphnia pulex and the development of a biotic ligand model. *Aquatic Toxicology (Amsterdam, Netherlands), 91*(3), 221–228. doi:10.1016/j.aquatox.2008.11.005 PMID:19111357

Kulka, M. (2016). A review of paraoxonase 1 properties and diagnostic applications. *Polish Journal of Veterinary Sciences, 19*(1), 225–232. doi:10.1515/pjvs-2016-0028 PMID:27096809

Kumar, M., Sawhney, N., & Lal, R. (2021). Chemistry of heavy metals in the environment. In V. Kumar, A. Sharma, & A. Cerdia (Eds.), *Heavy metals in the environment*. Elsevier. doi:10.1016/B978-0-12-821656-9.00002-X

Kupper, M., Weinbruch, S., Skaug, V., Skogstad, A., Thornér, E., Benker, N., Ebert, M., Chashchin, V., Odland, J. Ø., & Thomassen, Y. (2015). Electron microscopy of particles deposited in the lungs of nickel refinery workers. *Analytical and Bioanalytical Chemistry, 407*(21), 6435–6445. doi:10.100700216-015-8806-z PMID:26077746

Lamas, G., Goertz, C., Boineau, R., Mark, D., Rozema, T., Nahin, R. L., Drisko, J. A., & Lee, K. L. (2012). Design of the trial to assess chelation therapy (TACT). *American Heart Journal, 163*(1), 7–12. doi:10.1016/j.ahj.2011.10.002 PMID:22172430

Lentini, P., Zanoli, L., Granata, A., Signorelli, S., Castellino, P., & Aquila, R. (2017). Kidney and heavy metals-the role of environmental exposure [review]. *Molecular Medicine Reports, 15*(5), 3413–3419. doi:10.3892/mmr.2017.6389 PMID:28339049

Liaw, J., Marshall, G., Yuan, Y., Ferreccio, C., Steinmaus, C., & Smith, H. (2008). Increased childhood liver cancer mortality and arsenic in drinking water in northern Chile. *Cancer Epidemiology, Biomarkers & Prevention, 17*(8), 1982–1987. doi:10.1158/1055-9965.EPI-07-2816 PMID:18708388

Liu, K., & Shi, X. (2001). In vivo reduction of chromium (VI) and its related free radical generation. *Molecular and Cellular Biochemistry*, *222*(1/2), 41–47. doi:10.1023/A:1017994720562 PMID:11678610

Lloyd, R., Hanna, P., & Mason, R. (1997). The origin of the hydroxyl radical oxygen in the Fenton reaction. *Free Radical Biology & Medicine*, *22*(5), 885–888. doi:10.1016/S0891-5849(96)00432-7 PMID:9119257

Magalhaes, D., Marques, M., Baptista, D., & Buss, D. (2015). Metal bioavailability and toxicity in freshwaters. *Environmental Chemistry Letters*, *13*(1), 69–87. doi:10.100710311-015-0491-9

Magos, L., & Clarkson, T. (2006). Overview of the clinical toxicity of mercury. *Annals of Clinical Biochemistry*, *43*(4), 257–268. doi:10.1258/000456306777695654 PMID:16824275

Marchetti, C. (2014). Interaction of metal ions with neurotransmitter receptors and potential role in neurodiseases. *Biometals*, *27*(6), 1097–1113. doi:10.100710534-014-9791-y PMID:25224737

Martinez, D., Vucic, A., Becker-Santos, D., Gil, L., & Lam, L. (2011). Arsenic exposure and the induction of human cancers. *Journal of Toxicology*, *2011*, 1–13. doi:10.1155/2011/431287 PMID:22174709

Messner, B., Knoflach, M., Seubert, A., Ritsch, A., Pfaller, K., Henderson, B., Shen, Y. H., Zeller, I., Willeit, J., Laufer, G., Wick, G., Kiechl, S., & Bernhard, D. (2010). Cadmium is a novel and independent risk factor for early atherosclerosis mechanisms and in vivo relevance. *Arteriosclerosis, Thrombosis, and Vascular Biology*, *29*(9), 1392–1398. doi:10.1161/ATVBAHA.109.190082 PMID:19556524

Milton, A., Hussain, S., Akter, S., Rahman, M., Mouly, T., & Mitchell, K. (2017). A review of the effects of chronic arsenic exposure on adverse pregnancy outcomes. *International Journal of Environmental Research and Public Health*, *14*(6), 556. doi:10.3390/ijerph14060556 PMID:28545256

Miranda, L., Wijesiri, B., Ayoko, G., Egodawatta, P., & Goonetilleke, A. (2021). Water-sediment interactions and mobility of heavy metals in aquatic environments. *Water Research*, *202*(1), 117386. doi:10.1016/j.watres.2021.117386 PMID:34229194

Mishra, K., & Singh, S. (2020). Heavy metals exposure and risk to autoimmune diseases: A review. *Archives of Immunology and Allergy*, *3*(2), 22–26.

Mitra, S., Chakraborty, A., Tareq, A., Emran, T., Nainu, F., Khusro, A., Idris, A. M., Khandaker, M. U., Osman, H., Alhumaydhi, F. A., & Simal-Gandara, J. (2022). Impact of heavy metals on the environment and human health: Novel therapeutic insights to counter toxicity. *Journal of King Saud University. Science*, *34*(3), 101865. doi:10.1016/j.jksus.2022.101865

Miura, S., Takahashi, K., Imagawa, T., Uchida, K., Saito, S., Tominaga, M., & Ohta, T. (2013). Involvement of TRPA1 activation in acute pain induced by cadmium in mice. *Molecular Pain*, *9*, 1744-8069-9-7. Advance online publication. doi:10.1186/1744-8069-9-7 PMID:23448290

Mondal, M., Begum, W., Nasrollahzadeh, M., Ghorbannezhad, F., & Antoniadis, V. (2021). A comprehensive review on chromium chemistry along with detection, speciation, extraction and remediation of hexavalent chromium in contemporary science and technology. *Vietnam Journal of Chemistry*, *59*(6), 711–732. doi:10.1002/vjch.202100048

Mustafa, S., & AlSharif, M. (2018). Copper (Cu) an Essential Redox-Active Transition Metal in Living System—A Review Article. *American Journal of Analytical Chemistry*, *9*(1), 15–26. doi:10.4236/ajac.2018.91002

Naz, B., & Batool, S. (2017). Infertility related issues and challenges: Perspectives of patients, spouses, and infertility experts. *Pakistan Journal of Clinical Sociology and Psychology*, *15*, 3–11.

Nordstrom, S., Beckman, L., & Nordenson, I. (1978). Occupational and environmental risks in and around a smelter in northern Sweden. *Hereditas*, *88*(1), 43–46. doi:10.1111/j.1601-5223.1978.tb01600.x PMID:649423

Nyika, J. (2021). Tolerance of microorganisms to heavy metals. In S. Dey & B. Acharya (Eds.), *Recent advancements in bioremediation of metal contaminants*. IGI Global. doi:10.4018/978-1-7998-4888-2.ch002

Nyika, J., & Dinka, M. (2022). Heavy metal pollution in soils and vegetables from suburban regions of Nairobi, Kenya and their community health implications. *Pollution*, *8*(4), 1434–1447. doi:10.22059/POLL.2022.341522.1440

Nyika, J., Onyari, E., Dinka, M., & Mishra, S. (2019). Heavy metal pollution and mobility in soils within a landfill vicinity: A South African case study. *Oriental Journal of Chemistry*, *35*(4), 1286–1296. doi:10.13005/ojc/350406

O'Day, P. (2006). Chemistry and mineralogy of arsenic. *Elements (Quebec)*, *2*(2), 77–83. doi:10.2113/gselements.2.2.77

Obone, E., Chakrabarti, S., Bai, M., Malick, M., Lamontagne, L., & Subramanian, K. (1999). Toxicity and bioaccumulation of nickel sulfate in Sprague-Dawley rats following 13 weeks of subchronic exposure. *Journal of Toxicology and Environmental Health. Part A.*, *57*(6), 379–401. doi:10.1080/009841099157593 PMID:10478821

Oloniran, A., Balgobind, A., & Pillay, B. (2013). Bioavailability of heavy metals in soil: Impact on microbial biodegradation of organic compounds and possible improvement strategies. *International Journal of Molecular Sciences*, *14*(5), 10197–10228. doi:10.3390/ijms140510197 PMID:23676353

Omobowale, O., Oyagbemi, A., Akinrinde, S., Saba, B., Daramola, T., Ogunpolu, S., & Olopade, J. O. (2014). Failure of recovery from lead induced hepatoxicity and disruption of erythrocyte antioxidant defense system in Wistar rats. *Environmental Toxicology and Pharmacology*, *37*(3), 1202–1211. doi:10.1016/j.etap.2014.03.002 PMID:24814264

Orr, S., & Bridges, C. (2017). Chronic kidney disease and exposure to nephrotoxic metals. *International Journal of Molecular Sciences*, *18*(5), 1039. doi:10.3390/ijms18051039 PMID:28498320

Packer, M. (2016). Cobalt cardiomyopathy: a critical reappraisal in light of a recent resurgence. *Circulation: Heart Failure* *9*(12). . doi:10.1161/CIRCHEARTFAILURE.116.003604

Park, Y.-H., Kim, D., Dai, J., & Zhang, Z. (2015). Human bronchial epithelial BEAS-2B cells, an appropriate in vitro model to study heavy metals induced carcinogenesis. *Toxicology and Applied Pharmacology*, *287*(3), 240–245. doi:10.1016/j.taap.2015.06.008 PMID:26091798

Pi, H., Xu, S., Reiter, J., Guo, P., Zhang, L., & Li, Y. (2015). SIRT3-SOD2- mROS-dependent autophagy in cadmium-induced hepatotoxicity and salvage by melatonin. *Autophagy, 11*(7), 1037–1051. https://doi.org/. 1052208 doi:10.1080/15548627.2015

Pi, J., Horiguchi, S., Sun, Y., Nikaido, M., Shimojo, N., Hayashi, T., Yamauchi, H., Itoh, K., Yamamoto, M., Sun, G., Waalkes, M. P., & Kumagai, Y. (2003). A potential mechanism for the impairment of nitric oxide formation caused by prolonged oral exposure to arsenate in rabbits. *Free Radical Biology & Medicine*, *35*(1), 102–113. doi:10.1016/S0891-5849(03)00269-7 PMID:12826260

Pierce, L., Kibriya, G., Tong, L., Jasmine, F., Argos, M., Roy, S., Paul-Brutus, R., Rahaman, R., Rakibuz-Zaman, M., Parvez, F., Ahmed, A., Quasem, I., Hore, S. K., Alam, S., Islam, T., Slavkovich, V., Gamble, M. V., Yunus, M., Rahman, M., & Ahsan, H. (2012). Genome-wide association study identifies chromosome 10q24.32 variants associated with arsenic metabolism and toxicity phenotypes in Bangladesh. *PLOS Genetics*, *8*(2), e1002522. doi:10.1371/journal.pgen.1002522 PMID:22383894

Prohaska, J. (2000). Long-term functional consequences of malnutrition during brain development: Copper. *Nutrition (Burbank, Los Angeles County, Calif.)*, *16*(7-8), 502–504. doi:10.1016/S0899-9007(00)00308-7 PMID:10906536

Rahimzadeh, M., Rahimzadeh, M., Kazemi, S., & Moghadamnia, A. (2017). Cadmium toxicity and treatment: An update. *Caspian Journal of Internal Medicine*, *8*, 135–145. doi:10.22088/acadpub.bums.8.2.67 PMID:28932363

Rani, A., Kumar, A., Lal, A., & Pant, M. (2014). Cellular mechanisms of cadmium-induced toxicity: A review. *International Journal of Environmental Health Research*, *24*(4), 378–399. doi:10.1080/09603123.2013.835032 PMID:24117228

Rattan, S., Zhou, C., Chiang, C., Mahalingam, S., Brehm, E., & Flaws, J. (2017). Exposure to endocrine disruptors during adulthood: Consequences for female fertility. *The Journal of Endocrinology*, *233*(3), R109–R129. doi:10.1530/JOE-17-0023 PMID:28356401

Reczek, R., & Chandel, N. (2017). The two faces of reactive oxygen species in cancer. *Annual Review of Cancer Biology*, *1*(1), 79–98. doi:10.1146/annurev-cancerbio-041916-065808

Renu, K., Chakraborty, R., Myakala, H., Koti, R., Famurewa, A., Madhyastha, H., Vellingiri, B., George, A., & Valsala Gopalakrishnan, A. (2021). Molecular mechanism of heavy metals (lead, chromium, arsenic, mercury, nickel and cadmium)-induced hepatoxicity-a review. *Chemosphere*, *271*, 129735. doi:10.1016/j.chemosphere.2021.129735 PMID:33736223

Rerkmitr, P., Kantikosum, K., Chottawornsak, N., Tangkijngamvong, N., Kerr, S., & Prueksapanich, P. (2019). Chronic occupational exposure to lead leads to significant mucocutanerous changes in lead factory workers. *Journal of the European Academy of Dermatology and Venereology*, *33*(10), 1993–2000. doi:10.1111/jdv.15678 PMID:31087433

Reyes, J., Molina-Jijón, E., Rodríguez-Muñoz, R., Bautista-García, P., Debray-García, Y., & Namorado, M. (2013). Tight junction proteins and oxidative stress in heavy metals-induced nephrotoxicity. *BioMed Research International*, *2013*, 1–14. doi:10.1155/2013/730789 PMID:23710457

Robson, A. (1993). The chemistry of zinc. In P. Barak & P. Helmke (Eds.), *Zinc in soils and plants. Developments in plant and soil sciences*. Springer. doi:10.1007/978-94-011-0878-2_1

Roy, J., Chatterjee, D., Das, N., & Giri, K. (2018). Substantial evidences indicate that inorganic arsenic is a genotoxic carcinogen: A review. *Toxicological Research, 34*(4), 311–324. doi:10.5487/TR.2018.34.4.311 PMID:30370006

Salnikow, K., & Zhitkovich, A. (2008). Genetic and epigenetic mechanisms in metal carcinogenesis and cocarcinogenesis: Nickel, arsenic, and chromium. *Chemical Research in Toxicology, 21*(1), 28–44. doi:10.1021/tx700198a PMID:17970581

Saturday, A. (2018). Mercury and its associated impacts on environment and human health: A review. *Journal of Environment and Health Sciences, 4*(2), 37–43. doi:10.15436/2378-6841.18.1906

Scerri, E. (2020). Recent attempts to change the periodic table. *Philosophical Transactions - Royal Society. Mathematical, Physical, and Engineering Sciences, 378*(2180), 20190300. doi:10.1098/rsta.2019.0300 PMID:32811365

Sengupta, P., Banerjee, R., Nath, S., Das, S., & Banerjee, S. (2015). Metals and female reproductive toxicity. *Human and Experimental Toxicology, 34*(7), 679–697. doi:10.1177/0960327114559611 PMID:25425549

Sevim, C., Dogan, E., & Comakli, S. (2020). Cardiovascular disease and toxic metals. *Current Opinion in Toxicology, 19*, 88–92. doi:10.1016/j.cotox.2020.01.004

Sharifi, A., Darabi, R., Akbarloo, N., Larijani, B., & Khoshbaten, A. (2004). Investigation of circulatory and tissue ACE activity during development of lead-induced hypertension. *Toxicology Letters, 153*(2), 233–238. doi:10.1016/j.toxlet.2004.04.013 PMID:15451554

Singh, H., Mahajan, P., Kaur, S., Batish, D., & Kohli, R. (2013). Chromium toxicity and tolerance in plants. *Environmental Chemistry Letters, 11*(3), 229–254. doi:10.100710311-013-0407-5

Skalnaya, M., & Skalny, V. (2018). *Essential trace elements in human health: A physician's view*. Publishing House of Tomsk State University.

Slobodian, M., Petahtegoose, J., Wallis, A., Levesque, D., & Merritt, T. (2021). The effects of essential and non-essential metal toxicity in the drosophila melanogaster insect model: A review. *Toxics, 9*(10), 269. doi:10.3390/toxics9100269 PMID:34678965

Tchounwou, P., Yedjou, C., Patlolla, A., & Sutton, D. (2012). Heavy metals toxicity and the environment. *EXS, 101*, 133–164. doi:10.1007/978-3-7643-8340-4_6 PMID:22945569

Tellez-Plaza, M., Guallar, E., Howard, V., Umans, G., Francesconi, A., Goessler, W., Silbergeld, E. K., Devereux, R. B., & Navas-Acien, A. (2013). Cadmium exposure and incident cardiovascular disease. *Epidemiology (Cambridge, Mass.), 24*(3), 421–429. doi:10.1097/EDE.0b013e31828b0631 PMID:23514838

Valko, M., Izakovic, M., Mazur, M., Rhodes, C., & Telser, J. (2004). Role of oxygen radicals in DNA damage and cancer incidence. *Molecular and Cellular Biochemistry, 266*(1/2), 37–56. doi:10.1023/B:MCBI.0000049134.69131.89 PMID:15646026

Vaziri, D. (2008). Mechanisms of lead-induced hypertension and cardiovascular disease. *American Journal of Physiology. Heart and Circulatory Physiology, 295*(2), H454–H465. doi:10.1152/ajpheart.00158.2008 PMID:18567711

Wang, B., Chiu, H., Lee, Y., Li, C., Wang, Y., & Lee, Y. (2018). Pterostilbene attenuates hexavalent chromium-induced allergic contact dermatitis by preventing cell apoptosis and inhibiting IL-1b-related NLRP3 inflammasome activation. *Journal of Clinical Medicine, 7*(12), 489. doi:10.3390/jcm7120489 PMID:30486377

Wang, B., & Du, Y. (2013). Cadmium and its neurotoxic effects. *Oxidative Medicine and Cellular Longevity, 898034*, 1–12. Advance online publication. doi:10.1155/2013/898034 PMID:23997854

Wang, J., Zhu, H., Yang, Z., & Liu, Z. (2013). Antioxidative effects of hesperetin against lead acetate-induced oxidative stress in rats. *Indian Journal of Pharmacology, 45*(4), 395–398. doi:10.4103/0253-7613.115015 PMID:24014918

Wang, Z., Sun, Y., Yao, W., Ba, Q., & Wang, H. (2021). Effects of cadmium exposure on the immune system and immunoregulation. *Frontiers in Immunology, 12*, 695484. doi:10.3389/fimmu.2021.695484 PMID:34354707

World Health Organization. (2022). *Lead poisoning*. WHO. https://www.who.int/news-room/fact-sheets/detail/lead-poisoning-and-health

Wu, X., Cobbina, S. J., Mao, G., Xu, H., Zhang, Z., & Yang, L. (2016). A review of toxicity and mechanisms of individual and mixtures of heavy metals in the environment. *Environmental Science and Pollution Research International, 23*(9), 8244–8259. doi:10.100711356-016-6333-x PMID:26965280

Xu, D., Shen, Z., Dou, C., Dou, Z., Li, Y., Gao, Y., & Sun, Q. (2022). Effects of soil properties on heavy metal bioavailability and accumulation in crop grains under different farmland use patterns. *Scientific Reports*, *12*(1), 9211. doi:10.103841598-022-13140-1 PMID:35654920

Yang, O., Kim, H., Weon, J., & Seo, Y. (2015). Endocrine-disrupting chemicals: Review of toxicological mechanisms using molecular pathway analysis. *Journal of Cancer Prevention*, *20*(1), 12–24. doi:10.15430/JCP.2015.20.1.12 PMID:25853100

Yildiz, Y. (2017). *General aspects of the cobalt chemistry*. Intech Open. doi:10.5772/intechopen.71089

Yoshihisa, Y., & Shimizu, T. (2012). Metal allergy and systemic contact dermatitis: An overview. *Dermatology Research and Practice*, *1-5*, 1–5. Advance online publication. doi:10.1155/2012/749561 PMID:22693488

Yoshizawa, K., Rimm, B., Morris, S., Spate, L., Hsieh, C., Spiegelman, D., Stampfer, M. J., & Willett, W. C. (2002). Mercury and the risk of coronary heart disease in men. *The New England Journal of Medicine*, *347*(22), 1755–1760. doi:10.1056/NEJMoa021437 PMID:12456851

Yu, H., Liao, W., & Chai, C. (2006). Arsenic carcinogenesis in the skin. *Journal of Biomedical Science*, *13*(5), 657–666. doi:10.100711373-006-9092-8 PMID:16807664

Yumoto, T., Tsukahara, K., Naito, H., Iida, A., & Nakao, A. (2017). A successfully treated case of criminal thallium poisoning. *Journal of Clinical and Diagnostic Research : JCDR*, *11*, od01–od02. doi:10.7860/JCDR/2017/24286.9494 PMID:28571191

Zheng, W., Aschner, M., & Ghersi-Egea, J. (2003). Brain barrier systems: A new frontier in metal neurotoxicological. *Toxicology and Applied Pharmacology*, *192*(1), 1–11. doi:10.1016/S0041-008X(03)00251-5 PMID:14554098

Zhou, C., Huang, C., Wang, J., Huang, H., Li, J., Xie, Q., Liu, Y., Zhu, J., Li, Y., Zhang, D., Zhu, Q., & Huang, C. (2017). LncRNA MEG3 downregulation mediated by DNMT3b contributes to nickel malignant transformation of human bronchial epithelial cells via modulating PHLPP1 transcription and HIF-1a translation. *Oncogene*, *36*(27), 3878–3889. doi:10.1038/onc.2017.14 PMID:28263966

Chapter 2
Global Industrialization and the Introduction of Heavy Metal Pollution to the Environment

ABSTRACT

In this chapter, chronological development of industrialization since the 18th century and through the first to the fifth industrial revolution is examined along with its effects on the environment. Findings showed that developments in the industrialization sector are growing exponentially. Similarly, production of solid wastes, emissions, particulates, and effluents from the associated activities are on the rise compared to the ability of the environment to self-cleanse itself. Eventually, air, soil, and water compartments have become polluted with heavy metals that are persistent, non-biodegradable and bioaccumulative in nature. The metals can be transferred from the environmental compartments to plants, animals, and humans to exert their toxic effects. To prevent these advances, industrialization efforts of modern day should reconsider incorporating environmental sustainability aspects as proposed in the principles of industrial revolution 5.0.

INTRODUCTION

The environment, which describes the immediate surroundings where living things dwell is composed of the hydrosphere, lithosphere, atmosphere and biosphere that all interrelate with one another (Briffa et al., 2020). The four spheres are vulnerable to chemicals and pollutants that can occur at excessive levels than the environment

DOI: 10.4018/978-1-6684-7116-6.ch002

can self-clear resulting to contamination and pollution. Contamination refers to the presence of high levels of substances in the surroundings beyond natural background levels. Pollution is the anthropogenic introduction of energy or substances directly or indirectly into the environment resulting to negative and harmful effects on human health and the environment as well as their ability to function properly (Masindi & Muedi, 2019). Contaminants can be transformed to pollutants exacerbating environmental pollution. During the last century, industrialization has risen and resulted to exploitation of natural resources of the earth and ultimately, worsening the environmental pollution trend (Walker et al., 2012; Ali et al., 2021). From the industrial activities, pollutants such as heavy metals, organic compounds, organometallic compounds, nanoparticles, gaseous pollutants and radioactive isotopes have been introduced to the environment causing concerns to the four spheres of the earth and their associated functions (Gautam et al., 2016).

Heavy metal pollution, transport and mobilization resulting from rapid industrialization is a growing environmental concern (Nyika et al., 2019; Nyika & Dinka, 2022). The growth in this kind of pollution dates back to the 1940s (Ali et al., 2019). Industrial processes such as emissions, smelting, mining, combustion of fossil fuels lead to the release of heavy metals to the environment. Their bio-accumulative, persistent and non-biodegradable nature enables them to contaminate food chains and worsen the health problems associated with their toxicity (As discussed in Chapter 1). Heavy metals have been found to pollute soils, water (Nyika et al., 2019) and vegetables (Nyika & Dinka, 2022). In countries such as China, the rise of urbanization and industrialization has been associated with heavy metal pollution in soils (Adnan et al., 2022; Su et al., 2022). In Poland, industrialization levels were found to be directly related to HM pollution in water resources (Hubeny et al., 2021). In South Korea, rapid growth of industries resulted to HM pollution in soils and sediments (Choi et al., 2020). The extent of pollution however depended on the type of industries and the activities they engaged in and the surroundings to which the pollutants were released. In this chapter, industrialization trends around the globe are interrelated with HM pollution and their environmental effects are discussed.

GROWTH IN INDUSTRIALIZATION

Human history is known to shape industrialization and in particular, the human structure and way of life in contemporary society. In respect to civilization, humanity has changed to best meet his needs, dominate his needs and modify his needs accordingly. Three great revolutions shaped man's society and economic model (Sengel, 2021). In the first revolution, man settled and began practicing agriculture to develop an agricultural economy before transitioning to the industrial revolution. Industrial revolution transformed the once agricultural economy to the industrial economy characterized by a rise in information communication technology (ICT) (Sengel, 2019). ICT brought a new, digital and information filled economy. The three evolutions have benefited and supported one another to shape modern society (Aksoy, 2016). The historical background of humanity particularly aspects such as mass production management, economics and marketing have an effect on consumption even today because they define the development of novel consumption patterns. According to Ocal and Altintas (2018), substitution of labor with machinery during the industrial revolution modified humanity's history and the associated socio-economic structure in his surroundings. This section therefore evaluates the development of industries globally to relate their emergence to the rise of environmental pollution and contamination.

The industrial revolution describes a period of great change in multidimensional identity particularly, in the production sector, which occurred from the 18th century (Saygili, 2013). In the period, production developed from previously primitive and simpler ways to professional, tech-savvy, mechanized and large-scale approaches. The machines in the period used steam power rather than brawn-enhanced and manual mass production capacity (Gunay, 2002). Consequently, industrial revolution steered up person-education, capital-labor, nature-human, time-place and production-consumption dual dimensions (Cetin, 2002). Today, advancements that aim at pin-point precision in production have been adopted following improvements in the communication sector. Logistics and transport sectors in addition to mechanization and mass production have also grown exponentially.

The chronological development of the industrial revolution from the 18th century is categorized into five periods: - industrial revolution 1.0, 2.0, 3.0, 4.0 and 5.0, respectively (Cevik, 2017). The four periods were and are as summarized in Figure 1 (Adopted from Rojko, 2017). In the first period (industrial revolution 1.0) between 1760-1830, mechanization production efforts began to take shape (Kharb, 2018). Although, the advances were met with resistance initially because of facilitating limited human labor role (job security), mechanization became accepted because of its enhanced workload efficiency. The focus during industrial revolution 1.0 (IR 1.0) was on transportation, agriculture, mining, paper, glass, lighting, gas, chemicals, cement, tools, iron, steam power and textile sectors (Akundi et al., 2022). The period reported major achievements such as growth in the sectors and employability though it brought the rise of pollution. Processes used took longer to meet targets because of using mathematical tools such as geometry and linear programming for production processes (Vinitha et al., 2020). During the period, steam power was used to revolutionize the textile industry, create flexible manufacturing mechanisms and begin the virtualization of the automobile sector. The first industrial revolution began in Britain. To have an advantage over other European countries, Britain banned export of manufacturing techniques, skilled workers and machinery. The monopoly did not last since export of industrial opportunities was profitable among the Britons. Soon industrialization spread to Belgium, France and finally, to Germany in the 1870s and in the US, in the early late 19th century. The Eastern European, Asian and African countries were left behind until the 20th century when India, China and Japan joined in the race for such advances.

Figure 1. Industrial revolutions 1.0 to 5.0
(Adopted from Rojko, 2017)

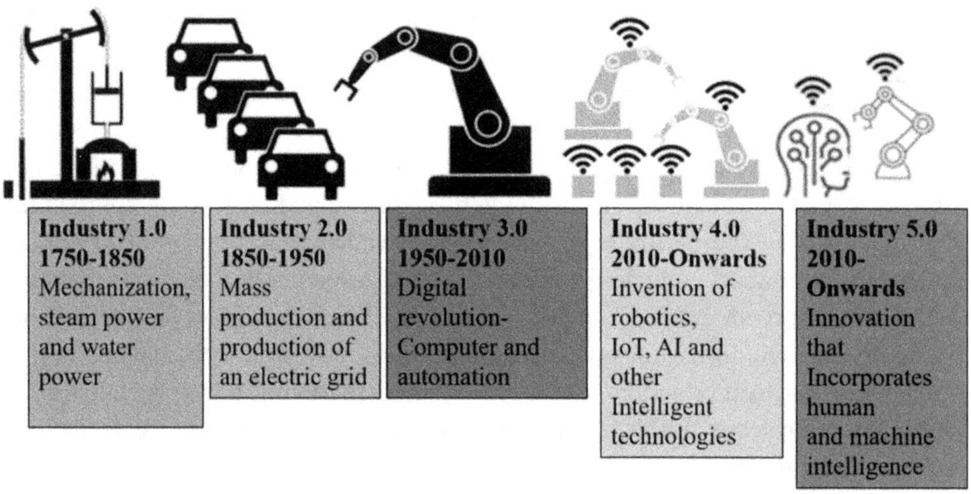

Industrial revolution 2.0 (IR 2.0) covers the period between 1850 and 1950 and saw the beginning of mass production of assorted industrial products particularly, in the automobile, electrical and electronic sectors. Mass production was facilitated by automation of assembly lines in industrial processes. The demand for the products in the period was based on diversity and volume (Yin et al., 2018). Sectors that were transformed during this period included business management, telecommunications, engine turbines, fertilizer production, applied sciences, automobile, rubber, maritime, chemical, petroleum, paper, machine tools, electrification, steel and iron industries (Akundi et al., 2022). During the period, technological revolution, which was characterized by the emergence of telegraph, telephones, electrical power grid, high power consumption and internal combustion engines prevailed (Taj & Jhanjhi, 2022). Mathematical models used in the period were based on linear and differential equations.

The period saw increased use of synthetic and natural resources including rare earth and lighter metals, plastics, alloys and new energy sources. The focus in industry 2.0 was volume and variety to ensure mass production of diversified products using efficient technology (Yin et al., 2018) The natural and synthetic resources combined with adoption of electricity as the energy source, invention of telegraph and railroads, use of computers, tools and machines brought the rise of automatic manufacturing. Additionally, a wider distribution and ownership of industries prevailed unlike in industry 1.0 where oligarchical ownership existed. Socioeconomic organization began to meet the needs of the industrializing societies unlike the previous revolution where laissez-faire politics prevailed. Agricultural revolution characterized by mechanized farming, experimentation of new crops, application of scientific breeding methods and land ownership changes predominated during industry 2.0, especially in Britain. The global north countries were more receptive to the revolution unlike global south countries that still struggled with such changes due to financial and technological limitations.

Industrial revolution 3.0 (IR 3.0) grew in the period between 1950-2010 and saw the development of computer software and hardware as well as integration of digital and network technologies in the production processes. Adoption of technology from analogue systems was revolutionary to the electronics industry since their architecture transformed to modular from integral products resulting to shortened life cycles. In addition to the variety and volume components of industry 2.0, delivery time was prioritized through adoption of technology and flexible manufacturing systems in industry 3.0 (Yin et al., 2018). Consequently, an information society with production value chains combined with dynamic data processing, automation, IT and cyber physical systems emerged (Sengel, 2021). Production at reduced cost was also realized since data, internet and cyber systems enabled a solid economic structure based on modern production tools and IT. According to Rojko (2017), flexible manufacturing that incorporated machine automation and microelectronics resulted to a variety of products during IR 3.0 period. Other sectors influenced during this period included renewable energy, telecommunication, robots and digital logic technology including the world wide web (WWW) (Kumar et al., 2021). Although the period enhanced flexible and customized manufacturing, such systems were not applicable in some situations and were cost-incremental to the production and operational costs in others (Vinitha et al., 2020). Zakoldaev et al. (2019) summarized the stages of industry 3.0 into six. The components of each stage were as summarized below.

i. The first stage was of company computerization where industries were supplied with the digital means to scale up production. An example is in the installation of some computers with computer numerical controls (CNC).

ii. The second stage involved the integration of digital net technologies to singular control systems excluding paperless documentation.
iii. In the third stage, company virtualization that involved the manufacture, testing and installation of cloud services in cyber and physical systems was done in digital factories at each industry. At this stage, documentation was administered in electronic form.
iv. The fourth stage involved the BigData technology implementation to collect massive engineering data, share, exchange and use it to transform production activities at workplaces. Additionally, remote cloud services and internet of things (IoT) were incorporated in industrial processes.
v. In the fifth stage, smart factories performed many virtual tests comparable to real scenarios for prognosticate purposes. The aim was to improve the industrial designs and model the function of various products to optimize their production systems.
vi. The last stage of industry 3.0 was to integrate physical and cyber technological equipment at industry level for adaptability (self-organization) using artificial intelligence (AI). Such advances were done using algorithms whose functions were to optimize industrial designs for effective functioning and more resilience to failure.

The period after 2010 characterizes the industrial revolution 4.0 (IR 4.0) aimed at advancing technology to transform production value chains to computer-based approaches (Sengel, 2021). In the period, inter-agency and intra-organizational production processes expanded through technological advancements such as additive manufacturing (3D printing), cloud technologies, augmented realities, deep learning, artificial intelligence and robotics (Sharma & Singh, 2020; Taj & Jhanjhi, 2022). Apart from the physical aspects of IR 4.0, the creation of artificial genetic engineering processes and internet of things are some biological and digital aspects predominant during the period, respectively (Apriliyanti & Ilham, 2022). According to Xu et al. (2018), IR 4.0 is manifested through the growth of businesses in the period that exploit horizontal and vertical chain digitalization to become central enterprises for the intricate industrial ecosystems of the near future. With this revolution period, resource efficiency, decentralized, flexible, on demand and on short-term development manufacturing and production processes are growing exponentially (Sengel, 2021). The most notable achievements of the period are intelligent and automated systems that work in uncertain conditions and machine learning (Akundi et al., 2022). Through network theory and optimization techniques, production processes of IR 4.0 period generate massive cloud data whose protection is limited (Vinitha et al., 2020). Industry 4.0 has some overlap with features of industry 3.0 such as the reliance of BigData, cloud computing, additive manufacturing and

IoT in industrial processes but at a more scaled-up level comparatively. The period was also marked by the use of autonomous robots to solve complex tasks beyond human ability, the use of software programs based on algorithms and mathematical models to simulate industrial processes and find solutions for challenges, horizontal and vertical integration of industrial processes for efficiency and effectiveness as well as augmented reality where humans interact with machines during industrial operations and maintenance (Erboz, 2017).

Most recently in April 2022, industrial revolution 5.0 (IR 5.0) was coined in a roundtable meeting at Brussels (Belgium) comprising of an expert group drawn from European countries (European Commission, 2022) though advances about it had been mentioned by Rada (2018). IR 5.0 is an emphasis of aspects in IR 4.0. The period is marked by use of collaborative robots to reduce and prevent risks during production. The robots learn, feel, understand and notice human operators and conceive their expectations and goals during production operations. In this period, it is expected that AI will penetrate human life to enhance capacity. Other advancements considered in this revolution include IoT, augmented reality and robotics (Skobelev & Borovik, 2017). IR 5.0 revolution is cognizant of the potential of industries to fulfill social needs of humans beyond development and employment and towards sustainable production and hence, enhanced human health. Therefore, workers' welfare is enhanced through technology upgrade in the manufacturing sector based on the principles of social benefit, environmental stewardship and human centricity in IR 5.0 (Akundi et al., 2022). By encouraging a human touch in industrialization, technology is tailored to nurture values and ethics that enhance novelty (Muller, 2020). With IR 5.0, manufacturing is resilient, human centric, sustainable and prioritizes on consumer preferences and capacity to pay. Although they are overlaps in the fourth and fifth industrial revolutions, the differences between them are as summarized in Table 1.

In all the industrial revolution periods, the global north countries were more advantaged with the technologies compared to the global south ones. The phenomenon across all periods is a leading source of environmental pollution by toxic substances such as pesticides, antibiotics and heavy metals (Hubeny et al., 2021). It is the onset of the revolution that led to increased heavy metal pollution by elements such as As, Cd, Cr, Cu, Hg and Pb (Proshad et al., 2018). Regions with advanced industrialization (developed countries) have higher concentrations of the pollutants resulting from high waste and wastewater production compared to poor developing nations of Africa and Asia (Cheng et al., 2014). Similarly, Anyanwu et al. (2018) noted that industrialized areas have increasing activities in production processes in residential areas, burning of noxious wastes, arbitrary dumping, production of airborne dust, use of fossil fuels, illegal refining and artisanal mining activities (Xu et al., 2022b). These activities contribute to heavy metal introduction to the environment.

Industrial emissions from chemicals, coal combustion, fossil fuel processing and cement plants as well as vehicular emissions in industrialized towns also induce heavy metal introduction to the environment (Proshad et al., 2018). The advances to evolve industries towards better economies globally must be balanced vis-à-vis the capacity to produce heavy metals whose effects on the environment are deadly.

Table 1. Differences between industrial revolutions 4.0 and 5.0 (Vaidya et al., 2018)

Characteristic	Industrial Revolution 4.0	Industrial Revolution 5.0
Objective	Smart manufacturing and optimization of systems	Production that encourages social benefit, human centricity, environmental stewardship and sustainability
System approaches	Integrated chain focused on end-of-life cycle phases and real-time data monitoring	Logistics efficiency incorporating the 6R-methodology, ethical technological use and social-centric decisions that advance production and human values
Human factors	Repeated movements, human reliability and human-computer interrelationship	Training and learning of employees as well as their safety and management
Enabling concepts and technologies	Cloud computing, IoT, bog data and analytics, cyber security, digitization	Cloud computing, IoT, bog data and analytics, cyber security, digitization
	Automation, cyber physical systems, vertical and horizontal value chains and additive manufacturing	Human-machine interaction, multi-lingual gesture and speech recognitions, collaborative robotics, tracking technologies, smart grids, predictive maintenance, decision support tools and equipment with bio-inspired support and safety
Environmental implications	Economic systems, waste prevention, extended product life cycle, increased material and energy consumption	Waste recycling and prevention, use of greener energy sources, smart and automatic sensors, energy efficiency during production processes.

ENVIRONMENTAL POLLUTION

Environmental pollution refers to the introduction of pollutants and contaminants in spheres of the earth, which results to undesirable and unwanted changes in land and water resources as well as living organisms (Masindi & Muedi, 2019). Industrial growth and urbanization are the drivers of pollution globally since they result to high energy use, production and use of chemical substances and generation of wastewater and solid waste discharge to the environment, which promotes natural resources pollution and emission of greenhouse gases (Kelishadi, 2012). The public health concern with environmental pollution is its ability to affect human health by resulting to mental disorders, cardiovascular diseases, malignancies, allergies, infant deaths, respiratory dysfunctions and perinatal disorders among others complications (Xu et al., 2022a) (As discussed in Chapter 1).

Environmental pollutants can be foreign or naturally occurring matter categorized as 1) organic, 2) inorganic and/or 3) biological. Organic pollutants are biodegradable substances resulting mainly from human manufacturing activities. They include organochlorine pesticides, petroleum containing products, pesticides, polycyclic aromatic hydrocarbons, polybrominated diphenyl ethers, polychlorinated biphenyls, food waste and human waste (El-Shahawi et al., 2010). The pollutants have high hydrophobicity, lipophilicity, stability and solubility to high lipid quantities, which eases their capacity to bioaccumulate to toxic levels (Lepp, 2012). Inorganic pollutants result from domestic, agricultural and industrial wastes and include substances such as artificial radionuclides, minerals, salts and heavy metals (Kepa & Zaborska, 2021). They enter bodies of living organisms during chemical, metallurgical, smelting and mine drainage processes and through natural sources such as volcanicity resulting to their accumulation and transfer to food chains (Masindi & Muedi, 2019). Biological pollutants describe pathogenic microbes such as parasites, viruses and bacteria as well as mold, pollen, cockroaches, mites, dust, cat saliva, animal dander, mildew. These biological components have a ubiquitous nature in the environment and induce mild to serious health complications in living things (Elliott, 2003; Postigo et al., 2018). Irrespective of the type, the pollutants have adverse effects on the environment. This chapter focuses on environmental pollution as a result of inorganic matter and in particular, heavy metals and relates the phenomenon to the global rise in industrialization trends.

HEAVY METAL POLLUTION

Heavy metals refer to elements whose density and atomic weight is at least five times compared to that of water, occur in minute levels unless augmented by human activities and are toxic even in small quantities. (Banfalvi, 2011). Common heavy metals include metals and metalloids such as Ag, As, Au, Cd, Co, Cr, Cu, Fe, Hg, Mn, Mo, Ni, Pb, Pt, Sn, Ti, V, Zn among others (Briffa et al., 2020). Metalloids covalently bond with other substances to manifest their toxic characteristics. They bond with other organic groups to form lipophilic compounds and ions that bind to non-metals of macromolecules in cells of living things resulting to toxic effects. The lipophilic nature associated with metalloids has varied toxic responses and distribution in living things. Examples of lipophilic compounds include methylated forms of arsenic and tributyltin oxide that bind various protein groups such as the sulfhydryl moiety. The metals and metalloids enter the biosphere through inhalation of contaminated gases and air, consumption of contaminated water and soil, contact of the pollutants with the skin or via ingestion of metal-contaminated foods that are grown in polluted soils (Walker et al., 2012; Xu et al., 2022b).

Heavy metals are non-biodegradable and hence not easy to metabolize. Living organisms usually take up metal ions and deposit them in their proteins and intracellular granules for either excretion or prolonged storage. Through continuous storage, the metals bioaccumulate causing physicochemical and biological complications in living things. Their bioaccumulation and non-biodegradable nature in living things is known to cause various forms of toxicity including immunological toxicity, reproductive and developmental toxicity, genotoxicity, neurotoxicity, nephrotoxicity, carcinogenicity, cardiotoxicity and hepatoxicity (Mitra et al., 2022). Apart from living organisms, heavy metals pollute soils and various plants such as vegetables (Nyika & Dinka, 2022) and also water resources (Postigo et al., 2018; Hubeny et al., 2021).

Industrial Sources of Heavy Metals

A number of industrial activities are responsible for heavy metal release to the environment. They are discussed under the following subheadings.

Mining and Milling Industries

Mining activities, which involve extraction of geological materials and precious minerals from the earth's crust release large quantities of toxic metals to the environment (Karn et al., 2021). Such metals include As, Hg, Cd, Cr, Ni, Pb and Zn. The operations involved in mining along with processes such as grinding, milling and clustering of ores as well as release of tailings that are transported via water and wind lead to environmental pollution (Coetzee et al., 2020). Exposure to natural rocks during mining of various minerals exposes one to compounds such as arsenopyrite (FeAsS), cinnabar (HgS), galena (PbS), cerussite ($PbCO_3$), chromite ($FeCr_2O_4$) and angesite ($PbSO_4$) that expose one to As, Hg, Pb and Cr metals (Martinez-Guijarro et al., 2021). Sedimentary rocks including sandstone, limestone and shale contain heavy metals such as Cd, Cr, Cu, Hg, In, Mn, Ni, Pb and Sn (Sharma & Agrawal, 2005).

The mining, grinding, transportation and combustion of coal was shown to introduce As, Cd, Cr, Cu, Hg, Ni and Pb in soils of Mongolia Autonomous region and vicinity in China (Bu et al., 2020). The mining, mineral processing and metallurgical extraction of gold breaks down the mineral ore's crystallographic bonds resulting to production of tailings. The tailings mainly contain pyrite (FeS_2) and heavy metals such as As, Cd, Co, Cu, Hg, Ni, Pb and Zn in the form of silicates (Fashola et al., 2017). Mine tailings and waste rock during extraction of Hg release the heavy metal as emissions where they are evaporated and later precipitated and deposited into water resources and soils resulting to pollution. Dust emissions during ore blasting and mining of Hg contain Cd and Pb metals. The acid mine drainage resulting from Hg processing also introduces As, Fe, Hg and Sb to the environment leading to its pollution (Dube et al., 2005). The mining of chromite ($FeCr_2O_4$), which contains Cr (III) and Fe (II) oxides and its phyrometallurgical processing to extract Cr results to dust and slag wastes (Coetzee et al., 2020). The wastes can be inhaled, ingested or come into contact with skin of living things resulting to Cr (III) and Cr (VI) introduction. The mining activities of Cd and Zn ores resulted to release of the metals in their tailings (Martinez-Guijarro et al., 2021). Therefore, mining and mineral processing is a major anthropic source of heavy metal pollution in the environment.

Foundry and Smelting Industries

Metal smelting and casting processes release heavy metals such as As, Cd, Co, Cr, Cu, Hg, Ni, Pb, Se and Zn to the environment. The heavy metals occur in ores used or maybe added as mixtures during the formation of alloys (Brusseau & Gerba, 2019. The smelting and casting wastes also contain the hazardous elements. During smelting and casting processes, high temperatures reduce metal rocks such as bauxite to aluminum and pyrite to iron. A case example is Al, Cu and Fe smelting and casting at 660, 1082 and 1536 °C, respectively. Within the temperature range (660-1536 °C), others metal from the ores such as Cd (765 °C) and Zn (906 °C) also melt and volatilize (Brusseau & Gerba, 2019). Consequently, smelted and casted stacks without gas and particulate scrubbers release heavy metals to the environment as bound or volatile particulates (Csavina et al., 2014). The chemical composition of the particulates depends on the particle size and source or the original ore from which smelting and casting is done. Smelting in various metallurgical industries also results to release of airborne heavy metal pollutants whose effects on the environment depend on the particular land properties (Kapusta & Sobczyk, 2015).

Oil and Petrochemical Refineries

Petroleum refineries work by separating crude oil to various products through physicochemical techniques such as cracking, coking, visbreaking, reforming, fractionation, blending, hydrotreating, hydrodesulfurization, acid removal, sweetening, desalting, manufacturing and transportation (Hazardous Substance Research Center, 2003). Resultant products include waxes, asphalt, motor oil, diesel fuel, kerosene and petroleum gas. Crude oil is a mixture of both inorganic components such as heavy metals and organic hydrocarbons and its processing to various products could lead to the release of trace metals and their subsequent environmental pollution. In a Northern Iraq oil firm, crude oil exposure among workers was associated with elevated levels of Al, Ba, Hg, Mn, Pb, V and Zn metals in their blood stream (Saleh et al., 2021). Accidental spillages, extraction, refining, storage and transportation of crude oil could also result to exposure to heavy metals (Chinedu & Chukwuemeka, 2018). In the Niger Delta region of Nigeria, crude oil enriched with heavy metals such as Cd, Co, Cr, Cu, Fe, Mn, Ni, Pb and Zn was reported to have been spilled resulting to soil and water pollution of the vicinity (Chinedu & Chukwuemeka, 2018). Companies also using petrochemicals and oils have exposed workers to heavy metals. For instance, Cu and Pb levels in the blood and hair of workers in oil production and distribution firms of Brazil were in elevated levels (Jaccob, 2020). In consumer fuel for vehicles, Pb is used as an additive in petrol particularly in the Middle East countries and as such, the metal can be introduced in the environment. Ahmed et al. (2020) reported that petrol station workers in Bangladesh's Dhaka were at risk of being exposed to Pb among other heavy metals in petroleum products they handled and sold.

Coal Industry

The mining and consumption of coal plays a crucial role in socio-economic development of many countries around the globe (Li et al., 2018). The countries rely on it as a source of electric power. Long-term and extensive mining of coal disturbs natural environments and is associated with water quality deterioration, soil erosion and pollution and subsidence. Toxic elements such as fly ash, particulate matter and refuse are byproducts of coal processing and they can be released to the environment through activities such as waste discharge, gangue accumulation, production of coal-electricity and emissions from coal transportation and coal-based power plants (Candeias et al., 2011). All these substances have the potential to pollute the environment through the release of heavy metals. In areas near coal mine facilities in Yanzhou and Zoucheng of Shandong province (China), soils were found to accumulate HM such as Cd, Cr, Cu, Ni, Pb and Zn (Li et al., 2018). In Liupanshui city of Guangxi province (China), coal mining was associated with accumulation and pollution of Minghu wetland with Cd, Ni and Pb metals (Yu et al., 2021). In China's Suzhou city near Sunan mining facility, groundwater pollution from elevated levels of As, Cr, Cu, Fe, Mn, Pb and Zn was associated with coal processing (Wang et al., 2022). Lead during coal processing is from the resultant particulate matter of coal-combustion of dry ash and emissions produced during coal transportation. In addition, thermal power generation results to coal ash dust with high levels of Pb.

Cadmium, Cr and Ni are sourced from coal mining in seams as well as coal gangue deposited in water bodies and soils (Yiu et al., 2016).

Chemical Industries

The agrochemical and petrochemical industries are growing in modern day and so is the demand for their products. Consequently, they are contributors to HM pollution in the environment. Chemicals that are introduced to soils and water resources from irrigation with sludge and wastewater, application of herbicides, sludge-based amendments and metallo-pesticides are rich in heavy metals that eventually enter food chains (Rai et al., 2019). Elevated levels of As, Cr, Cu, Hg and Pb in cocoyam grown in soils of East Africa were associated with the use of heavy metal-containing agrochemicals and soil additives (Mongi & Chove, 2020). Inadequate management of agrochemicals (overuse or misuse) was associated with the introduction of As, Cd, Cr, Cu, Ni, Pb and Zn in soils and vegetables (Jibrin et al., 2021). The use of chemical-containing wastewater from industries was associated with elevated levels of Cd, Cr, Co, Cu, Hg, Ni and Pb in soils of suburban regions of Nairobi, Kenya (Nyika & Dinka, 2022). Petrochemical-based sludge was assayed and had elevated levels of Cd, Cr, Cu, Ni and Pb metals, which negatively affected the germination and growth of mung after the sludge was used to irrigate the crop (Kumar et al., 2012). Production of inorganic fertilizers from mineral ores and their blending with recycled industrial effluents that contain mine tailings and flue dust from steel milling was associated with introduction of As, Cd and Pb to soils in Minnesota (Minnesota Department of Health, 2021) among other regions (Rai et al., 2019). Table 2 summarizes some of the agrochemicals used and the heavy metals that they introduce to soils.

Introduction of petrochemicals to the environment results to heavy metal pollution. This suggestion is best described by the Gulf War oil spill in 1991 on marine organisms of the Arabian Gulf. Lead and Zn were found in elevated levels compared to the permissible limits among fish of Bahrain, Kuwait, Oman and UAE who were part of the Gulf War oil spill of 1991 (Freije, 2014). The oil spill was also associated with introduction of Ba and Cr in soils of Khursania region of Saudi Arabia (Hussain & Gondal, 2008). In China's Shengli Oilfied (Fu et al., 2014), Exxon Valdez (USA) (Apeti & Hartwell, 2015), and in the Easten Mediterranean Sea (Barbour et al., 2009), oil spills were associated with introduction of HM such as Cd, Cr, Mn, Ni, Pb, V and Zn in soils, water and their transfer to living organisms.

Table 2. Chemicals used in agricultural activities and the heavy metals that they introduce to soils

Chemical Source	Specific Agrochemical	Heavy Metal Input	Reference
Atmospheric deposition	Particulates from manufacturing, transport, mining, refining, smelting and waste incineration processes	As, Cd, Cr, Cu, Hg, Ni, Pb, Zn	Deng et al., 2016
Wastewater	Irrigation using untreated municipal and industrial wastewater	As, Cd, Cr, Cu, Hg, Ni, Pb and Zn	Woldetsadik et al., 2017
Biosolids	Fly ash deposition and sewage sludge	As, Cd, Cr, Cu, Hg, Ni and Pb	Sharma et al., 2017
Pesticides	Fungicides, insecticides and herbicides	As, Cd, Cu, Pb and Zn	Toth et al., 2016
Fertilizers	Phosphatic, nitrate and potash fertilizers as well as lime	Cd, Cr, Cu, Mn, Ni, Pb and Zn	Sun et al., 2013

Cement Industry

The cement plants are significant environmental pollutants since such factories can emit gases, dust and particulate matter containing heavy metals. The pollutants are a result of limestone (raw material) crushing, raw material handling, clinker production, kiln processing and grinding of finished cement (Adeyanju & Okeke, 2019). Heavy metals affiliated with cement processing include Cd, Cr, Cu, Pb and Zn. The metals are contained in particulate matter and dust, which can be dispersed to long distances from their production centers causing extensive pollution (Ogunkunle & Fatoba, 2014). Majority of the pollutants are associated with raw materials. According to Achternbosch et al (2003), common cement raw materials contain 53, 20, 21 and 25 mg/kg of Zn, Pb, Cu and Cr, respectively in their constituent compositions. Cement clinkering processes and their associated raw materials, fuel use and waste from combustion activities are also major sources of heavy metals (Blois & Aime, 2021).

Soils of Obajana cement plant vicinity in Nigeria were found to have elevated levels of Cd, Co, Cr, Pb and Ni, which were associated with cement dust deposition from the factory (Akpambang et al., 2021). Similarly, the levels of Cd, Cr, Cu, Pb and Zn were found to be elevated beyond permissible levels in soils of Southwest Nigeria in the vicinity of Lafarge cement company (Ogunkunle & Fatoba, 2014). In Gubbio, Italy, cement industries were associated with introduction of As, Cd, Cr, Ni and Pb from the resultant atmospheric emissions during its processing (Blois & Aime, 2021). Elevated levels of As, Cd, Cr, Cu, Fe, Mn, Pb and Zn were reported in soils and barley around Abyssinia cement plant of Ethiopia and attributable to dust and particulate matter precipitation and deposition from the processing activities (Yitagesu & Bekele, 2019).

Manufacturing Industries

Manufacturing industries including the cosmetic and leather industries are sources of heavy metal pollution to the environment. Leather processing and tanning produces wastes such as effluents, fibers, particulate matter and dusts that are rich in Cr (III) (Junaid et al., 2017). Chromium release in leather tanning is associated with the use of chromium sulphate salts [$Cr(H_2O)(OH)SO_4$]. In addition, leather processing exposes individuals to hexavalent Cr, As and methyl Hg (Mikoczy & Hagmar, 2005). Owing to the exposure to HM, workers in leather manufacturing industries of Kanpur, India were reported to be highly susceptible to respiratory dysfunctions and morbidity due to ingestion of Cr (Rastogi et al., 2008). Similarly, exposure to As, Cr and Hg was associated with the high cancer prevalence in a leather tanning industry in Sweden after an epidemiological study from 1958 to 1999 was carried out (Mikoczy & Hagmar, 2005).

In the manufacture of cosmetics, HM are used and therefore, the products can be routes of exposure to the pollutants once are used and come in contact with the skin during application or through disposal of resultant wastes to the environment. According to Attard and Attard (2022), HM such as Cd are used as coloring agent in various cosmetic products while Ni is used in management of chronic and acute dermatitis. Mercury is used in various skin lightening products while As and Pb occur as contaminants in cosmetic manufacturing. Face care products such as face paints, face creams, make-up foundation, mascara, eyebrow pencils, eye shadow, eye liners and lipsticks are sources of As, Cd, Hg, Ni and Pb (Arshad et al., 2020). Sunblock cream, skin lightening creams, body lotions, hair conditioners and shampoos, cleansers, beauty, tonic and hair creams are also sources of As, Cd, Hg, Ni and Pb (Lim et al., 2018).

Paints and coatings manufacturing that involves the production of adhesives, inks and other decorative products are potential sources of heavy metals. Their manufacturing processes produce particulates (dust) and volatile organic compounds that comprise of Cd, Co, Hg and Pb whose effects to the environment are deleterious (US-Environmental Protection Agency, USEPA, 2016). Elevated levels of As, Cd, Cr, Cu, Hg, Ni, Pb and Zn in soils around a closed paint manufacturing company were reported in Yugoslavia and (Radomirovic et al., 2020) and China (Ji et al., 2012). In both cases, pollution of soils by heavy metals was associated with unsystematic management of wastes and effluents from the plants.

EFFECTS OF HEAVY METALS

Heavy metal pollution is a growing global environmental problem whose momentum is expected to rise as advances to industrial revolution continues. Soil, water and air are the main compartments that are affected by the pollutants. Industrial pollution sources release HM to riverine resources, land systems (soils and sediments) and the atmosphere. From these compartments, the pollutants then enter in food chains through plants (primary producers) and animals (secondary consumers) that are consumed by living organisms causing toxic effects. A schematic representation of the effects of heavy metals is a shown in Figure 2.

Figure 2. Schematic representation of heavy metal pollution effects to the environment

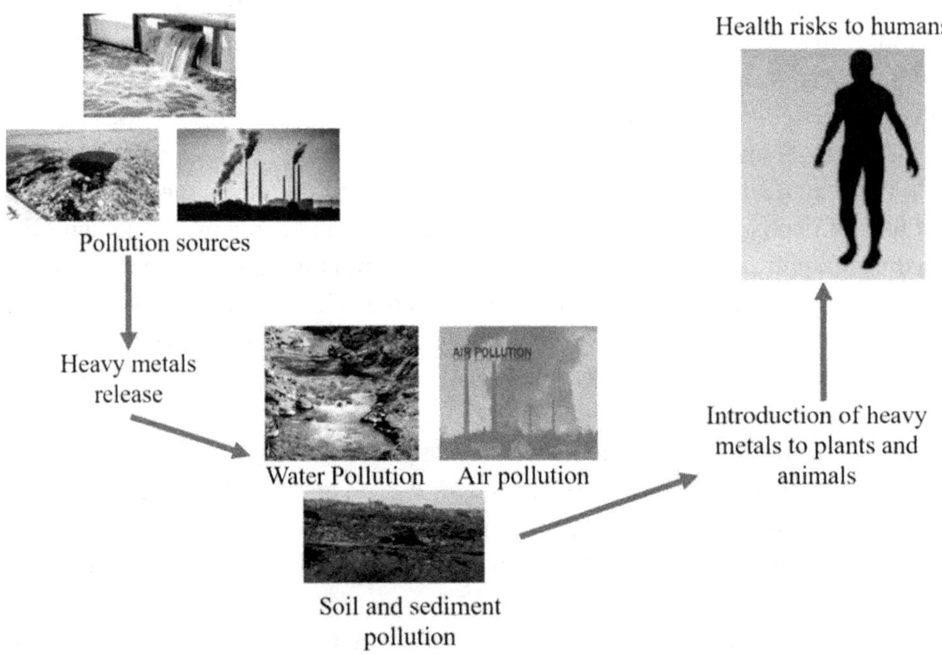

Soil Pollution

Deliberate soil pollution occurs when heavy metal containing solid wastes, coal combustion remains, petroleum distillate spills, mine tailings, sewage sludge, wastewater, mine ore waste, leaded paints, inorganic fertilizers and agrochemicals are introduced to soils. Since heavy metals are chemically and biologically non-biodegradable, their total concentrations are retained for a long time prior to their release to the environment, which is a consequence of bioaccumulation (Lepp, 2012). The nutrients in soils complexed with heavy metals are taken up by plants and animals before consumption of the polluted products by humans. Once in food chains, the pollutants ruin ecosystems and slow down the biodegradability of other organic pollutants. They can also be ingested from contaminated water and food as well as inhaled from the atmosphere. By altering the physicochemical characteristics of soils, heavy metals deteriorate their quality and ability to produce crops and also act as a natural water pollutant filter (Walker et al., 2012; Masindi & Muedi, 2018). Therefore, continued soil pollution increases the risk of groundwater pollution (Ali et al., 2021).

Air Pollution

The growing industrial revolution accompanied by urbanization and population rise has worsened the apparent air pollution globally. In particular, particulate matter ($PM_{2.5}-PM_{10}$) and dusts are released from various industrial emissions and processes causing health problems among living organisms. Atmospheric pollutants can also occur as droplets and gases that are in association with one another or with particulates. The particulates contain heavy metals among other forms of contaminants and travel long distances before their settling and based on their sizes. They can precipitate and be deposited in soils causing pollution in such environments. Reported health complications from exposure to heavy metals in humans include respiratory and cardiovascular dysfunctions, skin and eye irritations and premature death in extreme cases. Industrial and vehicular emissions particularly contain Cd, Cr, Cu, Ni, Pb, V and Zn metals (Soleimani et al., 2018).

Chimneys are the commonest way which industries release atmospheric pollutants (Briffa et al., 2020). Weather conditions and chimney heights influence the distance travelled by pollutants before their precipitation and deposition. In taller chimneys and when the weather is windier and warmer, air pollutants travel farthest as a result of convention and side currents. The trend is contrary when the weather is foggy and cold. Apart from emissions by chimneys, jet engines, internal combustion, poorly maintained diesel engines, production and use of pesticides are other sources of heavy metals released to the atmosphere as emissions or particulates and end up causing pollution (Masindi & Muedi, 2018; Walker et al., 2012).

Water Pollution

Water pollution by HM affects plants, animals, human and marine life (Ali et al., 2019, 2021). The pollution capacity of the metals is enabled by their ability to undergo biomagnification, bioaccumulation and their persistent nature in food chains (Rajaei et al., 2012). Various industrial processes release heavy metals that are transported by runoff to water resources. Release of untreated industrial effluents and solid wastes to water bodies also introduces HM in such environs. Depending on the organism exposed to the pollutants, its biological role, nature and the period of exposure, the metals can be toxic even at minute levels in water resources (Ali et al., 2021). Once the surface water bodies such as rivers, lakes, streams, seas and oceans are polluted, they interact with aquifer systems introducing the heavy metals in such groundwater systems. Soils especially those of coarse texture can leach heavy metals to aquifer systems once they bioaccumulate them and the natural land resources can no longer filter them.

TROPHIC TRANSFER OF HEAVY METALS

Following the introduction of HM to the environment (soil, air and water), their exposure and entry to the biosphere (plants, animals, fish, humans) depends on their trophic transfer in the food chain, their accumulation, bioavailability and metabolism rates (Ali et al., 2021). Environmental heavy metals enter an organism's body directly via non-living systems such as air, soil or via food/ prey systems. For instance, heavy metals can be introduced into a fish via sediment/soil and water, which it comes into contact with, the air it breaths, via skin contact with the pollutants or via feeding on a contaminated prey (Ali & Khan, 2018). The retention of the metals depends on an organism's detoxification and homeostasis capacity and speciation of the ingested heavy metals and their subsequent bioavailability. For instance, methyl-Hg has greater bioaccumulation and biomagnification capacity in living things due to its lipophilic nature. Some plants (metallophytes and hyperaccumulators) are adapted to concentrate metals with no physiological effects on their growth and development (Ali & Khan, 2018).

CONCLUSION

The growth in industrialization that began to shape in the 18th century has led to mass production, mechanization, automation and the use of both machine and human knowledge for innovative manufacturing of products. Consequently, natural resources have been exploited to manufacture products and chemicals that are required for mass production. Environmental pollution and heavy metals contamination has resulted from these trends. In this chapter, a detailed account of various heavy metals, their sources and effects in relation to the global industrial revolution is provided. Global industrial development is directly related to the release of environmental pollutants and in particular heavy metals. The metals can be in particulate matter, dust, effluents, solid wastes and residues of different industrial activities such as mining, manufacturing, metallurgical, chemical processing and petroleum refining processes. The bioacumulative, persistent and non-biodegradative nature of the elements enables their release to soils, water and air. From the environmental compartments, the pollutants are transferred to various trophic levels depending on their bioavailability, the nature of the organisms in a specific trophic level and the exposure period. Therefore, as advances to the fifth industrial revolution grow, there is need to consider approaches to minimize heavy metal pollution for environmental sustainability even as most nations prioritize on economic growth.

REFERENCES

Adeyanju, E., & Okeke, C. (2019). Exposure effect to cement dust pollution: A minireview. *SN Applied Sciences*, *1*(12), 1572. doi:10.100742452-019-1583-0

Adnan, M., Xiao, B., Xiao, P., Zhao, P., & Bibi, S. (2022). Heavy metal, waste, COVID-19, and rapid industrialization in this modern era—Fit for sustainable future. *Sustainability (Basel)*, *14*(8), 4746. doi:10.3390u14084746

Ahmed, S., Yesmin, M., Jeba, F., Hoque, S., Jamee, R., & Salam, A. (2020). Risk assessment and evaluation of heavy metals concentrations in blood samples of plastic industry workers in Dhaka, Bangladesh. *Toxicology Reports*, *7*, 1373–1380. doi:10.1016/j.toxrep.2020.10.003 PMID:33102140

Akpambang, V., Ebuzeme, G., & Oluwatobi, J. (2022). Heavy metal contamination of topsoil around a cement factory-a case study of Obajana cement plc. *Environmental Pollutants and Bioavailability*, *34*(1), 12–20. doi:10.1080/26395940.2021.2024090

Aksoy, A. (2016). Geleneksel devletten modern devlete: Sanayi devrimi ve kamu yönetimi düşüncesinde değişim. *Uluslararası Politik Araştırmalar Dergisi*, *2*(3), 31–37. doi:10.25272/j.2149-8539.2016.2.3.04

Akundi, A., Euresti, D., Luna, S., Ankobiah, W., Lopes, A., & Edinbarough, I. (2022). State of Industry 5.0—Analysis and Identification of Current Research Trends. *Applied System Innovation*, *5*(1), 27. doi:10.3390/asi5010027

Ali, H., & Khan, E. (2018). Assessment of potentially toxic heavy metals and health risk in water, sediments, and different fish species of River Kabul, Pakistan. *Human and Ecological Risk Assessment*, *24*(8), 2101–2218. doi:10.1080/10807039.2018.1438175

Ali, H., Khan, E., & Ilahi, I. (2019). Environmental chemistry and ecotoxicology of hazardous heavy metals: Environmental persistence, toxicity and bioaccumulation. *Journal of Chemistry, 6730305*, 1–14. doi:10.1155/2019/6730305

Ali, M., Hossain, D., Al-Imran, M., Khan, S., Begum, M., & Osman, M. (2021). *Environmental pollution with heavy metals: a public health concern*. Intech Open., doi:10.5772/intechopen.96805

Anyanwu, B., Ezejiofor, A., Igweze, Z., & Orisakwe, O. (2018). Heavy metal mixture and effects in developing nations: An update. *Toxics*, *6*(4), 65. doi:10.3390/toxics6040065 PMID:30400192

Apeti, D., & Hartwell, S. (2015). Baseline assessment of heavy metal concentrations in surficial sediment from Kachemak Bay, Alaska. *Environmental Monitoring and Assessment*, *187*(1), 1–11. doi:10.100710661-014-4106-x PMID:25394770

Apriliyanti, M., & Ilham, M. (2022). Challenges of the industrial revolution era 1.0 to 5.0: university digital library in Indonesia. *Library Philosophy and Practice (e-journal),* 6994.

Arshad, H., Mehmood, M., Shah, M., & Abbasi, A. (2020). Evaluation of heavy metals in cosmetic products and their health risk assessment. *Saudi Pharmaceutical Journal*, *28*(7), 779–790. doi:10.1016/j.jsps.2020.05.006 PMID:32647479

Attard, T., & Attard, E. (2022). *Heavy metals in cosmetics*. IntechOpen., doi:10.5772/intechopen.102406

Banfalvi, G. (2011). *Cellular effects of heavy metals*. Springer. doi:10.1007/978-94-007-0428-2

Barbour, K., Sabra, H., Bianu, G., Jaber, L., & Shaib, H. (2009). Oppositional dynamics of organic versus inorganic contaminants in oysters following an oil spill. *Journal of Coastal Research*, *25*(4), 864–869. doi:10.2112/08-1059.1

Blois, L., & Aime, L. (2021). Environmental impacts from atmospheric emission of heavy metals: A case study of a cement plant. *Measurement: Sensors*, *18*, 100313. doi:10.1016/j.measen.2021.100313

Briffa, J., Sinagra, E., & Blundell, R. (2020). Heavy metal pollution in the environment and their toxicological effects on humans. *Heliyon*, *6*(9), e04691. doi:10.1016/j.heliyon.2020.e04691 PMID:32964150

Brusseau, M., Gerba, C., & Pepper, I. (2019). *Environmental and pollution Science*. Academic Press Publisher.

Bu, Q., Li, Q., Zhang, H., Cao, H., Gong, W., Zhang, X., Ling, K., & Cao, Y. (2020). Concentrations, spatial distributions and sources of heavy metals in surface soils of the coal mining city Wuhai, China. *Journal of Chemistry*, *4705954*, 1–10. Advance online publication. doi:10.1155/2020/4705954

Candeias, C., Da Silva, F., Salgueiro, A., Pereira, H., Peis, A., & Patinha, C., &. (2011). Assessment of soil contamination by potentially toxic elements in the Aljustrel mining area in order to implement soil reclamation strategies. *Land Degradation & Development*, *22*(6), 565–585. doi:10.1002/ldr.1035

Cetin, H. (2002). Liberalizmin tarihsel kökenleri. *Cumhuriyet Üniversitesi İktisadi ve İdari Bilimler Dergisi*, *3*(1), 79–96. doi:10.25272/j.2149-8539.2018.4.1.02

Cevik, D. (2017). *Sanayi Devrimlerinin Süreci ve 4*. Sanayi Devrimi. https://www.alomaliye.com/2017/05/29/sanayi-devrimlerinin-sureci-4-sanayi-devrimi/

Cheng, M., Wu, L., Huang, Y., Luo, Y., & Christie, P. (2014). Total concentrations of heavy metals and occurrence of antibiotics in sewage sludges from cities throughout China. *Journal of Soils and Sediments*, *14*(6), 1123–1135. doi:10.100711368-014-0850-3

Chinedu, E., & Chukwuemeka, C. (2018). Oil spillage and heavy metals toxicity risk in the Niger Delta, Nigeria. *Journal of Health & Pollution*, *19*, 180905. doi:10.5696/2156-9614-8.19.180905 PMID:30524864

Choi, R., Kim, H., Yang, J., & Kim, J. (2020). Ecological impact of fast industrialization inferred from a sediment core in Seocheon, West Coast of Korean Peninsula. *Journal of Ecology and Environment*, *44*(24), 1–10. doi:10.118641610-019-0144-1

Coetzee, J., Bansal, N., & Chirwa, E. (2020). Chromium in environment, its toxic effect from chromite-mining and ferrochrome industries, and its possible bioremediation. *Exposure and Health*, *12*(1), 51–62. doi:10.100712403-018-0284-z PMID:33748533

Csavina, J., Taylor, M., Felix, O., Rine, K., Saez, A., & Betterton, E. (2014). Size-resolved dust and aerosol contaminants associated with copper and lead smelting emissions: Implication for emissions management and human health. *The Science of the Total Environment*, *493*, 750–756. doi:10.1016/j.scitotenv.2014.06.031 PMID:24995641

Deng, W., Li, X., An, Z., & Yang, L. (2016). The occurrence and sources of heavy metal contamination in peri-urban and smelting contaminated sites in Baoji, China. *Environmental Monitoring and Assessment*, *188*(4), 251. doi:10.100710661-016-5246-y PMID:27021694

Dube, M., MacLatchy, D., Kieffer, J., Glozier, N., Culp, J., & Cash, K. (2005). Effects of metal mining effluent on Atlantic salmon (*Salmo salar*) and slimy sculpin (*Cottus cognatus*): Using artificial streams to assess existing effects and predict future consequences. *The Science of the Total Environment, 343*(1-3), 135–154. doi:10.1016/j.scitotenv.2004.09.037 PMID:15862841

El-Shahawi, M., Hamza, A., Bashammakhb, A., & Al-Saggaf, W. (2010). An overview on the accumulation, distribution, transformations, toxicity and analytical methods for the monitoring of persistent organic pollutants. *Talanta, 80*(5), 1587–159. doi:10.1016/j.talanta.2009.09.055 PMID:20152382

Elliott, M. (2003). Biological pollutants and biological pollution—An increasing cause for concern. *Marine Pollution Bulletin, 46*(3), 275–280. doi:10.1016/S0025-326X(02)00423-X PMID:12604060

Erboz, G. (2017). *How to define industry 4.0: main pillars of industry 4.0. Managerial trends in the development of enterprises in globalization era conference*. Slovak University of Agriculture in Nitra.

European Commission (EC) (2022). *Industry 5.0 roundtable, Brussels, 27 April 2022 meeting report*. EC, Brussels, Belgium. doi:10.2777/982391

Fashola, M., Ngolo-Jeme, V., & Babalola, O. (2016). Heavy metal pollution from gold mines: Environmental effects and bacterial strategies for resistance. *International Journal of Environmental Research and Public Health, 13*(11), 1047. doi:10.3390/ijerph13111047 PMID:27792205

Freije, A. (2014). Heavy metal, trace element and petroleum hydrocarbon pollution in the Arabian Gulf: A review. *Journal of the Association of Arab Universities for Basic and Applied Sciences, 17*(1), 90–100. doi:10.1016/j.jaubas.2014.02.001

Fu, J., Wang, Q., Wang, H., Yu, H., & Zhang, X. (2014). Monitoring of non-destructive sampling strategies to assess the exposure of avian species in Jiangsu Province, China to heavy metals. *Environmental Science and Pollution Research International, 21*(4), 2898–2906. doi:10.100711356-013-2242-4 PMID:24154854

Gautam, P., Gautam, R., Chattopadhyaya, M., Banerjee, S., Chattopadhyaya, M., & Pandey, M. (2016). *Heavy metals in the environment: fate, transport, toxicity and remediation technologies Thermodynamic profiling of pollutants View project Materials for Solid oxide fuel cells View project Heavy Metals in the Environment: Fate*. Transport, Toxicity and Rem.

Gunay, D. (2002). Sanayi ve sanayi tarihi. *Mimar ve Mühendis Dergisi, 31*, 8–14.

Hazardous Substances Research Center. (2003). Environmental impact of the petroleum industry. CF Pub. https://cfpub.epa.gov/ncer_abstracts/index.cfm/fuseaction/display.files/fileID/14522

Hubeny, J., Harnisz, M., Korzeniewska, E., Buta, M., Zieliński, W., Rolbiecki, D., Giebułtowicz, J., Nałęcz-Jawecki, G., & Płaza, G. (2021). Industrialization as a source of heavy metals and antibiotics which can enhance the antibiotic resistance in wastewater, sewage sludge and river water. *PLoS One*, *16*(6), e0252691. doi:10.1371/journal.pone.0252691 PMID:34086804

Hussain, T., & Gondal, A. (2008). Monitoring and assessment of toxic metals in Gulf War oil spill contaminated soil using laser-induced breakdown spectroscopy. *Environmental Monitoring and Assessment*, *136*(1-3), 391–399. doi:10.100710661-007-9694-2 PMID:17406995

JaccobA. (2020). Evaluation of Lead and Copper content in hair of workers from oil product distribution companies in Iraq. *Brazilian Journal of Pharmaceutical Sciences*, 56. https://doi.org/ doi:10.1590/s2175-97902019000318061

Ji, W, Yang, T., Ma, S., & Ni, W. (2012). Heavy metal pollution of soils in the site of a retired paint and ink factory. *Energy Procedia, 16*(Part A), 21-26. doi:10.1016/j.egypro.2012.01.005

Jibrin, M., Abdulhameed, A., Nayaya, A., & Ezra, G. (2021). Health risk effect of heavy metals from pesticides in vegetables and soils: A review. *Dutse Journal of Pure and Applied Sciences*, *7*(3b), 24–33. doi:10.4314/dujopas.v7i3b.3

Junaid, M., Hashmi, Z., Tang, M., Malik, R., & Pei, D. (2017). Potential health risk of heavy metals in the leather manufacturing industries in Sialkot, Pakistan. *Scientific Reports*, *7*(1), 8848. doi:10.103841598-017-09075-7 PMID:28821790

Kapusta, P., & Sobczyk, L. (2015). Effects of heavy metal pollution from mining and smelting on enchytraeid communities under different land management and soil conditions. *The Science of the Total Environment*, *536*, 517–526. doi:10.1016/j.scitotenv.2015.07.086 PMID:26233783

Karn, R., Ojha, N., Abbas, S., & Bhugra, S. (2021). A review on heavy metal contamination at mining sites and remedial techniques. *IOP Conference Series. Earth and Environmental Science*, *796*(1), 012013. doi:10.1088/1755-1315/796/1/012013

Kelishadi, R. (2012). Environmental pollution: Health effects and operational implications for pollutants removal. *Journal of Environmental and Public Health, 341637*, 1–2. doi:10.1155/2012/341637 PMID:22619687

Kepa, P., & Zaborska, A. (2021). Sources, fate and distribution of inorganic contaminants in the Svalbard area, representative of a typical arctic critical environment-a review. *Environmental Monitoring and Assessment*, *193*(11), 724. doi:10.100710661-021-09305-6 PMID:34648070

Kharb, A. (2018). Industrial revolution–from industry 1.0 to industry 4.0. *Journal of Advances in Computational Intelligence and Communication Technologies*, *2*(1), 1–3.

Kumar, A., Bawge, G., & Kumar, V. (2021). An overview of industrial revolution and technology of industrial 4.0. *International Journal of Research in Engineering and Science*, *9*(1), 64–71.

Kumar, N., Bauddh, K., Kumar, S., Dwivedi, N., Singh, D., & Barman, S. (2012). Extractability and phytotoxicity of heavy metals present in petrochemical industry sludge. *Clean Technologies and Environmental Policy*, *15*(6), 1033–1039. doi:10.100710098-012-0559-1

Lepp, N. (2012). *Effect of heavy metal pollution on plants. metals in the environment, pollution monitoring series, applied science publishers. department of biology.* Liverpool Polytechnic.

Li, F., Li, X., Hou, L., & Shao, A. (2018). Impact of the coal mining on the spatial distribution of potentially toxic metals in farmland tillage soil. *Scientific Reports*, *8*(1), 14925. doi:10.103841598-018-33132-4 PMID:30297728

Lim, D., Roh, T., Kim, M., Kwon, Y., Choi, S., Kwack, S., Kim, K. B., Yoon, S., Kim, H. S., & Lee, B.-M. (2018). Non-cancer, cancer, and dermal sensitization risk assessment of heavy metals in cosmetics. *Journal of Toxicology and Environmental Health. Part A.*, *20*(81), 432–452. doi:10.1080/15287394.2018.1451191 PMID:29589992

Martínez-Guijarro, R., Paches, M., Romero, I., & Aguado, D. (2021). Sources, mobility, reactivity, and remediation of heavy metal(loid) pollution: a review. *Advances in Environmental and Engineering Research, 2*(4), 033. doi:10.21926/aeer.2104033

Masindi, V., & Muedi, K. (2019). *Environmental contamination by heavy metals.* Intech Open Publishers.

Mikoczy, Z., & Hagmar, L. (2005). Cancer incidence in the Swedish leather tanning industry: Updated findings 1958-99. *Occupational and Environmental Medicine*, *62*(7), 461–464. doi:10.1136/oem.2004.017038 PMID:15961622

Minnesota Department of Health. (2021). Heavy metal in fertilizers. Minnesota Department of Health. https://www.health.state.mn.us/communities/environment/risk/studies/metals.html

Mitra, S., Chakraborty, A., Tareq, A., Emran, T., Nainu, F., Khusro, A., Idris, A. M., Khandaker, M. U., Osman, H., Alhumaydhi, F. A., & Simal-Gandara, J. (2022). Impact of heavy metals on the environment and human health: Novel therapeutic insights to counter toxicity. *Journal of King Saud University. Science, 34*(3), 101865. doi:10.1016/j.jksus.2022.101865

Mongi, R., & Chove, L. (2020). Heavy metal contamination in cocoyam crops and soils in countries around the Lake Victoria basin (Tanzania, Uganda and Kenya). *Tanzanian Journal of Agricultural Sciences, 19*(2), 148–160.

Muller, J. (2020). *Enabling Technologies for Industry 5.0: Results of a Workshop with Europe's Technology Leaders*. European Commission.

Nyika, J., & Dinka, M. (2022). Heavy metal pollution in soils and vegetables from suburban regions of Nairobi, Kenya and their community health implications. *Pollution, 8*(4), 1434–1447. doi:10.22059/POLL.2022.341522.1440

Nyika, J., Onyari, E., Dinka, M., & Mishra, S. (2019). Heavy metal pollution and mobility in soils within a landfill vicinity: A South African case study. *Oriental Journal of Chemistry, 35*(4), 1286–1296. doi:10.13005/ojc/350406

Ocal, F., & Altintas, K. (2018). Dördüncü sanayi devriminin emek piyasaları üzerindeki olası etkilerinin incelenmesi ve çözüm önerileri. *OPUS Uluslararası Toplum Araştırmaları Dergisi, 8*(15), 35–35. doi:10.26466/opus.439952

Ogunkunle, C., & Fatoba, P. (2014). Contamination and spatial distribution of heavy metals in topsoil surrounding a mega cement factory. *Atmospheric Pollution Research, 5*(2), 270–282. doi:10.5094/APR.2014.033

Postigo, C., Martinez, D., Grondona, S., & Miglioranza, K. (2018). Groundwater pollution: sources, mechanisms and prevention. In A. Dominick & M. Goldstein (Eds.), *Encyclopedia of the Anthropocene*. Elsevier Publishers., doi:10.1016/B978-0-12-809665-9.09880-3

Proshad, R., Kormoker, T., Mursheed, N., Islam, M., Bhuyan, I., & Islam, S. (2018). Heavy metal toxicity in agricultural soil due to rapid industrialization in Bangladesh: A review. *International Journal of Advanced Geosciences, 6*(1), 83–88. doi:10.14419/ijag.v6i1.9174

Rada, M. (2018). *Industry 5.0-from Virtual to Physical*. LinkedIn. https://www.linkedin.com/pulse/industry-50-from-virtual-physical-michael-rada

Radomirovic, M., Cirovic, Z., Maksin, D., Bakic, T., Lukic, J., Stankovic, S., & Onjia, A. (2020). Ecological Risk Assessment of Heavy Metals in the Soil at a Former Painting Industry Facility. *Frontiers in Environmental Science*, *8*, 560415. doi:10.3389/fenvs.2020.560415

Rai, P., Lee, S., Zhang, M., Tsang, Y., & Kim, K. (2019). Heavy metals in food crops: Health risks, fate, mechanisms and management. *Environment International*, *125*, 365–385. doi:10.1016/j.envint.2019.01.067 PMID:30743144

Rajaei, G., Mansouri, B., Jahantigh, H., & Hamidian, A. (2012). Metal concentrations in the water of Chah Nimeh reservoirs in Zabol. *Bulletin of Environmental Contamination and Toxicology*, *89*(3), 495–500. doi:10.100700128-012-0738-0 PMID:22885539

Rastogi, K., Pandey, A., & Tripathi, S. (2008). Occupational health risks among the workers employed in leather tanneries at Kanpur. *Indian Journal of Occupational and Environmental Medicine*, *12*(3), 132. doi:10.4103/0019-5278.44695 PMID:20040972

Rojko, A. (2017). Industry 4.0 concept: Background and overview. *International Journal of Interactive Mobile Technologies*, *11*(5), 77–90. doi:10.3991/ijim.v11i5.7072

Saleh, M., Hamad, Z., & Hama, J. (2021). Assessment of heavy metals in crude oil workers from Kurdistan region, northern Iraq. *Environmental Monitoring and Assessment*, *193*(1), 49. doi:10.100710661-020-08818-w PMID:33415539

Saygili, S. (2013). Sanayi toplumundan bilgi toplumuna geçiş sürecinde eğitimde dönüştürücü bir entelektüel olarak öğretmenler. *Uşak Üniversitesi Sosyal Bilimler Dergisi, 6*(ÖYGE Özel Sayısı), 270-281.

Şengel, U. (2019). *Türkiye'nin Turizm Talebini Etkileyen Faktörlerin Sosyo-Ekonomik Açıdan Ampirik Olarak Değerlendirilmesi*. Basılmamış Doktora Tezi, Sakarya Uygulamalı Bilimler Üniversitesi Lisansüstü Eğitim Enstitüsü, Sakarya.

Sengel, U. (2021). Chronology of the interaction between the industrial revolution and modern tourism flows. *Journal of Tourism Intelligence and Smartness*, *4*(1), 19–30.

Sharma, A., & Singh, B. (2020). Evolution of industrial revolutions: A review. *International Journal of Innovative Technology and Exploring Engineering*, *9*(11), 66–73. doi:10.35940/ijitee.I7144.0991120

Sharma, R., & Agrawal, M. (2005). Biological effects of heavy metals: An overview. *Journal of Environmental Biology*, *26*, 301–313. PMID:16334259

Sharma, S., Dietz, J., & Mimura, T. (2017). Vacuolar compartmentalization as indispensable component of heavy metal detoxification in plants. *Plant, Cell & Environment*, *39*(5), 1112–1126. doi:10.1111/pce.12706 PMID:26729300

Skobelev, P., & Borovik, S. (2017). On the way from Industry 4.0 to Industry 5.0: From digital manufacturing to digital society. *Industry 4.0, 2*, 307–311.

Soleimani, M., Amini, N., Sadeghian, B., Wang, D., & Fang, L. (2018). Heavy metals and their source identification in particulate matter (PM2.5) in Isfahan City, Iran. *Journal of Environmental Sciences (China)*, *72*, 166–175. doi:10.1016/j.jes.2018.01.002 PMID:30244743

Su, C., Meng, J., Zhou, Y., Bi, R., Chen, Z., Diao, J., Huang, Z., Kan, Z., & Wang, T. (2022). Heavy metals in soils from intense industrial areas in South China: Spatial distribution, source apportionment, and risk assessment. *Frontiers in Environmental Science*, *10*, 820536. doi:10.3389/fenvs.2022.820536

Sun, C., Liu, J., Wang, Y., Sun, L., & Yu, H. (2013). Multivariate and geostatistical analyses of the spatial distribution and sources of heavy metals in agricultural soil in Dehui, Northeast China. *Chemosphere*, *92*(5), 517–523. doi:10.1016/j.chemosphere.2013.02.063 PMID:23608467

Taj, I., & Jhanjhi, N. (2022). Towards industrial revolution 5.0 and explainable artificial intelligence. Challenges and opportunities. *International Journal of Computing and Digital Systems*, *12*(1), 1–26. doi:10.12785/ijcds/120124

Toth, G., Hermann, T., Da Silva, M., & Montanarella, L. (2016). Heavy metals in agricultural soils of the European Union with implications for food safety. *Environmental Pollution*, *88*, 299–309. doi:10.1016/j.envint.2015.12.017 PMID:26851498

USEPA. (2016). Outdoor air-industry, business and home: paint and coating manufacturing. EPA. https://archive.epa.gov/airquality/community/web/html/paint_manuf_addl_info.html

Vaidya, S., Ambad, P., & Bhosle, S. (2018). Industry 4.0–A glimpse. *Procedia Manufacturing*, *20*, 233–238. doi:10.1016/j.promfg.2018.02.034

Vinitha, K., Prabhu, R., Bhaskar, R., & Hariharan, R. (2020). Review on industrial mathematics and materials at Industry 1.0 to Industry 4.0. *Materials Today: Proceedings*, *33*, 3956–3960. doi:10.1016/j.matpr.2020.06.331

Walker, C., Sibly, R., & Hopkin, D. (2012). *Principles of Ecotoxicology; Group, Taylor and Francis* (4th ed.). CRC Press.

Wang, L., Tao, Y., Su, B., Wang, L., & Liu, P. (2022). Environmental and health risks posed by heavy metal contamination of groundwater in the Sunan Coal Mine, China. *Toxics*, *10*(7), 390. doi:10.3390/toxics10070390 PMID:35878294

Woldetsadik, D., Drechsel, P., Keraita, B., Itanna, F., & Gebrekidan, H. (2017). Heavy metal accumulation and health risk assessment in wastewater-irrigated urban vegetable farming sites of Addis Ababa, Ethiopia. *International Journal of Food Contamination*, *4*(1), 9. doi:10.118640550-017-0053-y

Xu, D., Shen, Z., Dou, C., Dou, Z., Li, Y., Gao, Y., & Sun, Q. (2022b). Effects of soil properties on heavy metal bioavailability and accumulation in crop grains under different farmland use patterns. *Scientific Reports*, *12*(1), 9211. doi:10.103841598-022-13140-1 PMID:35654920

Xu, H., Jia, Y., Sun, Z., Su, J., Liu, Q., Zhou, Q., & Jiang, G. (2022a). Environmental pollution, a hidden culprit for health issues. *Eco-Environment and Health*, *1*(1), 31–45. doi:10.1016/j.eehl.2022.04.003

Xu, L., Xu, L., & Li, L. (2018). Industry 4.0: State of the art and future trends. *International Journal of Production Research*, *56*(8), 2941–2962. doi:10.1080/00207543.2018.1444806

Yin, Y., Stecke, E., & Li, D. (2018). The evolution of production systems from Industry 2.0 through Industry 4.0. *International Journal of Production Research*, *56*(1-2), 848–861. doi:10.1080/00207543.2017.1403664

Yitagesu, Y., & Bekele, E. (2019). Impacts of cement dust deposition on heavy metal pollution in soil and barley crop grown around Abyssinia cement factory, Ethiopia. *Chemistry and Materials Research*, *11*(2), 1–11. doi:10.7176/cmr

Yiu, N., Zhu, F., Wei, F., Miao, J., Zhou, M., & Guan, J. (2016). Characteristics and evaluation of heavy metal pollution in different functional areas of Luoyang City, Henan Province. *Environmental Sciences (Ruse)*, *37*(6), 2322–2328. doi:10.13227/j.hjkx.2016.06.041

Yu, L., Kaiyi, S., Jie, Y., & Qiyu, K. (2021). Evaluation of heavy metal pollutants from plateau mines in wetland surface deposits. *Frontiers in Environmental Science*, *8*, 5557302. doi:10.3389/fenvs.2020.557302

Zakoldaev, D., Shukalov, A., Zharinov, I., & Zharinov, O. (2019). Modernization stages of the industry 3.0 company and projection route for the industry 4.0 virtual factory. *IOP Conference Series. Materials Science and Engineering*, *537*(3), 032005. doi:10.1088/1757-899X/537/3/032005

Chapter 3
Environmental Pollution by Heavy Metals in Sub-Saharan Africa

ABSTRACT

This chapter discusses the rise in environmental pollution in sub-Saharan Africa (SSA) region, which was previously considered contamination free. Drivers such as industrialization, urbanization, and increased consumption due to a growing population enhance the pollution trend. Using named examples of SSA countries, the chapter reports that mining and smelting activities, unsystematic management of waste, agricultural and manufacturing sectors are the sources of heavy metal pollution in soils, air, and water in the region. Metals such as As, Cd, Cr, Cu, Hg, Ni, Pb, and Zn have been associated with these anthropic activities and in some cases found at elevated levels in different environmental compartments. With the apparent trend in pollution, SSA region should regularly review and monitor the buildup of heavy metals in the environment using appropriate legislations and policies as well as implement remediation initiatives in polluted areas to ensure sustainable development in addition to economic growth.

INTRODUCTION

The rise in urbanization and industrial revolution for the growing global population has resulted to extensive pollution. Technology advancements in contemporary society have improved the living standards of the global population but consequently, unwanted substances have been released into the environment inducing public health

DOI: 10.4018/978-1-6684-7116-6.ch003

concerns (Anyamwu et al., 2018). The ineffective regulation of the pollutants and their release owing to the anthropic activities has exposed humans to health etiology. Regions such as Sub-Saharan Africa (SSA) have been affected by environmental pollution sourced from the substances including chemicals and heavy metals gravely (Fasinu & Orisakwe, 2013). The toxicity by heavy metals extends to soils, water, animals, plants and humans once the pollutants enter in food chains (Nyika & Dinka, 2022). In SSA region, the situation is likely to worsen since economic growth has taken precedence over environmental conservation. Consequently, environmental pollution has become common in the region. According to Tindwa and Singh (2023), out of 80 world's most polluted countries, 36 of this total are from SSA region. Some of the countries include South Sudan, Democratic Republic of Congo, Seychelles, Somalia, Zambia, South Africa, Nigeria, Namibia, Sierra Leone, Mauritania, Malawi, Madagascar, Kenya, Lesotho, Ghana, Gabon and Eswatini among others (Nkonya et al., 2016). The region is also the least prepared to deal with pollution among other environmental challenges due to its economic and technological limitations.

Heavy metals exposure occurs via food, air, water, soil (as a result of geophagia) and during occupational exposure. The ecological importance of the pollutants is their toxicity even in minute concentrations, their translocation in food chains and their non-biodegradable nature, which is attributable to their persistence and bioaccumulation in the environment (Oruko et al., 2021). Activities such as industrial processes resulting to emissions and effluents, unscientific handling of solid waste, illegal refining, mining, smelting and foundry all lead to exposure to heavy metals (Ghane et al., 2022; Xu et al., 2022). Once the pollutants are introduced to environmental compartments, they are transferred through food chains to different environmental components increasing the risk of human exposure. Consequently, heavy metals have evolved to be a significant problem in dietary, natural and evolutionary perspectives and their insufficient reclamation and remediation has further exacerbated their unplanned exposure and negative effects to the environment (Jan et al., 2015; Anyamwu et al., 2018).

Public health concerns of heavy metals are affiliated to their ubiquitous nature from both natural and mainly anthropogenic sources. Although global industrialization induces most of the negative health effects of heavy metals, the severity of the outcomes is influenced by individual susceptibility defined by their immunity, age, exposure method and duration, dosage and specific metal one is exposed to (Kinuthia et al., 2020; Oruko et al., 2021). Voluntary or involuntary exposure to the pollutants especially Cd, Hg and Pb has increased due to industrialization and their ubiquity (Kinuthia et al., 2020). In this respect, it is essential to understand the specific sources and occurrence of heavy metals, which varies spatially and temporally. Such information is key in identifying pollution hotspots, assaying trends of pollutant distribution and sources and in planning for remediation and reclamation measures to avert extensive environmental pollution. The current book chapter provides an assessment of the drivers, status and extent of heavy metal exposure in different environmental compartments (soil, air and water) of SSA and with a focus on industrial-sourced pollution to soils of the region.

LEGAL FRAMEWORKS AND DRIVERS OF HEAVY METAL POLLUTION IN SUB SAHARAN AFRICA

The SSA region recognizes the problem of heavy metal pollution in soils and to this end, most of the countries have accepted, accessed and ratified some international conventions and/ or frameworks geared to ending soil pollution. Some of the conventions include the Minamata convention, Stockholm convention, Rotterdam convention and Basel convention. The Minamata convention seeks to prevent anthropogenic emissions that lead to the release of Hg in the atmosphere and eventually, to the soil (Minamata Convention on Mercury, 2019). The Stockholm convention seeks to safeguard environments of participating countries from release of toxic chemicals including heavy metals that have lethal effects on human health (Stockholm Convention, 2019). The Rotterdam convention protects the environment including soils from hazards including heavy metals, which result from irresponsible and unregulated trade and the use of such noxious chemicals (Rotterdam Convention, 2010). The Basel convention regulates transboundary movements of hazardous waste (including those containing heavy metals) to prevent their resultant environmental pollution (United Nations Environmental Program, UNEP, n.d). With the convention, export of pollutants should be minimized and once done, they should be treated according to the set standards before disposal. Although these milestones are progressive to regulating environmental pollution by heavy metals and streamlining the legal framework to do so, governance of environmental issues is poor in SSA due to non-enforcement of such legal frameworks (Tindwa & Singh, 2023). The observation is supported by the varied drivers of heavy metal pollution in SSA soils as detailed in subsequent sub-titles.

Manufacturing, Mining, and Smelting Industries

There is a rising demand for minerals globally as the industry strives for growth along with the manufacturing sector (Avkopashvili et al., 2022). Mining activities and the subsequent processing of minerals and their metallurgical removal facilitate physical isolation of minerals and ore concentrates in addition to breaking involved crystallographic bonds to get the desired quality of minerals. The open pit mining, clustering and grinding of ores affect natural ecosystems since they involve vegetation and soil removal and burial in addition to waste production (Fashola et al., 2016). In addition to physical disruption of landscapes, mining activities result to mine tailing production, accumulation of acidic drainage and emission dust during ore smelting and dressing (Sun et al., 2018). The tailings, dust and acid mine drainage all contain heavy metals and are a result of both small- and large-scale mining (Xu et al., 2022). The release of the substances induces pollution mainly in developing countries of South America, Asia and Africa where government oversight is poor and environmental awareness is limited (Fei et al., 2017). The mines of the regions are often illegal are extremely damaging to the environment in addition to inducing safety and health concerns in their vicinities.

Mining and associated processes are main heavy metal sources and pollute mainly the soils from where the pollutants are extracted (Sun et al., 2018). Pilling and uncovering of raw ore and its processing resulting to heavy metal containing wastes and their disposal to land and water pose danger to soils of farmlands, surface- and ground-water resources. For instance, 99% of extracted gold ore during the mineral processing is disposed off as waste to the environment and is associated with exposure to Ag, Cu and Pb heavy metals (Fashola et al., 2016). Similarly, the mining and processing of gold and copper resulted to Cd and Pb pollution of soils and water in Kazreti, south-east Georgia (Avkopashvili et al., 2022). Soil pollution resulted from the production of tailings and dust while water pollution was as a result of acid mine drainage during extraction and processing of the two minerals. In Qinghai-Tibet plateau of China, Cr and Ni pollution was reported in the water as a result of unsustainable coal, lead, copper and zinc mining and their resultant wastes (Wei et al., 2018). Dumped tailings in a polymetallic mine of Yaoposhan, China, farmland soils were found to be polluted by Cu, Cd, Pb and Zn beyond allowable limits (Sun et al., 2018). Soil and water resources of Kouh-e Zar mine of Iran were laden with As, Cd, Cr, Cu, Sb, Ni and Zn due to resultant tailings of gold and copper mining and their disposal on farmlands and waterways (Tahmasebi et al., 2020). In SSA, particularly southern, central and western Africa, mining activities are likely to induce heavy metal pollution due to the unsustainable management of the industry, its resultant wastes and the laxity of environmental regulators charged with environmental protection (Yabe et al., 2010; Tindwa & Singh, 2023).

Apart from the mining and metallurgy industry, refinery processes contribute to heavy metal pollution in the environment. Petroleum refineries, coal combustion power plants, high tension wires and nuclear energy plants emit heavy metals such as B, Cd, Cs, Cu, Ni, Se and Zn (Zhu et al., 2016). Processing industries for wood preservation, electronics, textile, paper and plastics also introduce heavy metals to the environment. The use and combustion of Pb-containing petrol pollute the environment through Pb-containing particulate matter. The release of Cd, Hg, Ni, Pb and Zn from inefficient engines and automobiles' anti-wear protectants also pollutes the environment (Srivastava et al., 2017). The resultant industrial solid waste and wastewater also comprises of heavy metals that can end up in soils and water resources especially if not treated adequately (Hubeny et al., 2021). Developing countries of SSA region, which are rapidly industrializing are prone to heavy metal pollution as they seek for economic development unless control measures are put in place to remediate such metals in the environment.

Agriculture

The use of heavy metal containing chemical fertilizers and pesticides is the main route of pollution in the agricultural sector. The pesticides and inorganic fertilizers are used to maximize the yields of coffee, tea, cotton and barley among other food crops but have negative effects on soils and farm laborers. Glyphosate-based pesticides were found laden with As, Cr, Co, Pb and Ni, which could be transferred to laborers as aerosols/ emissions and soils (after precipitation and deposition) during farm application (Defarge et al., 2017). The use of iron sulphate, copper sulphate and calcium superphosphate fertilizers was associated with introduction of Cd, Co, Cu, Ni and Zn in soils (Garcia & Boluda, 1996). The use of phosphatic fertilizers processed from rock phosphate releases Cd to agricultural soils (Suciu et al., 2022). This trend was established in Morocco, South Africa, Senegal and Togo of SSA region that are large producers of phosphate rock (Mar & Okazaki, 2012).

Soils enriched with the heavy metals can then pass on the contaminants to crops cultivated on such environments via root uptake. Similarly, use of fungicides and herbicides resulted to elevated levels of Co, Mn, Pb and Zn in agricultural soils (Garcia & Boluda, 1996). In Tanzania's coffee farms applied Cu-based fungicides, the metal (Cu) was found to be elevated beyond the allowable limits (Ma et al., 2019). Agricultural-based introduction of heavy metals in SSA is exacerbated by shipping of obsolete and banned stocks of such heavy-metal containing agrochemicals (Semu et al., 2019). The banned and obsolete stock of agrochemicals (such as Dichlorodiphenyltrichloroethane, DDT) are shipped abroad from developed countries to developing ones for incineration but end up being used as agrochemicals in SSA countries. For instance, in Kenya, Ethiopia and Tanzania such organopesticides, which are laden with heavy metals have been reported to accumulate in some of their soils (Wandiga, 2001; Loha et al., 2018). The legal frameworks to prevent pollution from such agrochemicals are existent but, they are not well enforced leading to their ineffectiveness and hence, an opportunity for extended pollution in SSA soils.

Other anthropogenic sources of agricultural heavy metals include the application of sludge, biosolids (such as compost and manure) and wastewater for growing crops (Nyika & Dinka, 2022) and emission outputs resulting from fossil fuel combustion, traffic and industry gaseous wastes. The emissions are released to the atmosphere, dispersed to distant places before their precipitation and deposition to soils. The use of sewage sludge and fly ash to lime soils introduced high levels of Co and Pb to such farmlands (Vacha, 2021).

Waste Disposal

In addition to production of wastes (solid waste and wastewater) in mining, industrialization and agricultural activities, there are growing deliberate efforts to dispose them (Nyika et al., 2020). Solid wastes from domestic activities, manufacturing and construction industries with different heavy metal composition are disposed in open dumpsites, landfills and riversides. In addition, ponds and lagoons are set up to store and partially treat industrial, municipal and mining wastewater. The infrastructural developments pose as threats to the introduction of heavy metals to land and water resources.

Many developing countries of SSA such as Nigeria, Kenya and Tanzania have large dumpsites at Ibadan, Dandora and Tabata suburban regions, respectively which have existed since the 1960s. The dumpsites have no waste conversion means, integrate both municipal and hazardous wastes, rely on incineration to reduce waste levels and pollute the atmosphere through heavy metal containing emissions and particulates. In addition, they introduce heavy metals to soils and water resources through generation of leachate that can migrate to such compartments (Semu et al., 2019). Introduction of Cr, Cu, Mn, Ni, Pb and Zn in soils and groundwater from leachates resulting from unscientific management of solid waste at Iringa municipality, Tanzania was reported (Sanga et al., 2022). Similar results were obtained in South Africa's Eastern Cape region where groundwater and soils were reported to have elevated levels of As, Cd, Cr, Fe, Hg and Pb as a result of poor solid waste management and subsequent leachate pollution at Roundhill landfill (Nyika & Onyari, 2019; Nyika et al., 2020). Leachate collected from six different dumpsites of Nigeria showed elevated levels of As, Cd, Cr, Fe, Ni, Pb and Zn, which were transferred to water resources and soils of their vicinities (Essein et al., 2022).

Waste lagoons, which are recipients of wastewater have also been found to be laden with heavy metals. In three wastewater treatment plants of Easten Cape, South Africa, sludge and effluent had elevated levels of Cd, Cu, Pb and Zn, which were poorly removed during treatment and prior to environmental release (Agoro et al., 2020). In soils and vegetation of Korle lagoon of Ghana, elevated levels of As, Cd, Cr, Cu, Hg, Ni, Pb, Sn and Zn were assayed at levels beyond permissible limits (Fosu-Mensah et al., 2017). The heavy metals originated from the disposal of electronic gadgets and chips that are heavy-metal containing and subsequent attempts to incinerate it. The metals have the capacity to build up in soils and water bodies after release of the sludge and wastewater causing extensive pollution.

Transportation

The transportation industry is also a driver to environmental pollution by heavy metals. Singh et al. (2018) noted that globally, soils near urban areas and on busy roadsides are prime targets of heavy metals due to increased traffic activities. Lead is emitted from vehicular smoke during the burning of leaded petrol. Lead particulates from the vehicles are deposited on lands near roads, some of which are used for agriculture (Dignam et al., 2019; UNEP, 2022). In addition to polluting soils, the heavy metal dust could fall on cultivated crops such as vegetables introducing Pb on them (Semu et al., 2019). Apart from through fuel combustion, heavy metals such as Cd, Cu, Ni, Pb and Zn are a result of the use of metallic parts, corrosion of radiators and batteries, leakage of oils and tire wearing out of vehicles (Singh et al., 2018). The situation is more pronounced in populous countries of the world such as China and India where such anthropic activities are concentrated. Pollution of soils at Bhomoraguri and Balipara roadsides of India by Cd and Pb was associated with vehicular emissions and wastes (Malunguja et al., 2022).

CURRENT STATUS AND EXTENT OF HEAVY METAL POLLUTION IN SUB-SAHARAN AFRICA

For many years, Africa was known to be free of heavy metal pollution but with urbanization and population growth, many regions have become vulnerable to environmental pollution. Figure 1 shows a number of regions in Africa including SSA that have reported heavy metal pollution. This section discusses the trend in heavy metal pollution of soils in various regions of SSA.

West Africa

The extraction and processing of petroleum products is the main causative of heavy metal pollution in west African soils (Anyanwu et al., 2018). Others drivers of heavy metal pollution in the region include the rise in industries especially in metropolitan regions. The exposure of metal contaminants to soils occurs via crude oil discharge to the environment in the form of gas and oil industries effluents, damage of oil processing facilities or corrosive destruction of the oil pipeline during its transportation. In soils and sediments from the Bonny/ New Calabar river estuary of Nigeria, bioaccumulation of Cd, Cr, Pb and V was reported and affiliated to the introduction of petroleum-based wastes (Chindah et al., 2004). Petrogenic and hydrocarbon pollution in Nigeria was attributable to elevated levels of Ni and Pb in soils where oil spillage had occurred (Oloruntegbe et al., 2009). A study in Asa River and vicinity in Nigeria linked industrial processes to high levels of Cr, Cu, Fe and Mn in soils and sediments. Heavy metal pollution in this case was associated with detergent making, leather tanning and bottling activities from adjacent industries and their resultant wastes. (Adekola & Eletta, 2007). In urbanized regions of Ghana near Iture estuary, Kakum and Sorowie rivers were found with elevated levels of Cd and Pb compared to the allowed limits due to increased industrial activities (Fianko et al., 2007). The polluted water also introduced heavy metals in soils since it was used for irrigation.

In Kumasi Metropolis of Ghana, industrial activities such as mining operations, metallurgy, tanning, auto-mechanic activities, gas and fuel processing were attributable to elevated levels of As, Cd, Co, Cu, Cr, Ni and Pb in areal soils (Akoto et al., 2017). Mining activities in the vicinities of Oda, Offin and Pra rivers of Ghana was attributable to Cd, Cr, Fe, Ni and Pb pollution of the soils of the vicinity beyond allowable limits (Ebo et al., 2018). A similar trend was established in Saint Louis and Dakar coast estuaries of Senegal where soils and sediments were polluted with Cd, Cr, Cu and Pb (Diop et al., 2015). Using industrial effluents and contaminated environs for farming is another cause of heavy metal pollution. The farming of vegetables and food crops by small-scale farmers in peri-urban areas to increase yields and use the organic nutrients in solid waste and wastewater leads to accumulation of heavy metals in the environment especially in developing countries of SSA. According to Odai et al. (2008), Cd, Cu, Pb and Zn accumulated to dangerous levels, which were beyond allowable limits in vegetables, fish, food, animals, soils and water of Kumasi, Ghana due to the use of untreated wastewater and biosolids.

Southern Africa

Southern Africa's mining industry is the main driver of heavy metal pollution in the environment. Countries such as South Africa hold large deposits of Au among other minerals {such as coal, zirconium (Zr), rutile (TiO_2), palladium (Pd), ilmenite {$(Fe,Ti)_2O_3$}, vermiculite {$(Mg,Fe^{2+},Al)_3(Al,Si)_4O10(OH)_2 \bullet 4(H_2O)$}, V, platinum (Pt), Mn and chrome (Cr_2O_3)} while Zambia has Co and Cu deposits (Ebenebe et al., 2017). The countries depend on the minerals for their economy and development. Metalliferous and acid mine drainage during gold and coal processing introduces Cu, Fe, Mn, Ni, Pb and Zn in the atmosphere resulting to soil and water pollution. Mining activities in Southern Africa areas along Limpopo River basin such as Shashe, Motloutse, Chagane, Mwenezi, Mzingwane, Dati, Ngotwane and Bubye were reported to be adversely affected by heavy metals, which led to extensive soil pollution (Masindi & Muedi, 2019). The main pollutants include As, Co, Cr, Cu, Hg and Ni from gold, coal and base metal processing. Ebenebe et al. (2017) also reported that the mining industry in South Africa has the potential to release heavy

Figure 1. Regions of Africa that have reported environmental pollution due to heavy metals
(Yabe et al., 2010)

metals in resultant dusts and effluents during their processing resulting to extensive bioaccumulation of the pollutants in soils prior to their phyto-transfer. Owing to the increased mining activities, which are energy-intensive and the coal-dependent electricity sources, South Africa is one of the most carbon-intensive countries globally (Moeletsi & Tongwane, 2020).

In the Copperbelt province of Zambia, the dumping of copper mine tailings in water resources and soils introduces heavy metals such as Cu, Fe, Mn, Pb and Zn. The trend was established in Kitwe and Mufulira districts that had several mine waste dumpsites and soils of the area had elevated heavy metal concentrations (Dusengemungu et al., 2022). The areal air also had particulates with high levels of Co, Cu, Fe, Pb and Zn, which originated from the mine dust. Similar findings were established by Mwaanga et al. (2019) in an investigation on the potential of mine air pollution in North-Western and Copperbelt provinces of Zambia. The soils and water resources in the provinces were also heavy metal (Cu, Fe and Zn) laden resulting to bioaccumulation of the pollutants in nearby plants and fishes of Solwezi and Kafue Rivers (Hasimuna et al., 2022).

The mining of gold, bornite, copper and chalcopyrite in Oamites and Klein Aub mines of Namibia resulted to tailings with high concentrations of Cu, Ni, Pb and Zn (Uugwanga & Kgabi, 2020). The metals polluted areal soils and sediments posing as a danger to the environment. Diamond, nickel, copper, gold and soda ash mining in Botswana resulted to production of mine waste. In Selebi Phikwe area, Botswana that hosts the BCL Cu-Ni mine, areal groundwater, soils and air were found laden with Cd, Co, Cu, Ni, Pb and Zn heavy metals due to smelting activities and resultant mine tailings (Manyiwa et al., 2022). The coal mining and processing industry of Mozambique has the potential to introduce heavy metals to the environment. This was established in Moatize district, which is a coal mining hotspot (Marove et al., 2022). The study reported elevated levels of Cd, Cr, Mn and Pb in areal groundwater, soils and sediments and attributed the pollution to coal cleaning and processing, which produced tailings that resulted to leachates and acid drainage and their hazardous effects after environmental release. The mining of Au and resultant tailings during processing at Kadoma, Sanyati catchment of Zimbabwe was associated with elevated levels of As, Cr, Hg, Mo, Ni and Sb in soils and sediments of the vicinity (Mudimbu et al., 2022).

In addition to mining activities, industrial effluents also induce heavy metal pollution in the Southern Africa region. The effluents are from petroleum, oil, iron and steel industrial processes. In Vaal Triangle of South Africa that hosts battery, metal galvanizing and tanking industries, effluents introduced Cu, Pb and Zn metals to soils and water resources resulting to Vaal River pollution and water quality deterioration (Iloms et al., 2020). The effluents had been released to Leeuwkuil wastewater treatment plant prior to their introduction to the Vaal River. However, the treatment plant should handle municipal rather than industrial effluent hence, wastewater treatment was inadequate. Farmers using the river water also introduced heavy metals to soils leading to extensive pollution. Unscientific management of industrial solid waste and its dumping in poorly managed landfills also introduces heavy metals to the environment. This supposition was proven in Roundhill landfill vicinity of South Africa where mismanagement of the facility and its dumped solid waste from industries of Eastern Cape province resulted to elevated levels of Cr and Cd in soils (Nyika et al., 2020) and Fe and Mn in groundwater (Nyika & Onyari, 2019) of the vicinity. Industrialization and production of industrial wastewater was attributable to soil, sediment and water pollution by As, Cd, Cr, Fe, Mn and Pb at Karas (Namibia) (Pitiya et al., 2022) and Kwekwe river vicinity (Zimbabwe) (Chinhanga, 2010).

Eastern Africa

In Eastern Africa, industrial solid waste and effluent mismanagement is the main cause of heavy metal introduction to the environment. In the region, generation of solid waste and effluents supersedes the set-up collection and treatment facilities. As such, industrial wastes and effluents are indiscriminately dumped in uncontrolled dumpsites resulting to generation of leachate that is heavy metal containing (Sanga et al., 2022). In Kenya's Dandora open dumpsite, where more than 2,000 tons of solid waste from among other industrial wastes were disposed daily, elevated levels of Cd, Co, Cr, Cu, Fe, Mn, Pb and Zn were found in areal soils and in the banks of Nairobi River where the levels exceeded recommended limits (Tsuma et al., 2016). In Uganda, industrial activities resulting to the introduction of steel rolling mill waste at Jinja was attributable to soil pollution by Cd, Cr, Cu, Ni, Pb and Zn (Namuhani & Kimumwe, 2015). In Tanzania's Iringa municipality, unscientific management of solid waste was attributable to the introduction of Fe, Mn, Ni, Pb and Zn in soils and groundwater of the vicinity (Sanga et al., 2022). The metals whose levels were beyond allowable limits had been introduced after leachate leakage to soils and areal aquifer system. Landfill leachate also introduced Cd, Ni and Pb in soils and river water of Tepi town vicinity in Ethiopia (Mekonnen et al., 2020). Solid waste dumping in areas of Addis Ababa (Ethiopia) and resultant leachate generation, along with the mixing of untreated wastewater with freshwater of the Akaki River is also causing heavy metal introduction in the surrounding environment (Mekuria et al., 2021). Introduction of the metals to the ecology was attributable to indiscriminate dumping of industrial and municipal waste in open spaces such as riversides, drainage areas, roadsides and residential areas. Disposed wastes with metallic contents also emanate from coal, gold and copper mining in Tanzania among other Easten African countries (Semu et al., 2019).

Industrial effluents also result to introduction of heavy metals to the environment in Eastern Africa. This is especially so if the effluents are mixed with municipal wastewater, they are not adequately treated or they are reused for agricultural activities. In Nairobi's suburban regions, a mixture of industrial and municipal effluents polluted soils and their use in vegetable cultivation introduced Cd, Cr, Co, Cu, Fe, hg, Mn, Ni, Pb and Zn in soils and crops (Nyika & Dinka, 2022). Production of industrial effluents from Dar es Salaam's brewing, textile and soap-making companies and its partial treatment exposed the water and soils of the vicinity to Pb pollution (Kihampa et al., 2016). The partial treatment was not adequate to decontaminate the heavy metal and subsequent release of the effluent into the Msimbazi River located near the city only extended the pollution problem. A similar trend was evident in Uganda where irregular discharge and retention of industrial effluents was associated with Cr, Cu, Hg and Pb pollution in Nakivubo stream of Kampala (the most industrious town of Uganda) and soils of its vicinity (Dietler et al., 2019). In Ethiopia's Awash River basin, discharge of industrial effluents was attributable to the presence of Cr, Cu, Fe, Mn and Zn in the soils and water beyond allowable limits (Dessie et al., 2022). The heavy metals resulted from effluents of thirty-three factories that had semi-functional or non-functional wastewater treatment plants. Ethiopia and Uganda among other Eastern African countries have been reported to use obsolete manufacturing technology without functional effluent treatment systems and are therefore forced to discharge resultant wastewater to freshwater resources and nearby soils without treatment (Dessie et al., 2022). Table 1 summarizes some of the heavy metals responsible for environmental pollution in SSA and the environmental compartment affected by the contaminants.

The cottage industry in Eastern Africa also introduces heavy metals to the environment. This has been documented in both Kenyan and Tanzanian cottage industries, which are of varying scales (Semu et al., 2019). Some of the metals associated with cottage processes are Cd, Cr, Cu, Pb and Zn. In soils of the vicinity of Ngara and Gikomba cottage industries in Nairobi (Kenya), the levels of Cd, Cr, Cu and Pb were beyond permissible limits (Odero et al., 2000). Similarly, soils near Morogoro cottage industries of Tanzania had elevated levels of Cd, Pb and Zn (Mmbaga & Semu, 1999).

Evidently most of SSA is suffering from elevated levels of heavy metals in the environment even as the region strives for industrial and economic growth amidst population rise. The drivers of heavy metal pollution are unscientific management of industrial solid wastes, emissions and effluents from mining, cottage, agricultural and transportation processes, incineration of wastes, use of leaded petrol and particulate matter from various processing and manufacturing activities. The reasons as to why SSA is progressing in heavy metal pollution are not only attributable to the drivers but also the weak legislation regarding environmental pollution control by member countries of the region (Yabe et al., 2010). Bildirici (2022) also shared similar sentiments suggesting the need to improve governance and enforce laws that support environmental sustainability stringently since such plans are essential steerers to development. SSA region also prioritizes more on industrialization and economic growth while overlooking environmental management (Tindwa & Singh, 2023). Consequently, the region has overexploited its natural resources through mineral exploration while producing massive wastes that they are incapable of managing (Yabe et al., 2010). The result has been uncontrolled heavy metal pollution on soils, water and the atmosphere. As such, the extent of heavy metal pollution in SSA region as discussed in this chapter is a background to devising approaches to monitor, assay and predict pollution spread accurately as well as take up exposure

Table 1. Heavy metal pollution in Sub-Saharan Africa

Region/ Country	Pollution Source	Heavy Metals	References
Kaduna, Nigeria	Soil and vegetables	Cd, Cr, Cu, Fe, Pb and Zn	Mohammed & Folorunsho, 2015
Niger Delta, Nigeria	Water	Cd, Cr, Cu, Pb and Zn	Akporido & Onianwa, 2015
Benin, Nigeria	Soil	Cd, Cr, Pb and Zn	Akporido & Asagba, 2013
Akwa Ibom, Nigeria	Fish	As, Cd, Cr, Cu, Pb, Zn	Akpanyung et al., 2014
Pretoria, South Africa	Soil	Cd, Cr, Cu, Fe, Pb, Zn	Olowoyo et al., 2012
Nairobi, Kenya	Soils and vegetables	Cd, Cr, Cu, Ni, Pb	Mutune et al., 2014
Accra, Ghana	Soil	Cd, Cu, Ni, Pb	Fosu-Mensah et al., 2017
Phillippi horticultural area, South Africa	Soils and vegetables	Cd, Cr, Cu, Ni, Pb, Zn	Malan et al., 2015
Bulawayo, Zimbabwe	Fish, soils, sediments, vegetables and water	Cd, Cr, Cu, Pb	Teta et al., 2017
Yaounde, Cameroon	River sediment	Cd, Cr, Cu, Ni, Pb	Ekengele et al., 2008
Ethiopia	Vegetables	As, Cd, Cr, Cu, Zn	Essumang, 2009
Ghana	Soils, sediment, water and fish	As, Cd, Cu, Pb, Zn	Odai et al., 2008
Namibia	Sediment	Cu, Ni, Zn	Taylor & Kesterton, 2002
South Africa	Water	Cd, Cu, Pb, Zn	Fakayode & Olu-Owolabi, 2003
Lake Victoria, Uganda	Water, soils, sediments and vegetables	Cd, Cr, Fe, Ni, Pb Cd,	Muwanga & Barifaijo, 2006
Lake Victoria, Tanzania	Water	Cr, Cu Pb, Zn	Kishe & Machiwa, 2003
Zambia	Sediment	Cd, Cr, Fe, Ni, Pb, Zn	Heyden & New, 2004
Harare, Zimbabwe	Vegetables and water	As, Cd, Cr, Cu, Mn, Pb, Zn	Muchuweti et al., 2006

and contamination control measures. These efforts are based on the precognition that once the environment is polluted with heavy metals, it is exorbitant and nearly impossible to cleanse them off the pollutants (Jan et al., 2015; Oruko et al., 2021).

CONCLUSION

Although the larger part of Africa, SSA was previously pollution free, advances to industrialization, increased consumption and urbanization have propagated the phenomenon and in particular, heavy metal exposure. The levels of heavy metals in most regions are critical and beyond permissible levels in fishes, soils, water, food, animals and edible vegetables. Drivers such as mining, manufacturing industries, agricultural, transportation, solid waste and effluent production and release to the environment are set to exacerbate the trend. In reviewed studies of this chapter, pollution by toxic metals including Cd, Cr, Hg and Pb have been reported. The metals cause toxicity to food chains and have the capacity to bio-magnify and bioaccumulate in environs causing harm to the biosphere. To manage the apparent situation, there is need to regularly review and monitor the buildup of heavy metals in the environment. Such undertakings are recommended to enable advanced planning on pollution control through enacting and enforcing appropriate environmental conservation legislations specific to control of heavy metal pollution in SSA.

REFERENCES

Adekola, F., & Eletta, O. (2007). A study of heavy metal pollution of ASA River, Ilorin. Nigeria; trace metal monitoring and geochemistry. *Environmental Monitoring and Assessment*, *125*(1-3), 157–163. doi:10.100710661-006-9248-z PMID:17058013

Agoro, M., Adeniji, A., Adefisoye, M., & Okoh, O. (2020). Heavy metals in wastewater and sewage sludge from selected municipal treatment plants in Eastern Cape province, South Africa. *Water (Basel)*, *12*(10), 2746. doi:10.3390/w12102746

Akoto, O., Sam, N., Ikenaka, Y., Nakayama, S., Baidoo, E., & Yohannes, B. (2017). Contaminated levels and sources of heavy metals and a metalloid in surface soils in the Kumasi metropolis, Ghana. *Journal of Health & Pollution*, *7*(5), 28–39. doi:10.5696/2156-9614-7.15.28 PMID:30524828

Akpanyung, E., Akanemesang, U., Akpakpan, E., & Anodoze, N. (2014). Levels of heavy metals in fish obtained from two fishing sites in Akwa Ibom State, Nigeria. *African Journal of Environmental Science and Technology*, *8*(7), 416–421. doi:10.5897/AJEST2014.1730

Akporido, S., & Asagba, S. (2013). Quality characteristics of soil close to the Benin River in the vicinity of a lubricating oil producing factory, Koko, Nigeria. *International Journal of Soil Science*, *8*(1), 1–16. doi:10.3923/ijss.2013.1.16

Akporido, S., & Onianwa, P. (2015). Heavy Metals and Total Petroleum Hydrocarbon Concentrations in Surface Water of ESI River, Western Niger Delta. *Research Journal of Environmental Sciences*, *9*(2), 88–100. doi:10.3923/rjes.2015.88.100

Anyanwu, B., Ezejiofor, A., Igweze, Z., & Oriasakwe, O. (2018). Heavy metal mixture exposure and effects in developing nations: An update. *Toxics*, *6*(4), 65. doi:10.3390/toxics6040065 PMID:30400192

Avkopashvili, M., Avkopashvili, G., Avkopashvili, I., Asanidze, L., Matchavariani, L., Gongadze, A., & Gakhokidze, R. (2022). Mining-related metal pollution and ecological risk factors in South Eastern Georgia. *Sustainability (Basel)*, *14*(9), 5621. doi:10.3390u14095621

Bildirici, M. (2022). The impacts of governance on environmental pollution in some countries of Middle East and sub-Saharan Africa: The evidence from panel quantile regression and causality. *Environmental Science and Pollution Research International*, *29*(12), 17382–17393. doi:10.100711356-021-15716-2 PMID:34665419

Chindah, A., Braide, A., & Sibeudu, O. (2004). Distribution of hydrocarbons and heavy metals in sediment and a crustacean (shrimps- *Penaeus notialis*) from the Bonny/New Calabar River Estuary, Niger Delta. *African Journal of Environmental Assessment and Management*, *9*, 1–17.

Chinhanga, J. (2010). Impact of industrial effluent from an iron and steel company on the physicochemical quality of Kwekwe river water in Zimbabwe. *International Journal of Engineering Science and Technology*, *2*(7), 129–140. doi:10.4314/ijest.v2i7.63754

Defarge, N., Vendomois, J., & Seralini, G. (2017). Toxicity of formulants and heavy metals in glyphosphate-based herbicides and other pesticides. *Toxicology Reports*, *5*, 156–163. doi:10.1016/j.toxrep.2017.12.025 PMID:29321978

Dessie, B., Tessema, B., Asegide, E., Tibebe, D., Alamirew, T., Walsh, C., & Zeleke, G. (2022). Physicochemical characterization and heavy metal analysis from industrial discharges in Upper Awash River basin, Ethiopia. *Toxicology Reports*, *9*, 1297–1307. doi:10.1016/j.toxrep.2022.06.002 PMID:36518430

Dietler, D., Babu, M., Cisse, G., Halage, A., Malambala, E., & Fuhrimann, S. (2019). Daily variation of heavy metal contamination and its potential sources along the major urban wastewater channel in Kampala, Uganda. *Environmental Monitoring and Assessment*, *191*(52), 1–13. doi:10.100710661-018-7175-4 PMID:30617634

Dignam, T., Kaufmann, R., LeStourgeon, L., & Brown, M. (2019). Control of Lead Sources in the United States, 1970-2017: Public Health Progress and Current Challenges to Eliminating Lead Exposure. *Journal of Public Health Management and Practice*, *25*(1), S13–S22. doi:10.1097/PHH.0000000000000889 PMID:30507765

Diop, C., Dewaele, D., Cazier, F., Diouf, A., & Ouddane, B. (2015). Assessment of trace metals contamination level, bioavailability and toxicity in sediments from Dakar coast and Saint Louis estuary in Senegal, West Africa. *Chemosphere, 138*, 980–987. doi:10.1016/j.chemosphere.2014.12.041 PMID:25592460

Dusengemungu, L., Mubemba, B., & Gwanama, C. (2022). Evaluation of heavy metal contamination in copper mine tailing soils of Kitwe and Mufulira, Zambia for reclamation prospects. *Scientific Reports, 12*(1), 11283. doi:10.103841598-022-15458-2 PMID:35787645

Ebenebe, P., Shale, K., Sedibe, M., Tikilili, P., & Achilonu, M. (2017). South African mine effluents: Heavy metal pollution and impact on the ecosystem. *International Journal of Chemical Science, 15*(4), 1–13.

Ebo, A., De Vries, N., & Nyarko, K. (2018). Assessment of heavy metal pollution in the main Pra River and its tributaries in the Pra basin of Ghana. *Environmental Nanotechnology, Monitoring & Management, 10*, 264–271. doi:10.1016/j.enmm.2018.06.003

Ekengele, N., Myung, C., Ombolo, A., Ngatcha, N., Georges, E., & Lape, M. (2008). Metal pollution in freshly deposited sediments from river Mingoa, main tributary to the Municipal Lake of Yaounde, Cameroon. *Geosciences Journal, 12*(4), 337–347. doi:10.100712303-008-0034-5

Essien, J., Ikpe, D., Inam, E., Okon, A., Ebong, G., & Benson, N. (2022). Occurrence and spatial distribution of heavy metals in landfill leachates and impacted freshwater ecosystem: An environmental and human health threat. *PLoS ONE, 17*(2): e0263279. https://doi.org/. pone.0263279 doi:10.1371/journal

Essumang, D. (2009). Analysis and human health risk assessment of arsenic, cadmium, and mercury in Manta birostris (manta ray) caught along the Ghanaian coastline. *Human and Ecological Risk Assessment, 15*(5), 985–998. doi:10.1080/10807030903153451

Fakayode, S., & Olu-Owolabi, B. (2003). Heavy metal contamination of roadside topsoil in Osogbo, Nigeria: Its relationship to traffic density and proximity to highways. *Environmental Geology (Berlin), 44*(2), 150–157. doi:10.100700254-002-0739-0

Fashola, M., Ngolo-Jeme, V., & Babalola, O. (2016). Heavy metal pollution from gold mines: Environmental effects and bacterial strategies for resistance. *International Journal of Environmental Research and Public Health, 13*(11), 1047. doi:10.3390/ijerph13111047 PMID:27792205

Fasinu, P., & Orisakwe, O. (2013). Heavy metal pollution in sub-Saharan Africa and possible implications in cancer epidemiology. *Asian Pacific Journal of Cancer Prevention, 14*(6), 3393–3402. doi:10.7314/APJCP.2013.14.6.3393 PMID:23886118

Fei, C., Min, X., Wang, Z., Pang, Z., Liang, Y., & Ke, Y. (2017). Health and ecological risk assessment of heavy metals pollution in an antimony mining region: A case study from South China. *Environmental Science and Pollution Research International, 24*(35), 27573–27586. doi:10.100711356-017-0310-x PMID:28980103

Fianko, J., Osae, S., Adomako, D., Adotey, D., & Serfor-Armah, Y. (2007). Assessment of heavy metal pollution of the Iture Estuary in the central region of Ghana. *Environmental Monitoring and Assessment, 131*(1-3), 467–473. doi:10.100710661-006-9492-2 PMID:17171259

Foso-Mensah, B., Addae, E., Tawiah, E., & Nyame, F. (2017). Heavy metals concentration and distribution in soils and vegetation at Korle Lagoon area in Accra, Ghana. *Cogent Environmental Science, 3*(1), 1405887. doi:10.1080/23311843.2017.1405887

Garcia, E., & Boluda, V. (1996). Heavy metals incidence in the application of inorganic fertilizers and pesticides to rice farming soils. *Environmental Pollution, 92*(1), 19–25. doi:10.1016/0269-7491(95)00090-9 PMID:15091407

Ghane, E., Khanverdiluo, S., & Mehri, F. (2022). The concentration and health risk of potentially toxic elements (PTEs) in the breast milk of mothers: A systematic revies and metal analysis. *Journal of Trace Elements in Medicine and Biology, 73*, 126998. doi:10.1016/j.jtemb.2022.126998 PMID:35617722

Hasimuna, O., Maulu, S., & Chibesa, M. (2022). Assessment of heavy metal contamination in water and largescale yellowfish (*Labeobarbus marequensis*, smith 1841) from Solwezi river, north-western Zambia. *Cogent Food & Agriculture, 8*(1), 1–17. doi:10.1080/23311932.2022.2121198

Heyden, C., & New, G. (2004). Sediment chemistry: A history of mine contaminant remediation and an assessment of processes and pollution potential. *Journal of Geochemical Exploration, 82*(1-3), 35–57. doi:10.1016/j.gexplo.2003.11.001

Hubeny, J., Harnisz, M., Korzeniewska, E., Buta, M., Zieliński, W., Rolbiecki, D., Giebułtowicz, J., Nałęcz-Jawecki, G., & Płaza, G. (2021). Industrialization as a source of heavy metals and antibiotics which can enhance the antibiotic resistance in wastewater, sewage sludge and river water. *PLoS One, 16*(6), e0252691. doi:10.1371/journal.pone.0252691 PMID:34086804

Iloms, E., Ololade, O., Ogola, H., & Selvarajan, R. (2020). Investigating industrial effluent impact on municipal wastewater treatment plant in Vaal, South Africa. *International Journal of Environmental Research and Public Health, 17*(3), 1096. doi:10.3390/ijerph17031096 PMID:32050467

Jan, A., Azam, M., Siddiqui, K., Ali, A., Choi, I., & Haq, Q. (2015). Heavy metals and human health: Mechanistic insight into toxicity and counter defense system of antioxidants. *International Journal of Molecular Sciences, 16*(12), 29592–29630. doi:10.3390/ijms161226183 PMID:26690422

Kihampa, C., Kaisi, G., & Kihampa, H. (2016). Assessing the contribution of industrial wastewater to toxic metals contamination in receiving urban rivers, Dar es Salaam city, Tanzania. *Elixir Pollution, 93*, 39532–39541.

Kinuthia, G., Ngure, V., Lugalia, R., Wangila, A., & Kamau, L. (2020). Levels of heavy metals in wastewater and soil samples from open drainage channels in Nairobi, Kenya: Community health implication. *Scientific Reports, 10*(1), 8434. doi:10.103841598-020-65359-5 PMID:32439896

Kishe, M., & Machiwa, J. (2003). Distribution of heavy metals in sediments of Mwanza Gulf of Lake Victoria, Tanzania. *Environment International, 28*(7), 619–625. doi:10.1016/S0160-4120(02)00099-5 PMID:12504158

Loha, K., Lamoree, M., Weiss, J., & Boer, J. (2018). Import, disposal and health impacts of pesticides in East Africa Rift (EAR) zone: A review on management and policy analysis. *Crop Protection (Guildford, Surrey), 112*, 322–331. doi:10.1016/j.cropro.2018.06.014

Ma, M., Dong, S., Jin, W., Zhang, C., & Zhou, W. (2019). Fate of the organophosphorous pesticide profenofos in cotton fiber. *Journal of Environmental Science and Health. Part B, Pesticides, Food Contaminants, and Agricultural Wastes, 54*(1), 70–75. doi:10.1080/03601234.2018.1505036 PMID:30633718

Malan, M., Muller, F., Cyster, L., Raitt, L., & Aalbers, J. (2015). Heavy metals in the irrigation water, soils and vegetables in the Philippi horticultural area in the Western Cape Province of South Africa. *Environmental Monitoring and Assessment, 187*(1), 4085. doi:10.100710661-014-4085-y PMID:25380711

Malunguja, G., Thakur, B., & Devi, A. (2022). Heavy metal contamination of forest soils by vehicular emissions: Ecological risks and effects on tree productivity. *Environmental Processes, 9*(1), 11. doi:10.100740710-022-00567-x

Manyiwa, T., Ultra, V. Jr, Rantong, G., Opaletswe, K., Gabankitse, G., Taupedi, S., & Gajaje, K. (2022). Heavy metals in soil, plants and associated risk on grazing ruminants in the vicinity of Cu-Ni mine in Selebi-Phikwe, Botswana. *Environmental Geochemistry and Health, 44*(5), 1633–1648. doi:10.100710653-021-00918-x PMID:33855629

Mar, S., & Okazaki, M. (2012). Investigation of Cd contents in several phosphate rocks used for the production of fertilizer. *Microchemical Journal, 104*, 17–21. doi:10.1016/j.microc.2012.03.020

Marove, C., Sotozono, R., Tangviroon, P., Baltazar, C., & Igarashi, T. (2022). Assessment of soil, sediment and water contaminations around open pit coal mines in Moatize, Tete Province, Mozambique. *Environmental Advances, 8*, 100215. doi:10.1016/j.envadv.2022.100215

Masindi, V., & Muedi, K. (2019). *Environmental contamination by heavy metals*. Intech Open Publishers., doi:10.5772/intechopen.76082

Mekonnen, B., Haddus, A., & Zeine, W. (2020). Assessment of the effect of solid waste dump site on surrounding soil and river water quality in Tepi town, Southwest Ethiopia. *Journal of Environmental and Public Health, 5157046*, 1–9. doi:10.1155/2020/5157046 PMID:32587623

Mekuria, D., Kassegne, A., & Asfaw, S. (2021). Assessing pollution profiles among little Akaki river receiving municipal and industrial wastewaters, central Ethiopia: Implications for emvironmental and public health safety. *Heliyon, 7*(7), e07526. doi:10.1016/j.heliyon.2021.e07526 PMID:34337176

Minamata Convention on Mercury. (2019). *Home*. UN. https://www.mercuryconvention.org/

Mmbaga, T., & Semu, E. (1999). Contents of heavy metals in some soils of the Morogoro municipality, Tanzania, as a result of cottage-scale metal working operations. *The International Journal of Environmental Studies, 56*(3), 373–383. doi:10.1080/00207239908711211

Moeletsi, M., & Tongwane, M. (2020). Projected direct carbon dioxide emission reductions as a result of the adoption of electric vehicles in Gauteng province in South Africa. *Atmosphere (Basel), 11*(6), 591. doi:10.3390/atmos11060591

Mohammed, S., & Folorunsho, J. (2015). Heavy metals concentration in soil and Amaranthus retroflexus grown on irrigated farmlands in the Makera Area, Kaduna, Nigeria. *Journal of Geography and Regional Planning, 8*(8), 210–217. doi:10.5897/JGRP2015.0498

Muchuweti, M., Birkett, J., Chinyanga, E., Zvauya, R., Scrimshaw, M., & Lester, J. (2006). Heavy metal content of vegetables irrigated with mixtures of wastewater and sewage sludge in Zimbabwe: Implications for human health. *Agriculture, Ecosystems & Environment, 112*(1), 41–48. doi:10.1016/j.agee.2005.04.028

Mudimbu, D., Davies, T., Tagwireyi, D., & Meck, M. (2022). Application of geochemical indices evaluating potentially harmful element contamination at mining centers in the Sanyati catchment, Zimbabwe. *Frontiers in Environmental Science, 10*, 829900. doi:10.3389/fenvs.2022.829900

Mutune, A., Makobe, M., & Abukutsa-Onyango, M. (2014). Heavy metal content of selected African leafy vegetables planted in urban and peri-urban Nairobi, Kenya. *African Journal of Environmental Science and Technology, 8,* 66–74. http://dx.doi.org/. 1573 doi:10.5897/ajest2013

Muwanga, A., & Barifaijo, E. (2006). Impact of industrial activities on heavy metal loading and their physicochemical effects on wetlands of Lake Victoria basin (Uganda). *African Journal of Science and Technology*, *7*(1). doi:10.4314/ajst.v7i1.55197

Mwaanga, P., Silondwa, M., Kasali, G., & Banda, P. (2019). Preliminary review of mine air pollution in Zambia. *Heliyon*, *5*(9), e02485. doi:10.1016/j.heliyon.2019.e02485 PMID:31687579

Namuhani, N., & Kimumwe, C. (2015). Soil contamination with heavy metals around Jinja steel rolling mills in Jinja, Municipality, Uganda. *Journal of Health & Pollution*, *9*(9), 61–67. doi:10.5696/2156-9614-5-9.61 PMID:30524777

Nkonya, E., Johnson, T., Kwon, H., & Kato, E. (2016). Economics of land degradation in Sub-Saharan Africa. In E. Nkonya, A. Mirzabaev, & J. von Braun (Eds.), *Economics of land degradation and improvement – a global assessment for sustainable development*. Springer. doi:10.1007/978-3-319-19168-3_9

Nyika, J., & Dinka, M. (2022). Heavy metal pollution in soils and vegetables from suburban regions of Nairobi, Kenya and their community health implications. *Pollution*, *8*(4), 1434–1447. doi:10.22059/POLL.2022.341522.1440

Nyika, J., & Onyari, E. (2019). Hydrogeochemical analysis and spatial distribution of groundwater quality in Roundhill landfill vicinity of South Africa. *Air, Soil and Water Research*, *12*, 1–8. doi:10.1177/1178622119872771

Nyika, J., Onyari, E., Dinka, M., & Shivani, B. (2020). Assessment of trace metal contamination of soils in a landfill vicinity: A southern Africa case study. *Chemistry Letters*, *9*(4), 171–182. doi:10.46488/nept.2020v19i02.009

Odai, S., Mensah, E., Sipitey, D., Ryo, S., & Awuah, E. (2008). Heavy metals uptake by vegetables cultivated on urban waste dumpsites: Case study of Kumasi, Ghana. *Research Journal of Environmental Toxicology*, *2*(2), 92–99. doi:10.3923/rjet.2008.92.99

Odero, D., Semu, E., & Kamau, G. (2000). Assessment of cottage industry derived heavy metal pollution of soils within Ngara and Gikomba areas of Nairobi city, Kenya. *African Journal of science and Technology. Science and Engineering Series*, *1*, 52–62.

Oloruntegbe, K., Akinsete, M., & Odutuyi, M. (2009). Fifty years of oil exploration in Nigeria: Physicochemical impacts and implication for environmental accounting and development. *Journal of Applied Sciences Research*, *5*, 2131–2137.

Olowoyo, J., Okedeyi, O., Mkolo, N., Lion, G., & Mdakane, S. (2012). Uptake and translocation of heavy metals by medicinal plants growing around a waste dump site in Pretoria, South Africa. *South African Journal of Botany*, *78*, 116–121. doi:10.1016/j.sajb.2011.05.010

Oruko, R., Edokpayi, J., Msagati, T., Tavengwa, N., Ogola, H., Ijoma, G., & Odiyo, J. (2021). Investigating the chromium status, heavy metal contamination and ecological risk assessment via tannery waste disposal in sub-Saharan Africa (Kenya and South Africa). *Environmental Science and Pollution Research International*, *28*(31), 42135–42149. doi:10.100711356-021-13703-1 PMID:33797722

Pitiya, R., Jacob, L., & Emilinot, R. (2022). A pilot study on the concentration of heavy metals in sediments from the lower orange river, Karas region, Namibia. *Journal of Materials Science and Chemical Engineering*, *10*(3), 1–14. doi:10.4236/msce.2022.103001

Rotterdam Convention. (2010). *Overview.* UN. https://www.pic.int/TheConvention/Overview/tabid/1044/language/en-US/Default.aspx

Sanga, V., Fabian, C., & Kimbokota, F. (2022). Heavy metal pollution in leachates and its impacts on the quality of groundwater resources around Iringa municipal solid waste dumpsite. *Environmental Science and Pollution Research International*, 1–13. doi:10.100711356-022-22760-z PMID:36053421

Semu, E., Tindwa, H., & Singh, B. (2019). Heavy metals and organopesticides: Ecotoxicology, health effects and mitigation options with emphasis on sub-Saharan Africa. *HSOA Journal of Toxicology : Current Research*, *2*(3), 1–10. doi:10.24966/TCR-3735/100010

Singh, D., Iqbal, J., Bhat, A., Bhat, R., Dervash, M., & Ganei, S. (2018). Vehicular stress a cause for heavy metal accumulation and change in physico-chemical characteristics of roadside soils in Pahalgam. *Environmental Monitoring and Assessment*, *190*(6), 353. doi:10.100710661-018-6731-2 PMID:29785575

Srivastava, V., Sarkar, A., Singh, S., Singh, P., Araujo, A., & Singh, R. (2017). Agroecological Responses of Heavy Metal Pollution with Special Emphasis on Soil Health and Plant Performances. *Frontiers in Environmental Science*, *5*, 64. doi:10.3389/fenvs.2017.00064

Stockholm Convention. (2019). *Home.* UN. http://chm.pops.int/

Suciu, N., De Vivo, R., Rizzati, N., & Capri, E. (2022). Cd content in phosphate fertilizer: Which potential risk for the environment and human health? *Current Opinion in Environmental Science & Health*, *30*, 100392. doi:10.1016/j.coesh.2022.100392

Sun, Z., Xie, X., Wang, P., Hu, Y., & Cheng, H. (2018). Heavy metal pollution caused by small-scale metal ore mining activities: A case study from a polymetallic mine in South China. *The Science of the Total Environment*, *639*, 217–227. doi:10.1016/j.scitotenv.2018.05.176 PMID:29787905

Tahmasebi, P., Taheri, M., & Gharaie, M. (2020). Heavy metal pollution associated with mining activity in the Kouh-e Zar region, NE Iran. *Bulletin of Engineering Geology and the Environment*, *79*(2), 1113–1123. doi:10.100710064-019-01574-3

Taylor, M., & Kesterton, R. (2002). Heavy metal contamination of an arid river environment: Gruben River, Namibia. *Geomorphology*, *42*(3-4), 311–327. doi:10.1016/S0169-555X(01)00093-9

Teta, C., Ncube, M., & Naik, Y. (2017). Heavy metal contamination of water and fish in peri-urban dams around Bulawayo, Zimbabwe. *African Journal of Aquatic Science*, *42*(4), 351–358. doi:10.2989/16085914.2017.1392925

Tindwa, H., & Singh, B. (2023). Soil pollution and agriculture in sub-Saharan Africa: State of the knowledge and remediation technologies. *Frontiers in Soil Science*, *2*, 1101944. doi:10.3389/fsoil.2022.1101944

Tsuma, J., Wandiga, S., & Abong'o, D. (2016). Methane and heavy metal levels from leachates at Dandora dumpsite, Nairobi County, Kenya. *IOSR Journal of Applied Chemistry*, *9*(9), 39–46. doi:10.9790/5736-0909023946

UNEP. (n.d). *Basel Convention on the control of transboundary movements of hazardous wastes*. UNEP. https://www.unep.org/resources/report/basel-convention-control-transboundary-movements-hazardous-wastes (Accessed July 13th, 2023).

United Nations Environmental Program. (2022). *The Lead Campaign*. UN. https://www.unep.org/explore-topics/transport/what-we-do/partnership-clean-fuels-and-vehicles/lead-campaign

Uugwanga, M., & Kgabi, N. (2020). Assessment of metals pollution in sediments and tailings of Kelin Aub and Oamites mine sites, Namibia. *Environmental Advances, 2*, 100006. doi:10.1016/j.envadv.2020.100006

Vacha, R. (2021). Heavy Metal Pollution and Its Effects on Agriculture. *Agronomy (Basel), 11*(9), 1719. doi:10.3390/agronomy11091719

Wandiga, S. (2001). Use and distribution of organochlorine pesticides. The future in Africa. *Pure and Applied Chemistry, 73*(7), 1147–1156. doi:10.1351/pac200173071147

Wei, W., Ma, R., Sun, Z., Zhou, A., Bu, J., Long, X., & Liu, Y. (2018). Effects of mining activities on the release of heavy metals (HMs) in a typical mountain headwater region, the Qinghai-Tibet Plateau in China. *International Journal of Environmental Research and Public Health, 15*(9), 1987. doi:10.3390/ijerph15091987 PMID:30213099

Xu, D., Shen, Z., Dou, C., Dou, Z., Li, Y., Gao, Y., & Sun, Q. (2022). Effects of soil properties on heavy metal bioavailability and accumulation in crop grains under different farmland use patterns. *Scientific Reports, 12*(1), 9211. doi:10.103841598-022-13140-1 PMID:35654920

Yabe, J., Ishizuka, M., & Umemura, T. (2010). Current levels of heavy metal pollution in Africa. *The Journal of Veterinary Medical Science, 72*(10), 1257–1263. doi:10.1292/jvms.10-0058 PMID:20519853

Zhu, C., Tian, H., Cheng, K., Liu, K., Wang, K., Hua, S. et al. (2016). Potentials of whole process control of heavy metals emissions from coal-fired power plants in China. *Journal of Cleaner Production, 114*, 343–351. https://doi.org/. 05.008 doi:10.1016/j.jclepro.2015

Chapter 4
Methods of Assessing and Analyzing Heavy Metal Pollution in Soils

ABSTRACT

As heavy metal pollution in soils grows to be an environmental and public health concern, there is need to apply accurate, precise, cost effective, and highly sensitive analytical approaches to assay the contaminant levels. The results of such analyses can be used in making informed plans to curtail pollution by using greener, reclamation and/or remediation methods in affected land. In this chapter, direct and indirect analytical methods to assay heavy metals in soils and their working approaches are discussed. The sampling of targeted soils and before-analysis preparations of analytical samples are also elaborated. Spectrometric techniques were found to be the most superior analytical methods due to their low detection limits, high sensitivity and contaminant identification power and extensive lineal ranges for multi-elemental analysis. However, equipment uses and sample introduction approaches require optimization in the techniques to prevent physical, chemical, and spectral interferences, which compromise analytical accuracy.

INTRODUCTION

The growth of technology affiliated with industrial revolution is a transformative characteristic of contemporary society. Its influence in human lives is phenomenal although its resultant environmental effects are understated and even overlooked in public arenas. Consequently, polluting agents from industrialization activities

DOI: 10.4018/978-1-6684-7116-6.ch004

and other related anthropogenic activities have added to the pollutants load from geogenic sources. The agents have adversely affected land and water resources inducing pollution. It is for this reason that 9 million deaths annually were affiliated to pollution of the natural resources, according to the Global Burden of Diseases, Injuries and Risk Factors Study of 2019 (Fuller et al., 2022). The growing pollution of soils in warfare affected areas, agricultural fields, industrial, mining and urban areas has been correlated to poor human health and inability of soils to provide their ecological goods and services (Brevik et al., 2020). Similarly, global water resources are vulnerable to contamination from agricultural, industrial and municipal activities through the generation and disposal of solid waste as well as release of effluents in such bodies (Chen et al., 2022). The soil and water pollution situation is dire in regions of Latin America, Asia and Africa where rapid population growth, urbanization and industrialization is evident and little attention is paid to environmental conservation efforts and initiatives (Chen et al., 2022). Of the growing pollution problems, exposure to heavy metals is more prominent because of their multiple applications in tech-savvy industries and the continued expansion of such industries (Sharma, 2014). The metals are toxic to ecosystems and humans even in minute concentrations owing to their ability to bioaccumulate and persist in the environment and their subsequent transfer to trophic systems where they are biomagnified and cause harmful effects.

Owing to their potency to cause environmental harm, heavy metal detection and analysis in environmental samples such as soils, sediments, water and plants is of paramount interest to analytical chemists. The analysis facilitates accurate and precise quantification of the heavy metal elements in environmental samples prior to using the data to derive information on current pollution trends and extent, compare resultant patterns with historical trends and predict future scenarios on pollution (Nyika et al., 2019a). These undertakings are key in planning on preventative measures and remediation/reclamation approaches to implement. Apart from assaying the total concentration of heavy metals in environmental samples, it is imperative to determine concentration levels of specific metal species. Bakirdere (2013) suggested that quantifying specific heavy metal ions provides information on their differential transport, bioavailability and toxicity mechanisms in the environment, which is dependent on their oxidation states. Concentrations of heavy metals occur in environmental samples in the tune of parts per billion (μg/l or μg/kg) and million (mg/l or mg/kg). Therefore, appropriate selection of the analytical methods used is key to high sensitivity and accuracy during quantification of total metal content and specific heavy metal species (Crompton, 2015). In this chapter, analytical methods used in determination of heavy metals present in soil samples are discussed. The aim is to understand recent analytical chemistry developments whose focus is to improve the sampling, sensitivity, precision and accuracy of heavy metal analysis in soils and other environmental samples.

TYPES OF HEAVY METALS ANALYSIS

The assay of heavy metals from soils has advanced with the development of novel analytical techniques. With these advancements, heavy metal analysis is categorized into two: 1) total concentration analysis and 2) speciation analysis. The former defines the identification, isolation and quantification of the summative concentration of a specific metal element. Many studies have used total concentration analysis of heavy metals in soil samples of sub-Saharan Africa (SSA) to assess their pollution levels (Nyika et al. 2019b; Asamoah et al., 2021; Yahaya et al., 2021 Mohajane & Manjoro, 2022; Nyika & Dinka, 2022; Shezi et al., 2022). Speciation analysis on the other hand, is the assay of the distribution of heavy metals of varied forms (species) (Templeton & Fujishiro, 2017). Heavy metal speciation research provides pertinent information to differentiate particular trace metals from total elemental composition in environmental samples (e.g. soils) based on their mineral contents, precipitates, complexes, redox species and organometallic compounds. Both total concentration and speciation analysis require highly sensitive and selective analytical approaches for accurate qualification and determination of the heavy metal's toxicity. According to Clough et al. (2018), contemporary analytical chemistry requires such analytical techniques owing to the increase of suspected polluted environs in soils. Similarly, Achterberg et al. (2019) pointed out that heavy metal speciation and assessment of their dissociation kinetics and stability is essential using sensitive analytical techniques since the phenomenon influences biodegradation, mobility, bioaccumulation, bioavailability and toxicity characteristics of the pollutants. A case example is the toxic heavy metal tellurium (Te), whose toxicity in the environment (including soils) is associated with solar panels use. The toxicity of the metal's tetravalent species is ten times more compared to the hexavalent species (Zare et al., 2017).

In soil pollution, speciation is essential and gives information on the toxicity and bioavailability of a given heavy metal (Cornelis & Nordberg, 2007). Quantifying varied chemical forms of heavy metals that are present in soil samples is imperative to obtain accurate results of the exposure risk associated with toxic elements and also assess the nutritional needs for elements of importance at minute concentrations (Bakirdere, 2013). A case example is chromium that exists in two stable oxidation states: 1) Cr (III) and Cr (VI). The two species have different physiological and biological characteristics so that the former, is important in metabolic enzymology in living things (synthesis of fatty acids and cholesterol) while the latter is noxious and associated with carcinogenicity in such organisms (Jozsef et al., 2019).

In SSA, several metal speciation studies on soils have been reported for varied reasons. A metal speciation analysis in soils of Calabar municipality, Nigeria reported that Ni exists in its residual or inert fraction while Fe and Pb were in their reducible forms unlike Cu and Cd in their oxidizable and acid extractable forms, respectively (Ebong et al., 2019). A metal speciation study in tail mines mixed with soils in Cameroon showed that their Cd, Co, Fe, Mn, Pb and Zn cations were highly mobile and likely to pollute environments of the vicinity of the study area (Yiika et al., 2023). In polluted soils of Struibult gold mine, South Africa, a speciation analysis reported that As, Bi, Cr, Cu, Mo, Ni, Pb, Sb, Se, Ti and Zn fractions of metals existed in their residual forms (Mngadi et al., 2020). Based on the computed individual contamination factors and risk assessment codes, Co, Ni and Zn species were highly likely to cause soil pollution compared to the other metal species.

Developments and application of advanced analytical techniques of soil samples therefore entails appropriate sampling and sample preparation coupled with sensitive detection and quantification mechanisms. The advancements enable quality heavy metal total concentration and speciation analyses in soils where pollutant concentration is below detection levels of µg/l (Kocot et al., 2016). In this section, sampling approaches and sample preparation methods are discussed prior to exploring on the analytical methods used in assaying heavy metal total and species concentrations in soils.

Sampling

Collection of samples (sampling) is a crucial initial step of quantifying heavy metals in environmental samples including soils. The sample collection regime used must be cultured to fit the purpose of a particular study in addition to being representative. Such considerations reduce the uncertainty associated with sampling when making overall inferences using obtained measurements and data. Studies have established that even with suitable sampling approaches, quantification of heavy metals in soils has measurement uncertainties (Kurfurst et al., 2004; Gustavsson et al., 2006).

The purpose of a particular study dictates the sampling regime to be used. For instance, in a study whose goal is to quantify the mean metal concentration, the field targeted for sampling is subdivided to segments (for an area of 1-2 ha, 20-25 segments can be divided) (Davidson, 2013). In such a case, the herring bone or regular grid patterns are used as sampling guides (Figure 1a). Using a hand-held GPS, the sampling points can be located and revisited easily if sampling is done severally. Sampling of the average heavy metal concentration can also be done along a zigzag path (Davidson, 2013). In this case, regular gridding of a land plot is done from a selection point and samples are collected to form a W-pattern as shown in Figure 1b. In studies establishing the spatial distribution patterns of heavy metals, regular gridding is used but more than one sample is collected from each segment for analysis. The hypothesis-guided soil sampling approach (Figure 1c) can be used in areas with or near known pollution sources (Scholz et al., 1994). For example, if the purpose of a study was to evaluate metal pollution from a point source such as an open dumpsite, the sampling approach maybe skewed to the direction of leachate movement considering that contamination is reasonably likely to be greater there. In a closed and/or abandoned industrial site, appropriate sampling requires prior knowledge on the industrial activities involved, raw materials used, approaches used to dispose resultant waste and previous locations of buildings (Davidson, 2013).

Figure 1. Different approaches of sampling soils for heavy metal analysis (Davidson, 2013)

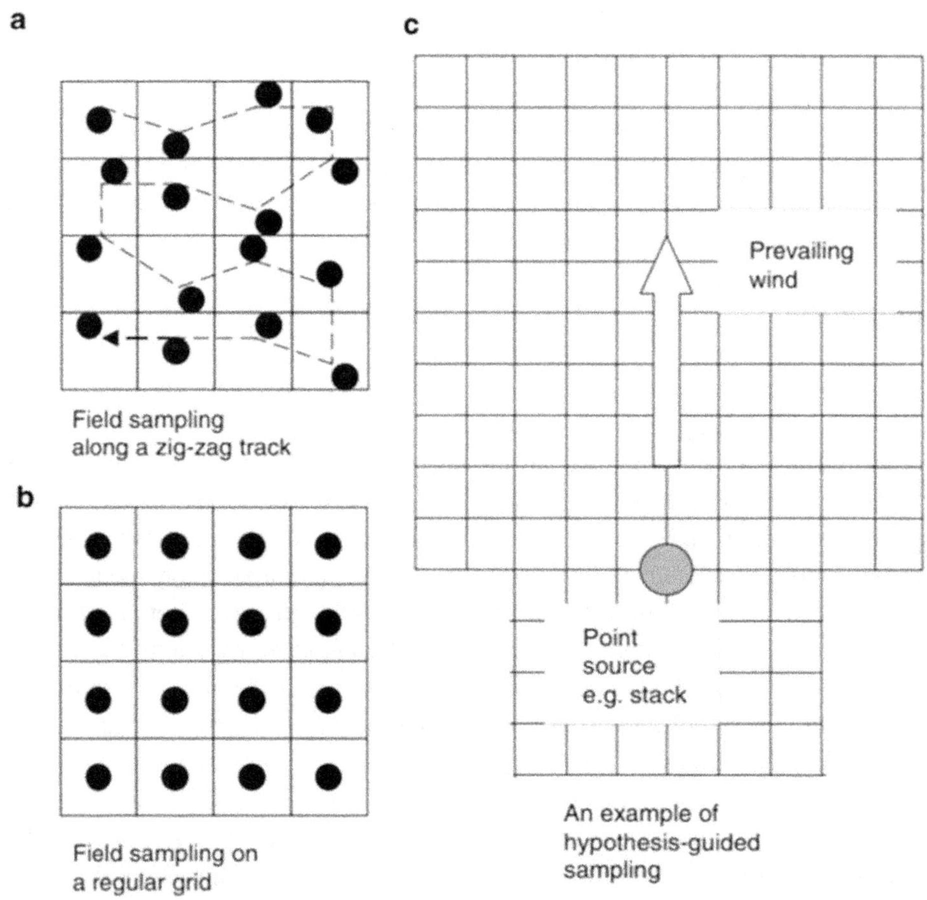

Using a corer or auger, the sample soils are collected to uniform depth. Depending on previous use of soils, in grassland and arable land, soil is collected at a depth of 7.5-10 and 15-20 cm, respectively (Davidson, 2013). At areas suspected to be contaminated, various depths can be used to assess the vertical movement of pollutants. For example, in assessing vertical migration of leachate from the vicinity Roundhill landfill, South Africa, Nyika et al. (2019b) collected soils at 0, 30, 60 and 100 cm depths. Obtained samples are put in suitable containers. Polyethylene bags are the most suitable for soil samples targeted for heavy metal analysis. However, in samples targeted for Hg analysis, separating the samples is advised to prevent pollution transfer of the element and cross-contamination of adjacent samples (Wagner et al., 2001). Aluminum and carbon steel sampling tools are preferred during sampling rather than stainless steel. This is because stainless steel could incorporate plastics, Cr, Mn, Mo and Ni among other trace metals leading to contamination of soils samples. After each sampling round, the tools should be wiped clean or rinsed with distilled water to prevent any form of cross-contamination. Using a permanent marker, samples should be labeled using unique codes or numbers on the outside of the bag. Details such as the sampling location, time and date can be included in the labels. The transportation of the samples from the field to the laboratory should also be done cautiously to avoid any external contamination. For samples collected in the field in readiness for transportation, a portable refrigerator can be used to store them at 4°C (Bailey et al., 2022). Once at the laboratory, the samples can be transferred to a freezer awaiting analysis (Bailey et al., 2022). Freezing helps to preserve the levels of organic nitrogen in the soils.

Sample Preservation and Preparation

After successful transportation of soils samples to the laboratory, they are then spread out to dry in aluminum trays accessorized with plastic sheets. Drying breaks aggregates of soil samples to smaller particles. Drying at room temperature (20°C) or air drying is highly preferred. In preparation of reference soils to compare with those targeted for analysis of metal contaminants, higher drying temperatures are preferred to enhance their long-term stability. After drying, sieve analysis is done using nylon or aluminum sieves of <2 mm diameter. Resultant fine earth particles are stored until they are required for analysis. A smaller diameter (50 μm) of the soil particles allows for more surface area for procedures such as acid digestion. Sand particles of >2 mm diameter are discarded. Sub-samples are pulverized to finer homogenous particles that can be used for direct method analysis. To obtain representative sub-samples, dried soils can be quartered, coned or mechanically riffled. In soils intended for speciation analysis, care should be taken during their preparation to preserve the inherent forms of the heavy metals (Dubiella et al., 2007).

During sample preparation, storage and preservation, caution should be taken to prevent contact with contaminants. According to Hoenig (2001), contaminants can occur at any step of the analytical process. The laboratory (plastics and paints used), instrumentation such as inductively coupled plasma mass spectrometer (ICP-MS), grinding and milling apparatus and areas used in acid digestion of soil samples are potential sources of contaminants. Therefore, extra caution should be taken to prevent contamination. Such considerations include avoiding stainless steel contact when assaying for Cr in soils and soaking all glassware to be used in the analysis in 5% nitric acid overnight before rinsing them in ultra-pure (deionized) distilled water (Davidson, 2013). Reagents of high purity are also recommended for use during soil analysis and to confirm the absence of contamination, they should be run as blanks. It is crucial to crosscheck reagents for those that have the lowest concentrations based on the targeted analytes due to variations by different chemical suppliers.

METHODS USED IN DETERMINING HEAVY METALS

The techniques applied in heavy metal analysis are of two categories: -1) direct and 2) indirect methods. The latter require the analytes from soil samples to be assayed by an instrument in solution form while the former uses solid substrates of soils and in some cases, analysis can be done in the field. In this section, direct and indirect analyses techniques and their application in heavy metal analyses of soils are discussed.

Direct Methods

X-ray Based Techniques

X-ray fluorescence (XRF) spectrometry describes a multi-element analytical technique that assays for total elemental concentrations in their solid phase (Madden et al., 2022). The technique is non-destructive, has relatively low sensitivity and detects elements whose atomic number is above 8. For successful analysis using XRF, primary X-rays irradiate soil samples from a radioactive source (usually an X-ray tube) that expels electrons from atoms (Madden et al., 2022). The resultant energy of electron ejection and fall results to an X-ray fluorescence spectrum whose specific wavelengths correspond to identified elements in a soil sample (Figure 2). XRF instruments used are of two types: 1) energy dispersive XRF (ED-XRF) and wavelength dispersive XRF (WD-XRF) (Byers et al., 2019). In the latter, sample emitted radiation is diffracted in varied directions and detection of X-rays based on their wavelengths is done by a mobile silicon/lithium (Si/Li) detector. In the former, an analyzer with many channels is combined with a detector to determine elemental composition of the soil sample. WD-XRF is expensive and has more superior resolution compared to ED-XRF whose parts are all immobile. Recent developments have seen the introduction of a portable XRF (PXRF) instrument that initially used excitation sources from radioactive isotopes (Cm^{244}, Am^{241}, Cd^{109}, Fe^{55} and Co^{57}) and Si-pin diode detectors (David & Somsubhra, 2020). The improved version of PXRF uses miniaturized tube-based X-ray sources including Ag, Ta/ Au or Rh for excitation and silicon drift in detection for enhanced analytical precision and accuracy (David & Somsubhra, 2020).

Figure 2. A description of how an XRF instrument works
(Byers et al., 2019)

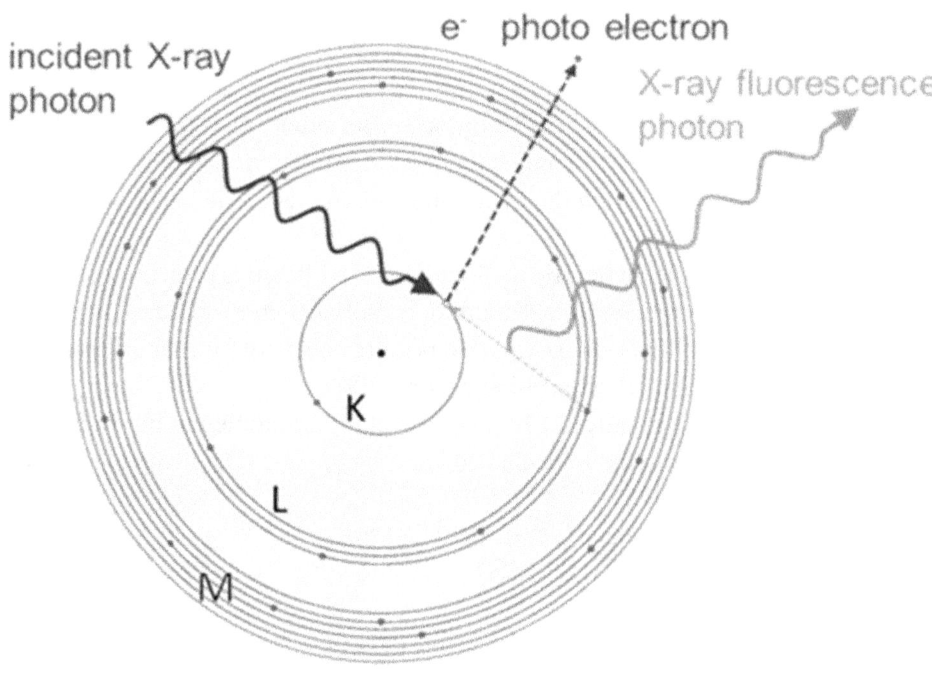

In both ED-XRF and WD-XRF, the soil samples are ground finely ground and fused to a glass for analysis. Alternatively, the samples are pressed and compressed to powder pellets. During analysis matric effects (ion enhancement or suppression as a result of modified ionization of target analytes when co-eluting elements are present) occur interfering with analytical precision (Davidson, 2013). They can be corrected mathematically. The PXRF is a pre-configured gun-shaped and handheld equipment that is taken to the field, powered through charging and scanned on soil for 60-90 seconds to display results (David & Somsubhra, 2020).

The advantage of XRF technique in heavy metal quantification of soils is the availability of the portable instrument usable in field settings. The PXRF was used to analyze heavy metals from suspected polluted fields such as industrial soils (Cr, Cu, Mn, Ni, Pb, V and Zn (Madden et al., 2022), wood preservative polluted soils (As, Cu and Cr) (Makinen et al., 2006), and solid waste polluted soils (As, Cu, Pb and Zn (Carr et al., 2008). Although the detection limits of the equipment were lower, obtained results in all cases were comparable to laboratory-based techniques. In preliminary studies that require rapid soil screening and identification of heavy metals of interest, the technique can be used and later, quantification can be done with much sensitive analytical methods. XRF analysis is influenced by soil moisture content and therefore soils targeted for analysis especially using PXRF should hold a moisture threshold of 20%. In addition, the analytical technique is vulnerable to inter-elemental interferences (especially from As and Pb) due to spectral peak sharing, which results to poor identification and isolation of the elements (David & Somsubhra, 2020). Additionally, the technique is not suitable for speciation analysis since it cannot distinguish Fe^{2+} from Fe^{3+}. To reduce the matric and inter-elemental interferences, advancements on the techniques are in place. These include

i. Total reflection XRF whereby the incident X-ray beam is projected on the soil sample at a reflected acute angle to reduce bulk penetration (Davidson, 2013)
ii. Using synchrotron-based analysis that is more powerful for speciation analysis of heavy metals in soils (Lombi & Susini, 2009)
iii. Using diamond as a source of light to prevent inter-elemental interferences of Pb and Zn and allow their speciation analysis in soils (Davidson et al., 2013).

Laser Ablation Inductively Coupled Plasma Mass Spectrometry (LA-ICP-MS)

LA-ICP-MS is used in compositional analysis of soils in their solid phase with high precision and sensitivity despite the conceptual simplicity of the technique (Koch & Gunther, 2017). Just like in XRF, soil samples are first ground to fine powders of <1 μm diameter prior to analysis, pressed to pellets in the presence of a binder and positioned in an ablation chamber. The LA-ICP-MS apparatus shown in Figure 3 consists of a laser source, sampling cell, which is air tight and atmospheric pressure controlled, optics to deliver the beam, a transport system and an ICP-MS detector. Once the soil sample is released in the ablation chamber, matrix- coordinated calibration standards are used to ensure a representative sample is used. Inert gases such as helium and argon are used in the ICP aerosol carriers to prevent molecular interferences of elements prior to MS detection and enhance accuracy of identified heavy metals (Koch & Gunther, 2017). The technique has been used in analysis of heavy metals (Cd, Cr, Cu, Pb and Zn) in soils of Kafr El-Zayat city, Egypt (Shaheen et al., 2021) and D industrial park, Korea (Kim et al., 2020).

Figure 3. The components of a laser ablation inductively coupled plasma mass spectrometry
(Davidson, 2013)

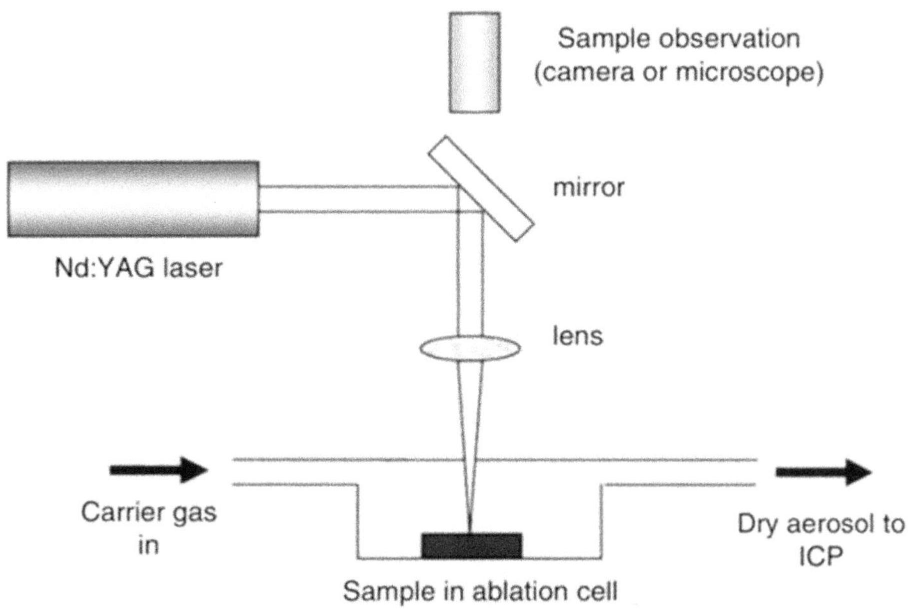

Laser -Induced Breakdown Spectroscopy (LIBS)

LIBS is a relatively new atomic emission spectrometry technique for analyzing soils, which requires minimal sample processing (Ding et al., 2019). The LIBS system uses a pulsed laser (usually Nd:YAG laser) to ablate soil sample surface and induce a high temperature micro-plasma where ablation plume constituents are transformed to atoms and ions. Thereafter, they undergo excitation and emit light based on their wavelength characteristics. Light is quantified following a short delay of 500 nanoseconds (ns) to allow plasma decays and facilitate broadband emission (Davidson, 2013). The schematic representation of the LIBS system is as shown in Figure 4. The technique has been used to analyze Cr, Cu, Ni and Zn from soils polluted with oil (Ding et al., 2019), Cd from field soils (Fu et al., 2018) and Cu, Ni and Zn from industry polluted soils (Meng et al., 2017). Compared to other techniques, the analytical procedure was effective in heavy metal assay of soils but required optimization to improve the limits of detection and accuracy as well as understand analyte processing in the system's plasma (Davidson, 2013). Fu et al. (2018) also made similar suggestions in an analysis of Cd in soils enriched with resins. Despite these limitations, LIBS just like XRF has portable instruments that enable in-situ heavy metal analysis in soils (Meng et al., 2017).

Figure 4. Schematic representation of the laser-induced breakdown spectroscopy (LIBS)
(Shen et al., 2018)

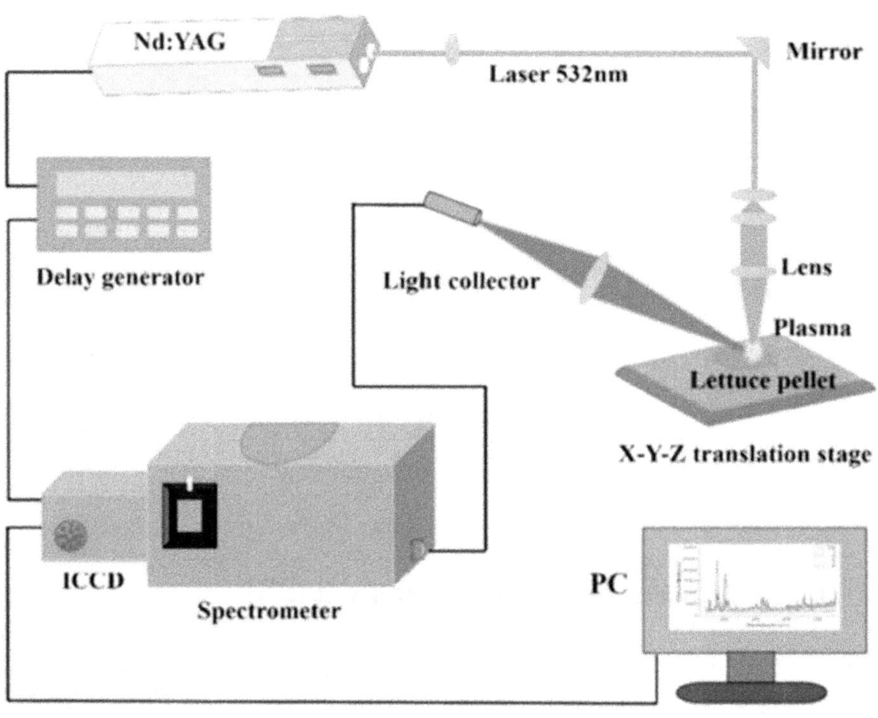

Instrumental Neutron Activation Analysis (INAA)

The Instrumental Neutron Activation Analysis performs multi-elemental and non-destructive analysis of solid samples (50-100 mg). It has a detection limit of <1 mg/kg (part per million). In INAA, a remote handling system puts a soil sample in a nuclear reactor prior to using thermal neutron flux to irradiate it. The resultant neutrons are captured by stable isotopes and changed to radionuclides. After this procedure, the sample is extracted prior to quantification of the activated products via ϒ-spectrometry. The technique has been applied to assay heavy metals in urban soils of Mexican cities (Mireles et al., 2012) and in a coastal oasis of Egypt (Wael et al., 2015). Some of the assayed heavy metals in the studies included As, Ba, Co, Cr, Fe, Mn, Ni, Ta, V and Zn. In soils polluted by soil amendments and agrochemicals, the technique was used to quantify the concentrations of Cr, Fe, Mn and Zn (Ogiyama et al., 2006; Nunes et al., 2009). The analytical technique requires highly trained personnel and specialized facilities for irradiation, which limits its wide use. The usability of the technique is also challenged by etiology to nuclear radiation exposure, unavailability of facilities and advancements of alternative analytical techniques that are more sensitive and accurate.

Indirect Methods

Indirect analysis of heavy metals in soils requires the transformation of analytes from their solid to liquid phases. As such, leaching or digestion is done prior to elemental analysis. The specificities of the extractant used in digestion of soils is essential for accurate analysis of heavy metals (Zeiner et al., 2007). After digestion, analytes are separated and preconcentrated to prevent analytical errors before the use of instrumental techniques to quantify heavy metals. In this section, indirect heavy metal analysis in soils will be discussed under three procedures: 1) sample digestion, 2) preconcentration and separation of analytes and 3) instrumental analysis of heavy metals.

Digestion of Soil Samples

The extraction of analytes through digestion occurs in four approaches. First is the total digestion usually applied in total concentration analysis of heavy metals in geochemical surveys to establish geogenic sources of the pollutants and relate them with anthropogenic sources (Morrison et al., 2009). In such samples, total digestion is done using hydrofluoric (HF) acid, which dissolves silicates. Soil samples are solubilized in the acid and total metal content obtained for quantification. HF produces corrosive vapors and therefore, caution when handling it should be observed by: 1) using fume hoods and wearing protective clothing when handling the acid, 2) avoiding using and storing the acid in glass containers and 3) availing calcium gluconate, which is a gel with antidote effects in events of HF burns. The dissolution of sulfides and organic matter is done in addition to silicate dissolution using oxoacids such as perchloric ($HClO_4$) and nitric (HNO_3) acids.

For soil samples containing persistent minerals such as chromite, fusion can be used as an alternative approach to total digestion (Morrison et al., 2009). In the process, a soil sample of 0.1 to 0.2 g with fine particles of <100μm diameter is put in a crucible containing approximately 0.8 g of lithium metaborate and subjected to 1000 °C heat. The soil sample dissolves in the resultant molten flux, which is allowed to cool before adding nitric acid and water to the suitable volume for analysis.

The second approach to sample digestion is pseudo-total digestion using a mixture of H_2O_2, HNO_3 and HCl. This digestion process is commonly used in soil pollution research to evaluate the anthropogenic input of heavy metals and approximate the levels of the pollutants that affect a receptor (e.g. soils) following their mobilization and transport (Thestorf & Makki, 2021). The digestion process does not dissolve primary silicates but liberates heavy metals associated with soil components for analysis. The procedure involves pre-weighing 0.5–2 g of soil into a vessel, moistening it with water prior to aqua regia addition. In this case, aqua regia refers to a mixture of 1:3 parts of HNO_3 and HCL. After an overnight allowance to stand, the vessel is heated and put in reflux for 2 hours before cooling. The resultant digestate is filtered to a volumetric flask before washing off the residue with dilute HNO_3 and adding the washings to the filtrate. The volumetric flask with the digestate is them added dilute HNO_3 or distilled water up to the mark and refrigerated at 4 °C in readiness for analysis.

In the case of total and pseudo-total digestion methods, experiments can be done in fume cupboards using watch-glass topped breakers (if using HF) and a hot plate as the heat source. Such manual digestion processes require careful monitoring to add acids lost via evaporation or in the event where digested soils have high organic content and require additional HNO_3 during digestion (Davidson, 2013). Alternatively, microwave oven, bomb or block digestors can be used for the processes. Depending on the digestion approach used, varied recovery results are obtained. For instance, Mico et al. (2008) compared microwave to open-vessel block digestion and reported that the latter had poor precision and higher extraction efficiency in calcareous soils compared to microwave digestion. Aluminum was underestimated when microwave digestion was done in the presence of HF compared to other acid mixtures (Marin et al., 2008). Recoveries of various digestion approaches in a comparative study on Florida soils was reported in the order of hotplate (aqua regia) < microwave (aqua regia) < microwave (aqua regia + HF) (Chen & Ma, 2001). Specific considerations before choosing a digestion method and based on a soil sample's metal content levels based on background levels, carbonate and organic matter content have been outlined by Sastre et al. (2002).

Singular and sequential extraction of various soil contents are the other methods of digestion subjected to soils before analysis. The two methods utilize chemical extractants to isolate the mobile species of a soil solution (with metals and free ions), which is chemically bound to ligands and organic matter. In singular extraction, each component of the soil is removed exclusively. For instance, soluble species of soil solution are removed using water extraction, hot water (for soluble boron) or via dilute treatment with $CaCl_2$ among other dilute salts (Davidson, 2013). Exchangeable metals found in organic and inorganic soils components are extracted using acetate salts ($C_2H_7NO_2$), calcium salts {($Ca(NO_3)_2$, $CaCl_2$)} and nitrates {($Mg(NO_3)_2$, NH_4NO_3)} (Davidson, 2013). The use of HNO_3, H_2O_2, sodium pyrophosphate and chelating agents such as diethylenetriaminepentaacetic acid (DTPA) and ethylenediaminetetraacetic acid (EDTA) can release heavy metals from soil solutions for quantitative analysis (Quenea et al., 2009). The chemicals used in extraction are not specific for some minerals but target some heavy metal reservoirs in soils and hence may not be applicable in all soils based on their specific physicochemical characteristics.

Sequential extraction on the other hand uses multiple chemical reagents to isolate soil mineral phases following modification in their redox potential and pH. According to the Community Bureau of Reference (BCR), Europe as detailed by Rauret et al. (1999), a sequential extraction occurs in four steps involving 1) removal of carbonates and cations, 2) removal of reducible Mn and Fe oxides, 3) removal of oxidizable sulfides and organic matter and 4) removal of residual matter. The chemicals used in the steps are acetic acid, hydroxylammonium chloride, ammonium acetate and aqua regia in respective order. Sequential extraction of metals using the BCR protocol and prior to their quantification has had several applications including: -

i. Assessing the mobilization and transport of uranium in weapon manufacturing and testing centers (Oliver et al., 2008)
ii. Understanding the geochemistry of soils in urban centers (Kartal et al., 2006)
iii. Evaluating time-based changes in acid wastewater and pyritic sludge contaminated water in a mining area in Spain (Pueyo et al., 2008)
iv. Relating geogenic processes and parent material composition to heavy metal concentration in Iranian soils (Nael et al., 2009)
v. Examining the mobility of heavy metal in agricultural soils of Greece (Golia et al., 2009)

To ensure success in sequential extractions, standard protocols must be adhered to including the application of quality control measures, application of comparative studies and publication of results based on operationally-specified fractions and absolute concentrations. According to Bacon and Davidson (2008), such considerations prevent incomplete extraction of heavy metals from soils and the possible analyte redistribution.

Preconcentration and Separation of Analytes

When subjecting digested soil samples for instrumental analysis, two challenges can occur. First the analytes present could be of low concentrations to levels nearing or below the limits of detection for the analytical instrument available. Second, the analysis process could be vulnerable to matric interferences, which could compromise the accuracy of heavy metal quantification (Nyika et al., 2019a). The two challenges can be controlled via analyte preconcentration before analysis. Preconcentration is done at a specific factor that is accounted for after analysis to determine the concentration of the original analyte. The three most common preconcentration procedures include cloud-point extraction, co-precipitation and solid phase extraction (Butler et al., 2009). In cloud-point extraction, a soil extract is mixed with a chelating agent and a surfactant to make dilute aqueous and surfactant-rich phases (Davidson, 2013). Using a solid carrier, analytes are trapped in the preconcentration process of co-precipitation. In solid phase extraction, samples containing targeted metal contaminants are concentrated to heavy metal complexes using specific sorbents (Butler et al., 2009). This is done in a column containing the sorbents where soil extracts are allowed to go through so that analytes are retained for later elution and analyses and interferents are excluded. Preconcentration is achieved by using a smaller eluant volume compared to the loaded extract. A complexing agent can also be added to the soil extract to facilitate preconcentration.

Instrumental Techniques for Heavy Metal Analysis in Soils

Digested and preconcentrated soil samples are subjected to instrumental analysis to measure their heavy metal content. Some of the chemicals used during digestion and concentration processes could be counter reactive to the instrument used for assay. Therefore, prior to instrument calibration using standard solutions of pre-defined concentrations is essential to avoid such analytical challenges. Calibration that is reagent-matched using standards reduces matric interferences during analysis and enhances accuracy. Some of the instrumental techniques used in heavy metal quantification in soils are discussed in the following sub-sections.

Ion Chromatography

In the many chromatographic techniques available, the use of ion chromatography to assay anions and cations in solution is of growing interest (Michalski, 2018). The technique is multi-elemental, cost and time efficient. Additionally, the technique requires a small sample, produces results of high resolution, selectivity and sensitivity (Liu et al., 2015). The premise in its working is isolating ionic species (including heavy metals) by applying ionic equilibrium and chromatographic principles together. These principles include ionic-exclusion, ionic pairing and ion exchange (Weiss, 2016). In this case, a column packed with resin is used as the stationary phase while the mobile phase is an eluent. The stationary phase binds to ionic species from soil samples digestate because it has ion exchange characteristics.

During the injection of samples to be analyzed in a column, the resin surface retains analytes before adding elution solution to form eluent ions that displace analytes (Liu et al., 2015). The targeted ions have different sizes, charges and elution time, which is the premise for their separation. Consequently, temporal separation of individual ions via conductimetric detection is achieved (Weiss, 2016). Apart from doing total metal analysis, ion chromatography can be used in speciation analysis as reported by Shaw and Haddad (2004) in a speciation analysis for As (V) and As (III) from environmental samples.

Surface -Enhanced Raman Spectroscopy

Surface -enhanced Raman spectroscopy (SERS) describes a technique using Raman scattering signals from target species adsorbed on nanoparticles and rough metal surfaces. Silver and gold nanostructures are used to enhance electromagnetic fields of surfaces near or next to targeted samples and increase their analytical sensitivity to limits of detection below ng/l (Borah et al., 2015). SERS can assay heavy metals in soils traditionally as a direct method where fine particles along with a suitable substrate in a dielectric environment are used. Using sensors, the technique can also be used in speciation analysis of heavy metals (in digestates) whose levels are trace in soils. Unlike the direct approach that uses substrates, the indirect method uses anchors specific for targeted heavy metals and positions them near the signal-enhancing surface to allow accurate quantification and speciation (Shaban & Galay, 2016). The use of the technique in indirect heavy metal analysis is growing due to its associated high sensitivity and accuracy (Borah et al., 2015).

Capillary Electrophoresis

Capillary electrophoresis (CE) is another method of heavy metal analysis in soils that is preferred for its low electrolyte and sample consumption, simplicity, high analysis speed and efficiency especially in speciation analysis (Imran & Hassan, 2002). Quantification and separation of heavy metal species in the technique is a result of their electrophoretic mobility such that the movement of metals is controlled.

After soil digestates are treated with chelating agents, species detection is done via ultraviolet detection. Heavy metals such as Ba, Pb and Zn have been assayed using the technique (Valsecchi & Polesello, 1999). CE has also has been combined with ICP-MS for platinum (Pt) speciation analysis from soil samples (Valsecchi & Polesello, 1999).

Atomic Spectrometry

The principle in using atomic spectrometry is quantifying electromagnetic radiation emitted or absorbed by unbound atoms once their valence electrons are transformed as they move in between energy levels (Sahin, 2019). During atomic absorption, transfer of energy to an atom from a photon occurs resulting to electron excitation while during atomic emission, excited electrons de-excite and release their excess energy as photons. With temperature increases, atomic excitation is favored and therefore, spectroscopic techniques should occur at varied temperatures. During atomic absorption, the combustion flame should be at relatively low temperatures while during emission, higher temperatures should be availed in plasma. The unique electronic configuration of each atom allows light absorption and emission at specific wavelength. Therefore, during atomic spectrometry, the light absorbed and subsequently emitted corresponds to the concentration levels of a given element in this case, the specific heavy metals in the soil digestates (Nyika et al., 2019a). They are different approaches to atomic spectrometry as discussed below.

Flame Atomic Absorption Spectrometry

The Flame Atomic Absorption Spectrometry (FAAS) just as any other AAS technique comprises of a source of light that provides photos, a detector as a photomultiplier tube, a wavelength isolator (monochromator) and an atom cell that releases free analyte ions (Figure 5i) (Davidson, 2013). A hollow cathode lamp (Fig 5ii) is used as the source of light but it can be substituted with an electrodeless discharge lamp (Davidson, 2013). An electrical discharge is generated in the used noble gas (argon and neon) after voltage is applied between the cathode and anode. The cathode is made of either the analyte or coated with it. Noble gas ions induce sputtering at the cathode and get excited. Thereafter, a line emission spectrum is produced after de-excitation where light enters via the atom cell to absorb targeted elements. Through the doppler and collision effects, spectral lines widen compared to natural widths (Mico et al., 2007). In a FAAS used in heavy metal quantification in soils, the combustion flame is found in the atom cell and mainly uses air acetylene flame of about 2,200 °C. In the spray chamber, fuel and air are mixed before flowing to the nebulizer that takes the sample to be analyzed at a rate of 5 ml/min. The resultant aerosol (about 10%) is directed to the bottom of the burner before aspiration to a flame. The alternative to an air-acetylene flame is a nitrous oxide-acetylene flame of about 2900 °C with high burning speed and atomization efficiency to use a shorter burner head. The flame is used for elements such as Al and Mo that have thermal oxides, which are stable. Irrespective of the flame used the burner height and fuel flow should be optimized for the targeted heavy metal to be assayed to enable atomization efficiency.

Figure 5. The basic functioning of atomic absorption spectrometer
(Davidson, 2013)

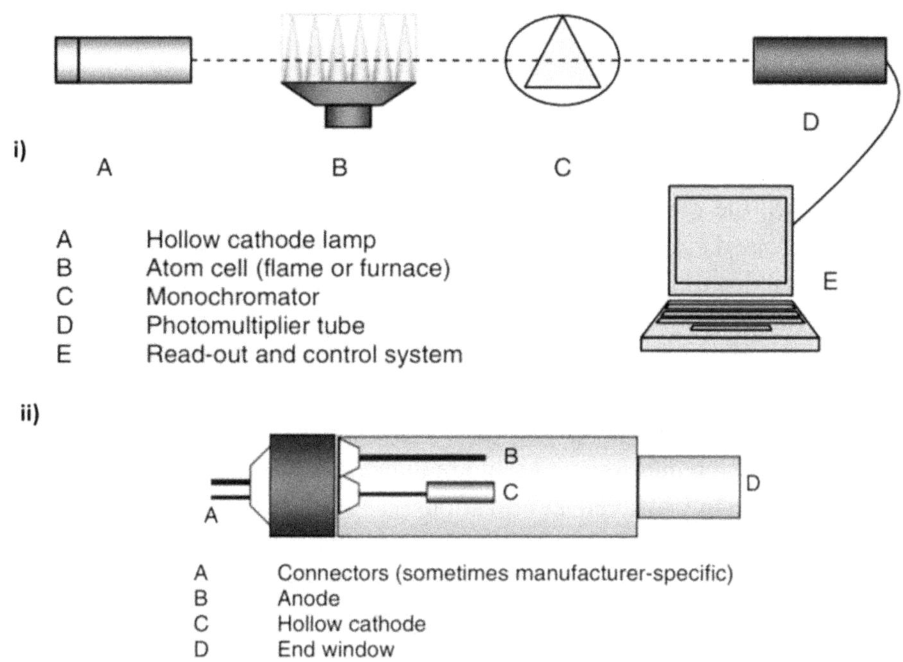

During FAAS technique, chemical, physical and spectral interferences can occur and compromise the accuracy of assayed heavy metals. In situations where multiple metals absorb light in an analogous wavelength as the analyte, spectral line interference occur. Although rare due to the narrow width of the cathode lamp, the errors can be eliminated via chemical isolation of the interferent before analysis (Santos et al., 2009). Using a detector to modulate light from the cathode lamp can also eliminate flame emission interferences. Background correction is also advised for FAAS instruments to prevent light scattering and molecular absorption of particles within the flame. Reagent matching is essential to prevent physical interferences that occur if standards and sample solutions are aspirated at varied rates due to differences in their surface tension and viscosity. Chemical interference, which disrupts atomization efficiency occurs during analysis and is overcome by adding a reagent such as Cesium (Cs), which binds preferentially to the interferents.

FAAS has been widely applied in assessing heavy metal content of soils. It was used to assay heavy metals in calcareous soils of Valencian Mediterranean region (Mico et al., 2007), quantify heavy metal uptake from soils to coffee (Santos et al., 2009) and assess Cd and Pb levels from roadside soils of Oromia region in Ethiopia (Sisay et al., 2019). The efficiency of the technique in terms of limits of detection is up to parts-per-million. The trend is however dependent on pneumatic nebulization of the sample, time taken by analytes in the cathode lamp and the dilution effect of the flame gas used.

Electrothermal Atomic Absorption Spectrometry

Unlike FAAS, the electrothermal AAS (ETAAS) uses a tube atomizer made of electrically-heated graphite. A soil sample digestate of 20 µl is put in the tube using a pipette via a dosing hole and heated to produce atoms of the analyte. The atoms pass along the tube axis of the cathode lamp for analysis to detection limits of parts-per-billion. By using chemically resistant and low permeable pyrolytic graphite and optimizing the heating system, chemical interferences during analysis can be overcome. Matrix modifiers such as nitrates of ammonium and magnesium can also be used to overcome interferences that are induced during ashing of volatile heavy metal species and in inefficient atomization. ETAAS is used in assessing individual heavy metal toxicity in soils especially if their levels are minute. It has been used in speciation analysis (Davidson, 2013), analysis of Cd (Petit & Rucandio, 1999), Cr (Figueiredo et al., 2007) and Sb in soil samples. The limitation of ETAAS just like FAAS is that commercially available hollow cathode lamps can only assay one heavy metal at a time, unlike many toxicity assessment studies that require multi-elemental analysis of soils.

Inductively Coupled Plasma- Atomic/Optical Emission Spectrometry

Inductively coupled plasma- atomic/optical emission spectrometry (ICP-A/OES) is a technique that uses spectra emissions to identify and assay elemental constituents (Nyika et al., 2021). The ICP-A/OES system has two parts: 1) the ICP, which holds three quartz tubes that make plasma and 2) the optical spectrometer that results to electromagnetic radiation, which is wavelength specific to the analyte. The samples are introduced just as in FAAS where they are dissolved, ionized and excited into a nebulizer and spray chamber (Nyika et al., 2021). They are then passed to a photo multiplier tube for quantification depending on the emission lines and specific intensity of the targeted analytes. The functioning and signal processing of the ICP-A/OES system is regulated by a microprocessor. The schematic presentation of the ICP-A/OES equipment is as shown in Figure 6.

Figure 6. Schematic representation of an ICP-OES
(Nyika et al., 2019a)

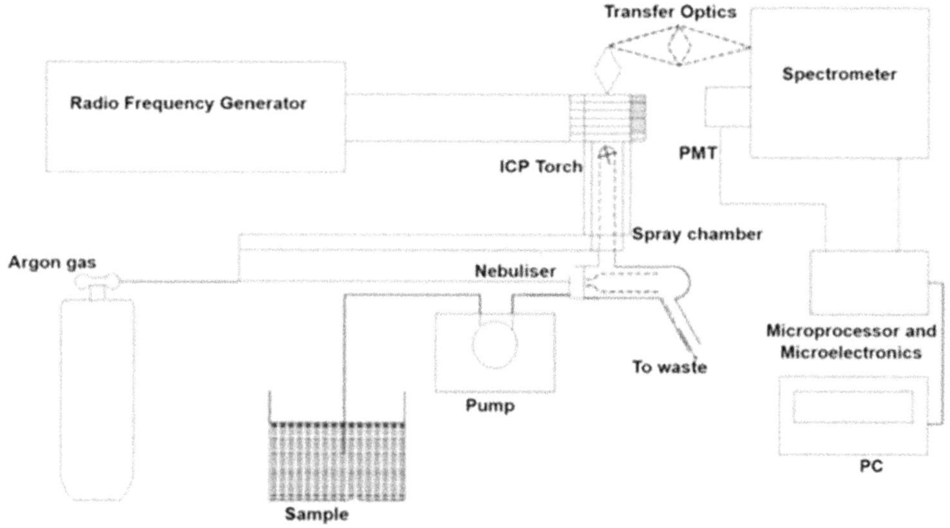

The technique has similar chemical and spectral interferences as FAAS since sample introduction is analogous. However, it is a more superior technique than ETAAS and FAAS in trace metal analysis due to its multi-elemental analysis and wide dynamic range. It has been used to identify and quantify heavy metals in leachate polluted soils of Roundhill landfill vicinity, South Africa (Nyika et al., 2019b), agricultural soils of Kafr El-Zayat region of Egypt (Shaheen et al., 2021) and mine tailings polluted soils of Slovakia (Stofejova et al., 2021). ICP-A/OES technique is limited in that its limits of detection (parts-per-million) are unrealistic to achieve if soils have relatively low pollution levels. In such a case, the trace metals are in the analytes are undetected since their concentrations are too low to quantify. The use of ICP-mass spectrometry (ICP-MS) with higher sensitivity could overcome this challenge.

Inductively Coupled Plasma Mass Spectrometry

ICP-MS is a highly sensitive and relatively new technique of heavy metal analysis. The system comprises of the ICP, which is an ion source, an interface system with a MS, detector, ion lens, skimming and pumping cores. Soil digestates are introduced in the injection system, excited to ions at low pressure and high temperature (Nyika et al., 2019a). Afterwards, heavy metal elements are detected using a quadrupole system that involves peak hopping to concentrations under parts-per-billion. A schematic representation of the ICP-MS system is a shown in Figure 7.

Figure 7. Schematic representation of an ICP-MS
(Nyika et al., 2019a)

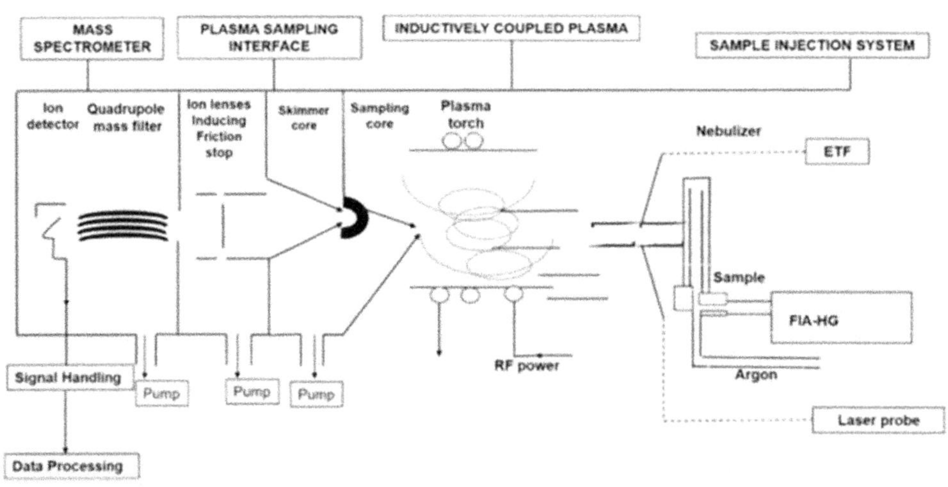

The challenges of using ICP-MS analytical technique are similar to those of ETAAS but it is more superior and highly sensitive from its ability to conduct multi-elemental analysis compared to the other spectrometric techniques. It has been used in heavy metal assay of polluted soils in South Africa (Nyika et al., 2019b), wastewater irrigated soils in suburban regions of Nairobi (Nyika & Dinka, 2022) and polluted soils of Baoji in China (Zhang et al., 2021) and Hamadan in Iran (Kharazi et al., 2021). Common interferences from using the technique in soil analysis include solid material deposition at the cone tips during sample injection. This especially occurs if the sample analyzed has high dissolved solids but can be overcome through sample dilution (Davidson, 2013; Nyika et al., 2019b). In polyatomic ions and ions of similar atomic number (e.g. ^{40}Ca and ^{40}Ar), isobaric interferences can occur and therefore, sample preparation precautions such as avoiding aqua regia soil digestion and instead, using only HNO_3 must be considered (Davidson, 2013).

Hydride Generation Atomic Absorption Spectrometry (HGAAS)

Heavy metals such as As, Bi, Ge, Pb, Sb, Se, Sn and Te that are reduced by sodium borohydride in acidic conditions to produce volatile hydrides can be assayed via HGAAS (Maleki et al., 2005). The analytical technique involves decomposition of hydride containing soils to an atomic vapor in an atomic cell made as a tubular quartz glass. Afterwards, the sample is subjected to heating in an FAAS instrument. The method is highly sensitive but vulnerable to matric interferences from transition metals, which disrupt the formation of hydrides in soil digestates. Heavy metals such as Sb and Pb as well as the metalloid, As have been quantified from soil extracts using the technique (Davidson, 2013).

Cold Vapor Atomic Absorption Spectrometry

Cold vapor atomic absorption spectrometry (CVAAS) is used to assay for organic and inorganic Hg in soils. Through reduction of soil digestates with tin (II) chloride, monoatomic Hg is generated, sucked into the light beam of the CVAAS equipment via a columnar cathode lamp. The sample does not require heating but it can be preconcentrated using gold. The technique has been successfully used in assaying Hg levels in paddy (Marcelo et al., 2017) and field (Katarzyna et al., 2016) soils.

CONCLUSION AND FUTURE PROSPECTS

Techniques used in analysis of heavy metals in soils play a crucial role in controlling environmental pollution since they identify and quantify the pollutants and can be used to assess the progress of remediation advances. Due to growing industrialization and technological developments, the demand for heavy metal use and their subsequent release to the environment is on the rise. Subsequently, the need to develop cost effective, highly sensitive, precise and accurate analytical techniques and continually improve existing ones is indispensable. This chapter has highlighted several challenges associated with assaying heavy metals in soils. Some soils have low trace metal concentrations, which require highly sensitive methods to quantify and such analytical techniques are not always available. The complication is worsened in speciation analysis where a particular species only represents a fraction of total element levels. As such, the analytical chemistry field will evolve in the near future to achieve accurate and low limits of detection.

Apart from indirect methods of heavy metal analysis in soils such as XRF, INAA and LA-ICP-MS, direct methods such as capillary electrophoresis, ion chromatography and atomic spectrometry can also be used. Spectrometric techniques demonstrated importance in determining a number of heavy metals in soils. However, the techniques require advancements to improve atomization and sample introduction. Of the spectrometric techniques discussed herein, ICP-MS was more superior due to its low detection limits, highly sensitive and identificatory power and extensive lineal ranges. Irrespective of their sensitive, the discussed analytical techniques are prone to physical, chemical and spectral interferences. As such, sample preparation approaches, reagent matching, standardization and instrument calibration and optimization are key considerations to counter interferences and improve analytical accuracy. Additionally, numerical simulations of heavy metals can be done to quantify the accuracy of analyzed results.

Even with the many direct and indirect methods for heavy metal analysis in soils available, the ultimate success of analysis depends on the analyst's criterion. Factors to consider include the heavy metal targeted for analysis, availability and cost of analytical equipment, the merits and demerits of each analytical technique and the purpose for analysis. To improve the accuracy in the targeted metal and its possible location in a study area, geographical information systems and remote sensing techniques can be used. Analytical chemists or technicians should therefore consider the objectives of a particular assessment and the problem, which analysis is intended to solve before choosing a convenient analytical approach. With the evolution and advancements in the analytical chemistry field, it is expected that more advantages will be leveraged in the future to enhance accuracy and sensitivity of the applied heavy metal analytical methods in soils while reducing their associated costs.

REFERENCES

Achterberg, E., Browning, T., Gledhill, M., & Schlosser, C. (2019). Transition metals and heavy metal speciation. In: Cochran, J., Bokuniewicz, H, Yager, P. (eds) Encyclopedia of ocean sciences, 3rd edn. Academic, Oxford. . 11394-6 doi:10.1016/B978-0-12-409548-9.11394-6

Asamoah, B., Asare, A., Okpati, S., & Aidoo, P. (2021). Heavy metal levels and their ecological risks in surface soils at Sunyani magazine in the Bono region of Ghana. *Scientific African*, *13*, e00937. doi:10.1016/j.sciaf.2021.e00937

Bacon, R., & Davidson, M. (2008). Is there a future for sequential chemical extraction? *Analyst (London)*, *133*(1), 25–46. doi:10.1039/B711896A PMID:18087610

Bailey, T., Robinson, N., Farrell, M., MacDonald, B., Weaver, T., Antille, D., Chin, A., & Brackin, R. (2022). Storage of soil samples leads to over-representation of the contribution of nitrate to plant available nitrogen. *Soil Research (Collingwood, Vic.)*, *60*(1), 22–32. doi:10.1071/SR21013

Bakirdere, S. (2013). *Speciation studies in soil, sediment and environmental samples*. CRC Press., doi:10.1201/b15501

Borah, S., Bora, T., Baruah, S., & Dutta, J. (2015). Heavy metal ion sensing in water using surface plasmon resonance of metallic nanostructures. *Groundwater for Sustainable Development*, *1*(1), 1–11. doi:10.1016/j.gsd.2015.12.004

Brevik, E., Slaughter, L., Singh, B., Steffan, J., Collier, D., Barnhart, P., & Pereira, P. (2020). Soil and human health: Current status and future needs. *Air, Soil and Water Research*, *13*, 1–12. doi:10.1177/1178622120934441

Butler, T., Cook, J., Davidson, C., Harrington, C., & Miles, L. (2009). Atomic spectrometry update: Environmental analysis. *Journal of Analytical Atomic Spectrometry*, *24*(2), 131–177. doi:10.1039/b821579k

Byers, H., McHenry, L., & Grundl, T. (2019). XRF techniques to quantify heavy metals in vegetables at low detection limits. *Food Chemistry: X*, *1*, 100001. doi:10.1016/j.fochx.2018.100001

Carr, R., Zhang, S., Moles, N., & Harder, M. (2008). Identification and mapping of heavy metal pollution in soils from a sports ground in Galway City, Ireland, using a portable XRF analyzer and GIS. *Environmental Geochemistry and Health*, *30*(1), 45–52. doi:10.100710653-007-9106-0 PMID:17610027

Chen, M., & Ma, L. (2001). Comparison of three aqua regia digestion methods for twenty Florida soils. *Soil Science Society of America Journal*, *65*(2), 491–499. doi:10.2136ssaj2001.652491x

Chen, S., Kimirei, A., Yu, C., Shen, Q., & Gao, Q. (2022). Assessment of urban river water pollution with urbanization in East Africa. *Environmental Science and Pollution Research International*, *29*(27), 40812–40825. doi:10.100711356-021-18082-1 PMID:35083687

Clough, R., Harrington, C., Hill, S., Madrid, Y., & Tyson, J. (2018). Atomic spectrometry update: Review of advances in elemental speciation. *Journal of Analytical Atomic Spectrometry*, *33*(7), 1103–1149. doi:10.1039/C8JA90025F

Crompton, T. (2015). Determination of metals in natural waters, sediments, and soils (1st ed.). Elsevier Science. 1016/B978-0-12-802654-0.00001-5.

Das, K., & Chakraborty, R. (1997). Electrothermal atomic absorption spectrometry in the study of metal ion speciation. *Fresenius' Journal of Analytical Chemistry, 357*(1), 1–17. doi:10.1007002160050102

David, W., & Somsubhra, C. (2020). Portable X-ray fluorescence spectrometry analysis of soils. *Soil Science Society of America Journal, 84*(5), 1384–1392. doi:10.1002aj2.20151

Davidson, C. (2013). Methods for the Determination of Heavy Metals and Metalloids in Soils. In B. Alloway (Ed.), *Heavy Metals in Soils. Environmental Pollution* (Vol. 22). Springer. doi:10.1007/978-94-007-4470-7_4

Ding, Y., Xia, G., Ji, H., & Xiong, X. (2019). Accurate quantitative determination of heavy metals in oily soil by laser induced breakdown spectroscopy (LIBS) combined with interval partial least squares (IPLS). *Analytical Methods, 11*(29), 3657–3664. doi:10.1039/C9AY01030K

Dubiella, A., Wasik, A., Przyjazny, A., & Namiesnik, J. (2007). Preparation of soil and sediment samples for determination of organometallic compounds. *Polish Journal of Environmental Studies, 16*, 159–176.

Ebong, G., Dan, E., Inam, E., & Offiong, N. (2019). Total concentration, speciation, source identification and associated health implications of trace metals in Lemma dumpsite soil, Calabar, Nigeria. *Journal of King Saud University. Science, 31*(4), 886–897. doi:10.1016/j.jksus.2018.01.005

Figueiredo, E., Soares, E., Baptista, P., Castro, M., & Bastos, M. L. (2007). Validation of an electrothermal atomization atomic absorption spectrometry method for quantification of total chromium and chromium (VI) in wild mushrooms and underlying soils. *Journal of Agricultural and Food Chemistry, 55*(17), 7192–7198. doi:10.1021/jf0710027 PMID:17661487

Fu, X., Li, G., Tian, H., & Dong, D. (2018). Detection of cadmium in soils using laser-induced breakdown spectroscopy combined with spatial confinement and resin enrichment. *RSC Advances, 8*(69), 39635–39640. doi:10.1039/C8RA07799A PMID:35558063

Fuller, R., Landrigan, P., Balakrishnan, K., Bose, S., & Braver, M. (2022). Pollution and health: A progress update. *The Lancet. Planetary Health, 6*(6), 6537–6547. doi:10.1016/S2542-5196(22)00090-0 PMID:35594895

Golia, E., Tsiropoulos, G., Dimirkou, A., & Mitsios, I. (2007). Distribution of heavy metals of agricultural sols of central Greece using the modified BCR sequential extraction method. *International Journal of Environmental Analytical Chemistry*, *87*(13-14), 1053–1063. doi:10.1080/03067310701451012

Gustavsson, B., Luthbom, K., & Lagerkvist, A. (2006). Comparison of analytical error and sampling error for contaminated soil. *Journal of Hazardous Materials*, *138*(2), 252–260. doi:10.1016/j.jhazmat.2006.01.082 PMID:17030410

Hoenig, M. (2001). Preparation steps in environmental trace element analysis – facts and traps. *Talanta*, *54*(6), 1021–1038. doi:10.1016/S0039-9140(01)00329-0 PMID:18968324

Imran, A., & Hassan, A. (2002). Determination of metal ions in water, soil and sediment by capillary electrophoresis. *Analytical Letters*, *35*(13), 2053–2076. doi:10.1081/AL-120015519

Jozsef, P., David, N., Sandor, K., & Aron, B. (2019). A comparison study of analytical performance of chromium speciation methods. *Microchemical Journal, 149*, 103958. https://doi.org/. microc.2019.05.058 doi:10.1016/j

Kartal, S., Aydin, Z., & Tokalioglu, S. (2006). Fractionation of metals in street sediment samples by using the BCR sequential extraction procedures and multivariate statistical elucidation of the data. *Journal of Hazardous Materials*, *132*(1), 80–89. doi:10.1016/j.jhazmat.2005.11.091 PMID:16466857

Katarzyna, S., Grazyna, D., Lukasz, Z., Grazyna, P., Katarzyna, G., & Barbara, A. (2016). Determination of mercury in selected environmental components using cold vapor atomic absorption spectrometry. *Bulletin off the Maritime Institute in Gdansk, 31*(1), 73-79. https://doi.org/ doi:10.5604/12307424.1201262

Kharazi, A., Leili, M., Khazaei, M., Alikhani, M., & Shokoohi, R. (2021). Human health risk assessment of heavy metals in agricultural soil and food crops in Hamadan, Iran. *Journal of Food Composition and Analysis*, *100*, 103890. doi:10.1016/j.jfca.2021.103890

Kim, J., Park, J., Choi, J., & Kim, J. (2020). Determination of metal concentration in road-side trees from an industrial area using laser ablation inductively coupled plasma mass spectrometry. *Minerals (Basel)*, *10*(2), 175. doi:10.3390/min10020175

Koch, J., & Gunther, D. (2017). Laser ablation inductively coupled plasma mass spectrometry. In J. Lindon, G. Tranter, & D. Koppenaal (Eds.), *Encyclopedia of spectroscopy and spectrometry* (pp. 526–532). Academic Press., doi:10.1016/B978-0-12-803224-4.00024-8

Kocot, K., Pytlakowska, K., Zawisza, B., & Sitko, R. (2016). How to detect metal species preconcentrated by microextraction techniques? *Trends in Analytical Chemistry*, *82*, 412–424. doi:10.1016/j.trac.2016.07.003

Kurfurst, U., Desaules, A., Rehnert, A., & Muntau, H. (2004). Estimation of measurement uncertainty by the budget approach for heavy metal content in soils under different land use. *Accreditation and Quality Assurance*, *9*, 64–75. doi:10.100700769-003-0697-6

Liu, J., Liu, C., Fang, G., & Wang, S. (2015). Advanced analytical methods and sample preparation for ion chromatography techniques. *RSC Advances*, *5*(72), 58713–58726. doi:10.1039/C5RA10348G

Lombi, E., & Susini, J. (2009). Synchrotron-based techniques for plant and soil science: Opportunities, challenges and future perspectives. *Plant and Soil*, *320*(1-2), 1–35. doi:10.100711104-008-9876-x

Madden, C., Pringle, K., Jeffery, A., Wisniewski, K., Heaton, V., Oliver, I., Glanville, H., Stimpson, I. G., Dick, H. C., Eeley, M., & Goodwin, J. (2022). Portable X-ray fluorescence (pXRF) analysis of heavy metal contamination in church graveyards with contrasting soil types. *Environmental Science and Pollution Research International*, *29*(36), 55278–55292. doi:10.100711356-022-19676-z PMID:35318600

Makinen, E., Korhonen, M., Viskari, L., Haapamaki, S., Jarvinen, M., & Lu, L. (2006). Comparison of XRF and FAAS methods in analyzing CCA contaminated soils. *Water, Air, and Soil Pollution*, *171*(1-4), 95–110. doi:10.100711270-005-9017-6

Maleki, N., Safavi, A., & Doroodmand, M. (2005). Determination of selenium in water and soil by hydride generation atomic absorption spectrometry using solid reagents. *Talanta*, *66*(4), 858–862. doi:10.1016/j.talanta.2004.12.053 PMID:18970063

Marcelo, B., Isabel, D., & Mariela, P. (2017). A low-cost device for sample introduction and determination of mercury by cold vapor atomic absorption spectrometry-application for irrigation water and paddy soil. *Brazilian Journal of Analytical Chemistry*, *4*(14), 34.

Marin, B., Chopin, B., Jupinet, B., & Gauthier, D. (2008). Comparison of microwave-assisted digestion procedures for total trace element determination in calcareous soils. *Talanta*, *77*(1), 282–288. doi:10.1016/j.talanta.2008.06.023 PMID:18804634

Meng, D., Zhao, N., Ma, M., Fang, L., Gu, Y., Jia, Y., Liu, J., & Liu, W. (2017, June 20). Application of a mobile laser- induced breakdown spectroscopy system to detect heavy metal elements in soil. *Applied Optics*, *56*(18), 5204. doi:10.1364/AO.56.005204 PMID:29047571

Michalski, R. (2018). Ion Chromatography applications in wastewater analysis. *Separations*, *5*(1), 16. doi:10.3390eparations5010016

Mico, C., Peris, M., Sanchez, J., & Recatala, L. (2008). Trace element analysis via open-vessel or microwave-assisted digestion in calcareous Mediterranean soils. *Communications in Soil Science and Plant Analysis*, *39*(5-6), 890–904. doi:10.1080/00103620701881246

Mico, C., Recatala, L., Peris, M., & Sanchez, J. (2007). A comparison of two digestion methods for the analysis of heavy metals by flame atomic absorption spectroscopy. *Spectroscopy Europe*, *19*(1), 23–27.

Mireles, F., Davila, J., Pinedo, J., Reyes, E., Speakman, R., & Glascock, M. (2012). Assessing urban soil pollution in the cities of Zacatecas and Guadalupe, Mexico by instrumental neutron activation analysis. *Microchemical Journal*, *103*, 158–164. doi:10.1016/j.microc.2012.02.009

Mngadi, S., Sihlahla, M., Lekoadu, S., Moja, S., & Nomngongo, P. (2020). Evaluation of mobility, fractionation and potential risk of trace metals present in soils from Struibult gold mine dumps. *Journal of African Earth Sciences*, *127*, 104008. doi:10.1016/j.jafrearsci.2020.104008

Mohajane, C., & Manjoro, M. (2022). Sediment-associated heavy metal contamination and potential ecological risk along an urban river in South Africa. *Heliyon*, *8*(12), e12499. doi:10.1016/j.heliyon.2022.e12499 PMID:36643299

Morrison, M., Goldhaber, B., Lee, L., Holloway, M., Wanty, B., Wolf, E., & Ranville, J. F. (2009). A regional-scale study of chromium and nickel in soils of northern California, USA. *Applied Geochemistry*, *24*(8), 1500–1511. doi:10.1016/j.apgeochem.2009.04.027

Nael, M., Khademi, H., Jalalain, A., Schulin, R., Kalbasi, M., & Sotohian, F. (2009). Effect of geo-pedological conditions on the distribution and chemical speciation of selected trace elements in forest soils of Western Alborz, Iran. *Geoderma*, *152*(1-2), 157–170. doi:10.1016/j.geoderma.2009.06.001

Nunes, P., Munita, S., Vasconcellos, A., Oliveira, S., Croci, A., & Faleiros, F. (2009). Characterization of soil samples according to their metal content. *Journal of Radioanalytical and Nuclear Chemistry*, *281*(3), 359–363. doi:10.100710967-009-0016-8

Nyika, J., & Dinka, M. (2022). Heavy metal pollution in soils and vegetable from suburban regions of Nairobi, Kenya and their community health implications. *Pollution Research*, *8*(4), 1434–1447. doi:10.22059/POLL.2022.341522.1440

Nyika, J., Onyari, E., Dinka, M., & Mishra, S. (2019a). A comparison of reproducibility of inductively coupled spectrometric techniques in soil metal analyses. *Air, Soil and Water Research*, *12*, 1–13. doi:10.1177/1178622119869002

Nyika, J., Onyari, E., Dinka, M., & Mishra, S. (2019b). Heavy metal pollution and mobility in soils within a landfill vicinity: A South African case study. *Oriental Journal of Chemistry*, *35*(4), 1286–1296. doi:10.13005/ojc/350406

Nyika, J., Onyari, E., Dinka, M., & Shivani, B. (2021). A review on methods of assessing pollution levels from landfills in South Africa. *International Journal of Environment and Waste Management*, *28*(4), 436–455. doi:10.1504/IJEWM.2021.118859

Ogiyama, S., Sakamoto, K., Suzuki, H., Ushio, S., Anzai, T., & Inubushi, K. (2006). Measurement of trace metals in arable soils with animal manure application using INAA and the concentrated acid digestion method. *Soil Science and Plant Nutrition*, *52*, 114–121. doi:10.1111/j.1747-0765.2006.00013.x

Oliver, W., Graham, C., MacKenzie, B., Ellam, M., & Farmer, G. (2008). Distribution and partitioning of depleted uranium (DU) in soils at weapons test ranges – Investigations combining the BCR extraction scheme and isotopic analysis. *Chemosphere*, *72*(6), 932–939. doi:10.1016/j.chemosphere.2008.03.029 PMID:18457863

Petit, D., & Rucandio, I. (1999). Sequential extractions for determination of cadmium distribution in coal fly ash, soil and sediment samples. *Analytica Chimica Acta*, *401*(1-2), 283–291. doi:10.1016/S0003-2670(99)00487-0

Pueyo, M., Mateu, J., Rigol, A., Vidal, M., Lopez-Sanchez, F., & Rauret, G. (2008). Use of the modified BCR three-step sequential extraction procedure for the study of trace element dynamics in contaminated soils. *Environmental Pollution*, *152*(2), 330–341. doi:10.1016/j.envpol.2007.06.020 PMID:17655986

Quenea, K., Larny, I., Winterton, P., Bermond, A., & Dumat, C. (2009). Interactions between metals and soil organic matter in various particle size fractions of soil contaminated with waste water. *Geoderma*, *149*(3-4), 217–223. doi:10.1016/j.geoderma.2008.11.037

Rauret, G., Lopez-Sanchez, F., Sahuquillo, A., Rubio, R., Davidson, C., Ure, A., & Quevauviller, P. (1999). Improvement of the BCR three step sequential extraction procedure prior to the certification of new sediment and soil reference materials. *Journal of Environmental Monitoring*, *1*(1), 57–61. doi:10.1039/a807854h PMID:11529080

Sahin, D. (2019). Atomic spectroscopy. In *M, Khan, G, Nascimento & M, El-Azazy. Modern spectroscopic techniques and applications*. Intech Open. doi:10.5772/intechopen.77480

Santos, S., Santos, P., Conti, M., Santos, N., & Oliveira, E. (2009). Evaluation of some metals in Brazilian coffees cultivated during the process of conversion from conventional to organic agriculture. *Food Chemistry*, *115*(4), 1405–1410. doi:10.1016/j.foodchem.2009.01.069

Sastre, J., Sahuquillo, A., Vidal, M., & Rauret, G. (2002). Determination of Cd, Cu, Pb and Zn in environmental samples: Microwave-assisted total digestion versus aqua regia and nitric acid extraction. *Analytica Chimica Acta*, *462*(1), 59–72. doi:10.1016/S0003-2670(02)00307-0

Scholz, W., Nothbaum, N., & May, T. (1994). Fixed and hypothesis-guided soil sampling methods – Principles, strategies and examples. In B. Markert (Ed.), *Environmental sampling for trace analysis* (pp. 335–345). VCH. doi:10.1002/9783527615872.ch17

Shaban, M., & Galaly, A. (2016). Highly sensitive and selective in-situ SERS detection of Pb^{2+}, Hg^{2+}, and Cd^{2+} using Nanoporous membrane functionalized with CNTs. *Scientific Reports*, *6*(1), 25307. doi:10.1038rep25307 PMID:27143512

Shaheen, M., Tawfik, W., Mankoula, A., Gagnon, J., Fryer, B., & El-Mekawy, F. (2021). Determination of heavy metal content and pollution indices in the agricultural soils using laser ablation inductively coupled plasma mass spectrometry. *Environmental Science and Pollution Research International*, *28*(27), 36039–36052. doi:10.100711356-021-13215-y PMID:33686601

Sharma, S. (2014). *Heavy metals in water: presence, removal and safety*. Royal Society of Chemistry. doi:10.1039/9781782620174

Shaw, M., & Haddad, P. (2004). The determination of trace metal pollutants in environmental matrices using ion chromatography. *Environmental International*, *30*(3), 403–431. https://doi.org/. envint.2003.09.009 doi:10.1016/j

Shen, T., Kong, W., Liu, F., Chen, Z., Yao, J., Wang, W., Peng, J., Chen, H., & He, Y. (2018). Rapid determination of cadmium contamination in lettuce using laser induced breakdown spectroscopy. *Molecules (Basel, Switzerland)*, *23*(11), 2930. doi:10.3390/molecules23112930 PMID:30424009

Shezi, B., Street, R., Webster, C., Kunene, Z., & Mathee, A. (2022). Heavy metal contamination of soil in preschool facilities around industrial operations, Kuils river, Cape Town (South Africa). *International Journal of Environmental Research and Public Health, 19*(7), 4380. doi:10.3390/ijerph19074380 PMID:35410061

Sisay, B., Debebe, E., Meresa, A., & Abera, T. (2019). Analysis of cadmium and lead using atomic absorption spectrophotometer in roadside soils of Jimma town. *Journal of Analytical & Pharmaceutical Research, 8*(4), 144–147. doi:10.15406/japlr.2019.08.00329

Stofejova, L., Fazekas, J., & Fazekasova, D. (2021). Analysis of Heavy Metal Content in Soil and Plants in the Dumping Ground of Magnesite Mining Factory Jelšava-Lubeník (Slovakia). *Sustainability (Basel), 13*(8), 4508. doi:10.3390u13084508

Templeton, D., & Fujishiro, H. (2017). Terminology of elemental speciation – an IUPAC perspective. *Coordination Chemistry Reviews, 352*, 424–431. doi:10.1016/j.ccr.2017.02.002

Thestorf, K., & Makki, M. (2021). Pseudo-total antimony content in topsoils of the Berlin Metropolitan Area. *Journal of Soils and Sediments, 21*(5), 2102–2117. doi:10.100711368-020-02742-9

Valsecchi, M., & Polesello, S. (1999). Analysis of inorganic species in environmental samples by capillary electrophoresis. *Journal of Chromatography. A, 834*(1-2), 363–385. doi:10.1016/S0021-9673(98)00914-5 PMID:10189695

Wael, M., Khaled, A., Hussein, S., Marina, F., Svetlana, G., & Octavian, D. (2015). Instrumental neutron activation analysis of soil and sediment samples from Siwa Oasis, Egypt. *Physics of Particles and Nuclei Letters, 12*(4), 637–644. doi:10.1134/S154747711504007X

Wagner, G., Mohr, M. E., Sprengart, J., Desaules, A., Muntau, H., Theocharopoulos, S., & Quevauviller, P. (2001). Objectives, concept and design of the CEEM soil project. *The Science of the Total Environment, 264*(1-2), 3–15. doi:10.1016/S0048-9697(00)00608-2 PMID:11213187

Weiss, J. (2016). *Handbook of ion chromatography* (4th ed., Vol. 1). Wiley. org/doi:10.1002/9783527651610

Yahaya, S., Abubakar, F., & Abdu, N. (2021). Ecological risk assessment of heavy metal-contaminated soils of selected villages in Zamfara state, Nigeria. *SN Applied Sciences, 3*(2), 168. doi:10.100742452-021-04175-6

Yiika, L., Tita, M., Suh, C., Mimba, M., & Lavenir, N. (2023). *Heavy metal speciation by Tessier sequential extraction applied to artisanal gold mine tailing sin eastern Cameroon*. Chemistry Africa. doi:10.100742250-023-00652-0

Zare, B., Nami, M., & Shahverdi, A. (2017). Tracing tellurium and its nanostructures in biology. *Biological Trace Element Research*, *180*(2), 171–181. doi:10.100712011-017-1006-2 PMID:28378115

Zeiner, M., Rezic, I., & Steffan, I. (2007). Analytical methods for the determination of heavy metals in the textile industry. *Journal of Chemistry and Chemical Engineering*, *56*(11), 587–595.

Zhang, J., Gao, Y., Yang, N., Dai, E., Yang, M., Wang, Z., & Geng, Y. (2021). Ecological risk and source analysis of soil heavy metals pollution in the river irrigation area from Baoji, China. *PLoS One*, *16*(8), e0253294. doi:10.1371/journal.pone.0253294 PMID:34339446

Chapter 5
Methods of Rating Heavy Metal Pollution in Soils Using Indices

ABSTRACT

Heavy metal contamination in soils is from both lithologic and anthropogenic origin. Assessment of such contaminants in soils is done using empirical geochemical studies that involve soil sampling and their subsequent analysis using different laboratory techniques and equipment. Contamination degree rating is done thereafter using pollution indices. In this chapter, single and total complex pollution indices are studied and their purpose, strengths, and weaknesses discussed. Findings showed that pollution rating is purpose-driven and site specific. Important considerations to make when choosing a pollution index to use depends on the target soil use, data availability, purpose of the study and comparability of obtained results. Simultaneous use of multiple indices is recommended for enhanced pollution rating accuracy.

INTRODUCTION

Contamination of soils by heavy metals is a growing environmental concern due to its effects on food security and human health (Kong et al., 2021; Qin et al., 2021). The trace elements are usually as a result of both natural and anthropogenic sources. Natural sources of heavy metals, which pollute soils include volcanic eruptions, metal-laden rock weathering, continental dust transport and volcanic emissions (Chen et al., 2020). Anthropogenic sources of heavy metal pollution include the manufacture, use and disposal of electronics, military training equipment, combustion of fossil fuels, mining and smelting and the use of metal containing industrial and

DOI: 10.4018/978-1-6684-7116-6.ch005

municipal wastewater and sludge in agricultural activities (Palansooriya et al., 2020; Wang et al., 2021). The application of phosphatic fertilizers, organic manures and composts among other biosolids also enrich the heavy metal content of soils (Khan et al., 2021). The use of pesticides to manage diseases during farming and livestock-keeping activities is also associated with the introduction of trace metals such as As, Cu, Ni, Pb and Zn into soils (Khan et al., 2021). Industrial emissions from paints, batteries, gas- and coal-fired stations, smelting and refining processes as well as road traffic emissions contribute to heavy metal content in the atmosphere, which is later dispersed to far distances, precipitated and deposited in soils (Cai et al., 2015).

Globally, over 5 million sites have been reported to be contaminated with heavy metals (Palansooriya et al., 2020). The observation was made using results and conclusions of many empirical and geochemical studies, which have resulted to the creation of a database on trace metal background values that is used in assessment of environmental quality (Kowalska et al., 2018). In Sub-Saharan Africa (SSA) region, soil pollution is growing due to advances in industrialization for economic growth and increased agricultural activities to produce more food for the growing population (Tindwa & Singh, 2023). The main sources of the heavy metal pollution in the region include e-waste, landfills and refuse dumps, auto-mechanic workshops, roadside emissions, mining and the use of agrochemicals and soil additives. The sources of soil pollution by trace metals are also widespread in the SSA region through oil spills, which are predominant in oil-rich nations such as Angola and Nigeria (Tindwa & Singh, 2023). Introduction of Cu, Pb and Zn from e-wastes to soils is also a growing concern (Fayiga et al., 2018). The situation is worse as nations in the SSA region attempt to balance economic development demand and needs vis-à-vis advances to sustainable environmental conservation. Since ancient times, the priority was on short-term advantages of increased industrialization for economic gains and sustenance of livelihoods at the expense of pollution prevention initiatives through cleaner production.

In light of this knowledge, the soil pollution situation in SSA region will continue to worsen and may become a major environmental issue at its peak in 2100 (Hoornweg et al., 2015). The trend can however be reversed with the appropriate interventions. In highlighting the effects of heavy metal pollution in soils, Khan et al. (2021 suggested the need to adopt ecological solutions such as remediation and reclamation, which involves ex- or in-situ techniques used to clean, restore and remove such pollutants from the environs. Such ecological solutions based their undertakings from accurate analysis of total or individual heavy metal concentrations in soils for efficient results (Kowalska et al., 2018). However, conducting an assessment of a soil's heavy metal content is not always a sufficient indicator of the pollution extent due to its localized nature and inability to cover the entire targeted area (Kowalska et al., 2016). This trend is observed because most empirical geochemical studies rely on sampled soils to generalize on the pollution status of a given area. Additionally, laboratory assessments and procedures (Discussed in Chapter 4) done in such studies are expensive, which makes it unrealistic to cover an entire area suspected to be contaminated. Additionally, such assessments does not interrelate the presence of heavy metals and their enrichment to specific soil properties (Mazurek et al., 2017). Computational tools such as regression and correlation have also been applied to deduce the relationship of metal content to soil properties but just as geochemical analysis, they fail to detail comprehensive information on the degree and extent of soil contamination (Kowalska et al., 2016, 2018). Therefore, the key to understanding soil contamination extent and degree by heavy metal pollution lies in the use of pollution indices. The indices are guides and tools that explain and supplement geochemical analysis by explaining the nature of soil environs suspected to have heavy metal contamination. The indices are suitable indicators of the effects of bioaccumulation of heavy metals in soils and their resultant environmental risks (Weissmannova & Pavlovsky, 2017). The focus of this chapter is to explore on various pollution indices, their strengths and weaknesses in relation to heavy metal contamination assessment in soils.

ASSESSMENT OF SOIL HEAVY METAL CONTAMINATION DEGREE

Assessment of trace metal contamination degree in soils is done using two approaches: 1) chemical indicators and 2) pollution indices. Chemical indicators are a derivative of sequential extraction technologies that target geochemical fractions detailed in Table 1 (Weissmannova & Pavlovsky, 2017; Khan et al., 2021). Through chemical extraction, both exchangeable and water-soluble fractions can be used to estimate the particular heavy metal concentrations in soil samples unlike in chemical analysis. The BCR-701 procedure from the Community Bureau of Reference (Discussed in Chapter 5) also assays for the chemical indicators in soils and relates them to the environmental effects of heavy metals (Wan et al., 2017). Through sequential extraction, phase specificity is realized and extractants are categorized depending on their reaction mode as shown in Table 1. The methods however are poor indicators of bioavailability and mobility of heavy metals in soils, which makes the approaches less preferred compared to the use of pollution indices (Weissmannova & Pavlovsky, 2017; Khan et al., 2021).

Table 1. Chemical indicators of assessing soil pollution through extractable trace metal content (Wan et al., 2017; Khan et al., 2021)

Fractions	Extraction Techniques
Exchangeable	$AlCl_3$, $BaCl_2$, $CaCl_2$, $Ca(NO_3)_2$, KCl, KNO_3, $Mg(NO_3)_2$, $NaNO_3$, NH_4Cl, NH_4NO_3
Water soluble	Ultrafiltration, filtration, dialysis, displacement, centrifugation
Potentially exchangeable	NH_4NO_3, NH_4OAc, $Pb(NO_3)_2$
Mineral lattice	HNO_3, HF, $HClO_4$
Oxide bound	Bicarbonate, citrate, dithionite, hydroxylamine hydrochloride
Residual	HNO_3, HNO_3 + HCL
Carbonate bound	EDTA, HOAc, NaOAc, HOAc

Pollution indices establish the relationship between heavy metal bioaccumulation in soils and their associated environmental risk (Nyika et al., 2020a). They also infer such metal accumulation to either anthropogenic activities or geogenic factors and therefore are useful tools to monitor human involvement in soil pollution by such toxins (Weissmannova & Pavlovsky, 2017; Kowalska et al., 2016, 2018). Pollution indices are often used in assessing the extent of soils contamination compared to chemical indicators (Bali & Sidhu, 2021; Xiao et al., 2020). The following sections details the categories of pollution indices, their uses, strengths and weaknesses in assessing the extent and degree of soil pollution by heavy metals.

POLLUTION AND ECOLOGICAL RISK INDICES

Soil pollution indices are classified into two: 1) single and 2) total complex indices (Kowalska et al., 2016; Khan et al., 2021). Single indices are computed for each trace metal separately. Total complex indices on the other hand, determine pollution and ecological risk through an integrated approach by considering assayed concentrations of more than one heavy metal or by summing up the single indices. The two classes are discussed in the following sub-sections.

Single Pollution Indices

Single pollution indices used to assess heavy metal pollution in soils include the enrichment factor (EF), contamination factor (CF), pollution index (PI), threshold pollution index (PI_T), and geo-accumulation index (I_{geo}) among others (Antoniadis et al., 2019; Ekoa Bessa et al., 2020; Shaheen et al., 2020). The indices classify and rate soils into several groups based on the pollution degree and using individual trace metal concentrations. Table 2 summarizes some the single pollution indices while Table 3 shows their classifications based on the computed pollution degree.

Table 2. Single and total complex pollution indices and their formulas (Weissmannova & Pavlovsky, 2017; Khai et al., 2021)

Single Pollution Index	Formula	Explanation for Formula
Geoaccumulation index (I_{geo})	$I_{geo} = Log_2\left(\frac{c_n}{1.5B_n}\right)$	C_n is the heavy metal content in a soil sample while B_n is the geochemical background concentration of a given heavy metal. 1.5 is a constant compensating for B_n.
Enrichment factor (EF)	$EF = \frac{C_n}{C_{Ref}}$	C_{Ref} is the reference metal level in the specified environment (values vary depending on the target country/ region). C_n is a described.
Single pollution index (PI)	$PI = \frac{C_n}{B_n}$	B_n is the background metal level in the specified environment while C_n is as described.
Threshold pollution index (PI_T)	$PI_T = \frac{C_i}{C_{TL}}$	C_i is the metal concentration while C_{TL} is the tolerance level of a given metal.
Contamination factor (CF)	$CF = \frac{C_M}{C_{np}}$	C_M is the average metal concentration while C_{np} is the preindustrial era metal level.
Total Complex Pollution Indices	**Formula**	**Explanation for Formula**
PI_{sum}	$PI_{sum} = \sum_{i-1}^{m} PI_i$	PI_i is the single pollution index of a specified heavy metal
IPI	IPI= mean(PI_i)	PI_i is the single pollution index of a specified heavy metal
PI_{avg}	$PI_{avg} = \frac{1}{m}\sum_{i-1}^{m} PI_i$	m and/ or i is the number of assayed heavy metals
IPI_T	$IPI_T = \frac{1}{n}\left(\frac{1}{n}+\frac{1}{n}...+\frac{1}{n}\right)$	n is the number of heavy metals

continues on following page

Table 2. Continued

$PI_{Nemerow}$	$PI_{Nemerow} = \sqrt{\frac{\left(\frac{1}{n}\sum_{i=1}^{n} PI_i\right)^2 + P_{imax}^2}{2}}$	P_{imax} is the maximum single pollution index PI of all trace metals
PIN	$PIN = \sum_{i=1}^{m} W_i^2 PI_i$	W_i is the class of heavy metal
PLI	$PLI = \sqrt[n]{PI_1 \times PI_2 \times PI_3 \ldots PI_n}$	PI is the computed value for a single pollution index and n is the number of assayed trace elements
C_{deg}	$C_{deg} = \sum_{i=1}^{n} CF_i$	CF is the contamination factor
mC_{deg}	$mC_{deg} = \frac{1}{n}\sum_{i=1}^{n} CF_i$	CF is the contamination factor
PERI	$PERI = \sum_{i=1}^{n} E^i_r$	E^i_r is the single index of ecological risk factor
MERMQ	$MERMQ = \frac{\sum_{i=1}^{n} \frac{C_i}{ERM_i}}{n}$	C_i is the concentration of a heavy metal ERMi is the effects range median
CSI	$CSI = \sum_{i=1}^{n} W_{ia}\left[\left(\frac{C_i}{ERL_i}\right)^{\frac{1}{2}} + \left(\frac{C_i}{ERL_i}\right)^2\right]$	ERL_i is the effects range low while W_{ia} is the weight of heavy metals resulting from multivariate analysis (PCA/FA)

The EF standardizes assayed metals compared to their predetermined reference metal at low concentration and with low variability. The metals used as reference include Al, Ca, Fe, Mn, Sc and Ti. The strengths in using this index are its ability to estimate anthropogenic contribution to soil pollution by a particular heavy metal and hence establish its origin. The EF also standardizes assayed heavy metals to those of low occurrence variability in other elements making it a precise scale. The accuracy of its results depends on the selected background and reference values (Sutherland, 2000). Background or reference values are computed by considering the underlying rock composition and the weathering rate in a given area. They are pre-industrial, geochemical or baseline values whose role is to distinguish potentially toxic heavy metals of geogenic origin from anomalous concentrations, which result due to anthropic activities (Izah et al., 2021).

The CF describes a ratio of the average heavy metal level in the soil from more than five samples (C_m) and the metal levels in soils with no pollution (C_n). The reference value for soil pollution by trace elements uses the preindustrial concentrations of the heavy metals in the earth's crust (Loska et al., 2004). The CF is a widely used single index of pollution because it is direct and simple in addition to having a direct scale. It includes differences between the reference values and sample concentrations. Computation of the index requires the pre-industrial reference values, which sometimes are not available. In addition, computation of CF does not factor in variations in lithologic processes and background values and hence erroneous (Loska et al., 2004; Inengite et al., 2015; Kowalska et al., 2018).

The single PI uses threshold pollution values, national standards of pollution, baseline values, background level, average crust and preindustrial concentrations to compare to assayed heavy metal levels. PI_T just as PI compares assayed levels of metals (C_i) to their tolerance levels (C_{TL}) predetermined in national guidelines (which differ based on country/region) on trace metal concentrations that have hazardous effects. PI and PI_T indices are easy to calculate, apply and have a precise scale, which makes their application wide. However, both indices do not account for changes in natural processes and their accuracy is determined by the selected background values (Sayadi et al., 2015).

The I_{geo} is a ratio of trace metal concentrations to their background levels. The I_{geo} also applies a 1.5 constant to detect the anthropogenic contribution to soil pollution and show deviation from natural metal concentrations (Lu & Bai, 2010; Nyika et al., 2020b). The index compares previous to present contamination levels in specified sampling sites, is widely used due to its ease in computing. Additionally, the use of the 1.5 constant in its computation minimizes the variation due to geogenic effects. The I_{geo} is sensitive to the chosen local and/or reference backgrounds, which could lead to wrong inferences due to the index's omission of natural geochemical variability (Loska et al., 2004; Kowalska et al., 2018).

Total Complex Pollution Indices

Total complex indices determine pollution using many elements or as a total of single pollution indices of a given metal (Qingjie et al., 2008). The indices include the sum of pollution index (PI_{sum}), integrated pollution index (IPI), average pollution index (PI_{avg}), integrated threshold pollution index (IPI_T), Nemerow pollution index ($PI_{Nemerow}$), new pollution index (PIN), potential ecological risk index (PERI), contamination degree (C_{deg}), modified contamination degree (mC_{deg}), contamination severity index (CSI) and mean ERM quotient (MERMQ) (Hakanson, 1980; Muller, 1981; Al-Anbari et al., 2015; Weissmannova & Pavlovsky, 2017; Khai et al., 2021). The indices and the formulas used to compute them are summarized in Table 2.

Table 3. Classes of various single pollution indices based on computed pollution degree

Single Index	Pollution Degree	Classification	Reference
I_{geo}	$I_{geo} \leq 0$ $0 \leq I_{geo} < 1$ $1 \leq I_{geo} < 2$ $2 \leq I_{geo} < 3$ $3 \leq I_{geo} < 4$ $4 \leq I_{geo} < 5$ $I_{geo} > 5$	-Uncontaminated -Uncontaminated to moderately contaminated -Moderately contaminated -Moderately to strongly contaminated -Strongly contaminated -Strongly to extremely contaminate -Extremely high contamination	Loska et al., 2003; Nyika et al., 2020b
EF	EF < 2 EF = 2–5 EF = 5–20 EF = 20–40 EF > 40	-Deficiency to minimal mineral enrichment -Moderate enrichment -Significant enrichment -Very high enrichment -Extremely high enrichment	Sutherland, 2000
PI	PI < 1 $1 \leq PI \leq 3$ $3 \leq PI$	-Unpolluted, low level of pollution -Moderately polluted -Strongly polluted	Wu et al., 2015
PI_T	$PI_T < 1$ $1 \leq PI_T \leq 2$ $2 \leq PI_T \leq 3$ $3 \leq PI_T \leq 5$ $5 \leq PI_T$	-Unpolluted -Low pollution -Moderate pollution -Strongly polluted -Very strongly polluted	Lu et al., 2009
CF	CF < 1 $1 \leq CF \leq 3$ $3 \leq CF \leq 6$ $6 \leq CF$	-Low contamination factor -Moderate contamination factor -Considerable contamination factor -Very high contamination factor	Loska et al., 2004; Nyika et al., 2020b

PI_{sum} adds up while PI_{avg} computes the mean pollution contribution from all assayed metals to assess the quality of sediments and soils in reference to the presence of heavy metals (Qingjie et al., 2008). Soils that are highly polluted and of low quality have a PI_{avg} of 1. The average value of integrated pollution is IPI according to Weissmannova and Pavloxsky (2017). IPI_T evaluates the contamination caused by multi-metals compared to their common effects to the soil (Khai et al., 2021). Soil pollution by heavy metals is also computed and rated into five classes using the $PI_{Nemerow}$ and PIN (Inengite et al., 2015). Soil contamination is also assessed using the C_{deg} and mC_{deg} and classified into 4 categories (Hakanson, 1980). PLI as a geometric mean of PI is a simplified index used to show the deterioration of soil quality due to heavy metal accumulation (Varol, 2011).

Table 4. Classification of various total complex pollution indices (Weissmannova & Pavlovsky, 2017)

Pollution Index	Value	Classification/ Ecological risk
IPI	IPI < 1 1 ≤ IPI ≤ 2 IPI < 2	Low contamination/pollution Moderate contamination/pollution Strongly contamination/pollution
PIN	0–7 7–95.1 95.1–518.1 518.1–25486 25486–∞	Clean (class 1) Trace contamination (class 2) Lightly contaminated (class 3) Contaminated (class 4) Highly contaminated (class 5)
$PI_{Nemerow}$	$PI_{Nemerow} \leq 0.7$ $0.7 \leq PI_{Nemerow} \leq 1$ $1 \leq PI_{Nemerow} \leq 2$ $2 \leq PI_{Nemerow} \leq 3$ $3 \leq PI_{Nemerow}$	Clean Warning limit Slight pollution Moderate pollution Heavy pollution
C_{deg}	Cdeg < 8 8 ≤ Cdeg ≤ 16 16 ≤ Cdeg ≤ 32 32 ≤ Cdeg	Low degree of contamination Moderate degree of contamination Considerable degree of contamination Very high degree of contamination
mC_{deg}	mCdeg ≤ 1.5 1.5 ≤ mCdeg ≤ 2 2 ≤ mCdeg ≤ 4 4 ≤ mCdeg ≤ 8 8 ≤ mCdeg ≤ 16 16 ≤ mCdeg ≤ 32 32 ≤ mCdeg	Very low contamination Low contamination Moderate contamination High contamination Very high contamination Extremely high contamination Ultra-high contamination
PERI	PERI < 150 150 < PERI < 300 300 < PERI <600 ≥ 600	Low ecological risk Moderate ecological risk Considerable risk High risk

The ecological risk index (PERI) combines three indices; 1) singular ecological risk index (E^i_r), a specific element's pollution coefficient (C^i_f) and the individual metal toxic response factor (T^i_r) to rate pollution by heavy metals in soils. For heavy metals such as Cd, Cr, Cu, Mn, Ni, Pb and Zn, their predefined T^i_r values are 30, 2, 5, 1, 5, 5 and 1, respectively while preindustrial levels are 1, 1.5, 15, 150, 90, 0.25 and 70, respectively (Weissmannova & Paclovsky, 2017). The index considers the ecological, environmental chemistry and toxicological effects that result from soil pollution by heavy metals (Ke et al., 2017). The MERMQ determines the presence of toxic effects due to the presence of trace elements in sediments and soils. The index categorizes metal concentrations as percentiles and those with 10th and 50th percentiles have low range (ERL) and medium (EMR) effects, respectively (Khai et al., 2021). Beyond 50th percentile, the concentrations of the metals have adverse effects (ERM). By combining values of ERL and ERM with multivariate statistical analysis including factorial analysis and principal component analysis (PCA), the CSI of heavy metals in soils can be computed (Pejman et al., 2015). The index identifies the origin and sources of heavy metals in a given soil and its computed weighted values, which are a result of human-activity contribution to pollution. The classification of various total complex pollution indices is as shown in Table 4 while Table 5 summarizes the strengths and weaknesses of some of the indices.

Table 5. Strengths and weaknesses associated with specified total complex pollution indices

Pollution Index	Strengths	Weaknesses	Reference
PIN	-It is easy to compute and applies background values -Integrates contamination into a singular value	-Does not consider variations in various geochemical processes -It requires the determination of the PI_{class} -It is not commonly used	Qingjie et al., 2008
PI_{sum}	-Uses multiple elements to rate pollution -Is based on PI values and allows comparisons of pollution levels in soils from different ecosystems	-Lacks a precise scale -Accuracy is dependent on the chosen background values -The index does not consider changes in natural processes	Inengite et al., 2015; Kowalska et al., 2016, 2018
$PI_{Nemerow}$	-Reflects the soil pollution of a given environment directly -Highlights the most toxic trace elements -Can use background, threshold and baseline values in its computations -Takes into account the contribution of individual metals to total pollution -Based on PI values, is precise and widely used	- Ranks heavy metals -Does not consider the weighting factor of elements	Qingjie et al., 2008; Al-Anbari et al., 2015
PI_{avg}	-It is easy to apply and is based on PI values	-Its accuracy depends on the specified reference values - Has no precise scale -Takes into account the average rather than actual metal concentrations	Inengite et al., 2015; Kowalska et al., 2018
C_{deg}	-Can be computed with unlimited number of assayed heavy metals -Uses the sum of CF and therefore takes into account individual metal contribution to soil pollution	-It is not widely used - It does not factor in the geochemical changes in sampled environs and excludes background values in its computation	Hakanson, 1980; Inengite et al., 2015 Loska et al., 2004

continues on following page

Table 5. Continued

Pollution Index	Strengths	Weaknesses	Reference
mC_{deg}	-Factors in contribution of all assayed metals to total pollution -Its widely used and easy to apply	-Requires E^i_r ranking -Does not include variations in natural processes and background values in its calculation	Kowalska et al., 2018, Khai et al., 2021
PERI	-Provides comprehensive assessment by considering the synergy of metal toxic effects and their ecological sensitivity -Has a precise scale and is widely used	-Does not consider background values -Requires toxic response values that are sometimes unavailable for some metal contaminants -Requires E^i_r ranking	Pejman et al., 2015; Sayadi et al., 2015
MERMQ	-Reduce high levels of pollutants to a single index -Has a precise scale to assess biological effects of pollution by heavy metals in soils	-Not common -Requires ERM values to compute - Does not factor in changes in natural processes	Gao & Chen, 2012; Pejman et al., 2015
CSI	-Determines the toxicity limit and sums it up to a single value - Includes negative biological effects in its pollution classes	-It is not widely used -Weighting of every trace metal is a requirement -Requires the values of ERL and ERM	Pejman et al., 2015

DISCUSSION

The use of pollution indices is a growing approach to assay trace metal pollution in soils even in SSA (Neeraj et al., 2022). Simultaneous use of pollution indices increases their accuracy in evaluation soil pollution by heavy metals (Kowalska et al., 2018). The selection of the indices does not follow specific guidelines provided by a country or region (Ahivar et al., 2023) but depends on the purpose of a particular study such as evaluation of heavy metal origin, the ecological risk associated with pollution and assessment of pollution levels (Qingjie et al., 2008; Al-Anbari et al., 2015). For this reason, optimal use of indices for comparative studies is difficult to achieve. In some countries of SSA, data on geochemical background values of some heavy metal pollutants and the means to assay soil samples before applying pollution indices is limited. This challenge in addition to lack of specific guidelines to apply such chemical indicators prevent optimal use of indices for rating pollution. Based on the purpose, pollution indices can be divided into 4: 1) indices for evaluating individual trace metal concentrations (CF, PI, I_{geo}), 2) indices to assess pollution levels (PI_{sum}, $PI_{Nemerow}$, PLI, PI_{avg}, mC_{deg}, C_{deg}, PIN and CSI), 3) indices explaining the origin of heavy metals (EF) and 4) indices to assess the potential ecological risk (PERI and MERMQ). The comparison between soil pollution indices can be standardized and enhanced using multivariate statistical analysis such as PCA and hierarchical cluster analysis (HCA) (Wang et al., 2015).

Pollution indices (Table 2) have some similarities based on their computation methods (Inengite et al., 2015; Wiessmannova & Pavlovsky, 2017). For both single and total complex indices, descriptive statistics including the weighted geometric, geometric averages and geometric sums are used as evident from their formulas. Some of the pollution indices also use predetermined reference data such as the background/ baseline/ pre-industrial/ geochemical values (Gao & Chen, 2012; Wang et al., 2015). Pollution indices such as I_{geo}, PI, PI_{sum}, $PI_{Nemerow}$, PI_{avg}, PLI, PIN and EF use background values in their computation while CF, PERI, MERMQ, CSI and C_{deg} use other data rather than background values. Indices such as mC_{deg} are computed using trace metal concentrations in soil profile excluding the parent material. Soil pollution indices also possess some differences despite their similarities, which makes them not readily comparable (Gao & Chen, 2012; Kowalska et al., 2016). The differences are statistical and emanate from the data they use in their computation as established using multivariate analysis such as cluster analysis and PCA (Kowalska et al., 2018).

The indices also possess various advantages and disadvantages. The single pollution indices such as PI and I_{geo} have high accuracy, precision and are commonly used in evaluating soil pollution by heavy metals (Karim et al., 2015; Sayadi et al., 2015, Nyika et al., 2020a, b). The indices compare previous and current contamination. The I_{geo} index is more superior than the other singular indices since it reduces the metal accumulation resulting from anthropogenic activities degree using its 1.5 constant (Li et al., 2016). However, single indices computation needs the correct background values (Kowalska et al., 2016). Another weakness associated with the indices is their inability to factor in the effects of trace metals on xenobiont (as a result of organisms) behavior and edaphic (influenced by soil properties) characteristics of target soils (Sayadi et al., 2015). The EF is another single pollution index which helps in differentiating natural and anthropogenic sources of heavy metal pollution in soils through its standardization of elements during computation (Kowalska et al., 2016). Calculating EF however requires the reference values and enrichment levels of sampled soils of low occurrence variability (Omatoso & Ojo, 2015). With low occurrence variability, geochemical influence of soil pollution is normalized and contamination from natural and anthropogenic activities is differentiated (Sutherland, 2000). CF as a single pollution index compares assayed metal concentrations to preindustrial reference levels. The use of the reference levels in place of background values reduces inconsistencies in the computed values (Varol, 2011) though the index does not factor in the changes due to geogenic processes in its computation (Loska et al., 2004).

Unlike the single pollution indices, total complex indices are more common in rating soils pollution since they integrate and compute means of assayed data. Some of the indices even go further to compare and rate pollution in different sites using a common and specific scale (Qingjie et al., 2008). PI_{sum}, $PI_{Nemerow}$, PLI, PI_{avg} and PIN have similar purposes, use PI values in their computation and are easily comparable (Inengite et al., 2015). Additionally, the indices are simple, easy to interpret and distinguish acceptable from unacceptable pollution levels (Shu & Zhai, 2014). All the indices depend on computed PI values that use background values and hence selection of the latter determines their accuracy. PI_{sum} and PI_{avg} are single-scaled and have no precise scale and hence of low accuracy (Mazurek et al., 2017).

Total complex indices, which do not require background values include C_{deg}, MERMQ and CSI. CSI and MERMQ use the effects range median (ERM) and the effects range low (ERL) values in place of background values (Pejman et al., 2015). The ERM and ERL values are determined through multiple field studies and toxicity tests to differentiate trace metal concentrations and their associated ranges (Gao & Chen, 2012). Although both indices provide soil pollution patterns that are spatially representative, assessments involved are complex and grade-based (Wang et al., 2015). CSI uses computed weight for each trace metal in comparison to overall contamination hence, its accuracy while MERMQ calculates the possibility of toxicity and effects of trace elements to humans after exposure (Gao & Chen, 2012; Pejman et al., 2015). Determination of site-specific pollution is possible using C_{deg} (Hakanson, 1980) though the index relies on computed CF values. In addition, C_{deg} is simple to interpret and compute.

In rating the ecological risk associated with soil pollution by trace elements, PERI is used (Hakanson, 1980; Al-Anbari et al., 2015; Sayadi et al., 2015). The information provided by computed values of the index facilitate decision making and management of the environment including preventing the release of toxic heavy metals, protecting natural resources and the constant monitoring and evaluation of ecological synergies and the sensitivities of heavy metals to the changes (Mazurek et al., 2017). Computation of the index requires the toxic response factors, which are coefficients associated with toxicity levels of different trace metals. The index does not link well with other indices and comparison is difficult in addition to the fact that the use of a wide range of heavy metals in its computation is limited due to unavailability of their toxic response coefficients (Kowalska et al., 2016, 2018). To be able to make optimal use of soil pollution indices, it is imperative to develop guidelines for using them to enable comprehensive and comparative application at national and international levels (Ahivar et al., 2023). Such guidelines would enable appropriate selection of geochemical backgrounds for specific indices that are comparable among nations and that discern anthropogenic and natural sources of soil pollution.

CONCLUSION

The concentration of heavy metals globally has become an environmental concern due to their lithologic sources as well as anthropogenic contribution to their levels in soils. Consequently, different methods to assay and rate their pollution effects have emerged. In this chapter, the use of pollution indices to compute and rate the degree of pollution by heavy metals in soils was explored. Findings showed that soil pollution indices are categorized as single or total complex indices. The chapter also details the uses, strengths and weaknesses of the indices. The computation of pollution indices was shown to be key in rating the effects of various heavy metals in soils. Findings further showed that the simultaneous uses of the indices result to more accuracy due to their synergistic characteristics. Additionally, it was deduced that the selection of pollution indices is determined by a number of factors including: - 1) the specific use of the soil, 2) availability of accurate data including background values, ERM and ERL values, 3) the purpose of the study and 4) the comparability of computed values. As such, the choice of the index to use is study-specific and purpose driven during soil quality assessment.

REFERENCES

Ahivar, B., Das, P., Srivastava, V., & Kumar, M. (2023). Perspectives of heavy metal pollution indices for soil, sediment and water pollution evaluation: An insight. *Total Environment Research Themes*, *6*, 100039. doi:10.1016/j.totert.2023.100039

Al-Anbari, R., Al Obaidy, H., & Ali, F. (2015). Pollution loads and ecological risk assessment of heavy metals in the urban soil affected by various anthropogenic activities. *International Journal of Advanced Research*, *3*(2), 104–110.

Antoniadis, V., Shaheen, M., Levizou, E., Shahid, M., Niazi, K., Vithanage, M., Ok, Y. S., Bolan, N., & Rinklebe, J. (2019). A critical prospective analysis of the potential toxicity of trace element regulation limits in soils worldwide: Are they protective concerning health risk assessment? – a review. *Environment International*, *127*, 819–847. doi:10.1016/j.envint.2019.03.039 PMID:31051325

Bali, S., & Sidhu, S. (2021). Heavy metal contamination indices and ecological risk assessment index to assess metal pollution status in different soils. In V. Kumar, A. Sharma, & A. Cerda (Eds.), *Heavy Metals in the Environment*. Elsevier. doi:10.1016/B978-0-12-821656-9.00005-5

Cai, C., Xiong, B., Zhang, Y., Li, X., & Nunes, L. (2015). Critical comparison of soil pollution indices for assessing contamination with toxic metals. *Water, Air, and Soil Pollution*, *226*(10), 352. doi:10.100711270-015-2620-2

Chen, X., Kumari, D., Cao, C. J., Plaza, G., & Achal, V. (2020). A review on remediation technologies for nickel-contaminated soil. *Human and Ecological Risk Assessment*, *26*(3), 571–585. doi:10.1080/10807039.2018.1539639

Ekoa Bessa, Z., Ngueutchoua, G., Kwewouo Janpou, A., El-Amier, A., & Nguetnga, O. A. (2020). Heavy metal contamination and its ecological risks in the beach sediments along the Atlantic Ocean (Limbe coastal fringes, Cameroon). *Earth Systems and Environment*, *5*(2), 433–444. doi:10.100741748-020-00167-5

Fayiga, A., Ipinmoroti, M., & Chirenje, T. (2018). Environmental pollution in Africa. *Environment, Development and Sustainability*, *20*(1), 41–73. doi:10.100710668-016-9894-4

Gao, X., & Chen, A. (2012). Heavy metal pollution status in surface sediments of the coastal Bohai Bay. *Water Research*, *46*(6), 1901–1911. doi:10.1016/j.watres.2012.01.007 PMID:22285040

Hakanson, L. (1980). An ecological risk index for aquatic pollution control: A sedimentological approach. *Water Research*, *14*(8), 975–1001. doi:10.1016/0043-1354(80)90143-8

Hoornweg, D., Bhada-Tata, P., & Kennedy, C. (2015). Peak waste: When is it likely to occur? *Journal of Industrial Ecology*, *19*(1), 117–128. doi:10.1111/jiec.12165

Inengite, K., Abasi, Y., & Walter, C. (2015). Application of pollution indices for the assessment of heavy metal pollution in flood impacted soil. *International Research Journal of Pure and Applied Chemistry*, *8*(3), 175–189. doi:10.9734/IRJPAC/2015/17859

Izah, S., Richard, G., Aigberua, A., & Ekakitie, O. (2021). Variations in reference values utilized for evaluation of complex pollution indices of potentially toxic elements: A critical review. *Environmental Challenges*, *5*, 100322. doi:10.1016/j.envc.2021.100322

Karim, Z., Qureshi, A., & Mumtaz, M. (2015). Geochemical baseline determination and pollution assessment of heavy metals in urban soils of Karachi, Pakistan. *Ecological Indicators*, *48*, 358–364. doi:10.1016/j.ecolind.2014.08.032

Ke, X., Gui, S., Huang, H., Zhang, H., Wang, C., & Guo, W. (2017). Ecological risk assessment and source identification for heavy metals in surface sediment from the Liaohe River protected area, China. *Chemosphere, 175*(Supplement C), 473 – 481. https://doi.org/. chemosphere.2017.02.029 doi:10.1016/j

Khan, S., Naushad, M., Lima, E., Zhang, S., Shaheen, S., & Rinklebe, J. (2021). Global soil pollution by toxic elements: Current status and future perspectives on the risk assessment and remediation strategies-a review. *Journal of Hazardous Materials*, *417*, 126039. doi:10.1016/j.jhazmat.2021.126039 PMID:34015708

Kong, F., Chen, Y., Huang, L., Yang, Z., & Zhu, K. (2021). Human health risk visualization of potentially toxic elements in farmland soil: A combined method of source and probability. *Ecotoxicology and Environmental Safety*, *211*, 111922. doi:10.1016/j.ecoenv.2021.111922 PMID:33472110

Kowalska, J., Mazurek, R., Gasiorek, M., Setlak, M., Zaleski, T., & Waroszewski, J. (2016). Soil pollution indices conditioned by medieval metallurgical activity: A case study from Krakow (Poland). *Environmental Pollution*, *218*, 1023–1036. doi:10.1016/j.envpol.2016.08.053 PMID:27574802

Kowalska, J., Mazurek, R., Gasiorek, M., & Zaleski, T. (2018). Pollution indices as useful tools for the comprehensive evaluation of the degree of soil contamination-a review. *Environmental Geochemistry and Health*, *40*(6), 2395–2420. doi:10.100710653-018-0106-z PMID:29623514

Li, M., Yang, W., Sun, T., & Jin, Y. (2016). Potential ecological risk of heavy metal contamination in sediments and macrobenthos in coastal wetlands induced by freshwater releases: A case study in the Yellow River Delta, China. *Marine Pollution Bulletin*, *103*, 227–239. doi:10.1016/j.marpolbul.2015.12.014 PMID:26719069

Loska, K., Wiechula, D., Barska, B., Cebula, E., & Chojnecka, A. (2003). Assessment of arsenic enrichment of cultivated soils in southern Poland. *Polish Journal of Environmental Studies*, *12*(2), 187.

Loska, K., Wiechuła, D., & Korus, I. (2004). Metal contamination of farming soils affected by industry. *Environment International*, *30*(2), 159–165. doi:10.1016/S0160-4120(03)00157-0 PMID:14749104

Lu, G., & Bai, Q. (2010). Contamination and potential mobility assessment of heavy metals in urban soils of Hangzhou, China: Relationship with different land uses. *Environmental Earth Sciences*, *60*(7), 1481–1490. doi:10.100712665-009-0283-2

Lu, X., Wang, L., Lei, K., Huang, J., & Zhai, Y. (2009). Contamination assessment of copper, lead, zinc, manganese and nickel in street dust of Baoji, NW China. *Journal of Hazardous Materials, 161*(2–3), 1058–1062. doi:10.1016/j.jhazmat.2008.04.052 PMID:18502044

Mazurek, R., Kowalska, J., Gasiorek, M., Zadrozny, P., Jozefowska, A., Zaleski, T., Kępka, W., Tymczuk, M., & Orłowska, K. (2017). Assessment of heavy metals contamination in surface layers of Roztocze National Park Forest soils (SE Poland) by indices of pollution. *Chemosphere, 168*, 839–850. doi:10.1016/j.chemosphere.2016.10.126 PMID:27829506

Muller, G. (1981). The heavy metal pollution of the sediments of Neckars and its tributary: A stocktaking. *Chemiker-Zeitung, 105*, 157–164.

Neeraj, A., Hiranmai, R., & Iqbal, K. (2022). Comprehensive assessment of pollution indices, sources apportionment and ecological risk mapping of heavy metals in agricultural soils of Raebareli district, Uttar Pradesh, India, employing a GIS approach. *Land Degradation & Development, 34*(1), 173–195. doi:10.1002/ldr.4451

Nyika, J., Onyari, E., Dinka, M., & Mishra, S. (2020a). Comparative assessment of trace metal concentrations and their eco-risk analysis in soils of the vicinity of Roundhill landfill, Southern Africa. *Nature Environment and Pollution Technology, 19*(2), 539–548. doi:10.46488/NEPT.2020.v19i02.009

Nyika, J., Onyari, E., Dinka, M., & Mishra, S. (2020b). Assessment of trace metal contamination of soil in a landfill vicinity: A southern Africa case study. *Current Chemistry Letters, 9*, 171–182. doi:10.5267/j.ccl.2020.2.003

Omatoso, A., & Ojo, J. (2015). Assessment of some heavy metals contamination in the soil of river Niger floodplain at Jebba, central Nigeria. *Water Utility Journal, 9*, 71–80.

Palansooriya, N., Shaheen, M., Chen, S., Tsang, W., Hashimoto, Y., Hou, D., Bolan, N. S., Rinklebe, J., & Ok, Y. S. (2020). Soil amendments for immobilization of potentially toxic elements in contaminated soils: A critical review. *Environment International, 134*, 105046. doi:10.1016/j.envint.2019.105046 PMID:31731004

Pejman, A., Bidhendi, G., Ardestani, M., Saeedi, M., & Baghvand, A. (2015). A new index for assessing heavy metals contamination in sediments: a case study. *Ecological Indicators, 58*, 365–373. https://doi.org/. ecolind.2015.06.012. doi:10.1016/j

Qin, G., Niu, Z., Yu, J., Li, Z., Ma, J., & Xiang, P. (2021). Soil heavy metal pollution and food safety in China: Effects, sources and removing technology. *Chemosphere, 267*, 129205. doi:10.1016/j.chemosphere.2020.129205 PMID:33338709

Qingjie, G., Jun, D., Yunchuan, X., Qingfei, W., & Liqiang, Y. (2008). Calculating pollution indices by heavy metals in ecological geochemistry assessment and a case study in parks of Beijing. *Journal of China University of Geosciences*, *19*(3), 230–241. doi:10.1016/S1002-0705(08)60042-4

Sayadi, H., Shabani, M., & Ahmadpour, N. (2015). Pollution index and ecological risk of heavy metals in the surface soils of Amir-Abad Area in Birjand City, Iran. *Health Scope*, *4*(1), 121–137. doi:10.17795/jhealthscope-21137

Shaheen, M., Antoniadis, V., Kwon, E., Song, H., Wang, L., Hseu, Z.-Y., & Rinklebe, J. (2020). Soil contamination by potentially toxic elements and the associated human health risk in geo-and anthropogenic contaminated soils: A case study from the temperate region (Germany) and the arid region (Egypt). *Environmental Pollution*, *262*, 114312. doi:10.1016/j.envpol.2020.114312 PMID:32193081

Shu, Y., & Zhai, S. (2014). Study on soil heavy metals contamination of a lead refinery. *Chinese Journal of Geochemistry*, *33*(4), 393–397. doi:10.100711631-014-0703-1

Sutherland, A. (2000). Bed sediment-associated trace metals in an urban stream, Oahu, Hawaii. *Environmental Geology (Berlin)*, *39*(6), 611–627. doi:10.1007002540050473

Tindwa, H., & Singh, B. (2023). Soil pollution and agriculture in sub-Saharan Africa: State of the knowledge and remediation technologies. *Frontiers in Soil Science*, *2*, 1101944. doi:10.3389/fsoil.2022.1101944

Varol, M. (2011). Assessment of heavy metal contamination in sediments of the Tigris River (Turkey) using pollution indices and multivariate statistical techniques. *Journal of Hazardous Materials*, *195*, 355–364. doi:10.1016/j.jhazmat.2011.08.051 PMID:21890271

Wan, X., Dong, H., Feng, L., Lin, Z., & Luo, Q. (2017). Comparison of three sequential extraction procedures for arsenic fractionation in highly polluted sites. *Chemosphere, 178* (Supplement C), 402–410. https://doi.org/. chemosphere.2017.03.078 doi:10.1016/j

Wang, Q., Shaheen, M., Jiang, Y., Li, R., Slany, M., Abdelrahman, H., Kwon, E., Bolan, N., Rinklebe, J., & Zhang, Z. (2021). Fe/Mn- and P-modified drinking water treatment residuals reduced Cu and Pb phytoavailability and uptake in a mining soil. *Journal of Hazardous Materials*, *403*, 123628. doi:10.1016/j.jhazmat.2020.123628 PMID:32814241

Wang, Z., Wang, Y., Chen, L., Yan, C., Yan, Y., & Chi, Q. (2015). Assessment of metal contamination in coastal sediments of the Maluan Bay (China) using geochemical indices and multivariate statistical approaches. *Marine Pollution Bulletin*, *99*(1-2), 43–53. doi:10.1016/j.marpolbul.2015.07.064 PMID:26233304

Weissmannova, H., & Pavlovsky, J. (2017). Indices of soil contamination by heavy metals-methodology of calculation for pollution assessment (minireview). *Environmental Monitoring and Assessment*, *189*(12), 616. doi:10.100710661-017-6340-5 PMID:29116419

Wu, S., Peng, S., Zhang, X., Wu, D., Luo, W., Zhang, T., Zhou, S., Yang, G., Wan, H., & Wu, L. (2015). Levels and health risk assessments of heavy metals in urban soils in Dongguan, China. *Journal of Geochemical Exploration*, *148*, 71–78. doi:10.1016/j.gexplo.2014.08.009

Xiao, X., Zhang, J., Wang, H., Han, X., Ma, J., Ma, Y., & Luan, H. (2020). Distribution and health risk assessment of potentially toxic elements in soils around coal industrial areas: A global meta-analysis. *The Science of the Total Environment*, *713*, 135292. doi:10.1016/j.scitotenv.2019.135292 PMID:32019003

Chapter 6
Heavy Metal Pollution of Soils and Their Ecological Risk in Suburban Areas:
A Case Study From Eastern Africa

ABSTRACT

In this study, soils of five suburban areas of Nairobi, Kenya where vegetables were irrigated with wastewater were assayed for heavy metals and resultant concentrations compared to the predefined permissible levels. Using multivariate statistical analyses, relationships among the assayed metals were established. Furthermore, pollution and ecological risk indices were used to rate pollution levels from assayed metal contaminants. Findings showed that concentrations of As, Cd, Cr, Co, Hg and Mn were higher than permissible levels, which was indicative of pollution. All metals except Co and Ni came from similar sources based on their positive Pearson's correlation coefficient values. Based on the computed pollution and ecological risk indices, Cd and Hg had the greatest pollution contribution in the soils. Pollution in the study area was largely a result of anthropogenic activities in the vicinity. Affirmative action is imperative to regulate the release and reuse of wastewater for agricultural purposes.

DOI: 10.4018/978-1-6684-7116-6.ch006

INTRODUCTION

The rapid urbanization and industrialization trends have become increasingly associated with soil pollution globally (Yahaya & Abdu, 2021; Wieczorek & Baran, 2022). Through urbanization and industrialization, synthetic chemicals, pesticides, inorganic fertilizers, mining by products, industrial and domestic wastes and wastewater all threaten agroecological ecosystems through soil pollution (Kang et al., 2020). It for this reason that soil pollution is one of the greatest agricultural challenges that compromise advances to provide safe and healthy food (Lu et al., 2015). Regardless of whether heavy metals are essential nutrients to living organisms or not, their elevated levels in the environment have a propensity to result to soil pollution. This is because heavy metals are toxic, can migrate to long distances, enter trophic chains and biogeochemical cycles, and eventually human food chains resulting in their bio-magnification and bioaccumulation and subsequent, negative health effects (Shen et al., 2019; Nyika & Dinka, 2022; Wieczorek & Baran, 2022). Heavy metals are also being produced faster than their geogenic breakdown through human activities, and their movement and exposure, characterized by increased bioavailability, is on the rise. These conditions according to Lee et al. (2020) make the pollutants a great environmental and ecological threat. In suburban areas of Kenya, soil contamination has been reported and associated with exposure to heavy metal containing solid wastes, agricultural runoff, domestic and industrial effluents in such environments (Mungai & Wang, 2019; Kinuthia et al., 2020; Tomno et al., 2020; Nyika & Dinka, 2022).

The heavy metal problem is a global threat and it is therefore, imperative to study their sources, distribution, influence and concentrations to safeguard the environment and protect human health from their associated negative effects (Shen et al., 2019). Currently, a variety of approaches are used to quantify trace metals in soils and determine their pollution degree. The most common of the approaches are the predefined quality standards and geochemical indicators that assay the human-based impact on soil quality and reveal contamination possibilities (Kowalska et al., 2018). These standards and indicators are collectively known as soil quality indices. Ecological risk indices have also been used to assay the threat that soil organisms face following trace element pollution. The predefined quality standards and indices are essential tools to perform comprehensive geochemical assessment and report on the state of soils of a given region (Lu et al., 2021). This study was aimed at quantifying the heavy metal content of soils from suburban areas of Nairobi, Kenya. Additionally, their pollution threat was quantified using pollution and ecological risk indices.

MATERIALS AND METHODS

Sampling Sites and Sample Collection

Soil samples were collected from five suburban areas near informal settlements in Nairobi: - Saika, Kayole, Njiru, Chokaa and Ruai (Figure 1). In the selected areas whose georeferenced data is shown in Table 1, soils were collected in active farms where farmers used wastewater from open drains to irrigate their farms, and grow vegetables such as collard, coriander, tomato and spinach. The suburban regions, where sampling sites were located also had small and medium scale industries that processed chemicals, soaps and plastics. Previous studies in other informal settlements of Nairobi such as Mukuru (Opiyo et al., 2020) and in Dandora and Kariobangi (Ondayo et al., 2016), had been established heavy metal pollution in soils. In the study areas, the release of untreated wastes and wastewater to the environment, which can pollute areal soils was uncontrolled and the activities along with use of wastewater for agriculture informed the selection of the sampling sites on a convenience basis. Using an auger sampler, soil of about 500 g was collected from the sampling sites at a depth of 0-20 cm and emptied in polyethylene bags. The bags were labeled for transportation and further analysis. At the laboratory, the soils were air dried and stored for physicochemical analysis at room temperature.

Figure 1. Maps of Kenya and Nairobi County showing the location of the sampled sites

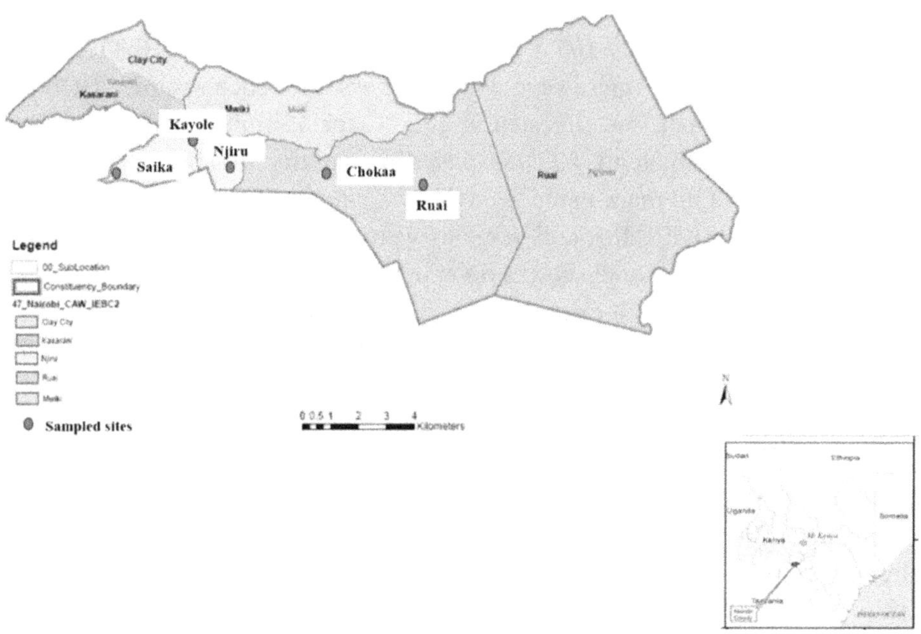

Soil Analysis

The air-dried soils were assayed for pH, EC, percentage organic matter (% OM), percentage moisture content (% MC), cation exchange capacity (CEC) and particle size distribution analysis. The respective meters were used to quantify the EC and pH of soils (International Institute of Tropical Agriculture, IITA, 1982). The particle size distribution analysis was done using a soil hydrometer (Nyika et al., 2019a) while the CEC was determined via extraction with the barium chloride method (Zeljka et al., 2018). Gravimetry was used to determine the moisture content in soils and the Walkey and Black method determined the OM content of the soil samples (Miyazawa et al., 2000).

Table 1. Name and georeferenced coordinates of the sampled sites to show their location

Name of the sampling point	Georeferenced data
Saika	-1.25863, 36.91451
Kayole	-1.26249, 36.91943
Njiru	-1.24686, 36.92547
Chokaa	-1.24910, 36.95144
Ruai	-1.26010, 36.97760

For heavy metal analysis, 0.5 g of pulverized soil, 9 ml 10 M HNO_3 and 3 ml 10 M HCL were put in a microwave digestion vessel for each sample. Thereafter, the vessels were capped and digested at 180 °C for 45 mins before filtering the digestate with a Whatman no. 42 filter paper in a volumetric flask. The flask was then filled to the 50 ml mark using 2% HNO_3. Using inductively coupled plasma-mass spectrometry (ICP-MS), soil samples were assayed for Cd, Cr, Co, Cu, Fe, Hg, Mn, Ni, Pb and Zn. Analysis was done in triplicate using the US-EPA method 6020B (US-EPA, 2014). Before analysis various instrument parameters such as plasma power, pump seed, coolant flow and nebulizer flow were optimized in the ICP-MS system. In addition, the instrument was calibrated using standards and blanks to enhance the accuracy of analyzed metal concentrations.

Computation of Pollution Indices

Pollution indices are important tools used in geochemical assessment of the conditions of soils while considering geogenic and anthropogenic contributors to pollution (Kowalska et al., 2018). They have been used to assess the pollution status in soils in several studies (Kang et al., 2020; Nyika et al., 2020; Wieczorek & Baran, 2022). In this study, the pollution load index (PLI), geoaccumulation index (I_{geo}) and contamination factor (CF) quantified heavy metal pollution extent in the study area while the ecological risk owing to the pollutants was assayed using the ecological risk factor and potential ecological risk index.

The geo-accumulation index (I_{geo}) by Muller (1969) compared current soil concentrations with pre-contamination levels. The index has been widely used (Zahra et al., 2014; Kang et al., 2020; Nyika et al., 2020) to compute pollution levels as shown in Equation (1).

$$I_{geo} = \log_2 \frac{C_n}{1.5B_n} \quad (1)$$

Where C_n was the quantified heavy metal level in sampled soil; B_n the geochemical background of the specified metal (n) and 1.5 is a constant used to account for variations in heavy metal background levels sourced from lithologic processes.

After computation of I_{geo} levels, the soils were graded based on the classes shown in Table 2.

The contamination factor evaluated the extent of contamination for individual heavy metal using the formula shown in Equation (2).

$$CF = \frac{\text{Concentration of the metal in soil samples}}{\text{Background value}} \quad (2)$$

In this case the background values are predefined levels of heavy metal content in soils. CF values that are <1 suggest the presence of contamination while those > 1 define pollution. The contamination and pollution grades for the index are as defined in Table 2.

The pollution load indices proposed by Tomlinson et al. (1980) was used to give the cumulative effect of pollution by heavy metals in a given sampling site. Levels of the index were computed using Equation (3).

$$PLI = (CF1 \times CF2 \times CF3 \times CF4 \times \ldots \times CFn)^{\frac{1}{n}} \quad (3)$$

Where n is the number of assayed heavy metals and CF is the contamination factor computed as shown in Equation (2). Classification groups using the PLI were as shown in Table 2.

In each of the assayed metals, the associated potential ecological risk was expressed using the ecological risk factor (ErF) computed as shown in Equation (4).

$$ErF = TR \times CF \quad (4)$$

Where TR is the toxic response factor and CF is as previously described. Based on availability, the TR for Cd, Cr, Hg, Ni, Pb and Zn were 30, 2, 40, 5, 5 and 1, respectively as noted by Yahaya and Abdu (2021).

The classification of the ErF was as described by Hakanson (1980) and as summarized in Table 2.

The potential ecological risk factor (RI) was used to compute the cumulative effects of the metal in the study area and was calculated as shown in Equation (5).

$$RI = (ErF1 + ErF2 + ErF3 + \ldots + ErFn) \quad (5)$$

Where ErF is as described in Equation (4) and n is the number of elements studied. The RI classification of computed values was as shown in Table 2.

Table 2. Classification of pollution indices based on computed values from assayed heavy metal concentrations

Index	Classification	Reference
I_{geo}	<0 Uncontaminated 0-1 Uncontaminated to moderately contaminated 1-2 Moderately contaminated 2-3 Moderately to heavily contaminated 3-4 Heavily contaminated 4-5 Heavily to extremely contaminated >5 Extremely contaminated	Kang et al., 2020
CF	<1 Low 1 ≤ CF <3 Moderate 3 ≤ CF <6 Considerable CF ≥ 6 Very high	Shen et al., 2019
PLI	<1 No Pollution 1.0 - < 2 Moderate pollution 2.0 - < 3 Heavy pollution ≥ 3 Extreme pollution	Yahaya & Abdu, 2021
ErF	< 40 Low risk 40 - <80 Moderate risk 80 - <160 Considerable risk 160 - <320 High risk ≥ 320 Very high risk	Hakanson, 1980
RI	< 150 Low risk 150 - <300 Moderate risk 300 - <600 Considerable risk ≥ 600 Very high risk	Hakanson, 1980

Statistical Analysis

All soil analyses were done in triplicate and Microsoft Excel 2016 used to check for any significant differences in the obtained data. The mean and standard deviation were computed using Microsoft Excel. Correlation and principal component analysis were used to check the source apportionment and factors contributing to heavy metal pollution in sampled soils. Their analysis was done using XLSTAT software.

RESULTS AND DISCUSSION

Physicochemical Properties of Soils

The sampled soils were assayed for physicochemical characteristics and results were as shown in Table 3. The soil pH ranged from 7.62 – 7.9 showing that soils of the study area were slightly alkaline. The obtained values were within the range of pH levels obtained in a study by Alghobar and Suresha (2017) who evaluated the accumulation of heavy metals in soils contaminated by sewage water in Karnataka (India). All soil samples had a clayey texture but at varying composition of silt, sand and clay. The EC of the soil samples ranged from 687.5 to 830.1 µS/cm. The levels are higher compared to those of a study by Alghobar and Suresha (2017) and lower compared to those of a study by Mekki and Sayadi (2017) who both evaluated heavy metal accumulation in soils exposed to wastewater. The EC levels were related to the clayey texture of soils since such soils are able to accumulate minerals. In addition, EC and soil alkalinity was associated with the evaporation rates of soils of the study area.

The OM content of the soils ranged between 2.04 and 2.31%. The levels were comparable to those obtained in a study by Sharma et al. (2018) evaluating the heavy metal contamination in Indian soils but considerably lower to minimum levels reported by Plunkett and Castle (2010) at 44.9%. Lower levels of OM could be associated with vulnerability of the soils to erosion and their over-cultivation leading to soil OC losses as Gebeyehu and Bayissa (2020) highlighted. The MC of sampled soils ranged between 19.66 to 25.27% while CEC ranged between 37.98 to 43.21 cmol (+)/ kg. Soil CEC levels are related to its capacity to retain nutrients and fertility and increase along with pH increases (Mukhopadhyay et al., 2019). In this case, CEC levels are associated with the clayey texture and organic matter of the soils, which enables their electrically charged portions to hold onto and attract unlike ions.

Table 3. Physicochemical characteristics of sampled soils and the values obtained after analysis

Sampling Point	pH	EC (µS/cm)	% OM	% MC	CEC (cmol (+)/kg)	Clay (%)	Silt (%)	Sand (%)	Soil Class
Saika	7.9	828.4	2.31	25.27	38.44	47.41	26.66	25.93	Clayey
Kayole	7.62	687.5	2.28	25.53	37.98	49.01	17.11	33.88	Clayey
Njiru	7.71	830.1	2.06	23.24	41.44	44.76	21.60	33.64	Clayey
Chokaa	7.83	775.2	2.09	19.92	43.21	50.01	19.23	30.76	Clayey
Ruai	7.62	695.6	2.04	19.66	42.10	39.49	23.41	37.1	Clayey
Mean	7.74	763.36	2.16	22.72	40.63	46.14	21.60	32.26	-
SD	0.13	69.23	0.02	2.82	2.31	4.21	3.70	4.19	-

Heavy Metal Concentration in Soils

Assayed heavy metals in the soil samples were quantified and their levels recorded as shown in Table 4. In all the samples, the assayed heavy metals were present, which was indicative of pollution. Further, the levels were compared to United Kingdom (UK) permissible limits from two sources (Abdi, 2010; Gebeyehu & Bayissa, 2020). This was because the permissible levels of the metals in soils of the locality (Kenya) were unavailable. The levels of arsenic surpassed the reference value of 14 mg/kg in all the sampled soils. The trend could be attributable to release of the pollutant or arsenic-based compounds along with effluents from industries of the sampled sites' vicinities to soils. Arsenic sources in the soils could also be geogenic in addition to being released from pesticides applied on the vegetables by the farmers. According to Shrivastava et al. (2017), elevated levels of arsenic in agricultural soils of West Bengal (India) are associated with both natural and anthropogenic sources.

Levels of Cd, Co and Cr exceeded the reference values of 0.8, 8 and 6.4 mg/kg, respectively while the levels of Cu, Ni, Pb and Zn were within the permissible values of 63, 50, 70 and 200 mg/kg, respectively in all sampled soils with a few exceptions. Although these heavy metals have natural sources, their elevation could be due to anthropogenic sources. According to Mahey et al. (2020), anthropogenic sources of the metal pollution to soils could be deposits of wastes and wastewaters with cosmetic chemical wastes, electronic wastes (such as computer monitors, televisions and mobile batteries), mining waste and paint residues in such resources. The soils of the study area could have absorbed these pollutants and hence the elevated heavy metal concentrations. The levels of the three heavy metals (Cd, Co and Cr) were in higher concentrations than levels reported by Sharma et al. (2018). The levels of Fe ranged between 26,939 to 36,505 mg/kg while those of Mn were between 2,756- 6,893 mg/kg. The elevated nature of the metals could be associated with their lithologic in addition to anthropogenic sources as noted by Nyika et al. (2020), in a heavy metal evaluation of South African soils located near a landfill facility.

The levels of Hg surpassed the reference values of 0.3 mg/kg in all sampled soils. The heavy metal is toxic if ingested by humans and causes negative health effects as noted by Kinuthia et al. (2020). Overall, sampled soils exhibited pollution presence with some of the assayed heavy metal levels exceeding the allowable limits. This could be a result of contamination by wastewater and agrochemicals used by farmers. Apart from parental material-sourced heavy metals, human activities such as discharge of effluents and wastes from industries, as well as, application of agrochemicals result in pollution of soils (Zhong et al., 2016). Sharma et al. (2018) also noted that soils accumulate heavy metals due to the discharge of effluents and wastes containing such pollutants to the environment in a study evaluating heavy metal levels in soils. The clayey nature of the soils could also be an enhancer of metal accumulation in sampled soils. According to Pikula and Stepien (2021), clay soils retain higher metal amounts compared to sandy soils because their particles act as binding or adsorption surfaces for the pollutants.

Table 4. Heavy metal concentrations of sampled soils in mg/kg and the threshold limits used to compare them

Heavy Metal Concentration (mg/kg)	Saika	Kayole	Njiru	Chokaa	Ruai	Mean	Standard Deviation	Allowable Limits (mg/kg)
As	29.81	20.92	31.37	27.67	26.49	27.25	4.01	14[b]
Cd	6.03	4.32	6.45	6.04	4.81	5.53	0.91	0.8[a]
Cr (vi)	60.73	49.21	50.0	48.15	53.24	52.27	5.10	6.4[a]
Co	21.7	18.94	15.9	13.52	15.11	17.03	3.27	8[b]
Cu	24.03	19.85	28.66	26.34	27.88	25.35	3.55	63[a]
Fe	28,108	26,939	36,505	27,349	33,286	30,437	4246.57	-
Hg	6.7	6.13	7.73	8.23	7.31	7.22	0.83	0.3[b]
Mn	3,666	3,203	6,893	2,756	3300	3,963	1669.37	2000[b]
Ni	50.73	44.32	42.57	35.02	41.29	42.79	5.66	50[a]
Pb	29.76	20.9	31.42	27.67	28.52	27.65	4.03	70[a]
Zn	126.67	97.64	135.91	108.26	99.43	113.58	16.97	200[a]

[a] Threshold set by UK (Adul, 2010)
[b] Threshold limits defined by Gebeyehu and Bayissa (2020)

Correlation and Principal Component Analyses

To establish a relationship between the observed heavy metal concentrations in the sampled sites with their origin, Pearson's correlation coefficient analysis was used as shown in Table 5. The approach has been used for this purpose in many studies (Malkoc & Yazici, 2017; Kurt, 2018; Nyika et al., 2019b). All heavy metals had positive correlations with the exception of Co and Ni. Negative relationships such as those between Co-As, Co-Cd, Cu-Cr, Fe and Cr and Co as well as Hg and Cr and Co inferred to the heavy metals having originated from different sources. Co-Ni, As-Cd, Cd-Zn, Cu-Pb and As-Pb had high correlations, which alluded to the source of each pair of heavy metals to be similar. A number of studies have reported that high Pearson's correlation values relate to common origin of heavy metals (Qu et al., 2012; Nyika et al., 2019b; Yahaya & Abdu, 2021). Therefore, As, Cd and Pb have common sources, which are largely from anthropogenic activities considering that all sampled areas were cultivated with polluted industrial wastewater from the vicinity. Cobalt and Ni that had the same principal component could be from geogenic sources. The two metals have been found to be dominant in some parent rocks as reported by Spahic et al. (2018).

Table 5. Pearson's correlation matrix for the assayed heavy metal concentrations showing significant relationships between them

Variables	As	Cd	Cr	Co	Cu	Fe	Hg	Mn	Ni	Pb	Zn
As	1										
Cd	**0.89**	1									
Cr	0.41	0.16	1								
Co	-0.03	-0.12	0.73	1							
Cu	0.75	0.60	-0.05	-0.62	1						
Fe	0.60	0.32	-0.08	-0.34	0.77	1					
Hg	0.53	0.65	-0.37	-0.81	0.82	0.37	1				
Mn	0.65	0.55	-0.07	-0.03	0.48	0.81	0.22	1			
Ni	0.17	-0.05	0.81	**0.95**	-0.39	-0.06	-0.74	0.16	1		
Pb	**0.97**	0.82	0.37	-0.17	**0.86**	0.64	0.62	0.56	0.05	1	
Zn	**0.88**	**0.87**	0.34	0.23	0.43	0.48	0.27	0.79	0.36	0.76	1
Values in bold are different from 0 with a significance level alpha=0.05											

Principal component analysis (PCA) was done on observed concentrations to validate their trend. In the analysis, four factor loadings (F1-F4) and respective loading scores for each heavy metal were obtained. The variability in the loading factors was 51.43, 31.58, 10.23 and 6.8% for F1-F4, respectively. However, the first loading factor (F1) whose aggregate contribution to data variation was 51.43% was considered. Obtained results were as shown in Table 6. As, Cd, Cu, Fe, Hg, Mn, Pb and Zn had high loading values that corresponded to significant contribution to the principal components unlike the case in the other factor loadings and in comparison, to Co and Ni. A similar trend was established in a biplot analysis of the active variables (heavy metals) and active observations (sampling sites) comparing F1 and F2 loading values as shown in Figure 2. Heavy metals that were within the same quadrant were most likely from the same origin and had a close interrelationship in their sources.

Table 6. Principal components of assayed heavy metals, their factor loading values and significance in relation to F1

	F1	F2	F3	F4
As	**0.874**	0.101	0.024	0.001
Cd	**0.733**	0.021	0.096	0.150
Cr	0.006	**0.721**	0.140	0.133
Co	0.116	**0.874**	0.000	0.009
Cu	**0.787**	0.092	0.002	0.118
Fe	**0.559**	0.004	0.323	0.114
Hg	**0.521**	0.375	0.099	0.005
Mn	**0.530**	0.045	0.380	0.045
Ni	0.018	**0.958**	0.015	0.009
Pb	**0.887**	0.033	0.042	0.039
Zn	**0.626**	0.250	0.003	0.121
***Values in bold represent loadings with the most significant values**				

High loading values could be associated to the heavy metals in the sampled sites being sourced from similar pollutants just as in the Pearson's correlation coefficient analysis. An analogous trend was established in a PCA of trace metal concentration in agricultural soils of Ethiopia (Duressa & Leta, 2015) and India (Singh et al., 2011). In both studies, high loading values were attributable to similar origin of heavy metals.

Figure 2. A biplot comparison of trends of heavy metals in sampled areas and based on F1 compared to F2 loading values

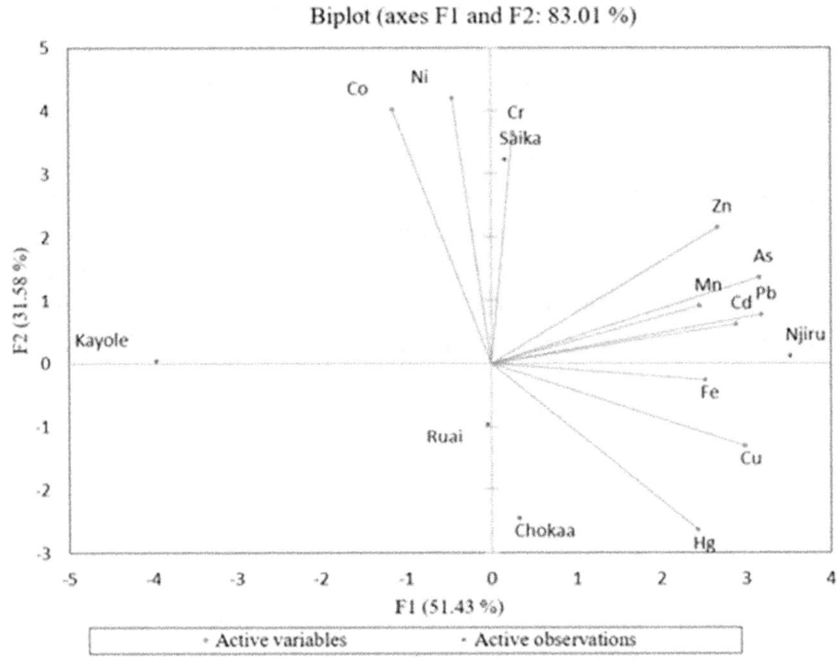

Pollution Assessment in Soils due to Heavy Metals

Computed I_{geo} values of various heavy metals in the sampled sites were as shown in Table 7. From the computed values, pollution by various heavy metals in increasing order was Cu> Pb> Zn> Ni> Cd> Mn>Cr> Co> Hg. In the case of Cu, Cd, Mn, Ni, Pb and Zn, computed I_{geo} values at all sampled sites were not indicative of pollution while in the case of Co and Cr, pollution was moderate. All sampled sites had heavy to extreme contamination from Hg and reported I_{geo} values were >4. High I_{geo} values corresponded to anthropogenic contribution to pollution and while low values alluded to the metal sources to be of lithologic origin (Weissmannova et al., 2019). Heavy to extreme pollution levels corresponded with irrigation of sampled soils with wastewater sourced from chemical industries in the regions. As noted by Qu et al. (2019) in a study on soil contamination in northeast Thailand, Hg can be introduced to soils through chemicals from chlor-alkali industries found in disposed solid wastes and effluents. The results showing elevated levels of Hg differ from a study by Jaworska and Klimek (2021) that reported low contamination from the heavy metal in agricultural soils adjacent to a motorway. Differences in computed I_{geo} values could emanate from variations in the reference values used. Some studies use the mean crust levels, pre-industrial references or mean shale heavy metal content reference values to compute the I_{geo} index (Hakanson, 1980; Wedepohl, 1995). For this research, reference studies were drawn from preexistent studies, which used a mixture of the three reference standards. However, based on the low permissible limits of Hg and its relation to anthropogenic activities in the study area, there is a likelihood that sampled soils were highly contaminated by the metal.

Table 7. Computed geo-accumulation index values for heavy metals assayed from the sampled soils

Heavy Metal	Saika	Kayole	Njiru	Chokaa	Ruai
Cd	0.43	0.30	0.45	0.38	0.38
Cr	1.51	1.08	1.61	1.51	1.20
Co	1.90	1.54	1.56	1.50	1.66
Cu	0.07	0.06	0.09	0.08	0.09
Hg	4.47	4.08	5.15	5.49	4.87
Mn	0.37	0.32	0.69	0.28	0.33
Ni	0.21	0.18	0.17	0.14	0.17
Pb	0.08	0.06	0.09	0.08	0.08
Zn	0.13	0.10	0.14	0.11	0.10

The calculated CF and PLI values for the metals in sampled soils were as shown in Table 8. Moderate to considerable contamination by Cd, Cr, Co, Hg and Mn was reported from the CF values unlike other metals that were indicative of no pollution. Higher values of CF corresponded to anthropogenic sources of the soil pollutants while low values show the geogenic contribution of the metals such as weathering of bedrock sources (Rahmanian & Safari, 2022). Irrigation of areal soils with industrial effluent could be related to elevated CF levels of Cd, Cr, Co and Hg. A related study in northwest Spain made similar conclusions in an evaluation of pollution levels in industrial soils and attributed high CF values for the metals to anthropogenic activities while Mn sources were concluded to be natural (Devanesan et al., 2017).

In Saika and Njiru, pollution was heavy compared to Kayole, Chokaa and Ruai that had moderate pollution based on the computed PLI values and the predetermined classes of the index detailed in Table 2. The results indicated that areal soils of the study area were generally polluted by heavy metals. Industrial activities conducted in the area and subsequent use of the resultant wastewater for irrigation had an impact on each of the assayed heavy metal. The results were similar to studies by Dankoub et al. (2012) and Dahri et al. (2018) who suggested that levels of assayed metals in

Table 8. Computed CF values for assayed heavy metals and PLI values for sampled sites

Metal	Saika	Kayole	Njiru	Chokaa	Ruai
Cd	7.54	5.4	8.06	7.55	6.01
Cr	9.49	7.69	7.81	7.52	8.32
Co	11.44	2.37	1.99	1.69	1.89
Cu	0.38	0.32	0.45	0.42	0.44
Hg	22.33	20.43	25.77	27.43	24.37
Mn	1.83	1.60	3.45	1.38	1.65
Ni	1.01	0.89	0.85	0.70	0.83
Pb	0.43	0.30	0.45	0.40	0.41
Zn	0.63	0.49	0.68	0.54	0.50
PLI	**2.45**	**1.71**	**2.20**	**1.82**	**1.86**

soils are a result of both geogenic and anthropogenic contribution. The computed values ranging from moderate to heavy pollution confirm this supposition. Similarly, Rahmanian and Safari (2022), noted that fluctuations in PLI values are indicative of the role of anthropogenic and geogenic factors in governing the distribution of heavy metals of a particular region.

Values of ErF and RI were computed based on the available geochemical background values of sampled soils and obtained results were as shown in Table 9. High pollution risk was associated with Cd and Hg, whose computed ErF values were >160 at all sampled sites while other metals showed low pollution risk (<40). A similar trend was found in the calculated RI (<100) where low ecological risk was associated with Cr, Ni, Zn and Pb, while Cd and Hg had very high ecological risk (RI >600). The results alluded to Cd and Hg as the leading potential ecological hazards. Similar results (high RI values for Cd and Hg) were found in a study on the potential ecological risk of trace metals in agricultural soils of North China (Yu et al., 2021). The results differ from a related study in Mexico that assessed the ecological risk associated with pollution by heavy metals in soils near an abandoned mine where computed RI values were <150 (Saha et al., 2022). Additionally, the findings show that the study is highly susceptible to contamination by the two metals and the risks they pose to both the surrounding living things and environment is high.

Table 9. Computed ErF and RI values of assayed heavy metals in the sampled soils

Metal	Saika	Kayole	Njiru	Chokaa	Ruai	RI
Cd	226.2	162	241.8	226.5	180.3	**1036.8**
Cr	18.98	15.38	15.62	15.04	16.64	**81.66**
Hg	893.2	817.2	1030.8	1097.2	974.8	**4813.2**
Ni	5.05	4.45	4.25	3.5	4.15	**21.4**
Pb	2.15	1.5	2.25	2.0	2.05	**9.95**
Zn	0.63	0.49	0.68	0.54	0.50	**2.84**

CONCLUSION

In this study, soils of five suburban regions of Nairobi County were assayed for heavy metals and the ecological risk associated with the metals assessed using pollution and ecological indices. All sampling sites showed the presence of the nine assayed heavy metals, which alluded to pollution in the soils. Some of the metals such as As, Cd, Cr, Co, Hg and Mn were above the allowable limits. The discharge of effluents in the soils from industries near the study area and the use of the wastewater for irrigation was attributable to the elevated levels of the heavy metals. The metals could have come from the same origin, which was either anthropogenic or lithologic based on the positive Pearson's correlation and the high loading values after PCA with exception of Co and Ni. Cd and Hg showed the highest pollution and ecological risk based on the computed I_{geo}, CF, ErF and RI indices. All the sampled areas were highly vulnerable to pollution based on the computed PLI values. This empirical evidence demonstrated the need to manage both industrial and municipal effluents prior to release and reuse for irrigation purposes. This is because observed pollution in soils is transferrable to plants through root uptake and consequently, to animals and humans following consumption. National (Kenyan) and local (Nairobi County) regulations on the release and reuse of wastewater should enacted and enforced to prevent extended pollution to soils.

REFERENCES

Abdi, N. (2010). *Availability, transfer and balances of heavy metals in urban agriculture of West Africa.* [PhD Dissertation, University of Kassel, Germany].

Alghobar, M., & Suresha, S. (2017). Evaluation of metal accumulation in soil and tomatoes irrigated with sewage water from Mysore city, Karnataka, India. *Journal of the Saudi Society of Agricultural Sciences, 16*(1), 49–59. doi:10.1016/j.jssas.2015.02.002

Dahri, N., Abdelfattah, A., Manel, E., & Habib, A. (2018). Assessment of streambed sediment contamination by heavy metals: The case of the Gabes Catchment, South-eastern Tunisia. *Journal of African Earth Sciences, 140*, 29–41. doi:10.1016/j.jafrearsci.2017.12.033

Dankoub, Z., Ayoubi, S., Khademi, H., & Lu, S. (2012). Spatial distribution of magnetic properties and selected heavy metals in calcareous soils as affected by land use in the Isfahan region, Central Iran. *Pedosphere, 22*(1), 33–47. doi:10.1016/S1002-0160(11)60189-6

Devanesan, E., Suresh-Gandhi, M., Selvapandiyan, M., Senthilkumar, G., & Ravisankar, R. (2017). Heavy metal and potential ecological risk assessment in sediments collected from Poombuhar to Karaikal Coast of Tamil Nadu using energy dispersive X-ray fluorescence (EDXRF) technique. *Beni-Suef University Journal of Basic and Applied Sciences, 6*(3), 285–292. doi:10.1016/j.bjbas.2017.04.011

Duressa, T., & Leta, S. (2015). Determination of levels of As, Cd, Cr, Hg and Pb in soils and some vegetables taken from Rive Mojo water irrigated farmland at Koka village, Oromia state, East Ethiopia. *International Journal of Sciences: Basic Applied Sciences, 21*(2), 352–372.

Gebeyehu, H., & Bayissa, L. (2020). Levels of heavy metals in soil and vegetables and associated health risks in Mojo area, Ethiopia. *PLoS One, 15*(1), e0227883. doi:10.1371/journal.pone.0227883 PMID:31999756

Hakanson, L. (1980). An ecological risk index for aquatic pollution control: A sedimentological approach. *Water Research, 14*(8), 975–100. doi:10.1016/0043-1354(80)90143-8

IITA. (1982). *Automated and semi-automated methods for soil and plant analysis*. International Institute of Tropical Agriculture.

Jaworska, H., & Klimek, J. (2021). Assessment of the impact of a motorway on content and spatial distribution of mercury in adjacent agricultural soils. *Minerals (Basel), 11*(11), 1221. doi:10.3390/min11111221

Kang, Z., Wang, S., Qin, J., Wu, R., & Li, H. (2020). Pollution characteristics and ecological risk assessment of heavy metals in paddy fields of Fujian province, China. *Scientific Reports, 10*(1), 12244. doi:10.103841598-020-69165-x PMID:32699372

Kinuthia, G., Ngure, V., Beti, D., Lugalia, R., Wangila, A., & Kamau, L. (2020). Levels of heavy metals in wastewater and soil samples from open drainage channels in Nairobi, Kenya: Community health implication. *Scientific Reports*, *10*(1), 8434. doi:10.103841598-020-65359-5 PMID:32439896

Kowalska, J., Mazurek, R., Gasiorek, M., & Zaleski, T. (2018). Pollution indices as useful tools for the comprehensive evaluation of the degree of soil contamination-a review. *Environmental Geochemistry and Health*, *40*(6), 2395–2420. doi:10.100710653-018-0106-z PMID:29623514

Kurt, M. (2018). Comparison of trace element and heavy metal concentrations of top and bottom soils in a complex land use area. *Carpathian Journal of Earth and Environmental Sciences*, *13*(1), 47–56. doi:10.26471/cjees/2018/013/005

Lee, P., Kang, M., Yu, S., & Kwon, Y. (2020). Assessment of trace metal pollution in roof dusts and soils near a large Zn smelter. *The Science of the Total Environment*, *713*, 136536. doi:10.1016/j.scitotenv.2020.136536 PMID:31955082

Lu, X., Gu, A., Huang, C., Wei, Y., Xu, M., Yin, H., & Hu, X.-F. (2021). Assessments of heavy metal pollution of a farmland in an urban area based on the environmental geochemical baselines. *Journal of Soils and Sediments*, *21*(7), 2659–2671. doi:10.100711368-021-02945-8

Lu, Y., Jenkins, A., Ferrier, R., Bailey, M., Gordon, I., Song, S., Huang, J., Jia, S., Zhang, F., Liu, X., Feng, Z., & Zhang, Z. (2015). Addressing China's grand challenge of achieving food security while ensuring environmental sustainability. *Science Advances*, *1*(1), e1400039. doi:10.1126ciadv.1400039 PMID:26601127

Mahey, S., Kumar, R., Sharma, M., Kumar, V., & Bhardwaj, R. (2020). A critical review on toxicity of cobalt and its remediation strategies. *SN Applied Sciences*, *2*(7), 1279. doi:10.100742452-020-3020-9

Malkoc, S., & Yazici, B. (2017). Multivariate analyses of heavy metals in surface soil around an organized industrial area in Eskisehir, Turkey. *Bulletin of Environmental Contamination and Toxicology*, *98*(2), 244–250. doi:10.100700128-016-1991-4 PMID:27942760

Mekki, A., & Sayadi, S. (2017). Study of heavy metal accumulation and residual toxicity in soil saturated with phosphate processing wastewater. *Water, Air, and Soil Pollution*, *228*(6), 215. doi:10.100711270-017-3399-0 PMID:28603317

Miyazawa, M., Pavan, M., Oliveira, E., Ionashiro, M., & Silva, A. (2000). Gravimetric determination of soil organic matter. Brazilian Archives of Biology and Biotechnology, 43(5), 475-478. https://doi.org/ doi:10.1590/S1516-89132000000500005

Mukhopadhyay, S., Masto, R., Tripathi, R., & Srivastava, N. (2019). Application of soil quality indicators for the phyto-restoration of mine spoil dumps. In V. Pandey & K. Bauddh (Eds.), *Phytomanagement of polluted sites* (pp. 361–388)., doi:10.1016/B978-0-12-813912-7.00014-4

Mungai, T., & Wang, J. (2020). Heavy metal pollution in suburban topsoil in Nyeri, Kapsabet, Voi, Ngong and Juja towns in Kenya. *SN Applied Sciences*, *1*(9), 960. doi:10.100742452-019-0996-0

Nyika, J., & Dinka, M. (2022). Heavy metal pollution in soils and vegetables from suburban regions of Nairobi, Kenya and their community health implications. *Pollution*, *8*(4), 1434–1447. doi:10.22059/POLL.2022.341522.1440

Nyika, J., Onyari, E., Dinka, M., & Mishra, S. (2019a). Analysis of particle size distribution of landfill contaminated soils and their mineralogical composition. *Particulate Science and Technology*, *38*(7), 843–853. doi:10.1080/02726351.2019.1635238

Nyika, J., Onyari, E., Dinka, M., & Mishra, S. (2019b). Heavy metal pollution and mobility in soils within a landfill vicinity: A South African case study. *Oriental Journal of Chemistry*, *35*(4), 1286–1296. doi:10.13005/ojc/350406

Nyika, J., Onyari, E., Dinka, M., & Mishra, S. (2020). Comparative assessment of trace metal concentrations and their eco-risk analysis in soils of the vicinity of Roundhill landfill, Southern Africa. *Nature Environment and Pollution Technology*, *19*(2), 539–548. doi:10.46488/NEPT.2020.v19i02.009

Ondayo, M., Simiyu, G., Raburu, P., & Were, F. (2016). Child exposure to lead in the vicinities of informal used lead-acid battery recycling operations in Nairobi slums, Kenya. *Journal of Health & Pollution*, *12*(12), 15–25. doi:10.5696/2156-9614-6.12.15 PMID:30524801

Opiyo, R., Osano, P., Mbandi, A., Apondo, W., & Muhoza, C. (2020). Commentary using citizen science to assess the cumulative risk from air and other pollution sources in informal settlements. *Clean Air Journal*, *30*(1), 1–4. doi:10.17159/caj/2020/30/1.8374

Pikula, D., & Stepien, W. (2021). Effect of the degree of soil contamination with heavy metals on their mobility in the soil profile in a microplot experiment. *Agronomy (Basel)*, *11*(5), 878. doi:10.3390/agronomy11050878

Plunkett, M., & Castle, J. (2010). *Soil organic matter and soil nutrient analysis soil organic matter*. Teagasc, Agriculture and Food Development Authority.

Qu, C., Ma, Z., Yang, J., Liu, Y., Bi, J., & Huang, L. (2012). Human exposure pathways of heavy metals in a lead-zinc mining area, Jiangsu Province, China. *PLoS One*, *7*(11), e46793. doi:10.1371/journal.pone.0046793 PMID:23152752

Qu, R., Han, G., Liu, M., & Li, X. (2019). The mercury behavior and contamination in soil profiles of Mun River basin, Northeast Thailand. *International Journal of Environmental Research and Public Health*, *16*(21), 4131. doi:10.3390/ijerph16214131 PMID:31717757

Rahmanian, M., & Safari, Y. (2022). Contamination factor and pollution load index to estimate source apportionment of selected heavy metals in soils around a cement factory, SW Iran. *Archives of Agronomy and Soil Science*, *68*(7), 903–913. doi:10.1080/03650340.2020.1861252

Saha, A., Gupta, B., Patidar, S., & Martínez-Villegas, N. (2022). Evaluation of Potential Ecological Risk Index of Toxic Metals Contamination in the Soils. Chemistry Proceedings, 10, 59. doi:10.3390/IOCAG2022-12214

Sharma, S., Nagpal, A., & Kaur, I. (2018). Heavy metal contamination in soil, food crops and associated health risks for residents of Ropar wetland, Punjab, India and its environs. *Food Chemistry*, *255*, 15–22. doi:10.1016/j.foodchem.2018.02.037 PMID:29571461

Shen, F., Mao, L., Sun, R., Du, J., Tan, Z., & Ding, M. (2019). Contamination evaluation and source identification of heavy metals in the sediments from the Lishui river watershed, Southern China. *International Journal of Environmental Research and Public Health*, *16*(3), 336. doi:10.3390/ijerph16030336 PMID:30691076

Shrivastava, A., Barla, A., Singh, S., Mandraha, S., & Bose, S. (2017). Arsenic contamination in agricultural soils of Bengal deltaic region of West Bengal and its higher assimilation in monsoon rice. *Journal of Hazardous Materials, 324*(Part B), 526–534. doi:10.1016/j.jhazmat.2016.11.022

Singh, V., Agrawal, H., Joshi, G., Sudershan, M., & Sinha, A. (2011). Elemental profile of agricultural soil by the EDXRF technique and use of the Principal Component Analysis (PCA) method to interpret the complex data. *Applied Radiation and Isotopes*, *69*(7), 969–974. doi:10.1016/j.apradiso.2011.01.025 PMID:21377884

Spahic, M., Sakan, S., Trbic, G., Tancic, P., Skrivanj, S., Kovacevic, J., & Manojlovic, D. (2018). Natural and anthropogenic sources of chromium, nickel and cobalt in soils impacted by agricultural and industrial activity (Vojvodina, Serbia). *Journal of Environmental Science and Health. Part A, Toxic/Hazardous Substances & Environmental Engineering*, *54*(3), 219–230. doi:10.1080/10934529.2018.1544802 PMID:30587075

Tomlinson, D., Wilson, J., Harris, C., & Jefrey, D. (1980). Problems in the assessment of heavy-metal levels in estuaries and the formation of a pollution index. *Helgoländer Wissenschaftliche Meeresuntersuchungen*, *33*(1), 566–575. doi:10.1007/BF02414780

Tomno, R., Nzeve, J., Mailu, S., Shitanda, D., & Waswa, F. (2020). Heavy metal contamination of water, soils and vegetables in urban streams in Machakos municipality, Kenya. *Scientific African*, *9*, e00539. doi:10.1016/j.sciaf.2020.e00539

United States-Environmental Protection Agency. US-EPA. (2014). Method 6020B: Inductively coupled plasma-mass spectrometry. Environmental Protection Agency, Washington DC, USA.

Wedepohl, K. (1995). The composition of the continental crust. *Geochimica et Cosmochimica Acta*, *59*(7), 1217–1232. doi:10.1016/0016-7037(95)00038-2

Wieczorek, J., & Baran, A. (2022). Pollution indices and biotests as useful tools for the evaluation of the degree of soil contamination by trace elements. *Journal of Soils and Sediments*, *22*(2), 559–576. doi:10.100711368-021-03091-x

Yahaya, S., & Abdu, F. (2021). Ecological risk assessment of heavy metals-contaminated soils of selected villages in Zamfara state, Nigeria. *SN Applied Sciences*, *3*(2), 168. doi:10.100742452-021-04175-6

Yu, L., Zhang, F., Zang, K., He, L., Wan, F., Liu, H., Zhang, X., & Shi, Z. (2021). Potential ecological risk assessment of heavy metals in cultivated land based on soil geochemical zoning: Yishui County, North China Case Study. *Water (Basel)*, *13*(23), 3322. doi:10.3390/w13233322

Zahra, A., Hashmi, Z., Malik, N., & Ahmed, Z. (2014). Enrichment and geo-accumulation of heavy metals and risk assessment of sediments of the Kurang Nallah-Feeding tributary of the Rawal Lake Reservoir, Pakistan. *The Science of the Total Environment*, *470–471*, 925–933. doi:10.1016/j.scitotenv.2013.10.017 PMID:24239813

Zeljka, Z., Branka, G., Aleksandra, P., Vlatka, J., Lola, G., & Nada, M. (2018). Comparison of two different CEC determination methods regarding the soil properties. *ACS. Agriculturae Conspectus Scientificus*, *84*(2), 151–158.

Zhong, T., Chen, D., & Zhang, X. (2016). Identification of potential sources of mercury (Hg) in farmlands soils using a decision tree method in China. *International Journal of Environmental Research and Public Health*, *13*(11), 1111. doi:10.3390/ijerph13111111 PMID:27834884

Chapter 7
Soil Pollution by Lead in Sub-Saharan Africa

ABSTRACT

This chapter discusses the anthropic sources of lead and its subsequent pollution to soils of sub-Saharan Africa (SSA) region. Using examples of east, west, and south African empirical studies, industrial activities characterized by cottage industries involved in paint use, electrical appliance and motor vehicle repair, hairdressing, and scrap metal recycling were associated with Pb introduction in soils leading to their contamination. The use of leaded petrol and manufacture and recycling of Pb-based batteries were additional causes of metal pollution in soils. The contaminant is introduced to soils via Pb-containing dusts, emissions, particulate matter, solid wastes, and effluents. Subsequent transfer to trophic levels is via ingestion, inhalation, and consumption of contaminated soils and food crops. It is essential to manage the sources of Pb to control and manage its accumulation in soil.

INTRODUCTION

Environmental pollution by heavy metals is on a growing trend in an era where the manufacturing and agricultural sectors are seeking to advance for economic development and dominance. With the advances, a lot of wastes are introduced in soils from mining and metallurgical processes as well as from using agrochemicals and in the chemical manufacturing sectors resulting to pollution. The disposal of such wastes is non-scientific due to lack of pre-treatment, sorting and separation prior to disposal, limited posttreatment facilities and equipment and lack of engineered

DOI: 10.4018/978-1-6684-7116-6.ch007

landfill disposal sites (O'Shea et al., 2021). Consequently, soils around the globe are vulnerable to heavy metal pollution.

The industrialization trend has led to the introduction of lead (Pb) in soils. The heavy metal found in nature and has no significant biological function is used widely in manufacture of batteries, paints, crystal glasses and electronics. Smelting and mining activities also introduce Pb in soils resulting to pollution of land resources (Bret et al., 2019). It has therefore been noted that many companies using Pb as a raw material such as smelting sites, battery producing and recycling firms have the soils of their surrounding having elevated levels of the pollutant (Gottesfeld et al., 2018). The metal is introduced to soils in the form of solid wastes, emissions from various processing companies, mine slag and tailings, dusts and particulate matter from the various industrial activities (Sun et al., 2022). The wastes can be leached to soils in the presence of water or deposited on the soil surfaces where they can be taken up by plants or introduced to groundwater systems. Lead pollution in soils is also associated with combustion of leaded gasoline (Bidar et al., 2020).

Although it is difficult to distinguish lead pollution from pedo-chemical effects with that of human activities, the entry and bioaccumulation of the metal in food chains through soil pollution is associated with a number of effects. These include elevated levels of the metal in blood and lead poisoning from soils polluted with Pb in and around mining areas (Shiomi, 2015), encephalopathy, brain dysfunctions among children and heart diseases among adults (Bret et al., 2019). Apart from inducing organ toxicity in humans, the metal acts as a neurobehavioral inhibitor, which induces performance and cognitive development damages (Markus & McBratney, 2001; O'Shea et al., 2021). The effects are associated with lead's ability to persist in the environment and in particular, soils compared to other media such as water and air (Hansson et al., 2019). According to Boldyrev (2018), the metal can remain in soils for hundreds to thousands of years without being degraded. In Africa and particularly, the sub-Saharan Africa (SSA) region, Pb pollution is on the rise as many nations seek to advance industrialization and technology with minimal attention on scientific solid waste and wastewater management approaches. Consequently, extensive pollution (formal and informal) is suspected in soils of the region though minimal testing has been done to validate the supposition (Gottesfeld et al., 2018). This chapter therefore seeks to enhance understanding on soil pollution by lead using the SSA case study. The characteristics and history on the use of lead in general will be evaluated prior to exploring the causes and effects of Pb pollution in soils of the region using a literature review study approach.

PROPERTIES OF LEAD

Atomic lead, which is found in the carbon group of the periodic table has 82 electrons whose configuration is $4f^{14}5d^{10}6s^{2}6p^{2}$. The pure form of the metal is bright and silver bluish but tarnishes when in contact with air to have a dull hue (Figure 1). Additionally, lead is resistant to corrosion, soft, ductile, malleable and has high density (Boldyrev, 2018). It has a boiling and melting point of 1749 and 327.5 °C, respectively. The metal consists of five short lived radioisotopes and four stable isotopes of masses 204, 206, 207 and 208). The metal has two oxidation states of +2 and +4 forming inorganic compounds of Pb(II) and Pb(IV), respectively. Lead occurs on earth in combination with sulfur in the mineral ore galena (PbS) although other forms such as anglesite ($PbSO_4$), boulangerite ($Pb_5Sb_4S_{11}$) and cerussite ($PbCO_3$) occur in smaller amounts (Boldyrev, 2018).

Figure 1. The physical characteristics of lead ore
(Boldyrev, 2018)

HISTORY OF LEAD USE AND CAUSES OF LEAD POLLUTION IN SOILS

Since the ancient Rome, Pb was used to manufacture products such as pipes, tubes and tableware. The trend was associated with the ability to manipulate the metal based on its malleability, corrosion resistance and low melting point characteristics (Shiomi, 2015). The metal was also used in making warfare materials (sling bullets), coins and in roofing (Boldyrev, 2018). Before the industrial revolution, Pb was mined in Congo Basin and South Africa regions of SSA, smelted and used as currency. Today, South Africa is one of the world's largest producers of Pb, with an output of about 40 tons in 2016 alone (Boldyrev, 2018).

It is in the industrial revolution era where extensive use of the metal for painting (lead paints) and in plumbing in many regions of the globe occurred. With extended use, Pb poisoning progressively increased. The 1970s also saw an increase in environmental pollution by Pb since the metal was used in gasoline. In particular, tetramethyl lead and/or tetraethyl lead were and still are used to protect car engines (Shiomi, 2015). Vehicles using leaded gasoline released volatile organic Pb, which is dispersed in the atmosphere, precipitated and deposited to further exacerbate soil and air pollution. In response to the potential of Pb-based gasoline to induce soil and air pollution, countries such as Japan and the USA, banned its use and modified their energy to Pb-free high-octane and alkylate gasoline. Despite these considerations to prevent Pb exposure by gasoline, developing countries mainly drawn from SSA and Asia are still using lead-based gasoline, which makes their environment, particularly soils vulnerable to pollution. In South Africa (Rollin et al., 2017) and West Africa (Obeng-Gyasi, 2022), use of and adoption of unleaded petrol remains elusive in preference to leaded petrol despite the foreknown environmental impacts of the latter. In SSA region as a whole, the use of leaded petrol was affiliated with Pb- contamination in soils (Anyanwu et al., 2018). Similarly, the use of Pb pipes to transport water has been banned in countries such as Japan and Uruguay since the 1970s (Shiomi, 2015). The trend was associated with the potential to introduce the metal to soils during irrigation or to human systems as a result of water consumption. Lead pipes however are still being used in many SSA regions such as West Africa for plumbing activities exposing the environment to potential contamination by the metal (Seltenrich, 2021; Obeng-Gyasi, 2022).

The use of Pb in SSA is growing and large users include manufacturers of lead-acid batteries used in industries and cars, production of inorganic chemicals including paints, crystal glass and polyvinyl chloride stabilizers. In soldering of ceramic glazes and electronic materials such as lagging materials used in underground cables, plating medical equipment and production of exhausting and draining tubes, the metal is also used (Farias et al., 2014). Smelting and mining are additional Pb poisoning sources. The recycling of Pb batteries is also a growing cause of environmental pollution. In SSA countries of Ghana, Nigeria, Mozambique, Cameroon, Kenya and Tanzania, battery recycling has been associated with soil pollution (Gottesfeld & Pokhrel, 2011; Gottesfeld et al., 2018). During the industrial processes, solid wastes, effluents, particulate matter and dust that contain Pb are produced and deposited in soils. The metal is thereafter introduced to food chains where it has harmful effects to humans.

LEAD POLLUTION IN SOILS OF SUB-SAHARAN AFRICA

Common sources of Pb in soils of SSA region are as shown in Figure 2. The focus in this chapter was on anthropogenic rather than lithologic sources of Pb. The human-based sources are mainly propagated by industrial activities such as the production and use of lead-based paints, water pipes, ceramics, cans, petrol and cosmetics (Obeng-Gyasi, 2022). Lead poisoning of soils also results from processes in cottage industries (World Health Organization, 2015, 2014). Mathee (2014) and Mathee et al. (2020) noted that cottage industries in SSA and more prevalent in urban South Africa, which engaged in recycling of scrap metal, vehicle spray painting, hair dressing, repair of motor vehicles and electrical appliances and painting services introduced Pb to the environment leading to soil pollution.

Figure 2. Common sources of lead

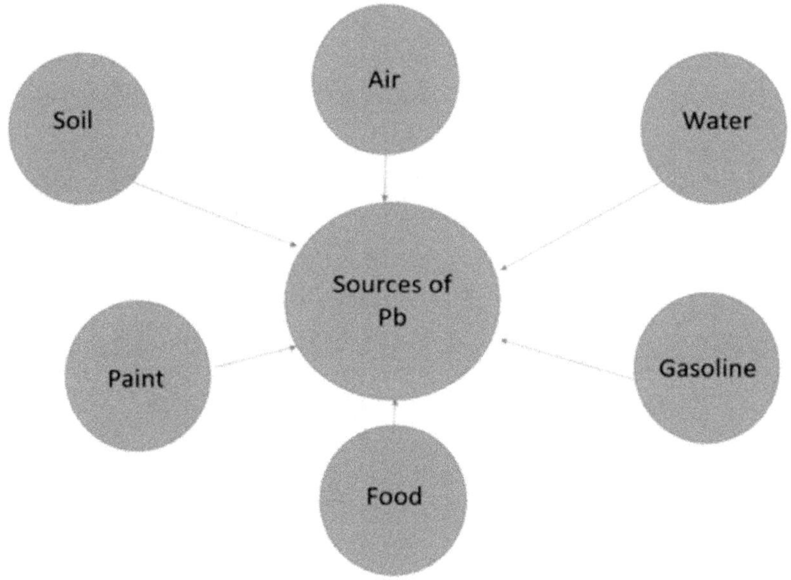

In majority of African countries, mostly drawn from SSA, Pb exposure is a growing concern as noted by WHO (2015) in a self-assessment excise. In African including SSA countries shown in Figure 3, public health concern on Pb was due to sources such as occupational exposure, petrol, toys and paints use as well as mining of the heavy metal and resultant wastes that were released to the environment. In various regions of SSA, exposure to Pb from varied sources and to soils is discussed in subsequent sections.

Figure 3. Countries in SSA that perceive lead as a public health concern
(Farias et al., 2014; WHO, 2014)

Lead Pollution in Soils of West Africa

Environmental pollution (air, soil and water pollution) by Pb has been reported in several countries of West Africa as detailed by Obeng-Gyasi (2022). Soil pollution in particular was reported in Benin (Bodeau-Livinec et al., 2016). Children and mothers ingested soils through eating (geophagia) and consequently, their blood had elevated levels of the metal. In another study in Burkina Faso, moderate soil pollution by Pb was reported and associated with disposal of mine tailings in the soils (Kagambega et al., 2014). In a battery recycling plant in Tema (Ghana), Pb levels were elevated in soils and subsequently, levels were found elevated in the blood of children who had been exposed to the metal via soil consumption or inhalation of polluted dust (Aboh et al., 2013). The use of leaded gasoline and its subsequent combustion in motor vehicles among other industrial activities in Warri, Nnewi, Port Harcourt, Aba, Onitsh, Enugu and Awka towns of Nigeria was attributable to Pb pollution of surface soils and roadside dust particles that were laden with Pb (Nduka & Orisakwe, 2010). Soils sampled near newly painted buildings in the cities of Ibadan and Lagos, Nigeria showed elevated levels of Pb compared to control samples due to paint spillage on them and deposition of Pb-laden particulate matter (Adeyi & Babalola, 2017). Soils in the vicinity of Mbeubeuss landfill of Senegal were found to be polluted by Pb (Cabral et al., 2012). The observation was attributed to leakage of Pb-containing leachate to soils from the facility. Additional exposure to the pollutant in children was through dust inhalation and ingestion of soils in the vicinity, which was associated with biomarkers of kidney injury. Other SSA studies focusing on soil pollution by lead and the sources of the metal are summarized in Table 1.

Table 1. Sources of lead pollution in soils of SSA countries and pathways of exposure to affected population

Country	Source of Lead in Soils	Exposure Pathway	Reference
West Africa			
Benin	-Unscientific solid waste and effluent disposal and management -Use of Pb-based petrol	Ingestion of soils and sediments	Yehouenou et al., 2013 Koumolou et al., 2013
Ghana	-Use of leaded petrol, recycling of Pb batteries, steel processing, disposal of Pb-based solid waste, scrap yard processes	Inhalation of particulate matter	Kylander et al., 2003; Aboh et al., 2013; Itai et al., 2014
Nigeria	-Use of leaded petrol	Inhalation of roadside dust	Amusan et al., 2003; Fakayode & Olu-Owalabi, 2003; Okunola et al., 2007

continues on following page

Table 1. Continued

Country	Source of Lead in Soils	Exposure Pathway	Reference
Nigeria	-Mining and Pb smelting	Ingestion and inhalation	Eze & Chukwe, 2011 Plumlee et al., 2013; Bartrem et al., 2014
Nigeria	-Disposal of electronic waste in open dumpsites	Ingestion	Olafisoye et al., 2013
Togo	-Mining of pyrite	Inhalation of dust in and near abandoned mine pits	Gnandi & Tobschall, 2002
South Africa			
Botswana	-Use of leaded petrol	Ingestion of Pb-poisoned roadside dust	Mbongwe et al., 2010
Namibia	-Mining of pyrite at Rosh Pinah mines	Inhalation of particulate matter	Kribek et al., 2014
South Africa	-Use of leaded petrol	Inhalation of roadside dust particles	Olowoyo et al., 2010
South Africa	-Production of lead-based paints, used of leaded petrol and mining activities	Inhalation of lead-poisoned particulate matter	De Villiers et al., 2010
South Africa	-Unscientific disposal of solid waste in open dumpsites.	Ingestion of Pb-containing plants and soils	Olowoyo et al., 2012
South Africa	-Use of lead-based petrol	Inhalation and ingestion of particulate matter with Pb	Okonkwo & Maribe, 2004
South Africa	-Cottage industries in Johannesburg that use Pb as a raw material or Pb-based products	Inhalation of particulate matter, emissions and dust	Mathee, 2014 Mathee et al., 2020
Zambia	-Use of lead-based petrol	Inhalation of roadside dust	Makondo et al., 2013
Zambia	-Mining activities at Kabwe mines	Inhalation and ingestion of particulate matter	Tembo et al., 2006; Ikenaka et al., 2010; Nakayama et al., 2011
Zambia	-Mining at the Copperbelt region	Inhalation of dust particles	Mihaljevic et al., 2011; Ettler et al., 2012
Eastern Africa			
Kenya Uganda Tanzania	-Use of leaded petrol	Ingestion of roadside dust and particulate matter	Makokha et al., 2011
Democratic Republic of Congo	-Geophagia among pregnant women, battery recycling and use of lead-based petrol in urban and rural pollution	Ingestion of Pb-poisoned soils and inhalation of polluted dust	Tuakuila et al., 2013
Kenya	-Improper management of electronic waste	Ingestion and inhalation of contaminated soils	Mungai & Wang, 2019
Kenya	-Use of Pb paints and improper management of solid waste	Ingestion of lead-containing dust and paint chips and inhalation of dust from aging surfaces that are painted	Owino et al., 2015
Tanzania	-Used of leaded petrol	Inhalation of roadside dust laden with Pb	Luilo & Othman, 2006

Lead Pollution in Soils of Southern Africa

Lead pollution in soils of southern Africa is also widely reported and is a result of the growing mining industry in the region. Exposure to lead poisoning in the country is at the formal and informal sectors. Mining and smelting of Pb-Zn ores in Kabwe, Zambia, which began operations in early eighteenth century has seen areal soils heavily polluted by the metals (Yabe et al., 2015). The situation was worsened by the closure of the mines in mid-1990s without taking precautionary measures to prevent extended pollution (WHO, 2014, 2015). Southern Africa region is rated as the top-ten most polluted globally and studies have shown elevated levels of Pb in children blood following ingestion of the polluted soils and inhalation of Pb-containing dusts (Blacksmith Institute, 2015; Yabe et al., 2013). In South Africa, the mining and smelting of Pb in the North West province was associated with soil pollution with the metal and transference of the effects to surrounding communities who had elevated levels of the metal in blood (Naicker & Mathee, 2015). In addition to mining activities, the use of Pb-based petrol and paints has been associated with soil pollution by Pb (Obeng-Gyasi, 2022). Other studies focusing on Pb pollution in soils of Southern African countries are summarized in Table 1.

Lead Pollution in Soils of Eastern Africa

Sources of Pb in soils of Eastern Africa are divergent and range from mining, use of leaded petrol and paints as well as disposal of electronic wastes. Soils of abandoned small-scale gold mines in Tanzania were quantified and found to have elevated Pb concentrations (Nyanza et al., 2014). Geophagic ingestion of such soils by pregnant women resulted to ill health due to Pb toxicity. In urban soils of Uganda exposed to Pb contamination from solid waste and industrial effluents, Pb pollution was detected and subsequently transferred to vegetables grown on affected soils (Nabulo et al., 2010). A similar trend was reported in Kenyan suburban regions where soils exposed to untreated industrial wastewater were polluted with Pb and the heavy metal was taken up by kales from the affected soils (Nyika & Dinka, 2022). East African countries of Kenya and Tanzania among other SSA countries were found to have 22-fold Pb polluted soils at sites of Pb battery manufacturing compared to control sites (Gottesfeld et al., 2018). The levels in soils were also beyond the permissible levels and posed as an environmental pollution threat. In Ethiopian towns, roadside soils were polluted with Pb as a result of high traffic density and the use of leaded gasoline (Teju et al., 2014). In other studies of the region, exposure to Pb was a result of geophagy of contaminated soils among pregnant women and children, unscientific management of solid waste, mining and smelting activities as summarized in Table 1.

CONCLUSION

The focus of the chapter was to evaluate soil pollution by Pb with the focus on SSA region. Findings showed that Pb is a growing environmental concern and a main causative of soil pollution. The concern is prompted and worsened by industrial activities including production and recycling of Pb-based batteries, mining, production and use of paints, smelting, glazing, use of leaded petrol and improper management of electronic solid waste. Specific empirical studies in west, eastern and southern Africa regions confirming soil pollution by the metal were evaluated and discussed. The exposure pathways of the heavy metal were found to be via ingestion of polluted soils and inhalation of particulate matter containing Pb from roadside soils or at workplaces. Geophagia and plant root uptake have been pointed out as common entry points of Pb to living things. With the potential of the metal to enter food chains (plants, animals and humans) via uptake and consumption of polluted soils and foods and cause negative health effects, there is need to manage the causatives of metal release to prevent extensive pollution spread to other resources.

REFERENCES

Aboh, K., Sampson, M., Nyaab, L., Caravanos, J., Ofosu, F., & Kuranchie-Mensah, H. (2013). Assessing levels of lead contamination in soil and predicting pediatric blood lead levels in Tema, Ghana. *Journal of Health & Pollution*, *3*(5), 7–12. doi:10.5696/2156-9614-3.5.7

Adeyi, A., & Babalola, B. (2017). Lead and cadmium levels in residential soils of Lagos and Ibadan, Nigeria. *Journal of Health & Pollution*, *7*(13), 42–55. doi:10.5696/2156-9614-7-13.42 PMID:30524813

Amusan, A., Bada, S., & Salami, A. (2003). Nigeria. *West African Journal of Applied Ecology*, *4*, 107–114.

Anyanwu, B., Ezejiofor, A., Igweze, Z., & Orisakwe, O. (2018). Heavy metal mixture exposure and effects in developing nations: An update. *Toxics*, *6*(4), 65. doi:10.3390/toxics6040065 PMID:30400192

Bartrem, C., Tirima, S., Lindern, I., Braun, M., Worrell, M., & Anka, S. (2014). Unknown risk: Co-exposure to lead and other heavy metals among children living in small-scale mining communities in Zamfara State, Nigeria. *International Journal of Environmental Health Research*, *24*(4), 304–319. doi:10.1080/09603123.2013.835028 PMID:24044870

Bidar, G., Pelfrene, A., Schwartz, C., Waterlot, C., Sahmer, K., Marot, F., & Douay, F. (2020). Urban kitchen gardens: Effect of the soil contamination and parameters on the trace element accumulation in vegetables-a review. *The Science of the Total Environment*, *738*, 139569. doi:10.1016/j.scitotenv.2020.139569 PMID:32516675

Blacksmith Institute. (2015). *Kabwe lead mines, 2015*. Blacksmith Institute. http:// www. blacksmithinstitute.org/projects/display/3

Bodeau-Livinec, F., Glorennec, P., Cot, M., Dumas, P., Durand, S., Massougbodji, A., Ayotte, P., & Le Bot, B. (2016). Elevated blood lead levels in infants and mothers in Benin and potential sources of exposure. *International Journal of Environmental Research and Public Health*, *13*(3), 316. doi:10.3390/ijerph13030316 PMID:26978384

Boldyrev, M. (2018). Lead: Properties, history and applications. *WikiJournal of Science*, *1*(2), 7. doi:10.15347/wjs/2018.007

Bret, E., Otieno, V., Nganga, C., Fort, J., & Taylor, M. (2019). Assessment of the presence of soil lead contamination near a former lead smelter in Mombasa, Kenya. *Journal of Health & Pollution*, *9*(21), 190307. doi:10.5696/2156-9614-9.21.190307 PMID:30931167

Cabral, M., Dieme, D., Verdin, A., Garcon, G., Fall, M., Bouhsina, S., Dewaele, D., Cazier, F., Tall-Dia, A., Diouf, A., & Shirali, P. (2012). Low-level environmental exposure to lead and renal adverse effects: A cross-sectional study in the population of children bordering the Mbeubeuss landfill near Dakar, Senegal. *Human and Experimental Toxicology*, *31*(12), 1280–1291. doi:10.1177/0960327112446815 PMID:22837546

De Villiers, S., Thiart, C., & Basson, N. (2010). Identification of sources of environmental lead in South Africa from sur - face soil geochemical maps. *Environmental Geochemistry and Health*, *32*(5), 451–459. doi:10.100710653-010-9288-8 PMID:20848346

Ettler, V., Kribek, B., Majer, V., Knesl, I., & Mihaljevic, M. (2012). Differences in the bioaccessibility of metals/metalloids in soils from mining and smelting areas (Copperbelt, Zambia). *Journal of Geochemical Exploration*, *113*, 68–75. doi:10.1016/j.gexplo.2011.08.001

Ezeh, B., & Chukwe, G. (2011). Small scale mining and heavy metals pollution of agricultural soils: The case of Ishiagu Mining District, South Eastern Nigeria. *Journal of Geology and Mining Research*, *3*, 87–104.

Fakayode, S., & Olu-Owolabi, B. (2003). Heavy metal contamination of roadside topsoil in Osogbo, Nigeria: Its relationship to traffic density and proximity to highways. *Environmental Geology (Berlin)*, *44*(2), 150–157. doi:10.100700254-002-0739-0

Farias, P., Alamo-Hernandez, U., Mancilla-Sanchez, L., Texcalac-Sangrador, J., & Carrizales-Yanez, L. (2014). Lead in school children from Morelos, Mexico: Levels, sources and feasible interventions. *International Journal of Environmental Research and Public Health*, *12*(11), 12668–12682. doi:10.3390/ijerph111212668 PMID:25493390

Gnandi, K., & Tobschall, H. (2002). Heavy metals distribution of soils around mining sites of cadmium-rich marine sedimentary phosphorites of Kpogamé and Hahotoé (southern Togo). *Environmental Geology (Berlin)*, *41*(5), 593–600. doi:10.1007002540100425

Gottesfeld, P., & Pokhrel, K. (2011). Lead exposure in battery manufacturing and recycling in developing countries and among children in nearby communities. *Journal of Occupational and Environmental Hygiene*, *8*(9), 520–532. doi:10.1080/15459624.2011.601710 PMID:21793732

Gottesfeld, P., Were, F., Adogame, L., Gharbi, S., San, D., Nota, M., & Kuepouo, G. (2018). Soil contamination from lead battery manufacturing and recycling in seven Africa countries. *Environmental Research*, *161*, 609–614. doi:10.1016/j.envres.2017.11.055 PMID:29248873

Hansson, S., Grusson, Y., Chimienti, M., Claustres, A., Jean, S., & Le Roux, G. (2019). Legacy Pb pollution in the contemporary environment and its potential bioavailability in three mountain catchments. *The Science of the Total Environment*, *671*, 1227–1236. doi:10.1016/j.scitotenv.2019.03.403

Ikenaka, Y., Nakayama, S., Muroya, T., Yabe, J., Konnai, S., Darwish, W., Muzandu, K., Choongo, K., Mainda, G., Teraoka, H., Umemura, T., & Ishizuka, M. (2012). Effects of environmental lead contamination on cattle in a lead/zinc mining area: Changes in cattle immune systems on exposure to lead in vivo and in vitro. *Environmental Toxicology and Chemistry*, *31*(10), 2300–2305. doi:10.1002/etc.1951 PMID:22821446

Itai, T., Otsuka, M., Asante, K., Muto, M., Ankomah, Y., Asare, O., & Tanabe, S. (2014). Variation and distribution of metals and metalloids in soil/ash mixtures from Agbogbloshie e-waste recycling site in Accra, Ghana. *The Science of the Total Environment*, *470-471*, 707–716. doi:10.1016/j.scitotenv.2013.10.037 PMID:24184547

Kagambega, N., Sawadogo, S., Bamba, O., Zombre, P., & Galvez, R. (2014). Acid mine drainage and heavy metals contamination of surface water and soil in southwest Burkina Faso–West Africa. *International Journal of Multidisciplinary Academic Research, 2*, 9–19.

Koumolou, L., Edorh, P., Montcho, S., Aklikokou, K., Loko, F., Boko, M., & Creppy, E. E. (2013). Health-risk market garden production linked to heavy metals in irrigation water in Benin. *Comptes Rendus Biologies, 336*(5-6), 278–283. doi:10.1016/j.crvi.2013.04.002 PMID:23916203

Kribek, B., Majer, V., Pasava, J., Kamona, F., Mapani, B., Keder, J., & Ettler, V. (2014). Contamination of soils with dust fallout from the tailings dam at the Rosh Pinah area, Namibia: Regional assessment, dust dispersion modelling and environmental consequences. *Journal of Geochemical Exploration, 144*, 391–408. doi:10.1016/j.gexplo.2014.01.010

Kylander, M., Rauch, S., Morrison, G., & Andam, K. (2003). Impact of automobile emissions on the levels of platinum and lead in Accra, Ghana. *Journal of Environmental Monitoring, 5*(1), 91–95. doi:10.1039/b211736c PMID:12619761

Luilo, G., & Othman, O. (2006). Lead pollution in urban roadside environment of Dar es Salaam city. *Tanzania Journal of Science, 32*(20), 61–67.

Makokha, O., Mghweno, L., Magoha, H., Nakajugo, A., & Wekesa, J. (2011). The effects of environmental lead pollution in Kisumu, Mwanza and Kampala. *The Open Environmental Engineering Journal, 4*, 133–140. doi:10.2174/1874829501104010133

Makondo, C., Mundike, J., & Mwaanga, P. (2013). Lead deposition from mobile sources: A case study of Ndola-Kitwe dual carriage highway. *American Journal of Environmental Protection, 2*(6), 128–133. doi:10.11648/j.ajep.20130206.12

Markus, J., & McBratney, A. (2001). A review of the contamination of soil with lead II. Spatial distribution and risk assessment of soil lead. *Environment International, 27*(5), 399–411. doi:10.1016/S0160-4120(01)00049-6 PMID:11757854

Mathee, A. (2014). Towards the prevention of lead exposure in South Africa: Contemporary and emerging challenges. *Neurotoxicology, 45*, 220–223. doi:10.1016/j.neuro.2014.07.007 PMID:25086205

Mathee, A., Street, R., Teare, J., & Naicker, N. (2020). Lead exposure in the home environment: An overview of risks from cottage industries in Africa. *Neurotoxicology, 81*, 34–39. doi:10.1016/j.neuro.2020.08.003 PMID:32835764

Mbongwe, B., Barnes, B., Tshabang, J., Zhai, M., Rajoram, S., & Mpuchane, S. (2010). Exposure to lead among children aged 1-6 years in the city of Gaborone, Botswana. *Journal of Environmental Health Research*, *10*, 17–26. doi:10.13140/RG.2.2.17136.20484

Mihaljevic, M., Ettler, V., Sebek, O., Sracek, O., Kribek, B., Kyncl, T., Majer, V., & Veselovský, F. (2011). Lead isotopic and metallic pollution record in tree rings from the Copperbelt mining-smelting area, Zambia. *Water, Air, and Soil Pollution*, *216*(1-4), 657–668. doi:10.100711270-010-0560-4

Mungai, T., & Wang, J. (2019). Heavy metal pollution in suburban topsoil of Nyeri, Voi, Ngong and Juja towns in Kenya. *SN Applied Sciences*, *1*(9), 960. doi:10.100742452-019-0996-0

Nabulo, G., Young, C., & Black, C. (2010). Assessing risk to human health from tropical leafy vegetables grown on contaminated urban soils. *The Science of the Total Environment*, *408*(22), 5338–5535. doi:10.1016/j.scitotenv.2010.06.034 PMID:20739044

Naicker, N., & Mathee, A. (2015). Trends in lead exposure in a rural mining town in South Africa, 1991-2008. *South African Medical Journal*, *105*(7), 515. doi:10.7196/SAMJnew.7809 PMID:26447246

Nakayama, S., Ikenaka, Y., Hamada, K., Muzandu, K., Choongo, K., Teraoka, H., Mizuno, N., & Ishizuka, M. (2011). Metal and metalloid contamination in roadside soil and wild rats around a Pb-Zn mine in Kabwe, Zambia. *Environmental Pollution*, *59*(1), 175–181. doi:10.1016/j.envpol.2010.09.007 PMID:20971538

Nduka, J., & Orisakwe, O. (2010). Assessment of environmental distribution of lead in some municipalities of South-Eastern Nigeria. *International Journal of Environmental Research and Public Health*, *7*(6), 2501–2513. doi:10.3390/ijerph7062501 PMID:20644686

Nyanza, E., Joseph, M., Premji, S., Thomas, D., & Mannion, C. (2014). Geophagy practices and the content of chemical elements in the soil eaten by pregnant women in artisanal and small scale gold mining communities in Tanzania. *BMC Pregnancy and Childbirth*, *14*(1), 144. doi:10.1186/1471-2393-14-144 PMID:24731450

Nyika, J., & Dinka, M. (2022). Heavy metal pollution in soils and vegetables from suburban regions of Nairobi, Kenya and their community health implications. *Pollution*, *8*(4), 1434–1447. doi:10.22059/POLL.2022.341522.1440

O'Shea, M., Toupal, J., Caballero-Gómez, H., McKeon, T., Howarth, M., Pepino, R., & Gieré, R. (2021). Lead Pollution, Demographics, and Environmental Health Risks: The Case of Philadelphia, USA. *International Journal of Environmental Research and Public Health*, *18*(17), 9055. doi:10.3390/ijerph18179055 PMID:34501644

Obeng-Gyasi, E. (2022). Sources of lead exposure in West Africa. *Science*, *4*(3), 33. doi:10.3390ci4030033

Okonkwo, J., & Maribe, F. (2004). Assessment of lead exposure in Thohoyandou, South Africa. *The Environmentalist*, *24*(3), 171–178. doi:10.100710669-005-6051-2

Okunola, J., Uzairu, A., & Ndukwe, G. (2007). Levels of trace metals in soil and vegetation along major and minor roads in metropolitan city of Kaduna, Nigeria. *African Journal of Biotechnology*, *6*, 1703–1709.

Olafisoye, O., Adefioye, T., & Osibote, O. (2013). Heavy metals contamination of water, soil and plants around an electronic waste dumpsite. *Polish Journal of Environmental Studies*, *22*, 1431–1439.

Olowoyo, J., Okedeyi, O., Mkolo, N., Lion, G., & Mdakane, S. (2012). Uptake and translocation of heavy metals by medicinal plants growing around a waste dump site in Pretoria, South Africa. *South African Journal of Botany*, *78*, 116–121. doi:10.1016/j.sajb.2011.05.010

Olowoyo, J., van Heerden, E., Fischer, J., & Baker, C. (2010). Trace metals in soil and leaves of Jacaranda mimosifolia in Tshwane area, South Africa. *Atmospheric Environment*, *44*(14), 1826–1830. doi:10.1016/j.atmosenv.2010.01.048

Owino, J., Ragama, P., Maghanga, C., & Njeru, N. (2015). Lead metal exposure to residents residing in informal settlement: A case study of residents in Nakuru municipality, Kenya. *Kabarak Journal of Research and Innovation*, *3*(1), 41–47.

Plumlee, G., Durant, J., Morman, S., Neri, A., Wolf, R., Dooyema, C., Hageman, P. L., Lowers, H. A., Fernette, G. L., Meeker, G. P., Benzel, W. M., Driscoll, R. L., Berry, C. J., Crock, J. G., Goldstein, H. L., Adams, M., Bartrem, C. L., Tirima, S., Behbod, B., & Brown, M. J. (2013). Linking geological and health sciences to assess childhood lead poisoning from artisanal gold mining in Nigeria. *Environmental Health Perspectives*, *121*(6), 744–750. doi:10.1289/ehp.1206051 PMID:23524139

Rollin, H., Olutola, B., Channa, K., & Odland, J. (2017). Reduction of in utero lead exposures in South Africans populations: Positive impact of unleaded petrol. *PLoS One*, *12*(10), e0186445. doi:10.1371/journal.pone.0186445 PMID:29036215

Seltenrich, N. (2021). A fix for fixtures: Addressing lead contamination in West African drinking water. *Environmental Health Perspectives, 129*(8), 084003. doi:10.1289/EHP9610 PMID:34402631

Shiomi, N. (2015). An assessment of the causes of lead pollution and the efficiency of bioremediation by plants and microorganisms. In N. Shiomi (Ed.), *Advances in bioremediation of wastewater and polluted soil*. IntechOpen. doi:10.5772/60802

Sun, X., Sun, M., Chao, Y., Shang, X., Wang, H., & Pan, H. (2022). Effects of lead pollution on soil microbial community diversity and biomass and on invertase activity. *Soil Ecology Letters*, 1–10. doi:10.100742832-022-0134-6

Teju, E., Megersa, N., Chandravanshi, B., & Zewge, F. (2014). Lead accumulation in the roadside soils from heavy metal density motor way towns of Eastern Ethiopia. *Bulletin of the Chemical Society of Ethiopia, 28*(2), 161–176. doi:10.4314/bcse.v28i2.1

Tembo, B., Sichilongo, K., & Cernak, J. (2006). Distribution of copper, lead, cadmium and zinc concentrations in soils around Kabwe town in Zambia. *Chemosphere, 63*(3), 497–501. doi:10.1016/j.chemosphere.2005.08.002 PMID:16337989

Tuakuila, J., Lison, D., Mbuyi, F., Haufroid, V., & Hoet, P. (2013). Elevated blood lead levels and sources of exposure in the population of Kinshasa, the capital of the Democratic Republic of Congo. *Journal of Exposure Science & Environmental Epidemiology, 23*(1), 81–87. doi:10.1038/jes.2012.49 PMID:22617721

World Health Organization. WHO. (2015). Lead exposure in African children, contemporary sources and concerns. WHO Regional Office for Africa, Brazzaville, Republic of Congo. .

WHO. (2014). Chemicals of public health concern in the African Region and their management – Regional Assessment Report. WHO, Geneva, Switzerland.

Yabe, J., Nakayama, S., Ikenaka, Y., Muzandu, K., Choongo, K., Mainda, G., Kabeta, M., Ishizuka, M., & Umemura, T. (2013). Metal distribution in tissues of free-range chickens near a lead–zinc mine in Kabwe, Zambia. *Environmental Toxicology and Chemistry, 32*(1), 189–192. doi:10.1002/etc.2029 PMID:23059509

Yabe, J., Nakayama, S., Ikenaka, Y., Yohannes, Y., Sam, N., & Oroszlany, B. (2015). Lead poisoning in children from townships in the vicinity of a lead-zinc mine in Kabwe, Zambia. *Chemosphere*, *119*, 941–947. doi:10.1016/j.chemosphere.2014.09.028 PMID:25303652

Yehouenou, E., Adamou, R., Azehoun, P., Edorh, P., & Ahoyo, T. (2013). Monitoring of heavy metals in the complex "Nokoué lake-Cotonou and Porto-Novo lagoon" ecosystem during three years in the Republic of Benin. *Research Journal of Chemical Sciences*, *3*, 12–18.

Chapter 8
Soil Pollution by Chromium in Sub-Saharan Africa

ABSTRACT

This chapter explores the physicochemical characteristics of chromium (Cr) and how they exert pollution in soils of sub-Saharan Africa. The metal is sourced mainly from ferrochrome- and chromite-based ores found in mafic and ultramafic rocks. The hexavalent form, Cr (VI), which is more soluble and bioavailable is more toxic compared to the trivalent form, Cr (III) that is immobile. Industrial activities including leather tanning that uses chromium basic sulfate, mining of ferrochrome and smelting activities, production of steel and manufacture, and use of agrochemicals introduce Cr to soils of SSA region. The unsystematic management of solid waste and wastewater containing the metal is also affiliated to its soil pollution. Evidence of Cr poisoning in soils of SSA region was reported in many studies, which is an environmental concern that requires corrective measures.

INTRODUCTION

The rise of human-based activities in contemporary society has seen a significant increase in pollution incidences from organic and inorganic contaminants following their environmental release (Gupta et al., 2020). The negative impact of such pollutants on human health and ecosystems is synergistic if their nature is mixed and their doses are high in such environs. The pollutants also have negative effects on soils and soil biota (Lacalle et al., 2020). Usually, more than one organic and/or inorganic pollutant causes modifications on soils and soil contents. For instance, due to the growth of industrialization, most of the soils have heavy metals that coexist with

DOI: 10.4018/978-1-6684-7116-6.ch008

other organic pollutants (Bakshi 2016). For instance, in some parts of the world, a mixture of chromium {Cr (VI)}, fertilizers and agrochemicals has been reported at levels beyond the permissible concentrations (Aparicio et al. 2018; 2019).

Chromium is found in water, air and soil. The heavy metal occurs in two stable states: 1) Cr (III), which is usually the natural form resulting from leaching of ores found in the earth's crust and 2) Cr (VI), which is anthropogenic in nature (Bakshi & Panigrahi, 2018). The trivalent form of Cr is essential in human metabolism (including the breakdown of fats, proteins, lipids and carbohydrates) as well as enhanced sensitivity to insulin (Lewicki et al., 2014). On the other hand, the hexavalent form is highly mobile, toxic, soluble and causes noxious effects on humans and animals in comparison to high levels of Cr (III) (Nakkeeran et al., 2018). Chromium (III) is immobile, insoluble and occurs bound to organic compounds in natural environs and hence, is less toxic (Jiang et al., 2019). Majority of environmental releases of Cr are from industries such as those dealing with manufacturing of stainless steel, chrome plating, tannery, electroplating, paint production, leather industries, rubber manufacturing, alloy cast irons, dyeing factories, chemical production, metallurgical works, cement production and wood treatment (Bakshi & Panigrahi, 2018; Lian et al., 2019). The involved industrial activities produce Cr-containing waste, solid sludge and wastewater whose release to the environment pollutes natural resources. Pollution also results from the releasing and leaching of chromite from mines to pollute both soil and water resources during its horizontal and vertical movement (Prasad et al., 2021).

The increasing concentrations of Cr in soils, water and air have resulted to a great consideration for the metal and its toxicity in various studies (Bakshi, 2016; Bakshi & Panigrahi, 2018). The increases in concentrations are mainly anthropogenic and a result of the two stable forms of the metal. Cr is found to accumulate in soils from different sources whereby, the metal is subsequently transferred to plants through root uptake. In studies in Bangladesh (Bakshi, 2016), China (Qu et al., 2015), Slovakia (Kulikova et al., 2019), India (Prasad et al., 2021), South Africa (Nyika et al., 2020) and Kenya (Mwamburi, 2016; Nyika & Dinka, 2022), Cr levels were found to be beyond allowable limits in soils with the sources being attributable to mainly industrial activities in the regions. The allowable limits of Cr in the studies differed based on the standards used to compare assayed concentrations of the metal. According to Nyika et al. (2019), allowable limit for Cr in soils of South Africa is 6.5 mg/kg while the World Health Organization (WHO) limits are 100 mg/kg (Nyika & Dinka, 2022). Allowable limits of Cr in Ethiopian soils were reported as 20 mg/kg according to the Ethiopian Environmental Protection Authority (EEPA, 2003). The allowable limits differ from the 180, 6.4 and 400 mg/kg allowable limits in Europe, United Kingdom and United States for the same metal in soils, respectively (Ediene & Umoetok, 2017). Chromium's permissible limits in soils according to the European Union standards is 150 mg/kg (Srivastava et al., 2017). Most studies on heavy metal assessment in SSA use the WHO and EU standards as specific countries do not have such predefined soil pollution standards. The accumulation of the metal in soils is dependent on the parent rock from which the soil originated, soil cation exchange capacity, organic content, texture and pH (Shahid et al., 2014). Additionally, Cr uptake from soils to plants is dependent on transpiration rates, root surface area, root type, plant type and absorption rates of the involved plants. To understand the occurrence and toxicity of Cr in soils, this chapter focuses on the characteristics of the metal, its sources and the extent of pollution effects in soils with a specific focus of SSA region.

Figure 1. The physical appearance of chromium metal
(Advanced Refractory Metals, 2023)

THE CHEMISTRY OF CHROMIUM METAL

Chromium is an abundant metal in the natural environment with an atomic mass of 24 and in the category of transition metals in the periodic table. It is steely gray lustrous in color (Figure 1), hard and an additive component of stainless steel. The metal is the 21st and the 7th most available element in rock ores and most abundant substance on earth respectively (Bakshi & Panigrahi, 2022). The physicochemical properties of the heavy metal are as summarized in Table 1.

In soils, both trivalent and hexavalent Cr forms have two ionic forms at normal electric potential and pH of such environs. For Cr (III), the two forms are Cr^{3+} and CrO^{2-} while for Cr (VI), the two forms are CrO_4^{2-} and $Cr_2O_7^{2-}$. The hexavalent ionic forms are soluble in soils while the trivalent forms are poorly soluble at pH of 4 and above and precipitate at acidic pH of 5.5 (Mohanty & Patra, 2011).

Table 1. Physicochemical properties of chromium (Bakshi & Panigrahi, 2022)

Characteristic	Value
Heat capacity	23.35 Joules/mole-Kelvin
Vaporization heat	347 kilojoules/ mole
Atomization heat	397 kilojoules/mole
Fusion heat	21 kilojoules/mole
Thermal conductivity	93.9 Joule per meter second kelvin
Electrical conductivity	77.52 megaohms-cm
Atomic radius	128 pm
Ionic radius	0.044 nm and 0.061 nm for Cr (IV) and Cr (III), respectively
Oxidation states	Cr (II), Cr (III), Cr (IV), Cr (V), Cr (VI) Cr (III) and Cr (VI) are the stable oxidation states
Electronic configuration	(Ar) $3d^5 4s^1$
Electronegativity	1.66 on a Pauling scale
Isotopes	6
Subgroup of periodic table	VIb positioned at the first series of transition metals
Mass number	51.996
Melting point	2130 K
Boiling point	2963 K
Hardness	8.5 Mohs
Density	7.19 g/cm^3 at 20 °C
Atomic mass	51.996 g/mol
Position	Group 6, period 4 in the d-block of the periodic table
Atomic number	24
State at standard temperature and pressure	Solid
Color	Silvery white
Electron per shell	Four shells with 2, 8, 13 and 1 electrons

Chromium is found in air, soil, water and volcanic dust and hence has a ubiquitous nature (Bakshi 2016; Bakshi & Panigrahi, 2018). Sources of the heavy metal are either natural or anthropogenic-based. In its natural form, it is found chemically bound to crystalized ores of iron oxide in the form of chromite ($FeCr_2O_4$) in ultra-mafic rocks (Bakshi & Panigrahi, 2022). Other compounds in which the heavy metal is found to be naturally occurring include crocoite ($PbCrO_4$), tarapacaite (K_2CrO_4), bentorite ($Ca_6(CrAl)_2SO_4)_3$) and vauquelinite ($CuPb_2CrO_4$-PO_4OH) (Babula et al., 2008). The Cr-containing chemicals undergo weathering and are transported to soils and sediments as components of their organic matter (Hsu et al., 2015). Most naturally occurring forms of Cr are found in mafic and ultra-mafic rocks although smaller quantities are found in sedimentary and igneous rocks (Kabata-Pendias, 2010). The concentrations of the metal on the earth's crust are varies spatially but range between 0.1 and 0.3 mg/kg. For this reason, the background levels of the heavy metal vary spatially and based on the percentage of clay in soils of a given area (Shahid et al. 2017).

INDUSTRIAL SOURCES OF CHROMIUM POLLUTION IN SOILS

The hexavalent form of Cr is used as an anti-corrosion agent in the form of spray coatings and chrome, in tanning of leather products, manufacture of stainless steel, as a wood preservative, in plastics, pigments, inks and paints as well as in the networking industries (Prasad et al., 2021). These industrial processes are sources of Cr that release the heavy metal in water, air and soils. In another study by Tumolo et al. (2020), industrial sources of Cr were mainly from the energy sector though other industries contributed to pollution by the heavy metal as shown in Table 2. In the energy sector, fossil fuels are combusted to contribute to release of 1723 metric tons of Cr mainly containing of dusts and particulate matter complexed with the metal, which precipitate and are deposited in the soils to pollute them (Tumolo et al., 2020). Out of the total amount of Cr, 0.2% is the hexavalent form. Activities such as repair and painting of aircrafts that involve spray painting, sanding, abrasive blasting and grinding also introduce Cr (VI) into soils (Rosaaen, 2017). In another study, manufacture of specialty chemicals, chromic acid, catalytic manufacture and production of cleaning agents and drilling muds are other industrial activities that introduce Cr-containing waste to soils (Ertani et al., 2017). Dhal et al. (2013) noted that most chrome use and pollution to the environment including soils with Cr comes from the metallurgical industry in addition to chemical industries, foundries and refractories with the share distribution shown in Figure 2.

Table 2. Industrial sources of chromium and their percentage share (Tumolo et al., 2020)

Industry Type	Percentage share of Cr pollution to the environment
Energy sector	83.9
Processing and production of metals	2.2
Mineral industry	0.5
Chemical industry	2.1
Wastewater and water management industry	10.9
Wood and paper processing industries	0.3

In soils, the concentration of Cr depends on sediments, rocks and parent material characteristics that make it up (Yadav et al., 2017). The aforementioned industrial activities lead to the production and dumping of Cr-containing solid and liquid wastes in the form of plating bath and/or slag, mining byproducts and particulate matter that is deposited to soils leading to their pollution with the metal (Shanker & Venkateswarlu, 2011). The unscientific management of the mine tailings, solid wastes and sludge from the industries by introducing them to soils enhances the concentrations of Cr in the soils. For instance, in India, the introduction of effluents from tannery industries to agricultural soils resulted to elevated levels of Cr metal (Ghosh et al. 2012). Similarly, surface soils of Wuhai in China had high concentrations of Cr beyond permissible levels due to coal mining activities and deposition of resultant tailings (Bu et al., 2020). Leather tanning industries globally have been identified as key introducers of Cr to the environment (Tumolo et al., 2020).

TOXICITY OF CHROMIUM IN SOILS

Chromium in the environment (air, soil and water) occurs in its particulate form. The metal's toxicity in soils depends on its concentrations and its speciation ability to different chemical forms (Rafiq et al., 2017). Chromium has a valence of 0-6 although its trivalent and hexavalent forms are the stable ones (Table 1). Cr (III) and Cr (VI) forms of the metal exist in nature and their redox reactions in soils are thermodynamically impulsive in the presence of pollution (Ding et al., 2016). Speciation controls the toxicity of the metal in soils considering that Cr (VI) is more toxic and readily soluble in soils compared to Cr (III). Speciation and subsequently toxicity of soils by Cr is regulated by a number of factors (Shahid et al. 2014; Taghipour & Jalali, 2016). These include: -

Figure 2. Industrial sources of chromium and the percentage of the metal used (Dhal et al., 2013)

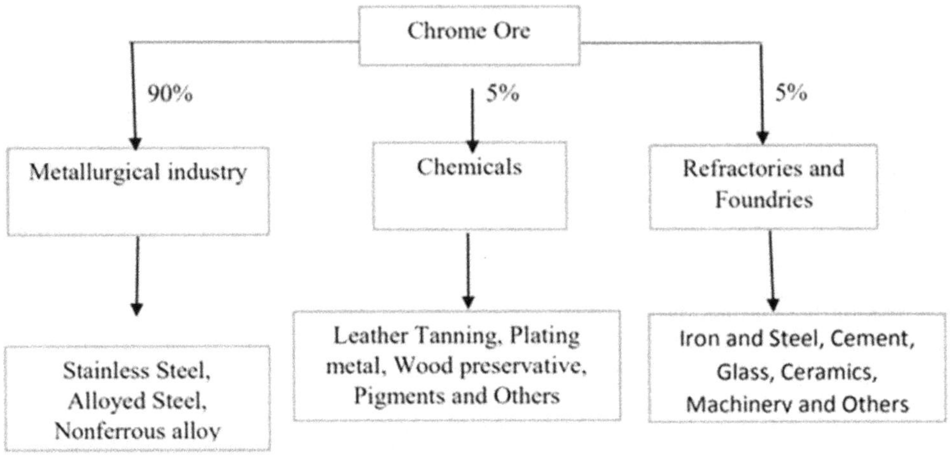

i. The concentration of the metal in the soils
ii. Availability of competing cations with a similar oxidation state
iii. The cation exchange capacity of a particular soil
iv. Soil pH
v. Redox potential
vi. The microbial environment of the soil

Soil pH has an indirectly proportional relationship to metal solubility and its ultimate toxicity to soils (Shahid et al., 2017). Cr (VI) adsorption is low at high pH while Cr (III) transformation from solid to solution is highly favored by low pH (Dias-Ferreira et al., 2015). Soil additives such as lime and organic matter decrease Cr (VI) sorption and increase Cr (III) sorption (Taghipour & Jalali, 2016). With pH increases, surface negative ions are affected by the increase of hydroxyl ions, which decreases soil desorption of Cr (VI) and promotes Cr (III) precipitation (Shahid et al., 2017). Toxicity of soils by Cr, which corresponds to increases in Cr (VI) is favored by acidity (pH of 1-6.5) while Cr (III) is favored by alkaline conditions (pH of 8-12) (Bakshi & Panigrahi, 2022).

The tendency to accept or donate protons also referred to us the redox potential influence the toxicity of soils by Cr. The observation is because the heavy metal exhibits various oxidation states (Xiao et al., 2015). Reduction conditions reduce soil toxicity by Cr allowing its transformation from the hexavalent to the trivalent form. In soils that are well aired, Cr (VI) is stored at neutral to alkaline pH because this form of Cr has high redox potential and is highly oxidizing at low pH unlike other forms (Mandina & Tawanda, 2013). The presence of organic matter content facilitates reduced toxicity in soils by favoring the formation of Cr (III) from Cr (VI) (Shahid et al., 2017). Using organic manure from compost, black carbon, farmyard manure and seaweed reduces toxicity of soil from Cr (VI) (Shahid et al., 2014). Some microorganisms promote formation of hexavalent Cr while others favor trivalent formation (Qian et al., 2016). The microbes and organic matter regulate the redox potential of soil matter and hence the level of Cr (VI), which is the determinant of toxicity levels. The influence of other factors on heavy metal bioavailability and toxicity is discussed in Chapter 1 of this book.

CHROMIUM POLLUTION OF SOILS IN SUB-SAHARAN AFRICA

Sub-Saharan Africa is growing and thriving economically due to the rise of industrialization. However, the economic advancements have come at a costly price due to environmental pollution by heavy metals. In particular, Cr pollution in SSA soils is a growing concern as noted by Oruko et al. (2021). In SSA, most Cr pollution of soils is from the leather tanning, mining and chemical industries as well as the application of agrochemicals in such environs.

Leather tanning activities have been identified as causatives of Cr introduction in soils of SSA. The processes of leather tanning use chromium basic sulfate ($Cr_2(OH(SO_4)_3$ to oxidize skin and/or hide organic materials hence tan the leather (Ertani et al., 2017). The compound is chemically reduced to chromium (III) oxide (Cr_2O_3) and/or chromium (III) hydroxide $\{Cr(OH)_3\}$ to enable leather softening and subsequent reshaping. Some of the Cr basic sulfate salts remain unused and become waste due to diffusion difficulties (Gezahegn et al., 2021). The chromium tanning process is the most preferred by 80-90% of the firms although it results to large quantities of Cr-containing waste (usually 3.5-4.5% w/w Cr_2O_3) (Oruko et al., 2020). These wastes are in the form of chrome leather trimmings, chrome-tanned leather shavings and chromium sludge in solid and liquid forms (Nigam et al., 2015). Once the wastes are introduced to soils, bioaccumulation of Cr can occur.

In SSA countries of East, South and Central Africa such as South Africa, Mozambique, Zimbabwe, Uganda, Ethiopia, Kenya, Malawi, Botswana, Tanzania, Zambia, Cameroon and Nigeria, such wastes from leather manufacturing activities are reported to pollute the environment including soils of the vicinity where such processes occur (Oruko et al., 2018, 2020, 2021). In the vicinity of tanneries at Dongo Bonde (Kenya) and Bath Ore (pty), South Africa, soils had Cr levels that exceeded the WHO, FAO and local permissible levels due to open dumping of Cr-tanned leather wastes and their subsequent pollution of such environs (Oruko et al., 2021). In many countries of SSA including Ethiopia, Kenya, South Sudan,

Namibia, Nigeria, South Africa, Uganda and Zimbabwe, the leather industry has been expanding since the 2000s (Hira et al., 2022). Consequently, Cr-containing waste produced during its processing are disposed off as sludge or solid wastes at landfills and open dumpsites resulting to environmental pollution including soil pollution. Hira et al. (2022) noted that the situation is worsened by rudimentary approaches to Cr-containing waste management in the SSA countries including disposal in freshwater bodies and wetlands in addition to open-dumping leading to extended pollution of soils and other environmental resources. In the vicinity of Colba tannery in Modjo, Ethiopia, areal soils were reported to have elevated Cr levels due to disposal of tannery solid waste and wastewater introducing the metal to the environment (Gezahegn et al., 2021). Similar results were also found in Bahir Dar City of Ethiopia, where disposal of tanning solid waste and wastewater was attributable to elevated Cr concentrations in agricultural soils compared to the allowable limits of 20 mg/kg (Alemu & Tegegne, 2022).

In West African countries such as Nigeria and Ghana, the tannery industry is a source of organic and inorganic wastes disposed off as sludge into soils. In such countries, the demand for leather products is growing along with population rise and so is the disposal of chrome leather tanning wastes, which is largely unregulated (Doka & Mohammed, 2020). Consequently, soil pollution by Cr has been established in agricultural soils of areas such as Kano metropolis (Doka & Mohammed, 2020; Njoku & Agwu, 2017), where most tannery industries are concentrated. In Ghana's Aboabo artisanal tannery and the vicinity, soils were polluted with Cr among other heavy metals as a result of unscientific management of the wastewater from the tanning activities (Appiah-Brempong et al., 2022). Most of the water was released to soils and freshwater resources without any form of treatment and hence, a potential public health and environmental concern.

The mining industry is also another high contributor of Cr-pollution in soils of SSA region. In particular, Southern African countries including nations of Zimbabwe and South Africa are large global producers of ferrochrome (one of the major ores of Cr). The resultant Cr is mainly used in the metal industry to produce stainless steel and other alloys of steel in addition to refractory and foundry applications. South Africa alone hosts 70% of the global chrome reserves while Zimbabwe has 20% of the reserves usually derived from the Bushveld igneous ores (Dzvinamurungu et al., 2020). During the mining and smelting of ferrochrome and/or chromite, dusts and wastes containing Cr in its hexavalent form are produced. The wastes are hazardous, highly reactive and soluble with the potential to induce environmental pollution (Das et al., 2021).

Coetzee et al. (2020) noted that chromite-mining and ferrochrome industrial activities produced smelter dust and slag, which led to the formation and remobilization of Cr (VI), which is a major soil pollutant. In Sekhukhune district of South Africa with multiple ferrochrome smelters and chromite mines, resultant waste dumps, dusts and tailings were reported to pollute soils where concentrations of the heavy metal were reported to be higher than the permissible levels (Adhikari et al., 2022). In another study in Zimbabwe, production of slag during chrome mining and ferrochrome smelting at Gweru smelting firm and its unscientific disposal through open dumping was associated with higher concentrations of Cr (VI) in soils of the vicinity (Mandina & Tawanda, 2013). In Zambia and Southern Africa, a reconnaissance study at the Copperbelt Province established that mine tailings deposited in soils were the source of elevated Cr (VI) levels in agricultural soils of the vicinity (Hamilton et al., 2020).

In West Africa, the smelting, mining, metallurgical and chemical industries are also responsible for Cr pollution in soils. The trend is more predominant in industrial, urban and suburban regions where resultant solid wastes and wastewater from the activities is disposed. In Kumasi Metropolis of Ghana where multiple mining, and industrial activities as well as metal fabrication, auto-repair and mechanic workshops are situated, soils were found to be enriched with Cr among other heavy metals (Akoto et al., 2017). During auto-mechanic repairs, Cr could be released to soils during peeling of paints, from auto-catalysts, in degraded car parts that require replacement and in auto-tires. In another study in Sunyani region of Ghana, concentrations of Cr in surface soils were found to be 7-10 times higher compared to the permissible reference standards (Asamoah et al., 2021). The trend was attributable to industrial activities such as wood preservation, tanning, metal plating, anodizing, production of pigments, dyes and textiles that were concentrated in the region. In Southwest Nigeria at Ota, soils had high Cr levels beyond the WHO/ FAO allowable limits of 100 mg/kg (Kayode et al., 2022). The observations were attributable to the application of agrochemicals and inorganic fertilizers that contained the heavy metal.

In East Africa, in addition to tanning activities other industrial activities result to Cr introduction in soils. These include the chemical industries, auto-mobile and steel processing industries. In Kenya for instance, using untreated wastewater from plastic, textile, soap and other chemical industries of Nairobi's suburban regions for irrigation was attributable to elevated levels of Cr among other heavy metals in studied soils beyond the recommended reference standards (Nyika & Dinka, 2022). Surface soils of central Kenya were also found laden with Cr among other trace metals due to the use of agrochemicals in addition to ferti-irrigation with untreated wastewater (Ndungu et al., 2019). At Kombolcha town of Ethiopia, exposure of soils to wastewater and solid waste from steel processing, textile and brewing industries of the area was attributable to their elevated Cr levels above the allowable limits of 20 mg/kg according to the EEPA (Muhammed et al., 2021). Auto-mechanical activities including the change of automobile catalytic converters and brake lining were associated with elevated levels of Cr in car workshops located at Ngara in

Kenya (Katana et al., 2013). Mining of tungsten in Gifurwe mining site of Rwanda was associated with production of tailings whose improper disposal led to high concentrations of Cr in soils of the vicinity (Hirwa et al., 2019).

CONCLUSION

This chapter studied pollution by Cr in soils of SSA nations. The hexavalent form of the element was affiliated with toxicity in soils while the trivalent form was found to have metabolic uses. The degree and extent of Cr poisoning in soils is dependent on the metal concentration, availability of competing ions, redox potential of soils, its pH and cation exchange capacity. These factors vary depending on the location of the soil. Soils of Southern Africa that are mafic and ultramafic in nature were found to be susceptible to Cr pollution due to the presence of naturally occurring ferrochrome, which becomes available during weathering processes. Chromium mining and subsequent use in production of alloys, tanning of leather and production of catalysts and other chemicals is growing in the SSA region to correspond to the population needs of modern day. Consequently, solid wastes and effluents containing the metal are also being introduced to the environment including soils leading to pollution and toxicity from the metal as evident in reported empirical studies. The energy sector and its associated processes as well as agrochemical manufacture and use were found to be potential sources of Cr. Controlling the accumulation of Cr in soils requires avoidance of its sources, which are mainly industry-based.

REFERENCES

Adhikari, S., Marcelo-Silva, J., Beukes, J., Van Zyl, P., Coetsee, Y., Boneschans, R., & Siebert, S. J. (2022). Contamination of useful plant leaves with chromium and other potentially toxic elements and associated health risks in a polluted mining-smelting region of South Africa. *Environmental Advances*, *9*, 100301. doi:10.1016/j.envadv.2022.100301

Advanced Refractory Metals. (2023). *Five uses of chromium/ uses of chromium in industry and everyday life*. Refractory Metals. https://www.refractorymetal.org/uses-of-chromium/

Akoto, O., Bortey-Sam, N., Ikenaka, Y., Nakayama, S., Baidoo, E., Yohannes, Y., & Ishizuka, M. (2017). Contamination levels and sources of heavy metals and a metalloid in surface soils in the Kumasi Metropolis, Ghana. *Journal of Health & Pollution*, *7*(15), 28–39. doi:10.5696/2156-9614-7.15.28 PMID:30524828

Alemu, A., & Tegegne, A. (2022). Assessment of chromium contamination in the soil and khat leaves (Catha edulis Forsk) and its health risks located in the vicinity of tannery industries; a case study of Bahir Dar city, Ethiopia. *Heliyon*, *8*(12), e11914. doi:10.1016/j.heliyon.2022.e11914 PMID:36506399

Aparicio, J., Garcia-Velasco, N., Urionabarrenetxea, E., Soto, M., Alvarez, A., & Polti, M. (2019). Evaluation of the effectiveness of a bioremediation process in experimental soils polluted with chromium and lindane. *Ecotoxicology and Environmental Safety*, *181*, 255–263. doi:10.1016/j.ecoenv.2019.06.019 PMID:31200198

Aparicio, J., Raimondo, E., Gil, R., Benimeli, C., & Polti, M. (2018). Actinobacteria consortium as an efficient biotechnological tool for mixed polluted soil reclamation: Experimental factorial design for bioremediation process optimization. *Journal of Hazardous Materials*, *342*, 408–417. doi:10.1016/j.jhazmat.2017.08.041 PMID:28854393

Appiah-Brempong, M., Essandoh, H., Asiedu, N., Dadzie, S., & Momade, F. (2022). Artisanal tannery wastewater: Quantity and characteristics. *Heliyon*, *8*(1), e08680. doi:10.1016/j.heliyon.2021.e08680 PMID:35024490

Asamoah, B., Asare, A., Okpati, S., & Aidoo, P. (2021). Heavy metal levels and their ecological risks in surface soils at Sunyani magazine in the bono region of Ghana. *Scientific African*, *13*, e00937. doi:10.1016/j.sciaf.2021.e00937

Babula, P., Adam, V., Opatrilova, R., Zehnalek, J., Havel, L., & Kizek, R. (2008). Uncommon heavy metals, metalloids and their plant toxicity: A review. *Environmental Chemistry Letters*, *6*(4), 189–213. doi:10.100710311-008-0159-9

Bakshi, A. (2016). *Analysis of anthropogenic disturbances and impact of pollution on fish fauna of River Churni with special reference to chromium pollution*. [PhD Thesis, University of Kalyani, Shodhganga]. http://hdl. handle.net/10603/241694

Bakshi, A., & Panigrahi, A. (2018). A comprehensive review on chromium induced alterations in fresh water fishes. *Toxicology Reports, 5*, 440–447. https://doi.org/. toxrep.2018.03.007 doi:10.1016/j

Bakshi, A., & Panigrahi, A. (2022). Chromium contamination in soil and its bioremediation: An Overview. In J. A. Malik (Ed.), *Advances in bioremediation and phytoremediation for sustainable soil management*. Springer. doi:10.1007/978-3-030-89984-4_15

Bu, Q., Li, Q., Zhang, H., Cao, H., Gong, W., Zhang, X., Ling, K., & Cao, Y. (2020). Concentrations, spatial distributions, and sources of heavy metals in surface soils of the coal mining city Wuhai, China. *Journal of Chemistry, 4705954*, 1–10. doi:10.1155/2020/4705954

Coetzee, J., Bansal, N., & Chirwa, N. (2020). Chromium in environment, its toxic effect from chromite-mining and ferrochrome industries, and its possible bioremediation. *Exposure and Health, 12*(1), 51–62. doi:10.100712403-018-0284-z

Das, P., Das, B., & Dash, P. (2021). Chromite mining pollution, environmental impact, toxicity and phytoremediation: A review. *Environmental Chemistry Letters, 19*(2), 1369–1381. doi:10.100710311-020-01102-w

Dhal, B., Thatoi, N., Das, N., & Pandey, D. (2013). Chemical and microbial remediation of hexavalent chromium from contaminated soil and mining/metallurgical solid waste: A review. *Journal of Hazardous Materials, 250*, 272–291. doi:10.1016/j.jhazmat.2013.01.048 PMID:23467183

Dias-Ferreira, C., Kirkelund, G., & Ottosen, L. (2015). Ammonium citrate as enhancement for electrodialytic soil remediation and investigation of soil solution during the process. *Chemosphere, 119*, 889–895. doi:10.1016/j.chemosphere.2014.08.064 PMID:25240953

Ding, W., Stewart, D., Humphreys, P., Rout, S., & Burke, I. (2016). Role of an organic carbon-rich soil and Fe (III) reduction in reducing the toxicity and environmental mobility of chromium (VI) at a COPR disposal site. *The Science of the Total Environment, 541*, 1191–1199. doi:10.1016/j.scitotenv.2015.09.150 PMID:26476060

Doka, M., & Mohammed, M. (2020). Effects of sludge on some soil properties and heavy metals uptake by some vegetables grown on tannery sludge amended soils in Kano metropolis, Nigeria. *Dutse Journal of Pure and Applied Sciences, 6*(1), 173–181.

Dzvinamurungu, T., Rose, D., Vijoen, K., & Bafubiandi, A. (2020). A process mineralogical evaluation of chromite at the Nkomati nickel mine, Uitkomst complex, South Africa. *Minerals (Basel), 10*(8), 709. doi:10.3390/min10080709

Ediene, V., & Umoetok, S. (2017). Concentration of heavy metals in soils at the municipal dumpsite in Calabar metropolis. *Asian Journal of Environment and Ecology, 3*(2), 1–11. doi:10.9734/AJEE/2017/34236

EEPA. (2003). Guideline ambient environmental standards for Ethiopia. Ethiopian Environmental Protection Authority and United Nations Industrial Development Organization, UNIDO, Addis Ababa, Ethiopia.

Ertani, A., Mietto, A., Borin, M., & Nardi, S. (2017). Chromium in agricultural soils and crops: A review. *Water, Air, and Soil Pollution*, *228*(5), 190. doi:10.100711270-017-3356-y

Gezahegn, A., Feyessa, F., Tekeste, E., & Beyene, M. (2021). Chromium laden soil, water and vegetables nearby tanning industries: Speciation and spatial distribution. *Journal of Chemistry*, *5531349*, 1–10. doi:10.1155/2021/5531349

Ghosh, I., Chatterjee, S., & Mukherjea, K. (2012). Chromium (VI) in tannery effluents: Assessment, biodistribution, and environmental health impact. *Journal of the Indian Chemical Society*, *89*(4), 479–483.

Gupta, N., Yadav, K., Kumar, V., Krishnan, S., Kumar, S., & Nejad, D. (2020). Evaluating heavy metals contamination in soil and vegetables in the region of North India: Levels, transfer and potential human health risk analysis. *Environmental Toxicology and Pharmacology*, *103563*. doi:10.1016/j.etap.2020.103563 PMID:33310081

Hamilton, E., Lark, R., Young, S., Bailey, E., Sakala, G., Maseka, K., & Watts, M. J. (2020). Reconnaissance sampling and determination of hexavalent chromium in potentially-contaminated agricultural soils in Copperbelt province, Zambia. *Chemosphere*, *247*, 125984. doi:10.1016/j.chemosphere.2020.125984 PMID:32079057

Hira, A., Pacini, H., Wadee, K., Sikander, M., Oruko, R., & Dinan, A. (2022). Mitigating tannery pollution in sub-Saharan Africa and South Asia. *Journal of Developing Societies*, *38*(3), 360–383. doi:10.1177/0169796X221104856

Hirwa, H., Nshimiyimana, F., Ngendahayo, E., Akimpaye, B., Nahayo, L., & Ngamata, O. (2019). Evaluation of soil contamination in mining areas of Rwanda. *American Journal of Water Science and Engineering*, *5*(1), 9–15. doi:10.11648/j.ajwse.20190501.12

Hsu, L., Liu, Y., & Tzou, Y. (2015). Comparison of the spectroscopic speciation and chemical fractionation of chromium in contaminated paddy soils. *Journal of Hazardous Materials*, *296*, 230–238. doi:10.1016/j.jhazmat.2015.03.044 PMID:25935296

Jiang, B., Gong, Y., Gao, J., Sun, T., Liu, Y., Oturan, N., & Oturan, A. (2019). The reduction of Cr (VI) to Cr (III) mediated by environmentally relevant carboxylic acids: State-of-the-art and perspectives. *Journal of Hazardous Materials*, *365*, 205–226. doi:10.1016/j.jhazmat.2018.10.070 PMID:30445352

Kabata-Pendias, A. (2010). *Trace elements in soils and plants*. CRC Press. doi:10.1201/b10158

Katana, C., Murungi, J., & Mbuvi, H. (2013). Speciation of chromium and nickel in open air automobile mechanic workshop soils in Ngara, Nairobi, Kenya. *World Environment*, *3*(5), 143–154. doi:10.5923/j.env.20130305.01

Kayode, O., Ogunyemi, E., Odukoya, A., & Aizebeokhai, A. (2022). Assessment of chromium and nickel in agricultural soils: Implications for sustainable agriculture. *IOP Conference Series. Earth and Environmental Science*, *993*(1), 012014. doi:10.1088/1755-1315/993/1/012014

Kulikova, T., Hiller, E., Jurkovič, L., Filová, L., Šottník, P., & Lacina, P. (2019). Total mercury, chromium, nickel and other trace chemical element contents in soils at an old cinnabar mine site (Merník, Slovakia): Anthropogenic versus natural sources of soil contamination. *Environmental Monitoring and Assessment*, *191*(5), 263. doi:10.100710661-019-7391-6 PMID:30953219

Lacalle, R., Aparicio, J., Artetxe, U., Urionabarrenetxea, E., Polti, M., & Soto, M. (2020). Gentle remediation options for soil with mixed chromium (VI) and lindane pollution: biostimulation, bioaugmentation, phytoremediation and vermiremediation. *Heliyon* *6*(8), e04550. https://doi.org/. heliyon.2020.e04550 doi:10.1016/j

Lewicki, S., Zdanowski, R., Krzyzowska, M., Lewicka, A., Debski, B., Niemcewicz, M., & Goniewicz, M. (2014). The role of chromium III in the organism and its possible use in diabetes and obesity treatment. *Annals of Agricultural and Environmental Medicine*, *21*(2), 331–335. doi:10.5604/1232-1966.1108599 PMID:24959784

Lian, G., Wang, B., Lee, X., Li, L., Liu, T., & Lyu, W. (2019). Enhanced removal of hexavalent chromium by engineered biochar composite fabricated from phosphogypsum and distillers' grains. *The Science of the Total Environment*, *697*, 134119. doi:10.1016/j.scitotenv.2019.134119 PMID:32380611

Mandina, S., & Tawanda, M. (2013). Chromium, an essential nutrient and pollutant: A review. African. *Journal of Pure Applied Chemistry*, *7*, 310–317. doi:10.5897/AJPAC2013.0517

Mandina, S., & Tawanda, M. (2013). Speciation of chromium in soils, plants and wastewater at a ferrochrome slag dump in Gweru. *IOSR Journal of Environmental Science, Toxicology and Food Technology*, *7*(4), 43–49. doi:10.9790/2402-0744349

Mohanty, M., & Patra, H. (2011). Attenuation of chromium toxicity by bioremediation technology. In D. M. Whitacre (Ed.), *Reviews of environmental contamination and toxicology* (Vol. 210, pp. 1–34). Springer. doi:10.1007/978-1-4419-7615-4_1

Muhammed, A., Hussen, A., Redi, M., & Kaneta, T. (2021). Remote investigation of total chromium determination in environmental samples of the Kombolcha industrial zone, Ethiopia, using microfluidic paper-based analytical devices. *Analytical Sciences*, *37*(4), 585–591. doi:10.2116/analsci.20P325 PMID:33041309

Mwamburi, J. (2016). Chromium distribution and spatial variations in the finer sediment grain size fraction and unfractioned surficial sediments on Nyanza gulf, of Lake Victoria (East Africa). *Journal of Waste Management*, *7528263*, 1–15. doi:10.1155/2016/7528263

Nakkeeran, E., Patra, C., Shahnaz, T., Rangabhashiyam, S., & Selvaraju, N. (2018). Continuous biosorption assessment for the removal of hexavalent chromium from aqueous solutions using *Strychnos nux vomica* fruit shell. *Bioresource Technology Reports*, *3*, 256–260. doi:10.1016/j.biteb.2018.09.001

Ndungu, A., Yan, X., Makokha, V., Githiga, K., & Wang, J. (2019). Occurrence and risk assessment of heavy metals and organochlorine pesticides in surface soils, central Kenya. *Journal of Environmental Health Science & Engineering*, *17*(1), 63–73. doi:10.100740201-018-00326-x PMID:31321038

Nigam, H., Das, M., Chauhan, S., Pandey, P., Swati, P., & Yadav, M. (2015). Effect of chromium generated by solid waste of tannery and microbial degradation of chromium to reduce its toxicity: A review. *Advances in Applied Science Research*, *6*(3), 129–136.

Njoku, C., & Agwu, O. (2017). Assessment of the solid waste disposal on ground water quality in selected Tube wells and Bore holes in Kano metropolis North West Nigeria. *Journal of Applied chemistry*, *10*(7), 56–60. doi:10.9790/5736-1007035660

Nyika, J., & Dinka, M. (2022). Heavy metal pollution in soils and vegetables from suburban regions of Nairobi, Kenya and their community health implications. *Pollution*, *8*(4), 1434–1447. doi:10.22059/POLL.2022.341522.1440

Nyika, J., Onyari, E., Dinka, M., & Shivani, M. (2019). Heavy metal pollution and mobility in soils within a landfill vicinity: A South African case study. *Oriental Journal of Chemistry*, *35*(4), 1286–1296. doi:10.13005/ojc/350406

Nyika, J., Onyari, E., Dinka, M., & Shivani, M. (2020). Assessment of trace metal contamination off soil in a landfill vicinity: A southern Africa case study. *Current Chemistry Letters*, *9*(4), 171–182. doi:10.5267/j.ccl.2020.2.003

Oruko, R., Edokpayi, J., Msagati, T., Tavengwa, N., Ogola, H., Ijoma, G., & Odiyo, J. O. (2021). Investigating the chromium status, heavy metal contamination, and ecological risk assessment via tannery waste disposal in sub-Saharan Africa (Kenya and South Africa). *Environmental Science and Pollution Research International*, *28*(31), 42135–42149. doi:10.100711356-021-13703-1 PMID:33797722

Oruko, R., Odiyo, J., & Edokpayi, J. (2018). Chromium tanning, management challenges and environmental legislation in sub–Saharan African tanneries. *Proceedings of ICSMNR2018*, Polokwane, South Africa.

Oruko, R., Selvarajan, R., Ogola, H., Edokpayi, J., & Odiyo, J. (2020). Contemporary and future direction of chromium tanning and management in sub-Saharan Africa tanneries. *Process Safety and Environmental Protection*, *133*, 369–386. doi:10.1016/j.psep.2019.11.013

Prasad, S., Yadav, K., Kumar, S., Gupta, N., Cabral-Pinto, M., Rezania, S., Radwan, N., & Alam, J. (2021). Chromium contamination and the effect on environmental health and its remediation: A sustainable approach. *Journal of Environmental Management*, *285*, 112174. doi:10.1016/j.jenvman.2021.112174 PMID:33607566

Qian, J., Wei, L., Liu, R., Jiang, F., Hao, X., & Chen, G. (2016). An exploratory study on the pathways of Cr (VI) reduction in sulfate-reducing up-flow anaerobic sludge bed (UASB) reactor. *Scientific Reports*, *6*(1), 23694. doi:10.1038rep23694 PMID:27021522

Qu, M., Li, W., Zhang, C., Huang, B., & Zhao, Y. (2015). Assessing the pollution risk of soil chromium based on loading capacity of paddy soil at a regional scale. *Scientific Reports*, *5*(1), 18451. doi:10.1038rep18451 PMID:26675587

Rafiq, M., Shahid, M., Abbas, G., Shamshad, S., Khalid, S., Niazi, N., & Dumat, C. (2017). Comparative effect of calcium and EDTA on arsenic uptake and physiological attributes of Pisum sativum. International *Journal of Phytoremediation, 19*(7), 662–669. https://doi.org/. 2016.1278426 doi:10.1080/15226514

Rosaaen, D. (2017). *Hexavalent Chromium Exposure to Military Aircraft Painters*. Airverter. http://www.airverter.com/controlling-exposure-hexavalent-chromium-aerospaceair transport-painting/

Shahid, M., Austruy, A., Echevarria, G., Arshad, M., Sanaullah, M., Aslam, M., Nadeem, M., Nasim, W., & Dumat, C. (2014). EDTA-enhanced phytoremediation of heavy metals: A review. *Soil & Sediment Contamination*, *23*(4), 389–416. doi:10.1080/15320383.2014.831029

Shahid, M., Shamshad, S., Rafiq, M., Khalid, S., Bibi, I., Niazi, N., Dumat, C. & Rashid, M. (2017). Chromium speciation, bioavailability, uptake, toxicity and detoxification in soil-plant system: a review. *Chemosphere, 178*, 513–533. https://doi.org/. 2017.03.074 doi:10.1016/j.chemosphere

Shankar, K., & Venkateswarlu, B. (2011). *Chromium: environmental pollution, health effects, and mode of action. Encyclopedia of Environmental Health.* Elsevier., doi:10.1016/B978-0-444-52272-6.00390-1

Srivastava, V., Sarkar, A., Singh, S., Singh, P., Araujo, A., & Singh, R. (2017). Agroecological responses of heavy metal pollution with special emphasis on soil health and plant performances. *Frontiers in Environmental Science, 5*, 64. doi:10.3389/fenvs.2017.00064

Taghipour, M., & Jalali, M. (2016). Influence of organic acids on kinetic release of chromium in soil contaminated with leather factory waste in the presence of some adsorbents. *Chemosphere, 155*, 395–404. doi:10.1016/j.chemosphere.2016.04.063 PMID:27139119

Tumolo, M., Ancona, V., De Paola, D., Losacco, D., Campanale, C., Massarelli, C., & Uricchio, V. (2020). Chromium pollution in European water sources, health risk, and remediation strategies: An overview. *International Journal of Environmental Research and Public Health, 17*(15), 5438. doi:10.3390/ijerph17155438 PMID:32731582

Xiao, W., Ye, X., Yang, X., Li, T., Zhao, S., & Zhang, Q. (2015). Effects of alternating wetting and drying versus continuous flooding on chromium fate in paddy soils. *Ecotoxicology and Environmental Safety, 113*, 439–445. doi:10.1016/j.ecoenv.2014.12.030 PMID:25546832

Yadav, K., Gupta, N., Kumar, V., & Singh, K. (2017). Bioremediation of heavy metals from contaminated sites using potential species: A review. *Indian Journal of Environmental Protection, 37*, 65–84.

Chapter 9
Soil Pollution by Arsenic in Sub-Saharan Africa

ABSTRACT

In this chapter, the chemistry of arsenic (As) and the toxicity of the metalloid in soils of sub-Saharan Africa (SSA) region was studied. The pentavalent (arsenate) and trivalent (arsenite) forms of the metalloid were found to be the most common, with the latter being more mobile and toxic compared to the former. In soils, the mobility and toxicity of As was influenced by soil physicochemical characteristics such as the organic matter content, pH and redox potential as well as the dissolution and speciation capacity of the metalloid. Industrial activities such as production and use of electronics, agrochemicals, wood preservatives, ore mining and smelting, coal processing, production of cosmetics, alloy-making and disposal of their resultant wastes were associated with As contamination in soils. Evidently, As has the potential to contaminate soils in SSA region and transfer the toxicity to other resources and into trophic chains where it exerts negative effects.

INTRODUCTION

Soil pollution is a growing environmental challenge of modern day that poses serious threat to economic growth, human health and food security globally (Onyia et al., 2020; Nyika & Dinka, 2022). The trend is because environmental pollution is linked to ecosystem degradation and anthropogenic activities such as waste disposal, metallurgical processes, mining, energy generation, application on agrochemicals on farmlands and industrial revolution (Yan & Zhang, 2021). Of the kinds of environmental pollution, the introduction of heavy metals such as As, Cd, Co, Cr,

DOI: 10.4018/978-1-6684-7116-6.ch009

Cu, Hg, Mn, Ni, Pb and Zn is proving hazardous to the soils. The heavy metals have a persistence and non-biodegradable nature and hence tend to bioaccumulate in soils and eventually end up in food chains through uptake by plants (Nyika et al., 2020). Their entry in food chains results to irreversible and chronic health complications to living organisms.

Arsenic (As), a metalloid, which is naturally occurring is one of the elements that introduces pollution to the environment. Although a metalloid, As is categorized with other heavy metals due to its toxicity and potential to cause pollution as well as its metal-like characteristics. When found in significant concentrations, the metalloid enters water resources (through water flow to surface water bodies or via infiltration to groundwater from the vadose zone) and soils and can cause harm to living organisms (Shrivastava et al., 2015). In soils, introduction of the metalloid as particulate matter, dust, As-containing solid waste and wastewater can occur. Kayode et al. (2021) noted that the metalloid is found in agricultural soils and once it enters cells and tissues of living things in adequate quantities, it can cause harm. The average concentrations of As in soils is about 5 mg/kg though quantities vary spatially and depending on the source of the metalloid, which can be of geogenic and/or anthropogenic origin (Punshon et al., 2017). Some soils are naturally endowed with toxic levels of As from their parent material (Emilie et al., 2017). However, anthropogenic activities such as mining activities, use of agrochemicals (insecticides, pesticides and herbicides), soil amendments, use of coal (which is rich in the metalloid) and unsystematic management of industrial and municipal solid wastes and effluent also lead to As pollution in soils (Chung et al., 2014; Kayode et al., 2021; Yan & Zhang, 2021).

Owing to pollution vulnerability by As, different national and international guidelines and standards have been set to rate pollution by the metalloid in soils. The Department of Environmental Affairs (DEA, 2017) in South Africa set the permissible limit of As in soils at 5.8 mg/kg while the WHO and the United States' agency for toxic substances and disease registry (ASTDR) standards were at 20 mg/kg (Rahaman et al., 2013; Nyika et al., 2019; Singh & Srivastava, 2020). The limits for the metalloid according to the European Union are 14 mg/kg (Nyika & Dinka, 2022). In many SSA countries, the WHO and EU standards are used since many of the countries do not have such documented standards.

This book chapter explores on the toxicity of the metalloid in soil environment. The focus in particular is in gaining insight on the anthropogenic sources of As and understanding the chemistry of the metalloid, which enables its toxicity in soil environs. The chapter also narrows down the search to pollution in sub-Saharan Africa (SSA) whose heavy metal concentrations (including those of As) are on the rise due to anthropogenic factors despite inadequate preparedness to deal with the crises in the region as Anyanwu et al. (2018) noted.

CHEMISTRY OF ARSENIC

Arsenic is a derivative of a Greek word 'arsenikos', which means potent. It is the 12th most abundant element of the earth's crust, 12th in human body and 14th in the sea. Arsenic exists in two forms: its sulfides and oxides. Sulfide forms have been known since ancient times and include the red colored realgar (As_4S_4) and the bright yellow orpiment (As_2S_3) discovered by Theophrastus, a Greek philosopher (Swaran, 2015). In the 5th century AD, a Greek historian from Thebes known as Olympiodorus heated arsenic sulfide to form white arsenic (As_2O_3) (Swaran, 2015). In the 1200, a German philosopher, Albertus Magnus discovered arsenic as an element and explained its metal-like characteristics. There are two ways to obtain pure forms of the metalloid: 1) reacting orpiment with soap in the presence of heat and 2) reacting arsenic trioxide (As_2O_3), produced during copper refining with olive oil in the presence of heat (Swaran, 2015). In the 1500, Tsao Kan-Mu who was a Chinese scientist, studied the toxicity of the metalloid in soils of rice farmlands and attributed its source to pesticides applied in the farms. Nriagu (2002) noted that As was used as a curative and poison in ancient times. Compound forms of the metalloid were used for decoration, pigmentation, warfare, metallurgy and pyrotechnics. For instance, $As_2O_{3,}$ which is odorless and tasteless was used as a chemical warfare agent while copper acetoarsenate ($C_4H_6As_6Cu_4O_{16}$), which is green in color was used as a pigment in making wallpapers (Nriagu, 2002).

Arsenic is a member of the nitrogen family and the 33rd element of the periodic table. It has an atomic weight of 74.921, which makes it heavier compared Mn, Ni, Fe and lighter than Au, Pb and Ag. The metalloid has 12 isotopes (Table 1) and out of this total, only As-75 is non-radioactive and stable. As-75 has 33 electrons, 42 neutrons and 33 protons. The 11 unstable and radioactive isotopes of As can be converted to the stable form through internal transition, neutron emission, positron emission, electron emission or by electron capture (Audi, 2003). In the stable As form, its electronic configuration has two, eight, eighteen and five electrons in the first to fourth energy levels in respective order and an electronic configuration of $1s^2 2s^2 2p^6$, $3s^2$, sp^6, $3d^{10}$, $4s^2$,$4p^3$. From the unfilled p orbital in the fourth energy level, arsenic can have 0, +3, +5 and -3 redox states. The -3 state results from the acquisition of three electrons to fill in the 4p orbital. By acting as a non-metal, elemental arsenic (^0As) can share the three electrons in its 4p orbital with other arsenic atoms and forms a trigonal pyramidal of the brittle gray structure of the metalloid. When the three outermost electrons are attracted to a non-metal such as oxygen, the element has an oxidation state of +3. If the electrons of 4s and 4p orbitals react with a non-metal, the element exhibits an oxidation state of +5.

Table 1. Isotopes of arsenic (Swaran, 2015)

Isotope	Mass
^{68}As	67.937
^{69}As	68.932
^{70}As	69.931
^{71}As	70.927
^{72}As	71.927
^{73}As	72.924
^{74}As	73.924
^{75}As	74.922
^{76}As	75.922
^{77}As	76.921
^{78}As	77.922
^{79}As	78.921

Elemental As has a number of physical properties as shown in Table 2. The element has a high electronegativity compared to nitrogen and equivalent to phosphorous. As such, it has greater oxidation potential and easily loses electrons increasing its cationic/metallic features. A higher electronegativity enhances easy exhibition of the +3 and +5 oxidation states. The oxidation states enable As to combine with other elements chemically, especially sulfur and oxygen. Unlike other heavy metals, As allows the sharing of valence electrons with anti-bonding orbitals and hence act as a ligand. By acting both as a metal and as a non-metal, As can shift from electronegative to electropositive states forming metal arsenides and oxo-anions, respectively. The element therefore does not occur freely in nature but forms compounds with metals such as Mn, Al, Ni, Ag, Cu and Fe to form sulfides, hydroxides and oxides (Kayode et al., 2021).

The oxidation and reduction (redox) reactions of As just as other heavy metals determine its toxicity to the environment including soils (Jiang et al., 2009; Han et al., 2019). There are two common oxidation states for As, 1) arsenite (AsO_3^{-3}), which is mobile and toxic and 2) arsenate (AsO_4^{-3}). The lethal dose of As (III) is estimated at 15-42 mg/kg while that of As (V) is 20-800 mg/kg (Shrivasta et al., 2015). The pH and the redox potential conditions regulate the metalloids redox reactions. In natural soils, As exists in its pentavalent arsenate (As^{V}) and trivalent arsenite oxyanion (As^{III}) forms. The pentavalent form is favored by oxidizing conditions while the trivalent form is dependent on the pH levels. At pH levels higher than 6.9, $HAsO_4^{-2}$ oxyanions exist while at lesser pH $H_2AsO_4^{-}$ predominate. The trivalent species of

Table 2. Physicochemical properties of arsenic (Swaran, 2015)

Property	Value
Thermal conductivity	0.502 W/cmK
Electrical conductivity	0.0345 10^6/cmΩ
Enthalpy of fusion	24.44 KJ/mole
Enthalpy of vaporization	34.76 KJ/mole
Enthalpy of atomization	301.3 KJ/mole at a temperature of 25°C
Energy for 1st, 2nd and 3rd ionization energy	947, 1798 and 2736 KJ/mole
Electron shell	[Ar] $3d^{10}$ $4s^2$ $4p^3$
Ionic radius	0.222, 0.047, 0.058 for -3, +5 and +3 oxidation states
Van der Waal radius	0.139 nm
Physical appearance	Brittle, gray, non-metal flakes
Specific heat	0.33 J/g Kelvin
Molar volume	13.08 cm^3/mole
Density	5.7 g/cm^3 at 14 °C
Electronegativity	2.0 in a Pauling scale
Melting point	814 °C at 36 bars
Boiling point	615 °C (sublimation)
Atomic number	33
Atomic mass	74.922 g/mol

As (H_3AsO_3), which are neutral predominate at reducing environs and with a pH below 9.2. At higher pH, the species dissociate to form the trivalent anions. Figure 1 shows the different forms of As based on the prevalent redox potential and pH conditions (Swaran, 2015). The concentrations of the metalloid are higher at low redox potential compared to higher redox potential (Han et al., 2019).

INDUSTRIAL SOURCES OF ARSENIC

Arsenic occurs in sediments, rocks, soils and earth's crust ubiquitously in both organic and inorganic forms. Organic forms of As include trimethylarsine, dimethylarsine and monomethylarsine (Shrivastava et al., 2015). The inorganic or mineral forms of As include its metal alloys, arsenates, arsenides, sulfosalts, and sulfides. Examples of the common inorganic forms of As include yukonite, scorodite, beudantite, pyrite, arsenopyrite, realgar and leollingite among others (Shrivastava et al., 2015).

Figure 1. Various forms of arsenic based on pH and redox potential conditions (Swaran, 2015, pg. 5)

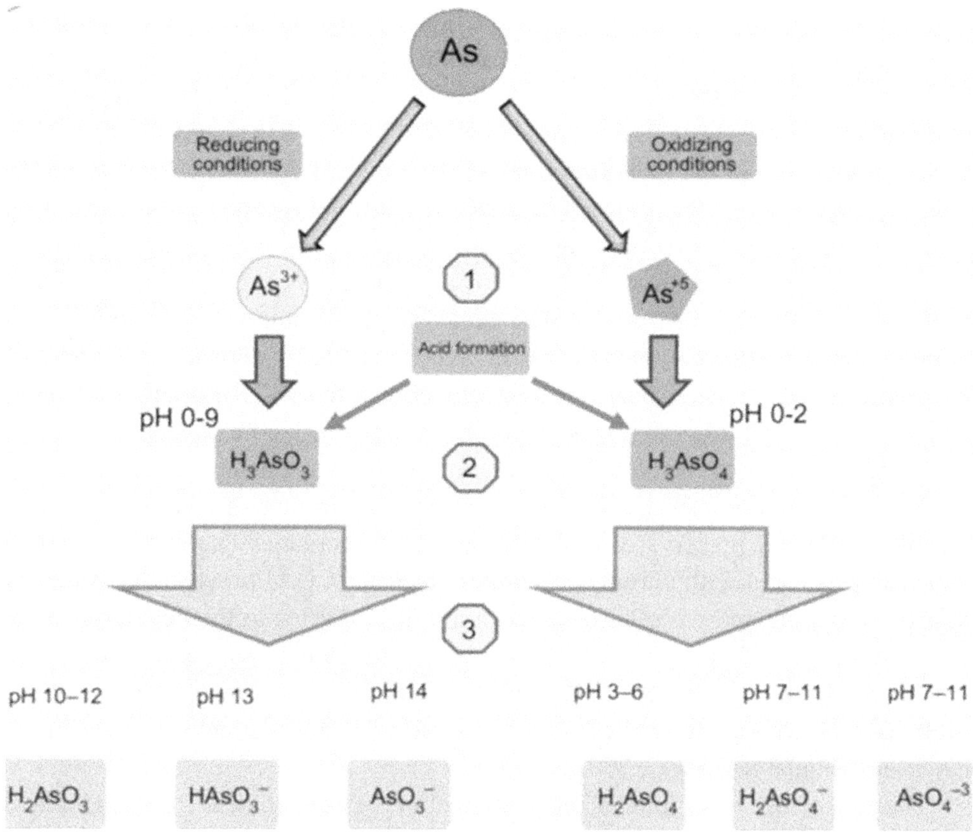

It is also present in the biosphere, atmosphere and hydrosphere. The sources of the metalloid are sulfides and hydroxides of Fe, Au, Pb and Cu complexed with arsenide and stored in sedimentary rocks or aquifer matrices rich in As (Jang et al., 2016). Anthropogenic activities are also adding on to the geogenic sources of As leading to rising concentrations of the metalloid in the lithosphere. According to Bhattacharya et al. (2002), human-based activities resulted to more than 82,000 metric tons of the metalloid additions annually around the worldwide.

The metalloid occurs with other metals and as such, their mining and smelting for industrial use result to its release to the environment. The mining of sulfide ores such as arsenopyrite and production of resultant mine tailings result to the release of the metalloid to soils (Swaran, 2015). The extraction of minerals such as dimethyl arsenic/cacodylic acid, disodium methyl arsenate, monosodium methyl arsenate from the earth's crust also introduces the metalloid to soils (Jang et al., 2016). During the production and processing of coal and other petroleum products, As is released into the soil through deposition of combusted particulate matter (Swaran, 2015). The resultant particulate matter in coal processing is either syngenetic (ore deposits formed during rock enclosure) or epigenetic (rock deposits formed after rock deposition) containing organic and inorganic forms of As (Yudovich & Ketris, 2005). During the smelting of arsenic ore, it is rapidly oxidized to As_2O in the atmosphere and cools down to solid particles deposited on soil to pollute it (Yan & Zhang, 2021).

The human use of the metalloid predates to the 315 B.C, where As was used to enhance the shiny characteristics of bronze. The alloys of As and Cu were also used in the manufacture of ornaments and tolls (Nriagu, 1994). Realgar and orpiment, which are alloys of As were used in the cosmetic industry as dyes (Charles, 1980). Solutions containing the metalloid such as arsphenamine, arsenic iodide and potassium arsenite were used in the medical industry to cure diabetes, tuberculosis, trypanosome infections, malaria, asthma, amoebic dysentery, rheumatism, syphilis and arthritis among other diseases (Merian & Clarkson, 1991).

Inorganic forms of the metalloid including sodium arsenate, lead arsenate and calcium arsenate are useful components in the manufacture of electronics, glass, weedicides, wood preservatives, pesticides and insecticides (Jang et al., 2016). Wood impregnated with chromated copper arsenate (CCA) prevents its attack by microbes, insects and other animals (Swaran, 2015). Prior to the development of Dichlorodiphenyltrichloroethane (DDT) and in the 1940s, As was introduced to soils in the form of pesticides containing the metalloid that were applied on cotton orchards and farms as well as herbicides applied to livestock to prevent insect-based diseases (Mandal & Suzuki, 2002).

The use of the metalloid in the manufacture of weedicides, insecticides and pesticides is currently under evaluation for a ban following the toxic effects of As once it is introduced to soils and ends up in food chains (Kirsti et al., 2022). In modern electronic industries, As is used in manufacture of corrosion resistant materials, metal adhesives, semi-conductors for computers and lead acid batteries. Arsenic trioxide is also used in the making of pharmaceuticals, preservatives and paints, Other industrial uses of As and the specific compound of the metalloid used are as shown in Table 3. The industrial activities result to products containing the metalloid. During use, after use and disposal, if the products come in contact with soils, they can introduce As in such environs resulting to pollution. Unsystematic management of As-containing wastes, their incineration to produce As-based particulate matter and the use of wastewater with As for ferti-irrigation was found to pollute soils with the pollutant (Kayode et al., 2021).

Table 3. Industrial uses of various arsenic compounds (Jang et al., 2016)

Industry	Chemical form of As	Uses
Agricultural farming	Monosodium arsenate- $NaCH_3HAsO_3)_3$	Manufacture of larvicides, weedicides, insecticides and pesticides
Livestock keeping	Sodium arsenite- NaH_2AsO_4	In manufacture of algaecides, in cattle deeps to prevent insect infestation and as feed additives to prevent heartworm infection and swine dysentery
Wood	CCA as Arsenic trioxide- As_2O_3	Preservation of wood and glass decolorization
Medicine	Arsenic acid- H_3AsO_4	In manufacture of medication to treat syphilis, sleeping sickness, amebiasis and trypanosomiasis
Metallurgy	Arsine-AsH_3	Manufacture of alloys used as hardening agents in battery plates, radiators and automotive body solders
Electronic	Gallium arsenide- GaAs	Manufacture of digital watches, light-releasing diodes, semiconductors, optoelectronic gadgets and solar panels
Other industries	Arsenic acid-H_3AsO_4	In the manufacture of defoliated cotton balls, ceramics, paper, textile, soaps, dyes, antifouling paints, pyrotechnics, catalysts, electro photography and glassware products

TOXICITY OF THE METALLOID TO SOILS

The mobilization of As in soils and its subsequent toxicity to soils is a function of four factors, which include (Swaran, 2015):

i. Interaction of the metalloid with soil components such as sulfide minerals, clay, metal oxides and organic matter
ii. Interaction of As with iron, zinc, fluoride and phosphate ions
iii. Dissolution of As in the soils and sediments
iv. Speciation of the metalloid due to its oxidation states

Arsenic occurs as organic arsenic, arsenate or arsenite in soils. In its natural form, As is complexed with aluminum oxides and amorphous iron. The toxicity of soils by the metalloid is regulated by several factors including the redox potential, pH, climate and soil texture (Bhattacharya et al., 2002). When soils are wet (during rainy seasons), redox potential of soils reduces, which enhances release of arsenic from its iron hydroxide complex. In the presence of sulfide, a precipitate of arsenic sulfide is formed but excess amounts maybe released into solution reducing it concentration in soils. During the dry season, arsenic is adsorbed on the soil matrix making its concentrations higher (Jang et al., 2016). In organic and alluvia soils, As concentrations are higher compared to sandy soils (Mandal & Suzuki, 2002). Organic matter traps the metalloid along with other heavy metals, which enhances its accumulation since movement is limited while sandy soils promote leaching of As from the vadose zone to the saturated zone. Clay also plays an important role in As fixation since the metalloid is sorbed in clay particles at neutral pH (Jang et al., 2016). Within soils, the presence of oxides of Mg, Fe and Al promote the metalloids mobility and sorption and hence, its toxicity. Anthropogenic activities such as application of pesticides and fertilizers, mining, combustion of fossil fuels and smelting increase the cationic component of soils and also promote high levels of As in soils. Dissolution and speciation of As is dependent on the prevailing pH and redox potential conditions as earlier discussed.

ARSENIC POLLUTION IN SUB-SAHARAN AFRICA

Many nations of SSA are endowed with minerals ores such as zinc, bauxite, iron, copper, cobalt, gold, silver, diamond and coal, which make the region one of the largest mineral producers worldwide (Irunde et al., 2022). At all scales, the mining industry is growing and inducing several disruptions in the geogenic conditions and eventually, release of toxic heavy metals to land (soil) and water resources. Survey studies reported by Ahoule et al. (2015) noted that the As levels were high in the region between 0.02 and 1760 μg/L and could even go as high as 10,000 μg/L in some soils, surface systems and groundwater systems. Some of the studies in SSA that have reported elevated levels of As in soils are summarized in Table 4.

In parts of Ghana, Kenya, South Africa and Tanzania, elevated levels of As in soils with a range of 0.001 to 20.298 μg/g were reported in several studies (Miller et al., 2018; Nyanza et al., 2014; Ramudzuli & Horn, 2014; Mensah et al., 2020; Kortei et al., 2020, Nyika & Dinka, 2022). Mining activities and resultant mine tailings in Kenya polluted soils with elevated levels of As, whereby Ogola et al. (2002) assayed levels ranging between 0.06 and 76 μg/g in the environments. Pollution of soils with As from gold mine tailings has since been shown to be transferred to cassava at Rwamagasa village of Tanzania where levels of 0.60-1.120 μg/g were found in roots and leaves of the plant (Nyanza et al., 2014). Consumption of As-contaminated soils at Geita gold mine of Tanzania via geophagia was attributable to high levels of the metalloid in urine of pregnant women (Nyanza et al., 2019). The authors noted that in some women, As levels were as high as 150 μg/l.

Table 4. Studies on soil pollution by arsenic in the SSA region and their assayed concentrations

Country	Study Area	Concentration range for Arsenic	Reference
Botswana	Kanye	10-49.2 μg/L	Sracek et al., 2021
Cote D'Ivoire	Boruko	0.1-5.8 μg/g	Kinimo et al., 2021
Mali	Sanson village	6-479 μg/g	Bokar et al., 2020
Ethiopia	Mojo river basin	24.5 ± 0.60 μg/g	Gebeyehu & Bayissa, 2020
Ghana	Abandoned mining spoils	1807-8401 μg/g	Mensah et al., 2020
	Volta region	1.63 ± 0.03 μg/kg	Kortei et al., 2020
South Africa	Limpopo	1033- 1369 μg/g	Okonkwo, 2007
		0.001-46.76 μg/g	Ramudzuli & Horn, 2014
	Krugersdorp game reserve	5-170.3 μg/l	Shapi et al., 2020
	Greater Giyani	0.1-172.52 μg/l	Mudzielwana et al., 2020
Tanzania	Geita, Rwamagasa village	0.183-20.298 μg/g	Nyanza et al., 2014
Kenya	Migori gold belt	0.06-76 μg/g	Ogola et al., 2002
	Kakamega	300 μg/g	Miller et al., 2018
	Suburban regions of Nairobi	20.92-31.37 mg/kg	Nyika & Dinka, 2022

Arsenic Pollution in Eastern African Soils

Evidence of As polluted soils was found in East Africa using different environmental samples including urine from women who engaged in geophagia (Miller et al., 2018), in soils mixed with mine tailings in Kenya (Ogola et al., 2002) and in Tanzania (Nyanza et al., 2014). Arsenic was also found in breastmilk of women who consumed soil and food grown in soils contaminated by the metalloid (Ghane et al., 2022). Activities such as mining and small-scale artisanal engagements in Tanzania also resulted to As-polluted soils (Nyanza et al., 2019). In Ethiopia, As pollution was attributed to both geogenic and anthropogenic sources. The presence of volcanic rocks such as fissural basalts, trachybasalt, rhyolite, pumice and ignimbrites at the Ethiopian rift valley is associated with As-polluted soils (Reimann et al., 2003). The use of freshwater contaminated with As from agricultural drains and industrial effluents resulted to elevated levels of the metalloid in soils. This trend was reported in soils of Mojo area of Ethiopia where Gebeyehu and Bayissa (2020) showed that As pollution of the soils was then transferred to tomato and cabbage plants through root uptake. In Kenya's Nairobi, Kisumu and Kakamega areas where mining activities, unsystematic release of industrial effluents and irrigation using untreated wastewater has occurred, soils were found to be laden with As beyond the 20 mg/kg limits set by WHO (Nyika, 2022; Nyika & Dinka, 2022; Irunde et al., 2022).

Arsenic Pollution in Southern African Soils

Geological and anthropogenic sources of As in southern Africa cause pollution in soils and the distribution of the metalloid is spatially varied. In Botswana and South Africa, several studies showed pollution of soils by arsenic as shown in Table 4 (Ramudzuli & Horn, 2014; Shapi et al., 2020; Sracek et al., 2021). Abiye and Bhattacharya (2019) noted that the release of As in South African environments including soils is a result of sulfide oxidation, industrial and agricultural activities. In particular, the manufacture and use of insecticides, wood preservatives, dyes, steel, semiconductors and glass are some of the industrial causatives of As release. In Johannesburg region, soils were found laden with the pentavalent form of As, which was eventually high in areal groundwater due to leaching and percolation (Abiye & Bharracharya, 2019). Farms located near four mine tailing dumps in Johannesburg had soils whose As levels ranged between 0.1 to 65.3 mg/kg, which was higher than the permissible levels (Mathee et al., 2018). The use of pesticides in a cattle dip at Limpopo was associated with elevated levels of the metal in soils of its vicinity (Okonkwo, 2007; Ramadzuli & Horn, 2014). A similar trend was evident in Venda tribal region, South Africa where soils of previously used cattle dip areas that were not rehabilitated had elevated As levels (Horn & Ramudzuli, 2020). Soils of the Okavango delta in Botswana had elevated levels of As from both geogenic and anthropogenic sources (Mladenov et al., 2014). The high organic matter and naturally occurring orpiment in the region favored the high concentrations of the metalloid. In Eastern Cape region of South Africa, unsystematic disposal of As-containing municipal and industrial wastes was attributable to high levels of the metalloid in soils of Roundhill landfill vicinity and beyond the 5.8 mg/kg allowable limits by DEA (Nyika et al., 2019). In mine tailings of Kanya, southeast Botswana, As levels ranged between 10 and 49 μg/l owing to the natural manganese ore and the mining activities that released the metalloid to the environment (Sracek et al., 2021).

Arsenic Pollution in Western African Soils

Pollution of soils, plants and water resources by As was reported in several western African countries including Nigeria, Togo, Mali, Burkina Faso, Ivory Coast and

Ghana at varied levels spatially (Irunde et al., 2022). In Ghana, As levels were elevated at top-soils of West Ghana (Mensah et al., 2020) and Volta region (Kortei et al., 2020) due to ongoing mining activities and its resultant tailings. A similar trend was also established in Obuasi area (Kumi-Boateng, 2007) and the vicinity of Damang gold mine (Petelka et al., 2019) of Ghana where assayed sediments and soils had high levels of As, which were attributable to gold mining activities. Elevated concentrations of As in fish from Ankobra river were associated with the pollution of areal soils and sediments, which the animals ingested (Gbogbo et al., 2017). The use of agrochemicals and gold mining activities at Agbaou and Boniko areas of Cote d'Ivoire was attributable to high As levels in soils and their eventual uptake by rice grown in the areas (Kinimo et al., 2021). Mining activities were also attributable to high As levels in soils of Sanso village of Mali, where the metalloid was also taken up by areal plants (Bokar et al., 2020). At Iwo area of Nigeria, As found at low levels was associated with anthropogenic activities since the geology of the study showed no lithologic presence of the metalloid (Atobatele & Olutona, 2015). Areas around Poura gold mine of Burkina Faso had an average of 753 ppm of As in soils due to the mine tailings in addition to the lithology of areal rocks (Kagambega et al., 2014).

CONCLUSION

Pollution by arsenic is a growing concern worldwide and more so, in the SSA region. In this chapter, it was established that the metalloid pollutes soils and other environmental resources from geogenic and anthropogenic sources. Anthropogenic sources are associated with the electronic, metallurgy, medicine, agriculture, mining and smelting industries and the unsystematic disposal of wastes from such industries. Arsenite, which is the trivalent form of As was found to be more toxic and affiliated with human-based activities. SSA regions of Western, Eastern and Southern Africa had reported cases of elevated levels of the metalloid in soils mainly due to mining and smelting activities, use of agrochemicals and cattle dip solutions and unscientific management of As-containing wastes. The human-activities introduced As to soils and the pollutant ended up in plants through root uptake, in humans through geophagia among expectant women and into water resources through water flow to surface water bodies or via infiltration to groundwater from the subsurface regions of soils. The lithologic nature of parent rocks in some regions also enhanced the levels of As in soils.

REFERENCES

Abiye, A., & Bhattacharya, P. (2019). Arsenic concentration in groundwater: Archetypal study from South Africa. *Groundwater for Sustainable Development, 9*, 100246. doi:10.1016/j.gsd.2019.100246

Ahoule, G., Lalanne, F., Mendret, J., Brosillon, S., & Maiga, H. (2015). Arsenic in African waters: A review. *Water, Air, and Soil Pollution, 226*(9), 302. Advance online publication. doi:10.100711270-015-2558-4

Anyanwu, B., Ezejiofor, A., Igweze, Z., & Orisakwe, O. (2018). Heavy metal mixture exposure and effects in developing nations: An update. *Toxics, 6*(4), 65. doi:10.3390/toxics6040065 PMID:30400192

Atobatele, O., & Olutona, G. (2015). Distribution of arsenic (As) in water, sediment and fish from a shallow tropical reservoir (Aiba reservoir, Iwo, Nigeria). *Journal of Applied Science & Environmental Management, 19*(1), 95. doi:10.4314/jasem.v19i1.13

Audi, G., Bersillon, O., Blachot, J., & Wapstra, A. (2003). The NUBASE evaluation of nuclear and decay properties. *Nuclear Physics. A., 729*(1), 3–128. doi:10.1016/j.nuclphysa.2003.11.001

Bhattacharya, P., Jacks, G., Frisbie, S., Smith, E., Naidu, R., & Sarkar, B. (2002). Arsenic in the environment: a global perspective. In B. Sarkar (Ed.), *Heavy Metals in the Environment* (pp. 147–215). Marcel Dekker. Inc. doi:10.1201/9780203909300.ch6

Bokar, H., Traore, Z., Mariko, A., Diallo, T., Traore, A., Sy, A., Soumare, O., Dolo, A., Bamba, F., Sacko, M., & Touré, O. (2020). Geogenic influence and impact of mining activities on water soil and plants in surrounding areas of Morila Mine, Mali. *Journal of Geochemical Exploration, 209*, 106429. Advance online publication. doi:10.1016/j.gexplo.2019.106429

Charles, J. (1980). The coming copper and copper-base alloys and iron: A metallurgical sequence. In T. Wertime & J. Muhley (Eds.), *The coming of the age of iron*. Yale University Press.

Chung, J., Yu, S., & Hong, Y. (2014). Environmental source of arsenic exposure. *Journal of Preventive Medicine and Public Health, 47*(5), 253–257. doi:10.3961/jpmph.14.036 PMID:25284196

Department of Environmental Affairs (DEA) (2013). National Environmental Management, Waste Act, 2008. National norms and standards for the remediation of contaminated land and soil quality in the Republic of South Africa. *Government Gazette, No. 36447.*

Emilie, E., Harue, M., Takahiro, S., Aki, N., Yusuke, S., Yusuke, M., & Hitoshi, C. (2017). Geochemical distribution and fate of arsenic in water and sediments of rivers from the Hokusetsu area. *Journal of Hydrology. Regional Studies*, *9*, 34–47. doi:10.1016/j.ejrh.2016.09.008

Gbogbo, F., Otoo, D., Asomaning, O., & Huago, Q. (2017). Contamination status of arsenic in fish and shellfish from three river basins in Ghana. *Environmental Monitoring and Assessment*, *189*(8), 400. doi:10.100710661-017-6118-9 PMID:28718096

Gebeyehu, R., & Bayissa, D. (2020). Levels of heavy metals in soil and vegetables and associated health risks in Mojo area, Ethiopia. *PLoS One*, *15*(1), 1–22. doi:10.1371/journal.pone.0227883 PMID:31999756

Ghane, E., Khanverdiluo, S., & Mehri, F. (2022). The concentration and health risk of potentially toxic elements (PTEs) in the breast milk of mothers: A systematic revies and metal analysis. *Journal of Trace Elements in Medicine and Biology*, *73*, 126998. doi:10.1016/j.jtemb.2022.126998 PMID:35617722

Han, Y., Park, J., Kim, S., Jeong, H., & Ahn, J. (2019). Redox transformation of soil minerals and arsenic in arsenic-contaminated soil under cycling redox conditions. *Journal of Hazardous Materials*, *378*, 120745. doi:10.1016/j.jhazmat.2019.120745 PMID:31203129

Horn, C., & Ramudzuli, M. (2020). Arsenic Contamination of soil in relation to water in Northeastern South Africa. In A. Fares & S. Singh (Eds.), *Arsenic water resources contamination. Advances in water security*. Springer. doi:10.1007/978-3-030-21258-2_7

Irunde, R., Ijumulana, J., Ligate, F., Maity, J., Ahmad, A., Mtamba, J., Mtalo, F., & Bhattacharya, P. (2022). Arsenic in Africa: Potential sources, spatial variability and the state-of-the-art arsenic removal using locally available materials. *Groundwater for Sustainable Development*, *18*, 100746. doi:10.1016/j.gsd.2022.100746

Jang, Y., Somanna, Y., & Kim, H. (2016). Source, distribution, toxicity and remediation of arsenic in the environment-a review. *International Journal of Applied Environmental Sciences*, *11*(2), 559–581.

Jiang, J., Bauer, I., Paul, A., & Kappler, A. (2009). Arsenic redox changes by microbially and chemically formed semiquinone radicals and hydroquinones in a humic substance model quinone. *Environmental Science & Technology*, *15*(10), 3639–3645. doi:10.1021/es803112a PMID:19544866

Kagambega, N., Sawadogo, S., & Gordio, A. (2014). High arsenic enrichment in water and soils from Sambayourou watershed-Burkina Faso (West Africa). *International Journal of Environmental Monitoring and Analysis, 2*(6-1), 6-12. doi:10.11648/j.ijema.s.2014020601.12

Kayode, O., Aizebeokhai, A., & Odukoya, A. (2021). Arsenic in agricultural soils and implications for sustainable agriculture. *IOP Conference Series. Earth and Environmental Science*, *655*(1), 012081. doi:10.1088/1755-1315/655/1/012081

Kinimo, C., Yao, M., Marcotte, S., Kouassi, B., & Trokourey, A. (2021). Trace metal(loid)s contamination in paddy rice (Oryza sativa L.) from wetlands near two goldmines in Cote d'Ivoire and health risk assessment. *Environmental Science and Pollution Research International*, *28*(18), 22779–22788. Advance online publication. doi:10.100711356-021-12360-8 PMID:33423204

Kirsti, L., Muller, I., Reichel, S., Jones, C., Brunet, F., & Guedard, M. (2022). Risk management for arsenic in agricultural soil-water systems: Lessons learned from case studies in Europe. *Journal of Hazardous Materials*, *424*, 127677. doi:10.1016/j.jhazmat.2021.127677 PMID:34774350

Kortei, K., Koryo-Dabrah, A., Akonor, P., Manaphraim, N., Akonor, M., & Boadi, N. (2020). Potential health risk assessment of toxic metals contamination in clay eaten as pica (geophagia) among pregnant women of Ho in the Volta Region of Ghana. *BMC Pregnancy and Childbirth*, *20*(1), 160. doi:10.118612884-020-02857-4 PMID:32169034

Kumi-Boateng, B. (2007). *Assessing the spatial distribution of arsenic concentration from goldmine for environmental management at Obuasi, Ghana*. [Msc Thesis, International Institute for Geo-information Science and Earth Observation, The Netherlands].

Mandal, B., & Suzuki, K. (2002). Arsenic round the world: A review. *Talanta*, *58*(1), 201–235. doi:10.1016/S0039-9140(02)00268-0 PMID:18968746

Mathee, A., Kootbodien, T., Kapwata, T., & Naicker, N. (2018). Concentrations of arsenic and lead in residential garden soil from four Johannesburg neighborhoods. *Environmental Research*, *167*, 524–527. doi:10.1016/j.envres.2018.08.012 PMID:30142628

Mensah, K., Marschner, B., Shaheen, M., Wang, J., Wang, L., & Rinklebe, J. (2020). Arsenic contamination in abandoned and active gold mine spoils in Ghana: Geochemical fractionation, speciation, and assessment of the potential human health risk. *Environmental Pollution*, *261*, 114116. doi:10.1016/j.envpol.2020.114116 PMID:32220748

Merian, E., & Clarkson, T. (1991). *Metals and their compounds in the environment: occurrence, analysis, and biological relevance*. VCH.

Miller, D., Collins, M., Omotayo, M., Martin, L., Dickin, L., & Young, L. (2018). Geophagic earths consumed by women in western Kenya contain dangerous levels of lead, arsenic, and iron. *American Journal of Human Biology*, *30*(4), e23130. doi:10.1002/ajhb.23130 PMID:29722093

Mladenov, N., Wolski, P., Hettiarachchi, M., Murray-Hudson, M., Enriquez, H., & Damaraju, S. (2014). Abiotic and biotic factors influencing the mobility of arsenic in groundwater of a through-flow island in the Okavango Delta, Botswana. *Journal of Hydrology (Amsterdam)*, *518*, 326–341. doi:10.1016/j.jhydrol.2013.09.026

Mudzielwana, R., Gitari, W., Akinyemi, A., Talabi, O., & Ndungu, P. (2020). Hydrogeochemical characteristics of arsenic rich groundwater in Greater Giyani municipality, Limpopo province, South Africa. *Groundwater for Sustainable Development*, *10*, 100336. doi:10.1016/j.gsd.2020.100336

Nriagu, J. (1994). *Arsenic in the environment*. Wiley.

Nriagu, J. (2002). Arsenic poisoning through the ages. In W. T. Frankenberger (Ed.), *Environmental chemistry of arsenic* (pp. 1–26). Marcel Dekker.

Nyanza, C., Bernier, P., Manyama, M., Hatfield, J., Martin, W., & Dewey, D. (2019). Maternal exposure to arsenic and mercury in small-scale gold mining areas of Northern Tanzania. *Environmental Research*, *173*, 432–442. doi:10.1016/j.envres.2019.03.031 PMID:30974369

Nyanza, C., Dewey, D., Thomas, K., Davey, M., & Ngallaba, E. (2014). Spatial distribution of mercury and arsenic levels in water, soil and cassava plants in a community with long history of gold mining in Tanzania. *Bulletin of Environmental Contamination and Toxicology*, *93*(6), 716–721. doi:10.100700128-014-1315-5 PMID:24923470

Nyika, J. (2022). Wastewater for agricultural production, benefits, risks, and limitations. In H. Chatoui, M. Merzouki, H. Moummou, M. Tilaoui, N. Saadaoui, & A. Brhich (Eds.), *Nutrition and human health*. Springer. doi:10.1007/978-3-030-93971-7_6

Nyika, J., & Dinka, M. (2022). Heavy metal pollution in soils and vegetables from suburban regions of Nairobi, Kenya and their community health implications. *Pollution*, *8*(4), 1434–1447. doi:10.22059/POLL.2022.341522.1440

Nyika, J., Onyari, E., Dinka, M., & Mishra, S. (2019). Heavy metal pollution and mobility in soils within a landfill vicinity: A South African case study. *Oriental Journal of Chemistry*, *35*(4), 1286–1296. doi:10.13005/ojc/350406

Nyika, J., Onyari, E., Dinka, M., & Mishra, S. (2020). Comparative assessment of trace metal concentrations and their eco-risk analysis in soils of the vicinity of Roundhill landfill, Southern Africa. *Nature. Environment and Pollution Technology*, *19*(2), 539–548. doi:10.46488/NEPT.2020.v19i02.009

Ogola, S., Mitullah, V., & Omulo, A. (2002). Impact of gold mining on the environment and human health: A case study in the Migori gold belt, Kenya. *Environmental Geochemistry and Health*, *24*(2), 141–157. doi:10.1023/A:1014207832471

Okonkwo, O. (2007). Arsenic status and distribution in soils at disused cattle dip in South Africa. *Bulletin of Environmental Contamination and Toxicology*, *79*(4), 380–383. doi:10.100700128-007-9255-y PMID:17701088

Onyia, P., Ozoko, D., & Ifediegwu, S. (2020). Phytoremediation of arsenic-contaminated soils by arsenic hyperaccumulating plants in selected areas of Enugu state, Southeastern, Nigeria. *Geology. Ecology and Landscapes*, *5*(4), 308–319. do i:10.1080/24749508.2020.1809058

Petelka, J., Abraham, J., Bockreis, A., Deikumah, J., & Zerbe, S. (2019). Soil heavy metal(loid) pollution and phytoremediation potential of native plants on a former gold mine in Ghana. *Water, Air, and Soil Pollution*, *230*(11), 267. doi:10.100711270-019-4317-4

Punshon, T., Jackson, P., Meharg, A., Warczack, T., Scheckel, K., & Guerinot, M. (2017). Understanding arsenic dynamics in agronomic systems to predict and prevent uptake by crop plants. *Science of the Total Environment,* (581 – 582), 209 - 220. doi:10.1016/j.scitotenv.2016.12.111

Rahaman, S., Sinha, A., Pati, R., & Mukhopadhyay, D. (2013). Arsenic contamination: A potential hazard to the affected areas of West Bengal, India. *Environmental Geochemistry and Health*, *35*(1), 119–132. doi:10.100710653-012-9460-4 PMID:22618763

Ramudzuli, R., & Horn, C. (2014). Arsenic residues in soil at cattle dip tanks in the Vhembe district, Limpopo Province, South Africa. *South African Journal of Science*, *110*(7/8), 1–7. doi:10.1590ajs.2014/20130393

Reimann, C., Bjorvatn, K., Frengstad, B., Melaku, Z., Tekle-Haimanot, R., & Siewers, U. (2003). Drinking water quality in the Ethiopian section of the east African Rift valley I- data and health aspects. *The Science of the Total Environment*, *311*(1-3), 65–80. doi:10.1016/S0048-9697(03)00137-2 PMID:12826384

Shapi, M., Jordaan, A., Nadasan, S., Davies, C., Chirenje, E., Dube, M., & Lekoa, M. R. (2020). Analysis of the distribution of some potentially harmful elements (PHEs) in the Krugersdorp game reserve, Gauteng, South Africa. *Minerals (Basel)*, *10*(2), 1–18. doi:10.3390/min10020151

Shrivastava, A., Ghosh, D., Dash, A., & Bose, S. (2015). Arsenic Contamination in Soil and Sediment in India: Sources, Effects, and Remediation. *Current Pollution Reports*, *1*(1), 35–46. doi:10.100740726-015-0004-2

Singh, S., & Srivastava, P. (2020). Bioavailability of arsenic in agricultural soils under the influence of different soil properties. *SN Applied Sciences*, *2*(2), 153. doi:10.100742452-019-1932-z

Sracek, O., Kribek, B., Mihaljevic, M., Ettler, V., Vanek, A., Penizek, V., Veselovský, F., Bagai, Z., Kapusta, J., & Sulovský, P. (2021). Mobility of Mn and other trace elements in Mn-rich mine tailings and adjacent creek at Kanye, southeast Botswana. *Journal of Geochemical Exploration*, *220*, 106–658. doi:10.1016/j.gexplo.2020.106658

Swaran, F. (2015). Arsenic: chemistry, occurrence and exposure. In S. Flora (Ed.), *Handbook of arsenic toxicology*. Academic Press, Elsevier. doi:10.1016/B978-0-12-418688-0.00001-0

Yan, B., & Zhang, X. (2021). Current status, causes and harm of soil arsenic pollution. *IOP Conference Series. Earth and Environmental Science*, *769*(2), 022034. doi:10.1088/1755-1315/769/2/022034

Yudovich, Y., & Ketris, M. (2005). Arsenic in coal: A review. *International Journal of Coal Geology*, *61*(3-4), 141–196. doi:10.1016/j.coal.2004.09.003

Chapter 10
Soil Pollution by Mercury in Sub-Saharan Africa

ABSTRACT

This chapter seeks to understand the sources and chemistry of mercury (Hg), which enable it to exert pollution in soils of Sub-Saharan Africa (SSA) region. Cinnabar was found to be the most common ore from which the heavy metal is found. In soils and as a result of anthropogenic activities, various forms of Hg are found including the particulate, gaseous reactive, and reactive gaseous compounds of the metal. Methyl Hg was found to be the most lethal form in soils due to its high absorption capacity and longer half-life allowing its cycling in different environmental compartments. In SSA, lethal forms of mercury were introduced to soil through Hg-amalgamation during artisanal gold mining. Industrial activities including coal processing and cement making and their associated wastes also polluted the soils of the region. From the growing concern with Hg pollution in soils, it is key to monitor the trends and devise remedial measures.

INTRODUCTION

Mercury (Hg) is one of the metals that are toxic to living organisms and the environment in addition being potential carcinogens (Ryzhenko et al., 2021). According to Liu et al. (2021) elemental Hg is classified as a class D carcinogen while methyl mercury is categorized as a class C carcinogen, which may trigger reproductive toxicity, neurotoxicity and nephrotoxicity. Class C carcinogens have the potential to cause cancers in humans while in class D, suspicions on cancer-causing potential of the element have been raised but are not yet proven by clinical tests. According to

DOI: 10.4018/978-1-6684-7116-6.ch010

the World Health Organization (WHO, 2017), Hg is a top ten chemical, affiliated with public health concerns due to its potential to induce environmental pollution. Mercury enters the environment and circulates around the world to pollute land and water resources, end up in plants, foods, animals and humans to become a global anthropogenic pollutant. The concentrations of Hg have increased by 300-500% over the past 100 years due to the rise in anthropogenic activities (Mason et al., 2012; AMAP/ UN Environment, 2019).

During the 19th century, the use and production of Hg has grown and so are its effects to the environment. Levels of the heavy metal in soils and sediments have risen 3-10 times more after the emergence of post-industrial anthropogenic activities (Zhang, 2019). Mercury concentrations in areas affected by a worldwide-pool atmospheric deposition range between 0.01 to 0.3 mg/kg (Agnan et al., 2016; Obrist et al., 2016) while anthropogenic-based contaminated sites have 2 to 4 times higher the concentration. In another study, Hg released to the environment (air, water and soil) amounted to 2 thousand tons in 2018 and an additional 260 tons of the heavy metal is estimated to pollute freshwater lakes and rivers from contaminated soils (US Environmental Protection Agency, USEPA, 2019). Gaseous Hg and particulates of the element are retained in the atmosphere for a period of 0.5 to 2 years and travel for long distances before their precipitation and deposition to soil surfaces (Li et al., 2020).

The effects associated with Hg-polluted soils include reduced seed germination chances, altered soil microbial activity and plant morbidity in addition to entry of the metal to food chains to induce negative health effects to humans. Methyl mercury taken up by plants and animals from polluted soils resulted to bio-amplification of the metal content in humans, which negatively affected their health (Tong et al., 2016; Abeysinghe et al., 2017). Specific effects of Hg polluted soils to plants, microorganisms, animals and humans are as summarized in Table 1. These effects necessitate an insight on the heavy metal and particularly its physicochemical features that enable its toxicity to soils. According to Gworek et al. (2020), understanding the cycling of heavy metals such as Hg is important because of two reasons. First, it is essential in predicting quantities of pollution levels to the environment and devising counter measures to reduce the concentrations in soils, water and air environmental compartments. Second, understanding heavy metal cycling is key in reducing the resultant impacts of the heavy metals in humans and the biota. In this chapter, the chemistry of Hg is discussed, its sources and extent of pollution in soils of sub-Saharan Africa (SSA) detailed. The aim is to related soil pollution by the metal to increased anthropogenic activities particularly, industrial activities in the region, which are on a growing trend.

Table 1. Noxious effects of soil mercury on living systems

Affected living organism	Targeted part/s	Signs	Reference
Humans and animals	DNA	Alteration of epigenetic and genetic factors	Basu et al., 2014
	Reproductive organs	Induces infertility, low sperm count and their reduced motility, increased likelihood for embryo defects and affects neonatal development	Fillion et al., 2006
	Heart	Causes myocardial infarction, hypertension, irregular pulse, palpitations and coronary dysfunction	Fillion et al., 2006
	Kidney	Hormonal imbalances and increased levels of renal Hg	Liu et al., 2021
	Nerves	Sleeping disorder, memory losses, bipolar disorder, schizophrenia, hearing losses	Cesarani et al., 2010
	In invertebrates	Causes reproductive disorders, and inhibits enzyme activity in insects, results to oxidative dysfunction	Tang et al., 2017
Microorganisms		Results to slow growth, low survival rates, reduces community diversity and inhibits soil enzyme activity, nitrification and respiration	Mahbub et al., 2017
Plants	Enzymes	Inhibits the activity of acid phosphatase, amylase, dehydrogenase, nitrate reductase and protease enzymes	Jiang & Zhao, 2001
	Antioxidant system	Enhances oxidative stress of plant cells and promote peroxidation of lipids	Israr & Shivendra, 2006
	Photosynthetic system	Inhibits chlorophyll synthesis and destroys its structure to reduce the rate of photosynthesis	Wang et al., 2008
	Leaves, stem, seed and roots	Stunted growth of all plant parts, browning/ blackening and eventual death of stems and leaves and reduced seed germination	Huang, 2011
	General plants	Suppresses plant growth and to the extreme can cause death	Qu et al., 2019

CHEMISTRY OF MERCURY

Mercury describes a fluid metal, which is silver colored (Figure 1), a liquid at 25 °C, in the d-block of heavy metals, solidifies at -38.83 °C and sometimes acts as a noble gas (Ariya et al., 2015). The metal is a rare element whose abundance is estimated at 0.08 mg/kg in the earth's crust. The element is associated with cinnabar (HgS), which is its commonest ore in addition to minerals such as corderoite ($Hg_3S_2Cl_2$), metacinnabar (β-HgS) and living stonite ($HgSb_4S_8$) (Rytuba, 2003; Beckers & Rinklebe, 2017). The metal can also be complexed with sulfides of Fe, Zn and other metals in the earth's crust. The metal ores occur near volcanic and hot spring regions in young orogenic belts, which contribute to half of its natural emissions. The rest of the fraction is from various industrial activities and the production of Hg-containing consumer products. Hg is the only metal that exists as a liquid at room temperature and has the highest surface tension at 466 ± 33 mN m^{-12}. The metal exists in various species and compounds found as vapor or particulates in combination with clouds and aerosols in the atmosphere. They include elemental mercury, (Hg^0), mercury (II) chloride, $HgCl_2$, mercury (II) bromide, $HgBr_2$ and mercury (II) hydroxide, $Hg(OH)_2$ (Li et al., 2020). The compounds are volatile, easily react with water and are known as the reactive gaseous mercury (RGM). Elemental Hg (Hg^0) is the commonest of the species and its levels in the atmosphere are approximated at 6000 tons. The physicochemical characteristics of various Hg species is a shown in Table 2.

Mercury in the soil exhibits different redox states: the metallic (Hg^0), mercurous (Hg^+ or Hg (I)) and mercuric (Hg^{++} or Hg (II) (WHO, 2000; Liu et al., 2012). The different redox forms in soils in addition to microbial activity, pH, and Cl^- regulate the speciation of the metal in soil solutions and its subsequent chemical reactions. The mercuric form is responsible for most of the organometallic and inorganic compounds of the metal. All compounds of Hg are toxic but methyl mercury (MeHg) is the most toxic (Liu et al., 2012). In organometallic forms of Hg, the metal occurs covalently bonded to one of two atoms of carbon. The elemental/ metallic form of Hg boils at 357 °C and is a dense and shiny metal (Figure 1). Additionally, it has a vapor pressure of 0.17 Pa at 20 °C and a saturated atmosphere of 14 mg/m^3 (WHO, 2000). At 25 °C the solubility of different mercury compounds varies such that mercuric chloride, mercurous chloride and elemental Hg have a solubility of 69, 2 and 60 g/liter, respectively.

Figure 1. The physical appearance of mercury
(Chemistry Explained, 2023)

Table 2. Physicochemical properties of mercury species and compounds (Ariya et al., 2015; Haynes et al., 2014)

Species	Species Name	Color/state	Standard enthalpy of formation (KJ/mol)	Gibbs energy change (KJ/mol)	Melting point (°C)	Water solubility at 25 °C	Density (kg/m³)	Ionization energy (eV)	Oxidation state	Atomic mass	Vapor pressure (Pa)
Hg^0	Elemental mercury	Silver colored liquid	0.0	0.0	-38.83	3 x 10^{-7}	13533.6	10.44	2, 1	200.59	1 at 42 °C
$HgCl_2$	Mercury (II) chloride	Crystalline white solid	-224.3	-178.6	277	0.27	5600	11.38	1	271.5	173.32 at 236°C
Hg_2Cl_2	Mercury (I) chloride	Crystalline white solid	-265.4	-210.7	383 as sublimation point	8 x 10^{-6}	7160	-	0	472.1	-
$Hg(CH_3)_2$	Dimethyl mercury	Colorless liquid	59.8	140.3	-43	Insoluble	3170	9.1	-	230.66	8305.98 at 25 °C
Hg_2Br_2	Mercury (I) bromide	Crystalline white solid	-206.9	-181.1	345 as decomposition point	7 x 10^{-7}	7307	-	0	560.99	-
$HgBr_2$	Mercury (II) bromide	Crystalline white solid	-170-7	-153.1	241	0.017	6050	10.56	1	360.40	16040 at 238 °C
HgO	Mercury (II) oxide	Red or yellow crystalline solid	-90.79	-58.5	500	2 x 10^{-4}	11140	-	0	216.59	-
HgS	Mercury (II) sulfide	Red or black crystalline solid	-58.2	-50.6	446	Insoluble	7700	-	0, -2	232.66	-
$HgSO_4$	Mercury (II) sulfate	Crystalline white solid	-707.5	-	-	Reacts with water	6470	-	0	296.65	-
$Hg(NO_3)_2$	Mercuric nitrate	Colorless crystalline solid	-	-	79	Soluble	4300	-	0	324.6	-

The mercuric form of the metal rarely occurs as a free cation (Hg^{2+}) due to its high reactivity and ability to form complexes in natural settings. In acidic solutions, Hg (II) has a 0.4 V redox potential, occurs as $HgCl_2^0$ complex and is stable. In basic solutions, the $Hg(OH)_2^0$ complex is formed. In the soil's humus, mercuric cations (Hg^{2+}) can be bound to reduce organic sulfur (Steinnes, 2013). When in reducing conditions, metallic mercury stabilizes in the presence of bisulfide (HS^-) and hydrogen sulfide (H_2S). However, at rising redox potential, HgS precipitates or forms hydrogen sulfide anions (HgS_2^{2-}) anions in soils that are strongly alkaline. An additional increase in the redox potential transforms sulfide to sulphate and enabled reduction of Hg^0, which can be oxidized to Hg (II) (Steinnes, 2013). In different environments including sediments, soils and water, Hg (II) is dominant while in the biota, MeHg is the commonest (Liu et al., 2012).

SOURCES AND USES OF MERCURY

A number of natural sources of Hg have been identified. These include 1) emissions of the metal from terrestrial environs including background soils and Hg-laden substrates, 2) geological activities such as geothermal and volcanic emissions and 3) volatilization of the metal from marine environs (Liu et al., 2012). Quantifying these kinds of Hg-emissions is difficult due to the uncertain nature in which they occur. Among the natural sources of Hg, gaseous elemental mercury (GEM) form is the most predominant while the reactive gaseous mercury (RGM) forms a minute portion.

The emission and use of Hg from various human activities globally has been associated with contamination of both terrestrial and aquatic ecosystems (Guedron & Acha, 2021). Some of the sources of the heavy metal and their affiliated quantities of Hg emissions are as shown in Figure 2. During the industrial era, the environmental burden of Hg has increased by a factor of 2 to 5 (Gworek et al., 2020). Although in the last four decades the emissions containing the metal have reduced, industrial activities still contribute more than half of the environmental emissions of the metal. Contrary to natural Hg sources, anthropogenic-based Hg can emit RGM, GEM and particulate Hg (PHg) species of the metal all of which are toxic in the environment (Liu et al., 2012).

Human-based Hg emissions emanate from coal-fired plants or via waste incineration and from unsystematic disposal of mine tailings and overburden rock, sewage sludge and landfill waste containing the metal (AMAP/UN Environment, 2019). During coal combustion by industrial boilers and in electrical power plants, Hg is emitted in the form of gaseous Hg^0, as particulate-bound Hg (Hg_p) or as oxidized mercury cations of the metal (Hg^{2+}) in very small quantities (Mathebula et al., 2020).

Figure 2. Anthropogenic sources of mercury
(Liu et al., 2012)

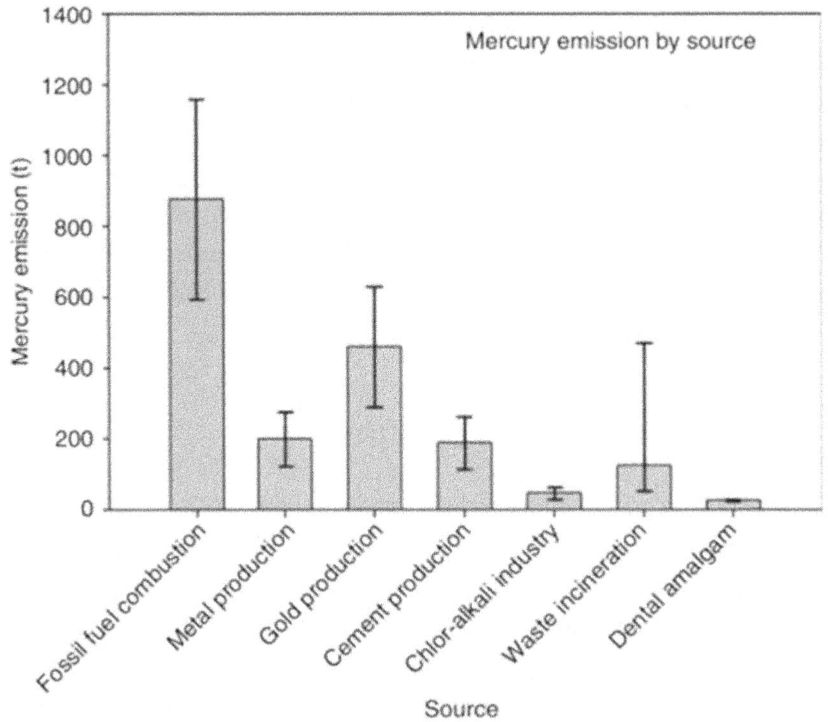

Point sources of the metal such as dental amalgams, manufacturing processes and combustion processes contribute more than 95% of all Hg anthropogenic emissions (Liu et al., 2012). Sources of the metal through combustive activities include burning oil and/or coal, municipal waste, medical waste incineration and sewage sludge all associated with industrial processes, residential and industrial heating using boilers as well as power generation. During combustive activities Hg (II) and Hg (0) are emitted in gaseous and particulate forms depending on the specific fuel being burned, the operating temperature and fuel gas cleaning at a specified atmosphere (Gworek et al., 2020). Figure 1 and Table 4 show that fuel combustion activities produce the largest portion of mercury emissions. Sources of Hg from the manufacturing sector include mining and large-scale production of gold, oil refining, production of chlor-alkali, paper and pulp manufacturing that produce and use mercury-based compounds as raw materials (Sundseth et al., 2017). In addition, production of cement, non-ferrous metals, Hg, steel and pig iron processing/ smelting industries contribute to the metal's emissions (Beckers & Rinklebe, 2017; Sundseth et al., 2017; Wang et al., 2020; Liu et al., 2021). Manufacturing processes unlike combustive ones, release Hg into the terrestrial and aquatic ecosystems in addition to the atmosphere. In the agriculture industry, some agrochemicals used in foliar sprays and seed dressing contain Hg and their application also results to soil pollution (Steinnes, 2013).

The use of Hg began during the ancient times based on archeological evidence that reports of the metal being used more than 3500 years ago. In the sixth century, Greece, China and India used elemental Hg and cinnabar in medicine while Egyptians used the metal to make copper and tin amalgamations (Steinnes, 2013). The Roman empire used the metal for production and commercialization of vermillion (HgS), a red pigment. It was not until the sixteenth century when the Spanish Americans introduced large scale mining of Ag and Hg for industrial uses such as the making of scientific instruments such as thermometers, vacuum pumps and barometers (Steinnes, 2013; WHO, 2000). Some of the industries that used the metal included the health industry for dentistry, in the agriculture, pharmaceutical, paint and chloralkaline firms. The production and mining of Hg in 1600 was rated at 400 tons and increased to 1000, 2000, 4000 and 8000 tons in the 1800, 1850, 1900 and after 1950s, respectively (Steinnes, 2013). As concerns on environmental pollution by Hg grew, the use and production of the metal in 1990s reduced to 6000 tons (Steinnes, 2013). Overall, more than 1,000,000 tons of Hg has been mined globally over the years and main producers are Spain, Slovenia, Italy, China, Australia, Kyrgyzstan, Tajikistan, Tunisia and Algeria (WHO, 2000).

Table 3. Annual emissions of Hg in selected parts of the globe in mg/year (Streets et al., 2019)

Region	Yearly emissions in 2000	Yearly emission in 2015	Percentage growth between 2000-2015	Percentage growth between 2010-2015
Southeast Asia	224.6	187.51	-16.5	2.2
South America	239	275.8	15.4	0.7
East Asia	532.7	1012.3	90	2.6
Eastern Africa	19.1	72.9	281.7	4.0
South Asia	120.8	191.6	58.6	4.6
Central America	33.5	34.3	2.4	5.4
Eastern Europe	49.4	33.6	-32	-1.3
Canada	12.7	8.3	-34.6	-3.2
USA	127.7	42.7	-66.6	-10.2
OECD Europe	106.2	38	-64.2	-5.8

In modern day, the applications of Hg have diversified even to the electrical, manufacturing and processing industries, which has led to extensive mining and consequently, negative effects on the environment. Regional distribution of Hg emissions in parts of the world between 2000 and 2015 are shown in Table 10. 3. Notable increases of over 280% in emissions of Hg in Eastern Africa were recorded between 2000 and 2010. The trend was recently observed in the region by Nipen et al. (2022) who noted that gaseous elemental mercury sources were growing especially in industry-concentrated urban Eastern Africa though its quantification was limited. The environmental increases of the element in the environment are a result of natural degassing of the earth's crust, which results to production of Hg vapor to the atmosphere and its deposition to soils. The emissions are either from natural or anthropogenic sources. Table 4 shows the global Hg budget according to the United Nations Environment Program, UNEP (2019). Evidently anthropogenic activities and subsequent deposition of emitted Hg is a great contributor to the quantities of the metal found in our environment including soils. Data also showed that yearly deposits exceeded emitted amounts of Hg, a trend attributable to enrichment of the metal in the environment as well as the prolonged retention of the metal in the atmosphere (Gworek et al., 2020).

Table 4. Global Hg budget in mg/ year (UNEP, 2019)

Type of emission	Amount (mg/ year)
Geogenic	500
Re-mobilized and re-emitted Hg Biomass burning Soil- and vegetation-sourced	600 1000
Anthropogenic/ industrial sources	2000-3000 (an average of 2500)
Oceanic emission of GEM	2900-4000 (an average of 3400)
Total	7000
Deposition To freshwater and land To ocean Total	3600 3880 7480

MERURY TOXICITY IN SOILS

Once different forms of Hg are introduced to soils, they can undergo several transformations. They can be 1) volatilized and escape to the atmosphere, 2) leached from the soil along with soil water, 3) retained and/ or 4) methylated.

Volatilization of Hg and Hg-based compounds makes them mobile and as such, they travel to long distances before they are precipitated and deposited in terrestrial and aquatic ecosystems including surface soils (Gworek et al., 2020). The retention and methylation of Hg is attributable to the toxicity of the metal in soils (Steinnes, 2013). Mercury retention in soils is facilitated by adsorption, a process influenced by the soil redox potential, soil pH, amount and nature of organic and inorganic soil colloids, soil grain size distribution and the chemical form of Hg (O'Connor et al., 2019). Through formation of selenide and sulfide complexes with Hg^{2+}, precipitates that have low solubility can be formed to enable the metal's fixation. Ion exchange also facilitates retention of Hg^{2+} in soils through formation of strong bonds such as ligands with soil humic substances or hydroxyligands formation with sesquioxides. In soils such as of lateritic nature, illitic clays, montmorillonite, kaolinite and organic soils, inorganic Hg (Hg^{2+}) is highly retained (Steinnes, 2013).

Methylation of Hg in soils is formed in the presence of fulvic acid and/or via the abiotic processes. The process is highly favored in water-saturated soils, under anaerobic conditions and when Hg (II) species are involved (Steinnes, 2013). Methyl Hg formation is favored by microorganisms and once the species are formed, they binds to proteins and can pass the metal's toxicity to biological membranes (Bigham et al., 2017). Unlike other Hg-species and compounds, MeHg bioaccumulates in soils, is transferred to foods and eventually food webs. As such, it is a key determinant of the effect of Hg on the environment and human health. Ryzhenko et al. (2021) made a similar observation suggesting that MeHg is absorbed six time better compared to Hg (II), which makes it toxic to ecosystems and living cells.

Mercury-pollution in soils is possible because of the long half-life of the metal and its recycling between the atmosphere and the surface environment. If the anthropogenic-based Hg is buried in mineral soils, then it can be removed from the biologically active portion of the environment. Soil therefore, is key in Hg circulation in that it enables the accumulation of the metal. In several studies, Hg accumulation in soils globally was estimated at 200 to 300 Gg (Mason et al., 2012; Hararuk et al., 2013; Amos et al., 2015). A study by Smith-Downey et al. (2010) estimated the levels to be 20% higher at 240 Gg due to a rise in industrial activities in modern day. As a result of industrial activities in the last century, levels of Hg emissions in the atmosphere increased by 300-500%, which translates to increases in soil deposition of the metal (Mason et al., 2012). The increase in concentrations of the metal translates to pollution rises considering all forms Hg are toxic. Bioaccumulation of the metal is more prone in soils of developing nations such as in SSA and Asia, where efforts and initiatives to remediate and reclaim polluted areas remain elusive. For this reason, the chapter focuses on the causes and extent of soil pollution by Hg in SSA in the following sub-sections.

SOIL POLLUTION BY MERCURY IN SUB-SAHARAN AFRICA

The SSA region has seen increasing Hg emissions unlike other parts of the world such as North America and Europe where such emissions are on a declining trend (Nipen et al., 2022). Consequently, a rising trend in deposition of the metal on surface soils and water bodies is evident. The situation is compounded by the lack of environmental measures to prevent pollution in general by regulating authorities in countries of the region. Monitoring of Hg emissions and their capacity to pollute soils in SSA is conducted through active sampling to provide temporal variations in the metal concentrations at different sites. Temporal variations provide information on the sources and changes in both emissions and depositions as a result of anthropogenic activities. Soil pollution in SSA region by Hg mainly results from artisanal gold mining activities (Bonzongo et al., 2004). Small-scale mining and artisanal processing of precious metals, building materials and gemstones is common in the region since it requires minimal to no skills and provides livelihoods to the rural poor who are mainly unskilled. For these reasons, millions of SSA residents engage in artisanal and mining activities, which is legal and an essential source of income. For instance, gemstones and gold mining alone result to more than $1 billion of income for the region annually (Bonzongo et al., 2004). With legalization of the activities, there is need to improve them to be environmentally friendlier and sensitive in addition to being economically productive.

In the artisanal gold mining activities in SSA region, Hg-amalgamation is used to isolate the native ore of gold or gemstone. The process involves addition of small concentrations of Hg to the grinding or stamping unit, where the resultant amalgam is contained on Hg-coated copper plates to extract the precious metal via scrapping and distilling off the Hg. The distilled Hg is then released to the environment to undergo volatilization and later deposition on soils in particulate form, where it causes pollution. Consequently, artisanal gold mining using Hg-amalgamation in addition to lack of knowledge on the functioning of ecosystems, diverse socio-economic variables and non-enforcement of or lack of legislations on environmental protection are attributable to Hg-pollution in SSA soils more so as the practices expand. SSA countries such as South Africa, Zimbabwe, Ghana, Democratic Republic of Congo, Mali, Ethiopia, Republic of Congo, Gabon, Tanzania, Rwanda and Central African Republic are known to engage in a lot of gold mining via Hg-amalgamation process (Bonzongo et al., 2004).

Soils of South African areas near gold mines of Inkomati water management regions were laden with Hg due to artisanal activities (Walters et al., 2011). In Witwatersrand (Kamunda et al., 2013) and West Rand (Lusilao-Makiese et al., 2013) areas of Johannesburg, South Africa, artisanal gold mining activities were attributable to Hg-laden soils. In both cases, Hg^{2+} and MeHg levels were found to be beyond the permissible levels. In Zimbabwe, Hg-pollution of the environment including soils due to artisanal and small-scale gold mining is a growing concern (Metcalf & Veiga, 2012) and areas near such mines such as the Farvic region of southeastern Zimbabwe have reported of soils laden with the heavy metal (Green, 2017). A similar trend was reported at gold-mining and smelting districts of Namibia including Kombat, Berg Aukas and Tsumeb as well as the Zambian Copperbelt regions of Mufulira and Kitwe, where assayed soils had Hg levels ranging between 0.0038- 4.39 mg/kg beyond the allowable limits of 0.0016 mg/kg in the two countries (Podolsky et al., 2015). Coal combustion by industrial boilers and at coal power plants in South Africa were attributable to the introduction of Hg gaseous, particulate and ionic forms in the atmosphere. Consequently, the forms of the trace metal cooled off after emission and were deposited on soils surrounding KwaZulu-Natal, Mpumalamga, Limpopo, Gauteng and Eastern Cape provinces of South Africa (Mathebula et al., 2020).

In West Africa, illegal small-scale gold mining commonly referred to us 'galamsey' is known to directly or indirectly introduce raw or residual forms of Hg to soils leading to pollution. In Western Ghana, 11 regions where galamsey practices were common had elevated levels of Hg in soils, which was attributable to unsystematic use, handling and disposal of the metal during mining and smelting activities (Mantey et al., 2020). Bonzongo et al. (2004) as well as Kuffour et al. (2018) also reported of Hg-polluted soils in most Ghanian mines due to the use of Hg-amalgamation during the gold mining. Soils of Nigeria's Itagunmodi region (Southwest area) (Eludoyin et al., 2016) and Maiwayo (Northern area) (Darma et al., 2016) were also reported to be laden with Hg due to artisanal gold mining. In Tanzania, related studies at Buzwagi (Gomezulu et al., 2018), North Mara and Geita (O'Sullivan et al., 2016) gold mines have confirmed that smelting and mining of gold at small-scale results to pollution of soils by Hg among other heavy metals. In soils of Migori-Transmara area of Kenya, artisanal gold mining was attributable to elevated Hg levels in soils at an average of 140 µg/kg owing to release of mine tailings to such environs (Odumo et al., 2014).

Apart from mining, soils of SSA region are polluted by Hg due to a number of anthropogenic activities. In named water management catchment areas of South Africa, Hg-pollution of soils was attributable to coal firing during energy production, cement manufacturing processes and contamination of soils by Hg-containing urban and agricultural effluents (Walters et al., 2011). In cement-making, Hg occurs as a complex in limestone, which is the main raw material or could be a result of coal-firing during the involved processes. In another study, Nipen et al. (2022) noted that soils of East Africa region were vulnerable to Hg-pollution due to smelting of other metals, incineration of medical wastes containing dental amalgams, cement production, burning of biomass and processing of electronic waste containing the metal. Increased use of mercury-containing lamps and landfill disposal of their waste after use has also been a concern due to its potential to increase the metal's emissions and deposition to soils, which pollutes the environment (World Bank, 2011). Incineration of Hg-containing household wastes that are hazardous also resulted to the metal's deposition and eventual pollution of soils in SSA region (Edokpayi et al., 2017). International trading of tons of Hg-containing waste from America and Britain to South Africa was attributable to high levels of the metal in the deposited environs including soils up to 1000-9000 times more than the WHO standards (Akpan & Olukanni, 2020).

CONCLUSION

Mercury pollution in soils among other environmental compartments is a growing concern in SSA. This study established that all forms of the metal are toxic in soils but methyl Hg has the greatest effect due to its longer half life and high absorption capacity in the environment. The point sources of Hg such as dental amalgams, manufacturing processes and coal combustion activities had the greatest contribution of the metal release in soils at 95%. Artisanal gold mining involving the use of Hg amalgams and subsequent distillation and release of the byproducts with high concentrations of the contaminant also polluted soils of SSA once emitted Hg was precipitate and deposited. Cement making using Hg-complexed limestone, incineration of medical wastes with dental amalgams and waste trading from developed to developing nations also resulted to soil pollution. Evidently, anthropogenic activities largely contributed to soil pollution by Hg. Therefore, SSA nations should invest on community awareness and devise laws to regulate regional industrial activities as well as waste trading and introduce the necessary environmental measures to control Hg pollution in soils and other environments.

REFERENCES

Abeysinghe, S., Qiu, G., Goodale, E., Anderson, N., Bishop, K., Evers, C., Goodale, M. W., Hintelmann, H., Liu, S., Mammides, C., Quan, R.-C., Wang, J., Wu, P., Xu, X.-H., Yang, X.-D., & Feng, X. (2017). Mercury flow through an Asian rice-based food web. *Environmental Pollution*, *229*, 219–228. doi:10.1016/j.envpol.2017.05.067 PMID:28599206

Agnan, Y., Le Dantec, T., Moore, C., Edwards, G., & Obrist, D. (2016). New constraints on terrestrial surface atmosphere fluxes of gaseous elemental mercury using a global database. *Environmental Science & Technology*, *50*(2), 507–524. doi:10.1021/acs.est.5b04013 PMID:26599393

Akpan, V., & Olukanni, D. (2020). Hazardous waste management: An African overview. *Recycling*, *5*(3), 15. doi:10.3390/recycling5030015

Amos, H., Sonke, J., Obrist, D., Robins, N., Hagan, N., Horowitz, H., Mason, R., et al (2015). Observational and modeling constraints on global anthropogenic enrichment of mercury. *Environmental Science and Technology, 49*, 4036– 4047. . doi:10.1021/es5058665

Arctic Monitoring and Assessment Program/ United Nations Environment (AMAP/ UN Environment) (2019). *Technical Background Report for the Global Mercury Assessment 2018*. Arctic Monitoring and Assessment Program, Oslo, Norway/UN Environment Program, Chemicals and Health Branch, Geneva, Switzerland

Ariya, P., Amyot, M., Dastoor, A., Deeds, D., Feinberg, A., Kos, G., Poulain, A., Ryjkov, A., Semeniuk, K., Subir, M., & Toyota, K. (2015). Mercury physicochemical and biogeochemical transformation in the atmosphere and at atmospheric interfaces: A review and future directions. *Chemical Reviews*, *115*(10), 3760–3802. doi:10.1021/cr500667e PMID:25928690

Basu, N., Goodrich, M., & Head, J. (2014). Ecogenetics of mercury: From genetic polymorphisms and epigenetics to risk assessment and decision-making. *Environmental Toxicology and Chemistry*, *33*(6), 1248–1258. doi:10.1002/etc.2375 PMID:24038486

Beckers, F., & Rinklebe, J. (2017). Cycling of mercury in the environment: sources, fate and human health implications: a review. *Critical Reviews in Environmental Science and Technology*, *47*(9), 693–794. doi:10.1080/10643389.2017.1326277

Bigham, G., Murray, K., Slowey, Y., & Henry, E. (2017). Biogeochemical controls on methylmercury in soils and sediments: Implications for site management. *Integrated Environmental Assessment and Management*, *13*(2), 249–263. doi:10.1002/ieam.1822 PMID:27427265

Bonzongo, J., Donkor, A., Nartey, V., & Lacerda, L. (2004). Mercury pollution in Ghana: a case study of environmental impacts of artisanal gold mining in sub-Saharan Africa. In L. Drude, R. Santelli, E. Duursma, & J. Abrao (Eds.), *Environmental Geochemistry in the tropical and subtropical environments*. Springer. doi:10.1007/978-3-662-07060-4_12

Cesarani, A., Minoia, C., Pigatto, D., & Guzzi, G. (2010). Mercury, dental amalgam, and hearing loss. *International Journal of Audiology*, *49*(1), 2. doi:10.3109/14992020902962439 PMID:20001448

Chemistry Explained. (2023). Mercury. *Chemistry Explained.* http://www.chemistryexplained.com/elements/L-P/Mercury.html

Darma, M., Kankara, I., & Abdullahi, S. (2016). Effect of artisanal gold mining at Maiwayo environ northern Nigeria: implication for environmental risk. *Sustainable Economic Development Conference*, Abuja, Nigeria.

Edokpayi, J., Odiyo, J., Durowoju, O., & Adetoro, A. (2017). Household hazardous waste management in sub-Saharan Africa. In D. Mmereki (Ed.), *Household hazardous waste management*. IntechOpen. doi:10.5772/66292

Eludoyin, A., Ojo, A., Ojo, T., & Awotoye, O. (2017). Effects of artisanal gold mining activities on soil properties in a part of southwestern Nigeria. *Cogent Environmental Science*, *3*(1), 1305650. doi:10.1080/23311843.2017.1305650

Fillion, M., Mergler, D., Carlos, J., Larribe, F., Melanie, L., & Jean, G. (2006). A preliminary study of mercury exposure and blood pressure in the Brazilian Amazon. *The Science of the Total Environment*, *5*(1), 29. doi:10.1186/1476-069X-5-29 PMID:17032453

Gomezulu, E., Mwakaje, A., & Katima, J. (2018). Heavy metals and cyanide distribution in the villages surrounding Buzwagi gold mine in Tanzania. *Tanzania Journal of Science*, *44*(1), 107–122.

Green, C. (2017). *A comparison of factors affecting the small-scale distribution of mercury contamination in a Zimbabwean stream system*. [MSc Thesis, Sam Houston State University, Texas, USA].

Guedron, S., & Acha, D. (2021). Mercury and methylmercury contamination of terrestrial and aquatic ecosystems. *Applied Sciences (Basel, Switzerland)*, *11*(11), 4807. doi:10.3390/app11114807

Gworek, B., Dmuchowski, W., & Dabrowska, A. (2020). Mercury in the terrestrial environment: A review. *Environmental Sciences Europe*, *32*(1), 128. doi:10.118612302-020-00401-x

Hararuk, O., Obrist, D., & Luo, Y. (2013). Modelling the sensitivity of soil mercury storage to climate-induced changes in soil carbon pools. *Biogeosciences*, *10*(4), 2393–2407. doi:10.5194/bg-10-2393-2013

Haynes, M., Bruno, T., & Lide, D. (2014). *CRC Handbook of Chemistry and Physics*. CRC Press. doi:10.1201/b17118

Huang, F. (2011). *Study on the toxic effects mercury on crops and the toxicity threshold value to crops of soil mercury*. Fujian Agriculture and Forestry University. doi:10.7666/d.y1878345

Israr, M., & Shivendra, V. (2006). Antioxidative responses to mercury in the cell cultures of *Sesbania drummondii. Plant Physiology and Biochemistry*, *44*(10), 590–595. doi:10.1016/j.plaphy.2006.09.021 PMID:17070690

Jiang, Y., & Zhao, F. (2001). Mechanism of heavy metal injury and resistance of plants. *Chinese Journal of Applied and Environmental Biology*, *7*(1), 92–99. doi:10.3321/j.issn:1006-687X.2001.01.022

Kamunda, C., Mathuthu, M., & Madhuku, M. (2016). Health risk assessment of heavy metals in soils from Witwatersrand gold mining basin, South Africa. *International Journal of Environmental Research and Public Health*, *13*(7), 663. doi:10.3390/ijerph13070663 PMID:27376316

Kuffour, R., Tiimub, B., & Agyapong, D. (2018). Impacts of illegal mining (Galamsey) on the environment (water and soil) at Bontefufuo area in the Amansie west district. *Journal of Environment and Earth Science*, *8*(7), 98–107.

Li, F., Ma, C., & Zhang, P. (2020). Mercury deposition, climate change and anthropogenic activities: A review. *Frontiers in Earth Science (Lausanne)*, *8*, 1–17. doi:10.3389/feart.2020.00316

Liu, G., Cai, Y., O'Driscoll, N., Feng, X., & Jiang, G. (2012). Overview of mercury in the environment. In G. Liu & Y. Cai (Eds.), *N. O'Driscoll. Environmental chemistry and toxicology of mercury*. John Wiley & Sons.

Liu, S., Wang, X., Guo, G., & Yan, Z. (2021). Status and environmental management of soil mercury pollution in China: A review. *Journal of Environmental Management*, *277*, 111442. doi:10.1016/j.jenvman.2020.111442 PMID:33069151

Lusilao-Makiese, J., Cukrowska, E., Tessier, E., Almouroux, D., & Weiersbye, I. (2013). The impact of post gold mining on mercury pollution in the West Rand region, Gauteng, South Africa. *Journal of Geochemical Exploration*, *134*, 111–119. doi:10.1016/j.gexplo.2013.08.010

Mahbub, R., Krishnan, K., Naidu, R., Andrews, S., & Megharaj, M. (2017). Mercury toxicity to terrestrial biota. *Ecological Indicators, 74*, 451–462. https://doi.org/. ecolind.2016.12.004 doi:10.1016/j

Mantey, J., Nyarko, K., Owusu-Nimo, F., Awua, K., Bempah, C., Amankwah, R., Akatu, W., & Effah, A. (2020). Mercury contamination of soil and water media from different illegal artisanal small-scale gold mining operations (galamsey). *Heliyon*, *6*(6), e04312. doi:10.1016/j.heliyon.2020.e04312 PMID:32637700

Mason, R., Choi, A., Fitzgerald, W., Hammerschmidt, C., Lamborg, C., Soerensen, A., & Sunderland, E. M. (2012). Mercury biogeochemical cycling in the ocean and policy implications. *Environmental Research*, *119*, 101–117. doi:10.1016/j.envres.2012.03.013 PMID:22559948

Mathebula, M., Panichev, N., & Mandiwana, K. (2020). Determination of mercury thermospecies in South African coals in the enhancement of mercury removal by pre-combustion technologies. *Scientific Reports*, *10*(1), 19282. doi:10.103841598-020-76453-z PMID:33159166

Metcalf, S., & Veiga, M. (2012). Using street theatre to increase awareness of and reduce mercury pollution in the artisanal gold mining sector: A case from Zimbabwe. *Journal of Cleaner Production*, *37*, 179–184. doi:10.1016/j.jclepro.2012.07.004

Nipen, M., Jorgensen, S., Nizzetto, P., Borga, K., Breivik, K., Mmochi, A., Mwakalapa, E., & … . (2022). Mercury in air and soil on an urban-rural transect in East Africa. *Environmental Science. Processes & Impacts*, *24*(6), 921–931. doi:10.1039/D2EM00040G PMID:35583028

O'Connor, D., Hou, D., Ok, Y., Mulder, J., Duan, L., Wu, Q., Wang, S., Tack, F., & Rinklebe, J. (2019). Mercury speciation, transformation and transportation in soils, atmospheric flux, and implications for risk management: A critical review. *Environment International*, *126*, 747–761. doi:10.1016/j.envint.2019.03.019 PMID:30878870

O'Sullivan, J., Lupakisyo, G., Purcell, J., Turner, N., & Mtalo, F. (2016). Assessing sediment and water quality issues in expanding African wetlands: The case of the Mara River, Tanzania. *The International Journal of Environmental Studies*, *73*(1), 95–107. doi:10.1080/00207233.2015.1116226

Obrist, D., Pearson, C., Webster, J., Kane, T., Lin, C., Aiken, G., & Alpers, C. N. (2016). A synthesis of terrestrial mercury in the western United States: Spatial distribution defined by land cover and plant productivity. *The Science of the Total Environment*, *568*, 522–535. doi:10.1016/j.scitotenv.2015.11.104 PMID:26775833

Odumo, B., Carbonell, G., Angeyo, H., Patel, J., Torrijos, M., & Martin, A. (2014). Impact of gold mining associated with mercury contamination in soil, biota sediments and tailings in Kenya. *Environmental Science and Pollution Research International*, *21*(21), 12426–12435. doi:10.100711356-014-3190-3 PMID:24943890

Podolsky, F., Ettler, V., Sebek, O., Jezek, J., Mihaljevic, M., Kribek, B., Sracek, O., Vaněk, A., Penížek, V., Majer, V., Mapani, B., Kamona, F., & Nyambe, I. (2015). Mercury in soil profiles from metal mining and smelting areas in Namibia and Zambia: Distribution and potential sources. *Journal of Soils and Sediments*, *15*(3), 648–658. doi:10.100711368-014-1035-9

Qu, R., Han, G., Liu, M., & Li, X. (2019). The mercury behavior and contamination in soil profiles in Mun river basin, Northeast Thailand. *International Journal of Environmental Research and Public Health*, *16*(21), 4131. doi:10.3390/ijerph16214131 PMID:31717757

Rytuba, J. (2003). Mercury from mineral deposits and potential environmental impact. *Environmental Geology (Berlin)*, *43*(3), 326–338. doi:10.100700254-002-0629-5

Ryzhenko, N., Zhavryda, D., Bokhonov, Y., & Ryzhenko, D. (2021). Mercury contamination in soil, water, plants and hydrobionts in Kyiv and the Kyiv region. *Polish Journal of Soil Science*, *2*(2), 185. doi:10.17951/pjss.2021.54.2.185

Smith-Downey, N., Sunderland, E., & Jacob, D. (2010). Anthropogenic impacts on global storage and emissions of mercury from terrestrial soils: Insights from a new global model. *Journal of Geophysical Research*, *115*(G3), G03008. doi:10.1029/2009JG001124

Steinnes, E. (2013). Mercury. In B. Alloway (Ed.), *Heavy Metals in Soils. Environmental Pollution* (Vol. 22). Springer., doi:10.1007/978-94-007-4470-7_15

Streets, D., Horowitz, H., Lu, Z., Levin, L., Thackray, C., & Sunderland, E. (2019). Global and regional trends in mercury emissions and concentrations. 2010–2015. *Atmospheric Environment*, *201*, 417–427. doi:10.1016/j.atmosenv.2018.12.031

Sundseth, K., Pacyna, J., Pacyna, E., Pirrone, N., & Thorne, R. (2017). Global sources and pathways of mercury in the context of human health. *International Journal of Environmental Research and Public Health*, *14*(1), 105. doi:10.3390/ijerph14010105 PMID:28117743

Tang, H., Liu, Z., Li, S., Sha, Y., & Qiu, P. (2017). Effects of mercury pollution stress on activity of several kinds of antioxidant enzymes in earthworms. *Journal of Shanghai Jiaotong University*, *35*(3), 17–23. doi:10.3969/J.ISSN.1671-9964.2017.03.003

Tong, D., Zhang, W., Deng, Y., & Wang, J. (2016). Pollution characteristics analysis and risk assessment of total mercury and methylmercury in aquatic. *Journal of Integrative Environmental Research*, *37*(3), 942–949. doi:10.13227/j.hjkx.2016.03.019 PMID:27337885

UNEP. (2019). Global mercury assessment 2018. Environment UN, Chemical and health branch, Geneva, Switzerland.

USEPA. (2019). *Mercury emissions: the global context (2019).* United States Environmental Protection Agency. https://www.epa.gov/international-cooperation/mercury-emissions-global-context

Walters, C., Somerset, V., Leaner, J., & Nel, J. (2011). A review of mercury pollution in South Africa: Current status. *Journal of Environmental Science and Health. Part A, Toxic/Hazardous Substances & Environmental Engineering*, *46*(10), 1129–1137. doi:10.1080/10934529.2011.590729 PMID:21806457

Wang, L., Hou, D., Cao, Y., Ok, Y., Tack, F., Rinklebe, J., & O'Connor, D. (2020). Remediation of mercury contaminated soil, water and air: A review of emerging materials and innovative technologies. *Environment International*, *134*, 105281. doi:10.1016/j.envint.2019.105281 PMID:31726360

Wang, P., Hua, C., Chen, Z., & Dai, H. (2008). Effects of water pollution caused by Hg^{2+} on the chlorophyll content and antioxidase activity in spinach seedlings. *Anhui Nongye Kexue*, *36*(14), 5738–5739. doi:10.3969/j.issn.0517-6611.2008.14.023

WHO. (2000). *Air quality guidelines for Europe (second edition).* WHO. https://wedocs.unep.org/20.500.11822/8681

WHO. (2017). *Mercury and health.* WHO. https://www.who.int/news-room/fact-sheets/ detail/mercury-and-health

World Bank. (2011). Study of mercury-containing lamp waste in sub–Saharan Africa. World Bank.

Zhang, J. (2019). Discussion on how to repair the soil environment of polluted sites. *China Resources Comprehensive Utilization*, *37*(2), 140–142. doi:10.1016/j.jcou.2018.12.002

Chapter 11
Soil Pollution by Cadmium in Sub-Saharan Africa

ABSTRACT

In this chapter, the sources and toxikinetics of cadmium (Cd) in soils of sub-Saharan Africa were identified and discussed. The metal was found to be toxic in soils due to its high mobility and ability to replace divalent cations of near similar or similar physicochemical characteristics. Natural sources of the heavy metal include its carbonate, sulfide, and selenide ores. Anthropic sources include the mining and smelting of gold, tin, and cobalt among other metals mainly by artisans, manufacturing of Cd-nickel batteries and resultant wastes laden with the metal and agricultural additives such as phosphatic fertilizers as well as the use of Cd-containing biosolids. Pollution by the activities in soils was worsened by their non-regulation and the laxity of SSA countries to take up environmental conservation initiatives and projects. Therefore, soils can only be used as temporary rather than permanent sinks of Cd due to its toxicity potential.

INTRODUCTION

Rapid expansion of agricultural, urbanization and industrialization activities in addition to population rise around the globe since the 18th century has prompted extensive environmental pollution, which deteriorates the ecological balance (Hocaoglu-Ozyigit & Genc, 2020; Bouida et al., 2022). Of the different forms of pollution, heavy metal contamination is a global problem due to the nonbiodegradable and bio-accumulative nature of the trace metals in addition to their hazardous effects to the biosphere, hydrosphere, lithosphere and atmosphere when they supersede the

DOI: 10.4018/978-1-6684-7116-6.ch011

permissible levels (Olowoyo et al., 2022). Among the heavy metals, cadmium (Cd) toxicity is of growing concern to the scientific community due to the highly toxic nature and mobility of the element to the environment where it has the potential to replace metals of analogous charge, ionic radius and chemical behavior (Kubier et al., 2019). The metal induces adverse effects to plants, animals and humans (Berglund et al., 2015). Apart from exerting carcinogenic effects to experimental animals and humans, Cd is toxic to gastrointestinal tract, placenta, skin, heart, prostrate, bones, kidneys and lungs of higher animals (Zhao et al., 2020).

Cadmium occurs naturally as a component of rocks of the earth's crust. The metal can them be transferred to the environment including soils through natural and human-based processes although the concentrations are generally low in such environs. Average levels of Cd in soils without anthropogenic enrichment range from 0.06 to 1.10 mg/kg (Kabata-Pendias, 2010). This observation alludes to natural enrichment of soils with Cd due to the nature of the parent rock and its associated weathering processes (Liu et al., 2013). In regions with karst covered by black shale such as Poland (Kabata-Pendias, 2010) and China (Zhang et al., 2018; Luo et al., 2019; Liu et al., 2013), soils have high concentrations of Cd naturally. Unlike other forms of shale, black shales are rich in Cd since they are laden with sulfides complexed with the metal and can be easily released once exposed to water and oxygen (Liu et al., 2017).

Anthropogenic activities can also prompt Cd emissions to the atmosphere and subsequent precipitation and deposition to soils. Activities such as smelting and mining of nonferrous metals for instance, account for more than 70% of all anthropogenic sources of the trace metal (Zhao et al., 2020). Most of such Cd emissions (80-90%) end up in soils through deposition (Zhang et al., 2015). The average concentrations of Cd in soils were ranked at 8.27, 1.91, 0.46 and 0.16 mg/kg in mining and smelting areas, where wastewater irrigation was carried out, in urban and suburban areas and in remote areas, respectively. The highest contribution of the metal in soils during mining and smelting activities occurred due to refining, ore transportation, mineral excavation and the disposal of mine wastewater, rock overburden and mine tailings that end up in soils. For this reason, soils near mining and smelting areas were laden with Cd compared to soils located distances away (Zhao et al., 2020). The inputs of both geogenic and anthropogenic activities make total concentrations of Cd in soils high and induce problematic effects that are transferrable to plants and animals. For this reason, assessing the anthropogenic sources of Cd, its toxi-kinetics in soils and spatial distribution changes is key in understanding and managing the effects of the metal in the environment. This chapter focused on the chemistry of Cd, the anthropogenic sources of the trace metal, its pollution in soils and particularly those of sub-Saharan Africa region during the modern-day industrialization era.

CHEMISTRY OF CADMIUM

Cadmium is a group 12, period 5 and d block post-transition element that was discovered in 1817 by a German chemist F. Strohmeyer from the zinc ore constituent, smithsonite ($ZnCO_3$) (Sharma et al., 2015). It has an atomic number of 48 and is a ductile, soft and silvery white metal with an electronic configuration of [Kr] $4d^{10}$ $5s^2$. Additionally, the metal has an atomic weight of 112.41, which results from eight stable isotopes (Genchi et al., 2020). The metal is mainly sourced from minerals such as Cadmium selenide (cadmoselite, CdSe), cadmium sulfide (Greenockite, CdS) and cadmium carbonate (otavite, $CdCO_3$). The metal is recovered as a constituent of copper, zinc and lead. The metal occurs in high concentrations in sedimentary rocks compared to igneous and metamorphic rocks. Some of the rocks that contain Cd and their mean concentrations are shown in Table 1 (Kubier et al., 2019). The metal whose distribution varies due to geogenic factors acts as a substitute for divalent cations including calcium, cobalt, iron, lead and zinc in minerals such as pyrite (FeS_2), sphalerite (ZnS) and smithsonite ($ZnCO_3$) because it has the same ionic radius (Tabelin et al., 2018). Apart from Cd sourced in soils, sediments and rocks, the metal is found in natural emissions of soil particles found in hydrothermal vents, volcanoes, biogenic materials, forest fires, sea sprays and from deserts (Kubier et al., 2019).

Table 1. Cadmium content in different types of sedimentary, igneous and metamorphic rocks (Kubier et al., 2019)

Rock type	Mean Cd concentration in PPM
Sedimentary Rocks	
Phosphorites	25
Oceanic manganese oxides	8
Organic sediment	0.5
Carbonate stone	0.012
Sandstone	0.028
Limestone	0.1
Shale and claystone	1
Red clay	0.56
Marlstone	2.6
Bentonite	1.4
Red shales	0.03
Black or bituminous shales	0.8
Metamorphic Rocks	
Schists	0.02
Gneisses	0.04
Igneous Rocks	
Basalt	0.22
Obsidian	0.25
Ultramafic rocks	0.02
Mafic rocks	0.11
Granitic rocks	0.12

Cadmium has an oxidation state of +2 and has similar physical characteristics as zinc. The metal is inflammable, water-insoluble and corrosion resistant. It reacts with oxygen from air to form cadmium oxide (CdO). The metal reacts with oxoacids such as nitric and sulfuric acid to form cadmium nitrate {$Cd(NO_3)_2$} and cadmium sulfate ($CdSO_4$), respectively. Some of the physicochemical and geochemical properties of the metal are as summarized in Table 2.

Table 2. Physical, chemical and geochemical characteristics of Cd (Sharma et al., 2015; Genchi et al., 2020)

Physical characteristics	Chemical characteristics	Geochemical characteristics
Soft, ductile and silvery-white	Atomic number of 48	
A density of 8.645 g/cm³ at 20°C	Atomic weight of 112.4	Occurs in high levels in phosphorites, lacustrine sediments, shales and oceanic sediments
Water insoluble	It is a transition metal of Group IIb in the periodic table	Low concentrations of the metal in igneous rocks
Inflammable	The presence of 14 electrons in the fourth orbital allows the formation of stable compounds	About 0.15-0.2 ppm in the earth's crust
Melting point of 321.07°C	Oxidation state +2	The metal is highly associated with calcium
Boiling point of 767.3°C	Forms covalent bonds with sulfur	
Heat of fusion at 6.21 kJ/mol	Forms soluble complexes with amines and cyanides	
Heat of vaporization of 99.6 kJ/mol	Has 8 isotopes Cd 106, 108, 111, 112, 118, 116, 0 and 1"4	
Electronegativity of 1.69 on a Pauling scale	An ionic radius of 1.03 and 0.88 A° for sixth- and fourth-fold coordination, respectively	
First and second ionization energy of 867.8 kJ/mol and 1631.4 kJ/mol, respectively		
A vapor pressure of 1.4 mm at 400°C		

Cadmium is one of the priority hazardous trace metals recognized by the United States Environmental Protection Agency (US-EPA) and non-governmental organizations. Metal exposure to human being occurs either via food chains (from polluted water, soils and contaminated food) or through the respiratory tract. The World Health Organization limited the tolerance to Cd exposure as 50 μg for a

0.007 mg/kg body weight every week (WHO, 2000). The entry of the metal in soil environs is a result of anthropogenic activities and geogenic sources. Some of the anthropogenic activities include the use of soil additives including inorganic fertilizers and pesticides (Ozyigit et al., 2016; Tabelin et al., 2018), use of wastewater for irrigation and atmospheric deposition of the trace metal due to a rise in industrial activities.

ANTHROPOGENIC SOURCES OF THE CADMIUM

Cadmium is naturally occurring in many soil systems but elevated levels are attributable to human activities. Temporal and spatial changes in minerals and rocks could sometimes induce elevated concentrations of the metal due to its easy mobility. For this reason, soils should not be used as permanent but rather temporary sinks of Cd-containing wastes since combination of such wastes with geogenic Cd could enhance bioaccumulation and leaching of the metal to pollute groundwater sources (Birke et al., 2017). Satarug (2019) noted that the rapid modernization of industries and technologies is associated with elevated levels of the trace metal in air, food, soils and water. In industries, Cd is used in manufacture of Nickel (Ni)-Cd batteries, pigments, coatings and stabilizers (Sharma et al., 2015). The main industrial uses of Cd globally and the amount of Cd they release to soils is as summarized in Table 3 (Kabir et al., 2012; Kubier et al., 2019). The manufacturing activities release solid wastes and wastewater, whose introduction to soils leads to elevated concentrations of the metal in such environs. Genchi et al. (2020) also noted that the metal is used in industrial electroplating and in paint manufacturing and its spraying introduces atmospheric depositions of the metal to the air, which are later deposited on soils. In addition to these activities, airborne Cd can result from combustion emissions containing the metal, the use sewage sludge in farms, the use of inorganic fertilizers with high cadmium phosphate $\{Cd_3(PO_4)_2\}$ levels, release of Cd-containing landfill leachate and in the mining and smelting industries (Bigalke et al., 2017). Abandoned mines were also sources of Cd into soils.

The manufacture of Ni-Cd batteries is one of the primary sources of Cd to the environment. Wastes from the batteries are disposed off at landfills in the form of municipal or industrial wastes based on previous use of the battery. Additionally, byproducts of alloys, polyvinyl chloride stabilizers, platings, coatings and pigments containing Cd also end up in landfills which can be released to soils to pollute them after the formation of leachate (Kubier et al., 2019). In various studies, phosphate fertilizers were found to have 77 mg (Azzi et al., 2017), 60 mg (Six and Smolders, 2014) and 36 mg (Grant, 2011) of Cd for every kilogram of phosphorous pentoxide fertilizers. The input of such fertilizers in soils and the effects of atmospheric deposition of Cd- containing particulate matter has been reported to pollute soils (Kubier et al., 2019). Mining and smelting of ores particularly zinc results to elevated levels of Cd in the atmosphere and soils. Ray and Datta (2016) noted that smelting of sphalerite (ZnS), which is the main mineral ore where Zn is sourced resulted to release of 0.2-0.4% of Cd as impurities that end up polluting the soils.

Table 3. Sources of Cd and concentrations of the metal in soils located near industrial areas (Kabir et al., 2012; Kubier et al., 2019)

Industrial Sources of Cd	Average Cd levels (mg/kg)	Maximum Cd levels (mg/kg)
Nonmetallic mineral products	25.8	72
Leathers	0.63	1.26
Textiles	42	83.6
Petroleum, chemicals and fertilizer production	0.51	2.13
Metal and mining industry	37.6	289

CADMIUM POLLUTION IN SOILS OF SSA

Cadmium (II) manifests high toxicity and persistence in soils and can easily contaminate food chains due to its easy uptake by plant roots and eventual accumulation in higher animals including human beings resulting to toxic effects (Subasic et al., 2022). The trace metal pollutes soils in addition to destroying its ecology and deteriorating their quality and fertility (Bouida et al., 2022). In addition to polluting soils and altering their physicochemical features, Cd through the soil plant interaction, the metal can be taken up by plant and animal systems contaminating them. In Japan for instance, chronic Cd poisoning was reported in the early 20th century and termed as the *itai-itai* disease that manifested in humans through symptoms such as osteoporosis, osteomalacia and renal tubular dysfunction where Cd replaced Ca and other divalent ions in bones and kidney (Khan et al., 2017; Kubier et al., 2019). In soils, various lithologic sources and anthropogenic activities result to the accumulation of Cd in soils, whose total content can be assayed. In the soils, toxicity of the metal is influenced by its bioavailability and mobility in such environs (Wieczorek et al., 2018). Processes of trace metal bioavailability and mobility are controlled by mechanisms such as adsorption, precipitation, ion exchange and surface complex formation with the solid phase of a specified soil (Puga et al., 2015; Venegas et al., 2015). Factors such as the amount of calcium carbonate, levels of manganese and ion oxides, organic matter content, clay mineral content, redox potential and pH that are specific to a given soil also influence the behavior, accumulation and toxicity of Cd to soils (Mazurek et al., 2017). These factors and their influence on heavy metal toxicity in soils are discussed in Chapter 1 of this book.

In alkaline soils, Cd has neutral solubility while the metal is relatively soluble in acidic conditions. In the presence of carbonates in soils, the solubility of Cd is limited. Organic matter binds onto Cd and reduces its bioavailability by transforming it to an organically bound complex (Chen et al., 2010). Cadmium promotes formation of halide, HS^-, S^{2-} complex ions with thiols and organic sulfides due to its soft Lewis acid nature. If organic or inorganic ligands are present in soil solutions, formation of dissolved complexes or soil adsorption may be reduced. Cd interacts with organic matter containing low molecular weight cations such as Cd, Co, Cu, Fe, Mn, Ni, Fe and Zn while organic acids with carboxyl groups promote Cd complexation (Bolan et al., 2003). The presence of nitrate and chloride anions prevents Cd sorption in soils since they form $H_2PO_4^-$ and HSO_4^- inorganic complexes that are soluble (Subasic et al., 2022). In soils, Cd competes for sorption sites with Zn because they have similar geochemical characteristics unless pH is optimized. After accumulation in soils, Cd is taken up by plant roots and shoots. The uptake depends on the presence or absence of organic acids and the pH of the soil. Forms of soil Cd that can be taken by plant roots from the soil at a pH of 6-7 include $CdCl^-$, $CdHCO_3^+$, $CdCl_n$ and $CdCO_3^+$ (Kabir et al., 2012).

Pollution of soils by Cd is a growing problem globally and more so in the SSA region (Tindwa & Singh, 2023). The situation in the region is dire because SSA is grappling to strike a balance between sustainable conservation of the environment and steady economic growth. Previously specific countries had short-term strategies to increase production while ignoring the resultant pollution effects. Assuming the status quo remains, industrial activities will overwhelm the region with Cd-containing waste and an environmental crisis by 2100 is expected (Hoornweg et al., 2015; Tindwa & Singh, 2023). The main sources of soil pollution by Cd in SSA are discussed in the following sections.

Industrial Production Activities

Industrial production in SSA region is growing and forecasted to grow exponentially though not as fast as the developed world (Tindwa & Singh, 2023). With the predisposition, the quantities of waste generated has increased and is further projected to increase exponentially (UN-Habitat, 2022). In 2016 for instance, Africa and a large portion being SSA generated approximately 174 million tons of wastes annually and the figure is expected to rise to 244 million tons annually by 2025 (Debrah et al., 2022; UN-Habitat, 2022). The World Bank estimates on the amount of waste generated in SSA region, as noted by Khan et al. (2022) are expected to be three times higher by 2050 and mostly constituting of organic wastes, plastics, metals, paper, glass and other forms of waste. Industrialization and population growth trends are attributable to the increased generation of such wastes, most of which are laden with heavy metals (Debrah et al., 2022; Khan et al., 2022; Kaza et al., 2018). The contaminants occur as liquid and solid wastes as well as emissions that end up in land resources such as soils and sediments. The situation is more likely in SSA region since regulations on waste management are less stringent and, in most cases, non-implemented and non-enforced. In industrial areas where manufacturing, petrochemical production and mining activities are concentrated, heavy metal contamination leads to stressful ecosystems and alteration of natural resources such as soils in SSA (Bekabil, 2020).

Small-scale industrial production including Ni-Cd battery recycling, cottage industries, auto-mechanical workshops and dry cleaning among others cause the introduction of heavy metals such as Cd (FAO & UNEP, 2021). Auto-mechanical workshops specializing in engine maintenance and repair, paint spraying and welding using Cd-containing chemicals and resultant wastes are released into surrounding soils and lead to environmental pollution. Such wastes include grease, lubricating oil, old hydraulic liquids, paint solvents, primers and paints (Ekeocha et al., 2017). Cottage industries, mainly home-based and which are usually unregulated and do not adhere to the occupational health and safety legislations also introduce Cd among other trace metals in soils owing to the chemicals used in such production processes and especially in urban areas (Mmbaga & Semu, 1999; Teare et al., 2015). Table 4 shows some of the industrial processes that result to pollution of soils by Cd in reported studies of SSA region.

Mining and Artisanal Activities

The mining of heavy metals such as cobalt, gold, lead, tin and zinc is a common practice and a significant economic contributor in SSA region done at small-scale by artisans. According to Bekabil (2020) millions of the region's residents rely on the proceeds realized from artisanal and small-scale mining activities. In another study, Hilson (2016) noted that skilled, semi-skilled and non-skilled SSA residents engage in artisanal and small-scale mining activities to make their lives sustainable. The activities are usually unregulated and informal, often characterized by illegality because of the bureaucracy and costs incurred in legitimization such activities (Tindwa & Singh, 2023). The absence of government controls and legislations to manage artisanal activities in addition to its widespread practice in the region has resulted to environmental pollution by Cd among other heavy metals and an ultimate destruction of natural resources and ecosystem stability.

Table 4. Studies on industrial production activities that have reported soil pollution by Cd in SSA region

Type of Industry	SSA Country	Reference
Cottage industries	South Africa	Teare et al., 2015
	Tanzania	Mmbaga & Semu, 1999
Welding services	Nigeria	Jimoh et al., 2020
	South Africa	Teare et al., 2015
Engine servicing and oil lubricant factories	Benin	Akporido & Asagba, 2013
	Kenya	Mutune et al., 2014
Recycling of scrap metal	Nigeria	Olatunji et al., 2018
Spray paint production and use	Kenya	Mutune et al., 2014
Metal smelting and mining	Tanzania, South Africa	Mkhize, 2020
Hair dressing and cosmetics use	Nigeria	Eneh, 2021
Soap, textile, plastic and chemical production	Kenya	Nyika & Dinka, 2022
	Ghana	Odai et al., 2008
Manufacturing of Pb, Cd-Ni batteries	Nairobi	Mutune et al., 2014
	Cameroon	Ekengele et al., 2008
	Ghana, Kenya, Tanzania, Nigeria, Cameroon, Mozambique	Gottesfeld et al., 2018
	Ghana	Odai et al., 2008
Production of electronics and their disposal	Ghana	Fosu-Mensah et al., 2017

Gold mining and the resultant tailings were reported to be complexed with Cd and hence release to the environment as emissions or particulates that end up polluting soils (Fashola et al., 2016). In a goldmine dump located at Ekurhuleni region of South Africa, soils had elevated levels of Cd among other trace metals such as mercury and lead that occurred during gold mining and where the trace metals resulted to soil pollution (Okereafor et al., 2019). In illegal gold mining sites of Ivory Coast (Kouadio et al., 2023), Nigeria (Adeyi & Babalola, 2017) and Niger (Ibrahim et al., 2019), elevated concentrations of Cd were also reported in soils of the vicinities. In Ghana's Bibiani mining district, artisanal and small-scale gold mining was associated with elevated Cd concentrations in soils of the vicinity (Hogarh et al., 2016).

Similar pollution problems occur in artisanal and small-scale mining of tin, which generates ex-tin mining tailing dumps and ponds in addition to creating large wastelands that are laden with heavy metals including Cd (Tindwa & Singh, 2023). The wastes also tend to form finer and dusty tailings that are released to the atmosphere and later deposited in soils. In SSA countries of Nigeria, Rwanda, Burundi, Uganda and Democratic Republic of Congo, tin mining at small-scale and industrial levels has been associated with heavy metal pollution leading to soil degradation (Omotehinse & Ako, 2019). The mining of coal, zirconium, rutile, palladium, ilmenite, vermiculite, vanadium, platinum, manganese and chrome in South Africa has also been associated with the introduction of Cd in soils among other environs (Ebenebe et al., 2017). The situation is more prevalent in Mpumalanga, Limpopo, Kwa-Zulu Natal, Eastern Cape and Free State provinces where many mines are located.

The mining of cobalt used in the manufacture of lithium batteries used in electronics such as laptops, computers, smartphones and for electric vehicles is also predominant in Democratic Republic of Congo (DRC) at the Katanga copper belt (Banza et al., 2018). The activities are done by artisanal miners and result to production of Cd among other heavy metals as by-products (Elenge & De Brouwer, 2011). Similar trends were reported in soils of Haut-Katanga province (Kalenga, 2019) and in neighboring Lubumbashi city (Van Brusselen, 2020) in DRC and attributable to the mining activities of the Katanga copper belt. In the latter study, the presence of Cd and other metals at elevated levels was associated with birth defects among neonates. In Zambia Copperbelt region, elevated levels of Cd in the soils were associated with its occurrence as a byproduct of other mineral rocks and disposal of mine tailings and overburden on non-polluted soils (Muimba et al., 2022; Tembo & Kwenga, 2006).

Agricultural Activities

The main contributor of soil pollution during agricultural activities are organic wastes used as soil additives, pesticides and fertilizers (Tindwa & Singh, 2023). Organic and chemical fertilizers used in agricultural activities in SSA region are laden with organic and inorganic pollutants including trace metals. Primarily, Cd is found in phosphatic fertilizers whose origin is phosphate rock used as a raw material during the manufacture of such fertilizers. According to Gupta et al. (2014), phosphate rock is the commonest raw material in production of phosphate fertilizers globally and application of such soil additives results to the buildup of Cd in such environs. Igneous rock deposits from countries such as South Africa, which are used in making inorganic fertilizers have high levels of Cd ranging from 1-4 mg/kg, which can be transferred to soils once such amendments are applied (Suciu et al., 2022). Such rocks contain phosphate minerals including hydroxyapatite, fluorapatite and carbonate hydroxyapatite all of which contain Cd as a natural contaminant (Umvoto Africa, 2014). The phosphate rocks are also found in East African regions of Mrima Hills (Kenya), Matongo and Minjingu in Burundi and Tanzania, respectively (Mwalongo et al., 2022). Once the rocks are used as raw materials for making inorganic fertilizers, they introduce Cd to soils. The import of Cd-containing phosphate fertilizers to SSA region from North Africa, USA, Japan and Israel and their use in agricultural activities also contributes to soil pollution by Cd (Suciu et al., 2022).

The application of Cd-containing compost, sludge or wastewater in soils also introduces the trace element if continuously used. Zhang et al. (2017) and Buta et al (2021) noted that sludge and wastewater contained non-essential trace metals including Cd and cautioned on the unregulated use by farmers. In Kenya's suburban regions, Cd-containing wastewater was used to irrigate vegetables and eventually, resulted to soil pollution by Cd among other heavy metals (Nyika & Dinka, 2022). A similar trend was evident in South Africa (Agoro et al., 2020), Eswatini (Tiruneh et al., 2014) and Benin (Suanon et al., 2016) where elevated levels of Cd were associated with irrigation of crops with metal-containing wastewater and use of sewage sludge as a fertilizer in agricultural activities.

CONCLUSION

In this chapter, anthropogenic sources and toxicity affiliated with Cd in soils of sub-Saharan Africa (SSA) region was discussed. Three main causes of pollution in soils by the metal were identified. The first source of the metal was in the manufacturing sector involving the production of Cd-Ni batteries, electroplating and the making of paints, pigments, stabilizers and coatings. The manufacturing activities use Cd and Cd-containing raw materials and produce wastes laden with the heavy metal. The mining of Co, Au and Sn usually done by artisans in SSA was another causative of Cd poisoning in soils. The metal was complexed with ores of the minerals and released during their processing. Agricultural activities particularly the use of phosphatic fertilizers, Cd-containing effluents and sewage sludge for farming also resulted to the metal introduction in soils and its subsequent bioaccumulation. The use of such soil additives is largely unregulated in SSA. Owing to the identified toxicity of Cd, soils cannot be used as permanent sinks of the metal to avoid extensive pollution to other environmental compartments.

REFERENCES

Adeyi, A., & Babalola, A. (2017). Lead and cadmium levels in residential soils of Lagos and Ibadan, Nigeria. *Journal of Health & Pollution*, *7*(13), 42–55. doi:10.5696/2156-9614-7-13.42 PMID:30524813

Agoro, M., Adeniji, A., Adefisoye, M., & Okoh, O. (2020). Heavy metals in wastewater and sewage sludge from selected municipal treatment plants in Eastern Cape province, South Africa. *Water (Basel)*, *12*(10), 2746. doi:10.3390/w12102746

Akporido, O., & Asagba, O. (2013). Quality characteristics of soil close to the Benin River in the vicinity of a lubricating oil producing factory, Koko, Nigeria. *International Journal of Soil Science*, *8*(1), 1–16. doi:10.3923/ijss.2013.1.16

Azzi, V., Kazpard, V., Lartiges, B., Kobeissi, A., Kanso, A., & El Samrani, A. (2017). Trace metals in phosphate fertilizers used in eastern Mediterranean countries. *Clean (Weinheim)*, *45*(1), 1–10. doi:10.1002/clen.201500988

Banza, L., Casas, L., Haufroid, V., De Putter, T., & Saenen, N. (2018). Sustainability of artisanal mining of cobalt in DR Congo. *Nature Sustainability*, *1*(9), 495–504. doi:10.103841893-018-0139-4 PMID:30288453

Bekabil, U. (2020). Industrialization and environmental pollution in Africa: An empirical review. *Journal of Resource Development and Management*, *69*, 1–4. doi:10.7176/JRDM/69-03

Bigalke, M., Ulrich, A., Rehmus, A., & Keller, A. (2017). Accumulation of cadmium and uranium in arable soils in Switzerland. *Environmental Pollution*, *221*, 85–93. doi:10.1016/j.envpol.2016.11.035 PMID:27908488

Birke, M., Reimann, C., Rauch, U., Ladenberger, A., Demetriades, A., Jähne-Klingberg, F., Oorts, K., Gosar, M., Dinelli, E., & Halamić, J. (2017). GEMAS: Cadmium distribution and its sources in agricultural and grazing land soil of Europe - original data versus clr-transformed data. *Journal of Geochemical Exploration*, *173*, 13–30. doi:10.1016/j.gexplo.2016.11.007

Bolan, S., Adriano, D., & Naidu, R. (2003). Role of phosphorus in mobilization and bioavailability of heavy metals in the soil-plant system. *Reviews of Environmental Contamination and Toxicology*, *177*, 1–44. doi:10.1007/0-387-21725-8_1 PMID:12666817

Bouida, L., Rafatullah, M., Kerrouche, A., Qutob, M., Alosaimi, A., Alorfi, H., & Hussein, M. (2022). Review on cadmium and lead contamination: Sources, fate, mechanism, health effects and remediation methods. *Water (Basel)*, *14*(21), 3432. doi:10.3390/w14213432

Buta, M., Hubeny, J., Zielinski, W., Harnisz, M., & Korzeniewska, E. (2021). Sewage sludge in agriculture - the effects of selected chemical pollutants and emerging genetic resistance determinants on the quality of soil and crops – a review. *Ecotoxicology and Environmental Safety, 214*, 112070. doi:10.1016/j.ecoenv.2021.112070 PMID:33652361

Chen, S., Huang, Y., Liu, L., Cai, P., Liang, W., & Li, M. (2010). Poultry manure compost alleviates the phytotoxicity of soil cadmium: Influence on growth of pak choi (*Brassica chinensis* L.). *Pedosphere, 20*(1), 63–70. doi:10.1016/S1002-0160(09)60283-6

Debrah, J., Teye, G., & Dinis, M. (2022). Barriers and challenges to waste management hindering the circular economy in Sub-Saharan Africa. *Urban Science (Basel, Switzerland), 6*(3), 57. doi:10.3390/urbansci6030057

Ebenebe, P., Shale, K., Sedibe, M., Tikilili, P., & Achilonu, M. (2017). South African Mine Effluents: Heavy Metal Pollution and Impact on the Ecosystem. *International Journal of Chemical Science, 15*(4), 198.

Ekengele, L., Myung, C., Ombolo, A., Ngatcha, N., & Georges, E. (2008, December). Metal pollution in freshly deposited sediments from river Mingoa, main tributary to the Municipal Lake of Yaoundé, Cameroon. *Geosciences Journal, 12*(4), 337–347. doi:10.100712303-008-0034-5

Ekeocha, C., Nwoko, I., & Onyeke, L. (2017). Impact of automobile repair activities on physicochemical and microbial properties of soils in selected automobile repair sites in Abuja, central Nigeria. *Chemical Science International Journal, 20*(2), 1–15. doi:10.9734/CSJI/2017/36065

Elenge, M., & De Brouwer, C. (2011). Identification of hazards in the workplaces of artisanal mining in Katanga. *International Journal of Occupational Medicine and Environmental Health, 245*(1), 7–66. doi:10.247813382-011-0012-4 PMID:21468903

Eneh, O. (2021). Health effects of selected trace elements in hairdressing cosmetics on hairdressers in Enugu, Nigeria. *Scientific Reports, 11*(1), 20352. doi:10.103841598-021-00022-1 PMID:34645821

Fashola, M., Ngole-Jeme, V., & Babalola, O. (2016). Heavy metal pollution from gold mines: Environmental effects and bacterial strategies for resistance. *International Journal of Environmental Research and Public Health, 13*(11), 1047. doi:10.3390/ijerph13111047 PMID:27792205

Food and Agricultural Organization (FAO) & United Nations Environmental Program (UNEP) (2021). *Global assessment of soil pollution: report.* FAO & UNEP. doi:10.4060/cb4894en

Fosu-Mensah, B., Addae, E., Tawiah, D., & Nyame, F. (2017). Heavy metal concentration and distribution in soils and vegetation at Korle Lagoon area in Accra, Ghana. *Cogent Environmental Science, 3*(1), 1405887. doi:10.1080/23311843.2017.1405887

Gottesfeld, P., Were, F., Adogame, L., Gharbi, S., San, D., Nota, M. M., & Kuepouo, G. (2018). Soil contamination from lead battery manufacturing and recycling in seven African countries. *Environmental Research, 161*, 609–614. doi:10.1016/j.envres.2017.11.055 PMID:29248873

Grant, C. (2011). Influence of phosphate fertilizer on cadmium in agricultural soils and crops. *Pedologist, 3*, 143–155. doi:10.18920/pedologist.54.3_143 PMID:26557096

Gupta, D., Chatterjee, S., Datta, S., Veer, V., & Walther, C. (2014). Role of phosphate fertilizers in heavy metal uptake and detoxification of toxic metals. *Chemosphere, 108*, 134–144. doi:10.1016/j.chemosphere.2014.01.030 PMID:24560283

Hilson, G. (2016). Artisanal and small-scale mining and agriculture: Exploring their links in rural sub-Saharan Africa. London: IIED. https://pubs.iied.org/sites/ default/files/pdfs/migrate/16617IIED.pdf

Hocaoglu-Ozyigit, A., & Genc, B. (2020). Cadmium in plants, humans and the environment. *Frontiers in Life Sciences and Related Technologies, 1*(1), 12–21.

Hogarh, J., Gyamfi, E., Nukpezah, D., Akoto, O., & Adu-Kumi, S. (2016). Contamination from mercury and other heavy metals in a mining district in Ghana: Discerning recent trends from sediment core analysis. *Environmental Systems Research, 5*(15), 1–9. doi:10.118640068-016-0067-0

Hoornweg, D., Bhada-Tata, P., & Kennedy, C. (2015). Peak waste: When is it likely to occur? *Journal of Industrial Ecology, 19*(1), 117–128. doi:10.1111/jiec.12165

Ibrahim, Z., Dan-Badjo, T., Guero, Y., Idi, M., & Feidt, C. (2019). Distribution spatiale des éléments traces métalliques dans les sols de la zone aurifère de Komabangou au Niger. *International Journal of Biological and Chemical Sciences, 13*(1), 557–573. doi:10.4314/ijbcs.v13i1.43

Jimoh, A., Agbaji, E., Ajibola, V., & Funtua, M. (2020). Application of pollution load indices, enrichment factors, contamination factor and health risk assessment of heavy metals pollution of soils of welding workshops at old Panteka market, Kaduna-Nigeria. *Open Journal of Analytical and Bioanalytical Chemistry, 4*(1), 011–019. doi:10.17352/ojabc.000019

Kabata-Pendias, A. (2010). *Trace elements in soils and plants*. CRC Press. doi:10.1201/b10158

Kabir, E., Ray, S., Kim, K., Yoon, H., Jeon, E., Kim, Y., Cho, Y.-S., Yun, S.-T., & Brown, R. J. C. (2012). Current status of trace metal pollution in soils affected by industrial activities. *TheScientificWorldJournal, 916705*, 1–18. doi:10.1100/2012/916705 PMID:22645468

Kalenga, J. (2019). Assessment of heavy metal concentrations in streams and economic effects in Haut-Katanga province of the Democratic Republic of Congo. *The Hosei University Economic Review, 86*(3-4), 1–21. doi:10.15002/00021806

Kaza, S., Yao, L., Bhada-Tata, P., & Woerden, V. (2018). *What is waste 2. a global snapshot of solid waste management to 2050*. World Bank Publications, The World Bank Group. doi:10.1596/978-1-4648-1329-0

Khan, I., Chowdhury, S., & Techato, K. (2022). Waste to energy in developing countries–a rapid review: Opportunities, challenges, and policies in selected countries of Sub-Saharan Africa and south Asia towards sustainability. *Sustainability (Basel), 7*(7), 3740. doi:10.3390u14073740

Khan, M., Khan, S., Khan, A., & Alam, M. (2017). Soil contamination with cadmium, consequences and remediation using organic amendments. *The Science of the Total Environment, 601*, 1591–1605. doi:10.1016/j.scitotenv.2017.06.030 PMID:28609847

Kouadio, L., Tillous, K., Coulibaly, V., Sei, J., & Martinez, H. (2023). Assessment of metallic pollution of water and soil from illegal gold mining sites in Kong 2, Hire and Degbezere (Ivory Coast). *Journal of Materials & Environmental Sciences, 14*(1), 41–61.

Kubier, A., Wilkin, R., & Pichler, T. (2019). Cadmium in soils and groundwater: A review. *Applied Geochemistry, 108*, 1–16. doi:10.1016/j.apgeochem.2019.104388 PMID:32280158

Liu, Q., Li, M., Duan, J., Wu, H., & Hong, X. (2013). Analysis on influence factors of soil Pb and Cd in agricultural soils of Changsha suburb based on geographically weighted regression model. *Nongye Gongcheng Xuebao (Beijing), 29*, 225–234.

Liu, Z., Xiao, F., Perkins, B., Zhu, M., Zhu, J., Xiong, Y., & Ning, P. (2017). Geogenic cadmium pollution and potential health risks, with emphasis on black shale. *Journal of Geochemical Exploration, 176*, 42–49. doi:10.1016/j.gexplo.2016.04.004

Luo, L., Mei, K., Qu, Y., Zhang, C., Chen, H., et al. (2019). Assessment of the geographical detector method for investigating heavy metal source apportionment in an urban watershed of Eastern China. *Science of the Total Environment, 653*, 714e722. doi:10.1016/j.scitotenv.2018.10.424

Mazurek, R., Kowalska, J., Gasiorek, M., Zadrozny, P., Jozefowska, A., Zaleski, T., Kępka, W., Tymczuk, M., & Orłowska, K. (2017). Assessment of heavy metals contamination in surface layers of Roztocze National Park Forest soils (SE Poland) by indices of pollution. *Chemosphere, 168*, 839–850. doi:10.1016/j.chemosphere.2016.10.126 PMID:27829506

Mkhize, T. (2020). *Assessment of heavy metal contamination in soils around Krugersdorp mining area, Johannesburg, South Africa.* [Msc Thesis, University of Kwa-Zulu Natal, South Africa].

Mmbaga, T., & Semu, E. (1999). Contents of heavy metals in some soils of the Morogoro municipality, Tanzania, as a result of cottage-scale metal working operations. *The International Journal of Environmental Studies, 56*(3), 373–383. doi:10.1080/00207239908711211

Muimba, K., Banza, L., Mwitwa, J., Kampemba, F., & Mulele, N. (2022). Impacts of trace metals pollution of water, food crops and ambient air on population health in Zambia and the DR Congo. *Journal of Environmental and Public Health, 4515115*, 1–14. doi:10.1155/2022/4515115

Mutune, A., Makobe, M., & Onyango, M. (2014). Heavy metal content of selected leafy vegetables planted in urban and peri-urban Nairobi, Kenya. *African Journal of Environmental Science and Technology, 8*(1), 66-74. https://doi.org/. 1573 doi:10.5897/AJEST2013

Mwalongo, D., Haneklaus, N., Lisuma, J., Kivevele, T., & Mtei, K. (2022). Uranium in phosphate rocks and mineral fertilizers applied to agricultural soils in East Africa. *Environmental Science and Pollution Research International, 30*(12), 1–9. doi:10.100711356-022-24574-5 PMID:36496520

Nyika, J., & Dinka, M. (2022). Heavy metal pollution in soils and vegetables from suburban regions of Nairobi, Kenya and their community health implications. *Pollution, 8*(4), 1434–1447. doi:10.22059/POLL.2022.341522.1440

Odai, N., Mensah, E., Sipitey, D., Ryo, S., & Awuah, E. (2008). Heavy metals uptake by vegetables cultivated on urban waste dumpsites: Case study of Kumasi, Ghana. *Research Journal of Environmental Toxicology*, *2*(2), 92–99. doi:10.3923/rjet.2008.92.99

Okereafor, G., Makhatha, M., Mekuto, L., & Mavumengwana, V. (2019). *Evaluation of trace elemental levels as pollution indicators in an abandoned goldmine dump in Ekurhuleni area, South Africa*. Intech Open.

Olatunji, A., Kolawole, T., Oloruntola, M., & Gunter, C. (2018). Evaluation of pollution of soils and particulate matter around metal recycling factories in Southwestern Nigeria. *Journal of Health & Pollution*, *8*(7), 20–30. doi:10.5696/2156-9614-8.17.20 PMID:30524846

Olowoyo, J., Lion, N., Unathi, T., & Oladeji, O. (2022). Concentrations of Pb and other associated elements in soil dust 15 years after the introduction of unleaded fuel and the human health implications in Pretoria, South Africa. *International Journal of Environmental Research and Public Health*, *19*(16), 10238. doi:10.3390/ijerph191610238 PMID:36011873

Omotehinse, A., & Ako, B. (2019). The environmental implications of the exploration and exploitation of solid minerals in Nigeria with a special focus on tin in Jos and coal in Enugu. *Journal of Sustainable Mining*, *18*(1), 18–24. doi:10.1016/j.jsm.2018.12.001

Ozyigit, I., Yilmaz, S., Dogan, I., Sakcali, S., Tombuloglu, G., & Demir, G. (2016). Detection of physiological and genotoxic damages reflecting toxicity in kalanchoe clones. *Global NEST Journal*, *18*(1), 223–232. doi:10.30955/gnj.001349

Puga, P., Abreu, A., Melo, A., Paz-Ferreiro, P., & Beesley, L. (2015). Cadmium, lead and zinc mobility and plant uptake in a mine soil amended with sugarcane straw biochar. *Environmental Science and Pollution Research International*, *22*(22), 17606–17614. doi:10.100711356-015-4977-6 PMID:26146374

Ray, P., & Datta, S. (2016). Solid phase speciation of Zn and Cd in zinc smelter effluent irrigated soils. *Chemical Speciation and Bioavailability*, *29*(1), 6–14. doi:10.1080/09542299.2016.1247656

Satarug, S. (2019). Cadmium sources and toxicity. *Toxics*, *7*(2), 25. doi:10.3390/toxics7020025 PMID:31064047

Sharma, H., Rawal, N., & Mathew, B. (2015). The characteristics, toxicity and effects of cadmium. *International Journal of Nanotechnology and Nanoscience*, *3*, 1–9.

Six, L., & Smolders, E. (2014). Future trends in soil cadmium concentration under current cadmium fluxes to European agricultural soils. *The Science of the Total Environment, 485*, 319–328. doi:10.1016/j.scitotenv.2014.03.109 PMID:24727598

Suanon, F., Tometin, L., Dimon, B., Agani, I., Mama, D., & Azandegbe, E. (2016). Utilization of sewage sludge in agricultural soil as fertilizer in the Republic of Benin (West Africa): What are the risks of heavy metals contamination and spreading? *American Journal of Environmental Sciences, 12*(1), 5–15. doi:10.3844/ajessp.2016.8.15

Subasic, M., Samec, D., Selovic, A., & Karalija, E. (2022). Phytoremediation of cadmium polluted soils: Current status and approaches for enhancing. *Soil Systems, 6*(3), 1–21. doi:10.3390oilsystems6010003

Suciu, N., De Vivo, R., Rizzati, N., & Capri, E. (2022). Cd content in phosphate fertilizer: Which potential risk for the environment and human health? *Current Opinion in Environmental Science & Health, 30*, 100392. doi:10.1016/j.coesh.2022.100392

Tabelin, C., Igarashi, T., Villacorte-Tabelin, M., Park, I., Opiso, E., Ito, M., & Hiroyoshi, N. (2018). Arsenic, selenium, boron, lead, cadmium, copper, and zinc in naturally contaminated rocks: A review of their sources, modes of enrichment, mechanisms of release, and mitigation strategies. *The Science of the Total Environment, 645*, 1522–1553. doi:10.1016/j.scitotenv.2018.07.103 PMID:30248873

Teare, J., Kootbodien, T., Naicker, N., & Mathee, A. (2015). The extent, nature and environmental health implications of cottage industries in Johannesburg, south Africa. *International Journal of Environmental Research and Public Health, 12*(2), 1894–1901. doi:10.3390/ijerph120201894 PMID:25664698

Tembo, B., & Kwenga, S. (2006). Distribution of copper, lead, cadmium and zinc concentrations in soils around Kabwe town in Zambia. *Chemosphere, 63*(3), 497–501. doi:10.1016/j.chemosphere.2005.08.002 PMID:16337989

Tindwa, H., & Singh, B. (2023). Soil pollution and agriculture in sub-Saharan Africa: State of the knowledge and remediation technologies. *Frontiers in Soil Science, 2*, 1101944. doi:10.3389/fsoil.2022.1101944

Tiruneh, A., Fadiran, A., & Mtshali, J. (2014). Evaluation of the risk of heavy metals in sewage sludge intended for agricultural application in Swaziland. *International Journal of Environmental Sciences, 5*(1), 197–216.

Umvoto Africa. (2014). *Role of fertilizers in trace metal (specifically cadmium) contamination of groundwater*. Water Research Commission.

UN-Habitat. (2022). *Africa's waste problem.* UN. https://unhabitat.org/ african-clean-cities-africas-waste-problems

Van Brusselen, D., Kitenge, T., Musanzayi, S., Kasole, T., & Ngombe, L. (2020). Metal mining and birth defects: A case-control study in Lubumbashi, Democratic Republic of the Congo. *The Lancet. Planetary Health*, *4*(4), e158–e167. doi:10.1016/S2542-5196(20)30059-0 PMID:32353296

Venegas, A., Rigol, A., & Vidal, M. (2015). Viability of organic wastes and biochar as amendments for remediation of heavy metal contaminated soil. *Chemosphere*, *119*, 190–198. doi:10.1016/j.chemosphere.2014.06.009 PMID:24995385

Wieczorek, J., Baran, A., Urbanski, K., Mazurek, R., & Pawlas, A. (2018). Assessment of the pollution and ecological risk of lead and cadmium in soils. *Environmental Geochemistry and Health*, *40*(6), 2325–2342. doi:10.100710653-018-0100-5 PMID:29589150

Zhang, Q., Hu, J., Lee, D., Chang, Y., & Lee, Y. (2017). Sludge treatment: Current research trends. *Bioresource Technology*, *243*, 1159–1172. doi:10.1016/j.biortech.2017.07.070 PMID:28764130

Zhang, S., Song, J., Cheng, Y., & McBride, M. (2018). Derivation of regional risk screening values and intervention values for cadmium-contaminated agricultural land in the Guizhou plateau. *Land Degradation & Development*, *29*(8), 2366–2377. doi:10.1002/ldr.3034

Zhang, Y., Chen, M., Zhong, Y., Zhang, M., Cheng, M., & Li, X. (2015). Assessment of cadmium (Cd) concentration in arable soil in China. *Environmental Science and Pollution Research International*, *22*(7), 4932–4941. doi:10.100711356-014-3892-6 PMID:25483971

Zhao, Y., Deng, Q., Lin, Q., Zeng, C., & Zhong, C. (2020). Cadmium source identification in soils and higher risk regions predicted by geographical detector method. *Environmental Pollution*, *263*, 114338. doi:10.1016/j.envpol.2020.114338 PMID:32304950

Chapter 12
Soil Pollution by Nickel in Sub-Saharan Africa

ABSTRACT

In this chapter, the sources and geochemical behavior of nickel (Ni) were studied and related to its mobility and toxicity in soils of sub-Saharan Africa region. The heavy metal, though important in plant growth, was toxic in high concentrations. Natural sources of the metal are from laterite and garnierite ores, though they are immobile and of low toxicity due to their solid phases. Exchangeable, mobile, and toxic Ni is of anthropic origin. Such sources include the mining of platinum group metals and Ni ores and their resultant wastes. Fumes, dust, tailings, rock overburden, and effluents were laden with the metal and once in soils, they caused Ni bioaccumulation. Electroplating, combustion of fuels, use of agricultural fertilizers and chemicals and unsystematic waste management also introduced the metal in soils of SSA. Managing industrial processes and resultant wastes is essential in controlling Ni pollution in soils of the region.

INTRODUCTION

Potentially toxic substances in soils emanate from anthropogenic and natural sources (Vischetti et al., 2022). Anthropogenic sources include the use of agricultural additives such as excess fertilizers, herbicides, weedicides, lime, biosolids and pesticides as well as industrial processes (Palansooriya et al., 2020). Other sources include unsystematic disposal of domestic and industrial waste, mining and smelting activities and oil or chemical spills, which occur during industrial processes. Once

DOI: 10.4018/978-1-6684-7116-6.ch012

the substances end up in soil environs, they introduce toxins in soils where they bioaccumulate beyond the maximum permissible levels to alter the physicochemical characteristics of soils and result to pollution and land degradation (Hou & Ok, 2019). Some of the potentially toxic substances include heavy metals such as antimony, arsenic, barium, cadmium, chromium cobalt, copper, lead, manganese, mercury, molybdenum, nickel, selenium, silver, tin, titanium, vanadium and zinc, whose negative environmental and human health effects are reported (Palansooriya et al., 2020; Liu et al., 2018; Wang et al., 2019) (see Chapter 1 of this book). The effects are a result of the substances ability to accumulate in soils and infiltrate in food chains before biomagnification to cause harmful effects to living things (Hou & OK, 2019; Vischetti et al., 2022).

Of the potentially toxic substances, is the micronutrient Ni, which is important to living things but is toxic at levels beyond the allowable limits (Jakubus & Graczyk, 2020). In plants and at low concentrations, Ni enhances the manufacture of carbohydrates, proteins and chlorophyll, improves fruit quality and yield, enhances growth of roots and shoots in addition to promoting seed germination (Gupta et al., 2017). The heavy metal pollutes soils from many human-based activities including combustion of fuel, diesel or oil, in industries dealing with chemicals, metals, glass and ceramics, application of organic amendments such as compost and sewage sludge containing the metal, mining and smelting activities (Ameen et al., 2019; Vischetti et al., 2022). The toxicity of Ni in soils is best presented in plants grown on soils polluted with the metal. According to Ameen et al. (2019), Ni is phytotoxic and causes reduced leaf area, necrosis, redundant shoot and root growth and chlorosis in addition to disrupting the enzymatic activity, uptake of other nutrients and the initiation of oxidative stress in the plants.

In animals, high concentrations of the metal (beyond 10 mg/kg) induce health effects such as lung dysfunctions, carcinogenicity and allergic reactions (Zambelli et al., 2016). Shaheen et al. (2020) also noted that the entry of Ni into food chains and its subsequent bioaccumulation in animals is a threat to their health and that of humans in addition to the potential to induce ecological instability even at global level. Similarly, Poznanovic Spahic et al. (2019) noted that the increasing levels of Ni in the environment globally and particularly, in developing countries such as those of sub-Saharan Africa (SSA), is a growing concern due to the metal's non-biodegradability and associated toxicity. In some reported areas, the levels of the trace metal in soils rose up to 26,000 mg/kg in the last decade due to the rise in industrialization among other anthropogenic activities (Rinklebe & Shaheen, 2017). Therefore, it is key to understand the physicochemical characteristics of the metal, its sources particularly those resulting from industrial activities and its toxicity to soils as focused in this chapter. By understanding the geochemical behavior of Ni, its hazard, mobility and eco-toxicity can be measured, rated based on available pollution standards, predicted, controlled and avoided. Therefore, the aim of this chapter is to detail the chemistry of Ni, its sources and toxicity in soils with a focus of SSA region.

CHEMISTRY OF NICKEL

About 0.008% of the earth's crust consists of Ni and hence the metal is widely distributed in nature. The distribution of the metal is at 5-50% in the meteorites, 1.5% in the nodules of the deep sea and 8.5% in the core. Nickel is mined from two types of ores: 1) laterites and 2) garnierites (National Center for Biotechnology Information, 2023). Laterites are mineral ores comprising of nickeliferous limonite with the formula [(Fe, Ni)O(OH)] while garnierite is hydrous nickel silicate whose chemical formula is $Mg,Ni)_3Si_4O_{10}(OH)_2{\cdot}nH_2O$. The metal is also sourced from pentlandite [$(Fe,Ni)_9S_8$] and pyrrhotite [Fe(1-x)S (x = 0 to 0.2)], which is an or-magmatic sulfide deposit. Ni can be substituted with Fe and Mg in the crystal lattices of the oxides and silicate minerals since the three metals are divalent with a near to ionic radius. The ultramafics, which are Fe- and Mg-rich rocks found in plutonic and volcanic environs are also rich in nickel sulfide (National Center for Biotechnology Information, 2023). Once the ultramafic rocks weather near the surface, they form laterites while at greater depths of the earth's core, high concentrations of nickel sulfide deposits occur.

Nickel is a ductile, odorless, malleable, corrosion resistant and strong hard silver white (Figure 1) heavy metal that occurs in the form of cubic crystals (Habashi, 2013; Helmenstine, 2019). The metal has oxidation rates of +1, +2, +3 and +4 but the most common of them is +2 where Ni (II) is the most toxic and readily available in its ionic form (Parades-Aguilar et al., 2021). Ni belongs to the iron family of the periodic table along with Co and Fe and in group 10 with an average atomic weight of 58.69. Ni has an electronic configuration of 2, 8, 16, 2 moving outwards from the nucleus (Habashi, 2013). The trace element is a transition metal with a boiling point of 2730 oC, a melting point of 1455 oC and a high density of 8.9 g/ cm^3 (Tsadilas & Rinklebe, 2018). Other physicochemical properties of the metal are as summarized in Table 1.

Nickel dissolves in ammoniacal solutions when oxygen is present to form ammine complexes as shown in Equation 1.

$$Ni + \frac{1}{2}O_2 + 2NH_3 + H_2O \rightarrow [Ni(NH_3)_2]^{2+} + 2OH^- \quad (1)$$

Figure 1. Physical appearance of Ni
(Helmenstine, 2019)

At high pressure and temperature and in the presence of hydrogen, Ni is precipitated to Ni powder as shown in Equation 2.

$$Ni^{2+} + H_2 \rightarrow Ni + 2H^+ \quad (2)$$

At ambient environs and in neutral medium, Ni sulfide is oxidized to nickel sulfate as shown in Equation 3.

$$NiS + 20_{2(aq)} \rightarrow NiSO_4 \quad (3)$$

In acidic environs, nickel sulfide forms the elemental form of the metal as shown in Equation 4.

$$NiS + \frac{1}{2}O_{2(aq)} + 2H^+ \rightarrow Ni^{2+} + S + H_2O \quad (4)$$

Table 1. The physical properties of nickel (Habashi, 2013; Helmenstine, 2019; National Center for Biotechnology Information, 2023)

Physicochemical Property	Value
Atomic number	28
Atomic weight	58.69
Color	Lustrous, silvery white solid
Melting point	1455 oC
Boiling point	2730 oC
Relative density at 25 oC	8.9 g/cm^3
The increase of volume after melting	4.5%
Heat of fusion at melting point	302 J/g
Heat of sublimation at 25 oC	7,317 J/g
Vapor pressure	0 mmHg
Stability	It is stable in air and at ordinary temperature and is not affected by water
Heat of vaporization	6,375 J/g
Brinell hardness	85
Elasticity modulus	199.5 GPa
Thermal expansion coefficient between 0 and 100 oC	13.3 x 10^{-6} K^{-1}
Electrical resistivity at 20oC	6.9 μΩ/cm
Heat capacity at 0 to 100oC	0.452 J/g/K
Thermal conductivity at 0 to 100oC	88.5 W/m/K
Solubility	Insoluble in water, soluble in dilute nitric acid and ammoniacal solution, slightly soluble in sulfuric and hydrochloric acids
Decomposition	The metal decomposes to release toxic vapors and gases
Corrosion	Excellent resistance to corrosion
Isotopes	The metal has five isotopes Ni^{58} at 67.76% Ni^{60} at 26.61% Ni^{61} at 1.25% Ni^{62} at 3.66% Ni^{64} at 1.16% There are also artificial isotopes including Ni 56, 57, 59, 63, 65-67

At high temperatures, nickel sulfide reacts with oxygen to form sulfur dioxide and nickel oxide (Equation 5).

$$NiS + 1\frac{1}{2} \rightarrow NiO + SO_2 \quad (5)$$

Elemental nickel is formed when molten nickel sulfide (Ni_3S_2) reacts with oxygen as shown in Equation 6.

$$Ni_3S_2 + 2O_2 3Ni + 2SO_2 \quad (6)$$

The basis for refining Ni is its reaction in the presence of carbon monoxide at 60 °C to form nickel tetracarbonyl, which is volatile (Equation 7). The reaction is reversible and the toxic nickel tetracarbonyl can be decomposed to elemental Ni and carbon monoxide at a higher temperature of 180 °C.

$$Ni + 4CO \leftrightarrow Ni(CO)_4 \quad (7)$$

SOURCES, USES, AND POLLUTION BY NICKEL

Nickel occurs in various parts of the ecosphere including the soil, marine and freshwater. Ni in soils from various pedogenic and anthropogenic processes is transferred to surface and groundwater resources where in high levels, it induces pollution. The weathering of minerals causes chemical modifications on surfaces of primary minerals to release soluble Ni-compounds in association with nitrates, sulphates and chlorides, which are used to form secondary minerals such as amorphous minerals, hydroxides, oxides, carbonates and alumina-silicates (El-Naggar et al., 2021). The lithogenic sources of Ni include parent material and volcanic ashes/ plume (Laribi et al., 2019), mafic and ultramafic parent rocks (Poznanovic Spahic et al., 2019). Some of the mafic and ultramafic rocks laden with Ni include ophiolites, pyroxenites, peridotites and serpentinites. In soils consisting of sedimentary rocks along with alluvial sediments, ophiolite and olivine complexes, Ni is found in high concentrations (Albanese et al., 2015). Pedogenic or lithologic sources of Ni are immobile and less toxic compared to anthropogenically sourced ones because they occur in the solid phase (Poznavic Spahic et al., 2019). However, with alternating redox conditions and changing pH, toxic natural occurring and labile Ni could be formed in soils and contaminate them.

Human activities are major sources of Ni and threaten the sustainability of ecosystems. Industrial activities such as cigarette making, glassblowing, lost-wax casting, manufacture of cement, ceramics and jewelry result to formation of fly ash, particulate matter and dust containing Ni (National Center for Biotechnology Information, 2023). Some studies reported that metal and cement processing industries rely heavily on fossil fuel combustion, which produces particulate matter such as fly ash resulting to elevated levels of Ni to the range of 91-1200 mg/kg in soils (Panagopoulos et al., 2015; Poznanovic Spahic et al., 2019). Such levels are beyond the metal's allowable limits according to WHO/FAO standards set at 50 mg/kg (Kayode et al., 2022). Industries engaging in stainless steel production, chemical making, manufacture of Ni-Cd batteries, oil refining, automobile servicing and repair also produce emissions and fumes to the atmosphere that are later deposited to soils and could introduce Ni in such environs (Albanese et al., 2015; Hernandez-Quiroz et al., 2012). The National Center for Biotechnology (2023) also listed that abrasive blasting, foundry processes, machining of metals, welding, electroplating, manufacture of semiconductors, thermal spraying of metals and coating of welded materials as industrial activities that introduce Ni to soils. The industrial activities produce solid wastes, wastewater and effluents that are laden with Ni and once such byproducts are unsystematically discharged, they introduce the metal to soils and other natural resources. In Peloponnese, Greece, industrial byproducts were attributable to elevated concentrations of Ni to levels more than 2.5 times the allowable limits in soils (Panagopoulos et al., 2015). Mining and smelting activities, production and use of agrochemicals such as herbicides, fungicides and insecticides as well as fertilizers also introduced Ni into soils (El-Naggar et al., 2018a; Khan et al., 2017; Palansooriya et al., 2020). Therefore, the introduction of highly mobile and toxic Ni is driven by anthropogenic activities, particularly industrial activities and hence the need to regulate them to prevent extended contamination by the trace metal to other spheres of the earth.

NICKEL TOXICITY IN SOILS

The toxicity of Ni to soils depends on its potentially mobile and already mobile fractions (El-Naggar et al., 2021). The mobility of Ni in soils and its subsequent phytoavailability is a function of several environmental factors including biological transformations, ion exchange, complexation, dissolution, precipitation, redox potential and pH (Rajapaksha et al., 2018; Shaheen et al., 2018). The factors promote geochemical fractionation of Ni and hence, various compounds containing the metal. Such compounds include oxides of manganese and iron as well as carbonates. The exchangeable Ni and the carbonate bound complex of the metal have the highest mobility while non-residual fractions of the metal become potentially mobile if there are variations in the soil redox potential, dissolved organic carbon levels and pH (El-Naggar et al., 2018a; Shaheen et al., 2020). The mobility of Ni enhances its uptake by plants and subsequent transfer to other environmental compartments using soil as the media and ultimately, its toxicity.

In soils, Ni has high affinity to metals such as Fe and Sb, metalloids such as As and nonmetals such as sulfur and selenium, which leads to its chemical complexation in minerals such as pentlandite [$(Fe, Ni)_9S_8$], gersdorffite (NiAsS), millerite (NiS), ullamannite (NiSbS), niccolite (NiAs) and kullerudite ($NiSe_2$) (Albanese et al., 2015; Hooda, 2010). The presence of Fe-Mn hydroxides and oxides, the redox potential of soils, soil pH, organic matter content, acidity levels, the content and type of clay all influence the biogeochemical characteristics of Ni in the soils and ultimately, its mobility and toxicity (Albanese et al., 2015; El-Naggar et al., 2018a; Panagopoulos et al., 2015). The metal has high affinity to organic matter and as such, organic matter composition and content controls its mobility in soils. Ni binds to organic ligands, which results to reduced reactivity and mobility of the metal and hence less toxicity in the soils. In the presence of humic and fulvic acids that have high chelation ability, the mobility of Ni is enhanced and so is its toxicity in soils (Kabata-Pendias, 2011). In clay minerals such as montmorillonite, Ni co-precipitates and becomes more mobile and toxic in soils. Complexation of Ni with Fe and Mn hydroxides and oxides such as hematite (Fe_2O_3) and magnetite (Fe_3O_4) in soils results to adsorption of the metal and its elevation in soils. For instance, levels of Ni were reported to be between 100-170 mg/kg and 39-4900 mg/kg in soils with Fe and Mn oxides and hydroxides, respectively (Kabata-Pendias, 2011). In acidic pH, the mobility of Ni in soils increases and the adsorption of Fe-Mn hydroxides and oxides is enhanced and so is the toxicity of the metal in soils (El-Naggar et al., 2018b). Mobilization of Ni can be favored or disfavored by high redox potential based on the specific soil pH and its related variations (Rinklebe et al., 2016).

NICKEL POLLUTION IN SUB-SAHARAN AFRICA

Sub-Saharan Africa (SSA) refers to Africa's geographical region south of the Sahara Desert and comprises of West Africa, Southern Africa, Eastern and Central Africa. The region has many mineral deposits and ranks second with the highest reserves of galena, zirconium, vermiculite, platinum-group metals, phosphate rock, industrial diamond, cobalt and bauxite (Ngure & Kinuthia, 2020). The mining of these resources is the second most useful economic activity after agriculture though it leaves soils laden with potentially toxic elements. The involved mining activities further lead to deposition of mine fumes and dust, processing tailings and wastes into soils leading to their pollution (Mwesigye et al., 2016). Mining of Ni containing ores occurs in many SSA regions whose geological characteristics have serpentinite that occurs along with trace metals such as Co, Cr, Mg, Ni, Pb and Zn (Prematuri et al., 2020). By releasing the trace metals to soils during mining activities, soils become polluted and transfer the contaminant to plants and water resources. The mining along with industrial activities using heavy metals pollute soils and other natural resources (Habib, 2017).

Electroplating processes and manufacture of catalysts, pigments, ceramics and alloys require the use of Ni. For this reason, the production and use of the metal is growing in SSA along with its diverse uses. In South Africa alone, more than 47, 000 tons of Ni were produced in 2011 alone during the production of platinum-group metals as a co-product (Young et al., 2018). The processing, refining and purification of the metal produces by-products that can lead to soil pollution by Ni. During milling and crushing of the metals, sulfidic nickel is released in the form of dust particles while oxidic nickel results during the refining of ores in furnaces as emissions. During nickel electrolysis in tank houses and at base metal refineries, soluble nickel sulfate is subjected to electrical current to form oxygen bubbles at the anode and producing Ni-containing aerosol and effluents (Young et al., 2018). The dust, effluents, aerosol and emissions laden with Ni are released to the atmosphere before being deposited on the soils at low temperatures. The mining and manufacturing activities have been associated with high levels of Ni in sediments of Burgersfort (Addo-Bediako et al., 2021) and Thohoyandou (Moyo et al., 2020) areas in Limpopo province, South Africa. A similar study in Zambia, Namibia and South Africa's Johannesburg city has shown that mining activities result to production of acid mine drainage, tailings and dust laden with Ni and other heavy metals that pollute areal soils (Kribek et al., 2014). Mining and smelting of base metals and gold results to Ni-containing effluents. These effluents were reported to pollute soils in Shashe, Motloutse and Mzingwane areas of South Africa (Ebenebe et al., 2017). Other human activities associated with Ni pollution in soils of South Africa include the combustion of fossil fuels, use of Ni-containing pesticides and chemical fertilizers (Addo-Bediako et al., 2021) and the unsystematic management of Ni-containing solid wastes and subsequent release of leachate (Nyika et al., 2019).

In Eastern Africa, a number of industrial processes have been associated with the introduction of Ni in soils. The importation, production and use of organochlorine pesticides such as hexachlorocyclohexane (HCH) and dichloro-diphenyl-trichloroethane (DDT) is known to introduce Ni among other heavy metals into agricultural soils of Kenya (Ndungu et al., 2019; Sun et al., 2016). The introduction of industrial effluents to soils through their use for irrigation has also been associated with elevated Ni levels in SSA region. The use of wastewater for irrigation in suburban areas of Nairobi (Nyika & Dinka, 2022) and introduction of industrial wastewater in open drainage channel soils of Nairobi's industrial area (Kinuthia et al., 2020) resulted elevated levels of Ni in sampled areas. The mining of gold and its associated artisanal activities in the Migori granite-green complex was associated with elevated levels of Ni in the soils and its transfer to grown plants (Ngure & Kinuthia, 2020). Copper mining in Uganda's Kilembe area was associated with cobaltiferous pyrite and cupriferous tailings containing Ni and other heavy metals being dumped into the soils and causing their pollution (Mwesigye et al., 2016). Other reasons associated with Ni contamination in soils of Eastern Africa include emissions, wastewater, solid waste and dust from manufacturing industries, metal processing industries, chemical processing firms, use of agrochemicals and soil additives and unsystematic disposal of Ni-containing solid waste (Mungai et al., 2016).

In west Africa, Ni pollution in soils has been associated with anthropogenic activities such as application of Ni-containing phosphatic fertilizers and manures, disposal of industrial and municipal wastes containing the metal, automobile emissions, metal mining and smelting and industrial processes such as electroplating and combustion of fossil fuels (Wiafe et al., 2022). In Ghana's Prestea Huni-Valley region, soils were laden with Ni and the trend was attributable to illegal artisanal mining activities and their resultant wastes that are introduced in such environs (Wiafe et al., 2022). Similar observations were made in Zamfara state of Nigeria, which is concentrated with artisanal miners where soils had elevated levels of Ni among other heavy metals (Yahaya et al., 2021). In automobile repair villages of Ibadan, Nigeria, soils had Ni levels ranging between 2 and 29 mg/kg due to combustion of fossil fuels and the use of Ni-containing products (Adelekan & Abegunde, 2011). In Challawa, Sharada and Bompai industrial areas of Nigeria's Kano state, Ni concentrations in soils were beyond the allowable limits (Egwuonwu et al., 2011; Hamidu et al., 2021; Idris et al., 2015; Musa et al., 2018). The observations were attributable to the concentration of chemical, plastic, food, fertilizer, textile and tannery industries in the areas, use of agrochemicals in nearby farmlands, the release of industrial effluents to soils of the vicinity and the disposal of metal containing solid wastes to the surroundings.

CONCLUSION

Evidence from empirical studies reviewed in this chapter show that the divalent form of Ni is toxic, highly exchangeable and mobile. It is replaced by Fe and Mg among other divalent cations of similar or near to ionic radius. In soils, the mobility of the heavy metal is a function of variation in the redox potential and pH of soils among other factors such as complexation, dissolution and ion exchange, which control the geochemical fractionation of Ni. Mobility of the metal in soils is enhanced by the presence of organic matter, some clays such as montmorillonite, fluvic and humic acids. Natural forms of the metal are immobile but can become labile with changes in pH and redox potential of soils to induce toxicity. Anthropogenic sources of Ni are mainly as a result of the mining of Ni-ores and platinum group metals, fuel combustion, excess use of agrochemicals and organic fertilizers, manufacture of chemicals such as pigments, ceramics, alloys and catalysts and unsystematic management of waste and wastewater that results from industrial activities. Therefore, pollution of Ni in soils is a function of anthropogenic activities rather than geogenic processes and their regulation is key in controlling extended contamination by the metal in the environment.

REFERENCES

Addo-Bediako, A., Nukeri, S., & Kekana, M. (2021). Heavy metal and metalloid contamination in the sediments of the Spekboom River, South Africa. *Applied Water Science*, *11*(7), 133. doi:10.100713201-021-01464-8

Adelekan, B., & Abegunde, K. (2011). Heavy metals contamination of soil and groundwater at automobile mechanic villages in Ibadan, Nigeria. *International Journal of Physical Sciences*, *6*(5), 1045–1058. doi:10.5897/IJPS10.495

Albanese, S., Sadeghi, M., Lima, A., Cicchella, D., Dinelli, E., Valera, P., Falconi, M., Demetriades, A., & De Vivo, B. (2015). GEMAS: Cobalt, Cr, Cu and Ni distribution in agricultural and grazing land soil of Europe. *Journal of Geochemical Exploration*, *154*, 81–93. doi:10.1016/j.gexplo.2015.01.004

Ameen, N., Amjad, M., Murtaza, B., Abbas, G., Shahid, M., Imran, M., Naeem, M. A., & Niazi, N. K. (2019). Biogeochemical behavior of nickel under different abiotic stresses: Toxicity and detoxification mechanisms in plants. *Environmental Science and Pollution Research International*, *26*(11), 10496–10514. doi:10.100711356-019-04540-4 PMID:30835069

Ebenebe, P., Shale, K., Sedibe, M., Tikilili, P., & Achilonu, M. (2017). South African mine effluents: Heavy metal pollution and impact on the ecosystem. *International Journal of Chemical Science*, *15*(4), 198.

Egwuonwu, G., Olabode, V., Bukar, P., Okolo, V., & Odunze, A. (2011). Characterization of topsoil and groundwater at leather industrial area, Challawa, Kano, Northern Nigeria. *Pacific Journal of Science and Technology*, *12*(1), 628.

El-Naggar, A., Ahmed, N., Mosa, A., Niazi, N., Yousaf, B., Sharma, A., Sarkar, B., Cai, Y., & Chang, S. X. (2021). Nickel in soil and water: Sources, biogeochemistry and remediation using biochar. *Journal of Hazardous Materials*, *419*, 126421. doi:10.1016/j.jhazmat.2021.126421 PMID:34171670

El-Naggar, A., Rajapaksha, U., Shaheen, M., Rinklebe, J., & Ok, S. (2018b). Potential of biochar to immobilize nickel in contaminated soils. In *Nickel in Soils and Plants* (pp. 293–318). CRC Press. doi:10.1201/9781315154664-13

El-Naggar, A., Shaheen, M., Ok, S., & Rinklebe, J. (2018a). Biochar affects the dissolved and colloidal concentrations of Cd, Cu, Ni, and Zn and their phytoavailability and potential mobility in a mining soil under dynamic redox-conditions. *The Science of the Total Environment, 624*, 1059–1071. doi:10.1016/j.scitotenv.2017.12.190 PMID:29929223

Gupta, V., Jatav, K., Verma, R., Kothari, L., & Kachhwaha, S. (2017). Nickel accumulation and its effect on growth, physiological and biochemical parameters in millets and oats. *Environmental Science and Pollution Research International, 24*(30), 23915–23925. doi:10.100711356-017-0057-4 PMID:28875293

Habashi, F. (2013). Nickel, Physical and Chemical Properties. In R. H. Kretsinger, V. N. Uversky, & E. A. Permyakov (Eds.), *Encyclopedia of Metalloproteins*. Springer. doi:10.1007/978-1-4614-1533-6_338

Habib, E. (2017). Heavy metals pollution of soil; toxicity and phytoremediation techniques. *International Journal of Advanced Research and Publications, 1*(1), 29–41.

Hamidu, H., Halilu, F., Yerima, K., Garba, L., Suleiman, A., Kankara, A. I., & Abdullahi, I. M. (2021). Heavy metals pollution indexing, geospatial and statistical approaches of groundwater within Challawa and Sharada industrial areas, Kano city, North-western Nigeria. *SN Applied Sciences, 3*(7), 690. doi:10.100742452-021-04662-w

Helmenstine, A. (2019). *Nickel element facts and properties*. ThoughtCo. https://www.thoughtco.com/nickel-facts-606565

Hernandez-Quiroz, M., Herre, A., Cram, S., de Leon, P., & Siebe, C. (2012). Pedogenic, lithogenic - or anthropogenic origin of Cr, Ni and V in soils near a petrochemical facility in Southeast Mexico. *Catena, 93*, 49–57. https://doi.org/.catena.2012.01.005. doi:10.1016/j

Hooda, S. (2010). *Trace elements in soils. trace elements in soils*. Blackwell Publishing Ltd. doi:10.1002/9781444319477

Hou, D., & Ok, S. (2019). Soil pollution — Speed up global mapping. *Nature, 566*(7745), 455. doi:10.1038/d41586-019-00669-x PMID:30809065

Idris, M., Khalid, D., & Abdullahi, Z. (2015). Comparative assessment of heavy metals concentration in the soil in the vicinity of tannery industries, Kumbotso old dump site and River Challawa, conference at Challawa industrial estate, Kano State, Nigeria. *International Journal of Innovative Research and Development, 4*(6), 122–128.

Jakubus, M., & Graczyk, M. (2020). Availability of nickel in soil evaluated by various chemical extractants and plant accumulation. *Agronomy (Basel), 10*(11), 1805. doi:10.3390/agronomy10111805

Kabata-Pendias, A. (2011). *Trace metals in soils and plants*. CRC Press.

Kayode, O., Ogunyemi, E., Odukoya, A., & Aizebeokhai, A. (2022). Assessment of chromium and nickel in agricultural soil: Implications for sustainable agriculture. *IOP Conference Series. Earth and Environmental Science, 993*(1), 012014. doi:10.1088/1755-1315/993/1/012014

Khan, N., Mobin, M., Abbas, K., & Alamri, A. (2017). Fertilizers and their contaminants in soils, surface and groundwater. In A. Dominick, S. Della, & M. Goldstein (Eds.), *The encyclopedia of the Anthropocene* (pp. 225–240). Elsevier. doi:10.1016/B978-0-12-809665-9.09888-8

Kinuthia, G., Ngure, V., Beti, D., Lugalia, R., Wangila, A., & Kamau, L. (2020). Levels of heavy metals in wastewater and soil samples from open drainage channels in Nairobi, Kenya: Community health implication. *Scientific Reports, 10*(1), 8434. doi:10.103841598-020-65359-5 PMID:32439896

Kribek, B., De Vivo, B., & Davies, T. (2014). Impacts of mining and mineral processing on the environment and human health in Africa. *Journal of Geochemical Exploration, 144*(part C), 387–390. doi:10.1016/j.gexplo.2014.07.018

Laribi, A., Shand, C., Wendler, R., Mouhouche, B., & Colinet, G. (2019). Concentrations and sources of Cd, Cr, Cu, Fe, Ni, Pb and Zn in soil of the Mitidja plain, Algeria. *Toxicological and Environmental Chemistry, 101*(1-2), 59–74. doi: 10.1080/02772248.2019.1619744

Liu, L., Li, W., Song, W., & Guo, M. (2018). Remediation techniques for heavy metal-contaminated soils: Principles and applicability. *The Science of the Total Environment, 633*, 206–219. doi:10.1016/j.scitotenv.2018.03.161 PMID:29573687

Moyo, B., Matodzi, V., Legodi, M., Pakade, V., & Tavengwa, N. (2020). Determination of Cd, Mn and Ni in fruits, vegetables and soil in the Thohayandou town area, South Africa. *Water S.A., 46*(2), 285–290. doi:10.17159/wsa/2020.v46.i2.8244

Mungai, T., Owino, A., Makokha, V., Gao, Y., Yan, X., & Wang, J. (2016). Occurrences and toxicological risk assessment of eight heavy metals in agricultural soils from Kenya, Eastern Africa. *Environmental Science and Pollution Research International*, *23*(18), 18533–18541. doi:10.100711356-016-7042-1 PMID:27291978

Musa, D., Sadiya, M., Yusuf, M., Garba, Y., & Gimba, E. (2018). Study of water and soil contamination by heavy metal from industrial effluents at Bompai industrial area Kano, Nigeria. *FUW Trends in Science and Technology Journal*, *3*(1), 234–238.

Mwesigye, R., Young, S., Bailey, E., & Tumwebaze, B. (2016). Population exposure to trace elements in the Kilembe copper mine area, Western Uganda: A pilot study. *The Science of the Total Environment*, *15*(573), 366–375. doi:10.1016/j.scitotenv.2016.08.125 PMID:27572529

National Center for Biotechnology Information. (2023). PubChem Compound Summary for CID 935, Nickel. NCBI. https://pubchem.ncbi.nlm.nih.gov/compound/Nickel

Ndungu, A., Yan, X., Makokha, V., Githaiga, K., & Wang, J. (2019). Occurrence and risk assessment of heavy metals and organochlorine pesticides in surface soils, central Kenya. *Journal of Environmental Health Science & Engineering*, *17*(1), 63–73. doi:10.100740201-018-00326-x PMID:31321038

Ngure, V., & Kinuthia, G. (2020). Health risk implications of lead, cadmium, zinc and nickel for consumers of food items in Migori gold mines, Kenya. *Journal of Geochemical Exploration*, *209*, 106430. doi:10.1016/j.gexplo.2019.106430

Nyika, J., & Dinka, M. (2022). Heavy metal pollution in soils and vegetables from suburban regions of Nairobi, Kenya and their community health implications. *Pollution*, *8*(4), 1434–1447. doi:10.22059/POLL.2022.341522.1440

Nyika, J., Onyari, E., Dinka, M., & Mishra, S. (2019). Heavy metal pollution and mobility in soils within a landfill vicinity: A South African case study. *Oriental Journal of Chemistry*, *35*(4), 1286–1296. doi:10.13005/ojc/350406

Palansooriya, N., Shaheen, M., Chen, S., Tsang, D., Hashimoto, Y., Hou, D., Bolan, N. S., Rinklebe, J., & Ok, Y. S. (2020). Soil amendments for immobilization of potentially toxic elements in contaminated Soils: A Critical review. *Environment International*, *134*, 105046. doi:10.1016/j.envint.2019.105046 PMID:31731004

Panagopoulos, I., Karayannis, A., Kollias, K., Xenidis, A., & Papassiopi, N. (2015). Investigation of potential soil contamination with Cr and Ni in four metal finishing facilities at Asopos industrial area. *Journal of Hazardous Materials*, *281*, 20–26. doi:10.1016/j.jhazmat.2014.07.040 PMID:25112552

Parades-Aguilar, J., Reyes-Martínez, V., Bustamante, G., Almendariz, J., Martínez-Meza, G. et al. (202)1. Removal of nickel (II) from wastewater using a zeolite-packed anaerobic bioreactor: bacterial diversity and community structure shifts. *Journal of Environmental Management, 279*, 111558 doi:10.1016/j.jenvman.2020.111558

Poznanovic Spahic, M., Sakan, M., Glavas, M., Tancic, P., & Skrivanj, B. (2019). Natural and anthropogenic sources of chromium, nickel and cobalt in soils impacted by agricultural and industrial activity (Vojvodina, Serbia). *Journal of Environmental Science and Health. Part A, Toxic/Hazardous Substances & Environmental Engineering, 54*(3), 219–230. doi:10.1080/10934529.2018.1544802 PMID:30587075

Prematuri, R., Turjaman, M., Sato, T., & Tawaraya, K. (2020). The impact of nickel mining on soil properties and growth of two fast-growing tropical tree species. *International Journal of Forestry Research, 8837590*, 1–9. doi:10.1155/2020/8837590

Rajapaksha, U., Alam, S., Chen, N., Alessi, S., Igalavithana, D., Tsang, D. C. W., & Ok, Y. S. (2018). Removal of hexavalent chromium in aqueous solutions using biochar: Chemical and spectroscopic investigations. *The Science of the Total Environment, 625*, 1567–1573. doi:10.1016/j.scitotenv.2017.12.195 PMID:29996453

Rinklebe, J., Shaheen, M., & Frohne, T. (2016). Amendment of biochar reduces the release of toxic elements under dynamic redox conditions in a contaminated floodplain soil. *Chemosphere, 142*, 41–47. doi:10.1016/j.chemosphere.2015.03.067 PMID:25900116

Rinklebe, J., & Shaheen, S. (2017). Redox chemistry of nickel in soils and sediments: A review. *Chemosphere, 179*, 265–278. doi:10.1016/j.chemosphere.2017.02.153 PMID:28371710

Shaheen, M., El-Naggar, A., Antoniadis, V., Moghanm, S., Zhang, Z., Tsang, D. C. W., Ok, Y. S., & Rinklebe, J. (2020). Release of toxic elements in fishpond sediments under dynamic redox conditions: Assessing the potential environmental risk for a safe management of fisheries systems and degraded waterlogged sediments. *Journal of Environmental Management, 255*, 109778. doi:10.1016/j.jenvman.2019.109778 PMID:32063315

Shaheen, M., El-Naggar, A., Wang, J., Hassan, E., & Niazi, K. (2018). Biochar as an (im)mobilizing agent for the potentially toxic elements in contaminated soils. In *Biochar from Biomass and Waste* (pp. 255–274). Elsevier. doi:10.1016/B978-0-12-811729-3.00014-5

Sun, H., Qi, Y., Zhang, D., Li, Q., & Wang, J. (2016). Concentrations, distribution, sources and risk assessment of organohalogenated contaminants in soils from Kenya, Eastern Africa. *Environmental Pollution*, *209*, 177–185. doi:10.1016/j.envpol.2015.11.040 PMID:26686059

Tsadilas, C., & Rinklebe, J. (2018). Nickel in soils and plants, first edition. CRC Press, Taylor & Francis Group, New York, USA. doi:10.1201/9781315154664

Vischetti, C., Marini, E., Casucci, C., & De Bernardi, A. (2022). Nickel in the environment: Bioremediation techniques for soils with low or moderate contamination in European Union. *Environments (Basel, Switzerland)*, *9*(10), 133. doi:10.3390/environments9100133

Wang, L., Cho, W., Tsang, C., Cao, X., Hou, D., Shen, Z., Alessi, D. S., Ok, Y. S., & Poon, C. S. (2019). Green remediation of As and Pb contaminated soil using cement-free clay-based stabilization/solidification. *Environment International*, *126*, 336–345. doi:10.1016/j.envint.2019.02.057 PMID:30826612

Wiafe, S., Yeboah, E., Boakye, E., & Ofosu, S. (2022). Environmental risk assessment of heavy metals contamination in the catchment of small-scale mining enclave in Prestea Huni-Valley district, Ghana. *Sustainable Environment*, *8*(1), 1–15. doi:10.1080/27658511.2022.2062825

Yahaya, S., Abubakar, F., & Abdu, N. (2021). Ecological risk assessment of heavy metal-contaminated soils of selected villages in Zamfara State, Nigeria. *SN Applied Sciences*, *3*(2), 168. doi:10.100742452-021-04175-6

Young, M., Van Der Merwe, C., Linde, S., & Du Plessis, J. (2018). A retrospective analysis of nickel exposure data at a South African base metal refinery. *Journal of Occupational and Environmental Hygiene*, *15*(3), 204–213. doi:10.1080/15459624.2017.1411596 PMID:29194021

Zambelli, B., Uversky, N., & Ciurli, S. (2016). Nickel impact on human health: An intrinsic disorder perspective. *Biochimica et Biophysica Acta (BBA)-. Biochimica et Biophysica Acta. Proteins and Proteomics*, *1864*(12), 1714–1731. doi:10.1016/j.bbapap.2016.09.008 PMID:27645710

Chapter 13
The Dynamics of Copper and Zinc Pollution in Soils:
The Case of Sub-Saharan Africa

ABSTRACT

In this chapter, the chemistry of copper (Cu) and zinc (Zn) and their physicochemical characteristics were studied and related to the toxicity of the metals in soils. The metals are essential nutrients in soils, but their bioaccumulation introduces toxic effects. Such effects include reduced respiration, low microbial, and microorganism activity and biodiversity and an imbalance of available essential nutrients leading to altered physicochemical characteristics, as well as low quality and fertility in the soils. The toxicity is controlled by factors such as pH and organic matter content that regulate sorption and desorption processes in soils. The toxicity in sub-Saharan Africa soils (SSA) is as a result of the mining and smelting of Cu and Zn ores where metals are introduced in soils as slug, waste rock, overburden materials, tailings, and acid mine drainage from the involved processes. The use of phosphatic fertilizers and fungicides also leads to accumulation of the metals in soils.

INTRODUCTION

Soils play an essential role in regulating ecological stability although the effectiveness of this function is regulated by their quality (Ngole & Ekosse, 2012). According to Genova et al. (2022), soil is a complex media whose functionality depends on many factors; it is a living, dynamic and non-renewable resource, which serves an importance role in sustaining animal and plant life as well as terrestrial ecosystems.

DOI: 10.4018/978-1-6684-7116-6.ch013

In the presence of potentially toxic metals (PTM), soils become contaminated, which results to an environmental problem. Datsenko et al. (2021) noted that PTM such as Cd, Cr, Cu, Fe, Hg, Pb Zn, which are of anthropogenic or lithologic origin are a growing environmental challenge in modern day. The concentrations of PTM in soils are growing due to the rise in urbanization, industrialization, infrastructural and economic development among other anthropic activities leading to ecological imbalances (Abhijith & Varghese, 2021). The application of fertilizers and agrochemicals, expansions of industries, petrochemical processing, disposal of solid wastes on soils and wastewater spillages all contaminate soils. Once soils are contaminated, their function and fertility are disrupted and the PTM are taken up by plants and/ or leach to surface and groundwater resources and eventually, enter food chains causing ill effects on biodiversity and hence, ecological imbalance (Genova et al., 2022). Consequently, the metals bioaccumulate and persist due to their non-biodegradable nature, which becomes a threat to the sustenance of life.

Pollution of soils by Cu and Zn is mainly a result of mining and disposal of resultant wastes as Ngole and Ekosse (2012) noted. Mining of the sulfidic ores and their pyrometallurgical processing results to wastes rich in the metals that infiltrate to the soils contaminating them. In the agricultural sector, Cu and Zn levels in soils increase due to application of minerals, wastewaters, manures, phosphate fertilizers and agrochemicals rich in the metals (Angelaki et al., 2022). Both Cu and Zn are essential micronutrients to plants in low concentrations but in high concentrations, they alter the soil chemistry and are toxic to plants (Gong et al., 2020; Genova et al., 2022). Elevated levels of Cu reduce the diversity and numbers of earthworms and springtails (important in cycling soil organic matter) in addition to altering the soil respiration, inhibiting metabolic activities by cyanobacterial and reducing microbial biomass (Karimi et al., 2021). Zn on the other hand slows down the activity of microbes (bacteria) and earthworms, retards organic matter breakdown and eventually, the biogeochemical cycling of nutrients and inhibits the activity of enzymes such as dehydrogenase, urease and phosphatase in the soils (Rajput et al., 2018; Garcia-Gomez et al., 2020). Elevated levels of the two metals also induce antagonistic and synergistic interactions among other soil elements in addition to causing imbalances in essential nutrients (Marastoni et al., 2019). Therefore, concentrations of the metals beyond the maximum allowable limits of 100 and 300 mg/kg for Cu and Zn, respectively presents a health risk to plants and secondary consumers including human beings. To this end, it is imperative to understand the occurrence of the two heavy metals in soils and their inherent characteristics, which influence their mobility, bioavailability and toxicity in the environment. Such advances will provide ideas to prevent and control further pollution by the metals. In this chapter, the chemistry of Cu and Zn is discussed to understand their sources, physicochemical characteristics and toxicity in soils. In addition, an evaluation on the causes and extent of pollution by the metals in Sub-Saharan Africa is discussed.

CHEMISTRY OF COPPER AND ZINC

Copper

Cu is a d-block element and a transition metal in Group Ib of the periodic table where it occurs along with silver (Ag) and gold (Au) (Mustafa & AlSharif, 2018). The metal has one s-orbital in addition to a filled electron shell used for chemical bonding. The electronic configuration for the metal is $3d^{10}4s^{1}$. Cu has three common oxidation states: 1) pure metal (Cu0), 2) cuprous (+1) and cupric (+2) forms used to react chemically and result to its compounds. The oxidation states also define the ionic state of the metal. Importance ores used as natural sources of the metal include copper pyrites ($CuFeS_2$ and $Cu_2S.Fe_2S_2$), copper glance also known as chalcocite (Cu_2S), cuprite (Cu_2O), azurite ($2\{CuCO_3.Cu(OH)_2\}$ and malachite $\{CuCO_3.Cu(OH)_2\}$. Most of these ores are found in Jharkhand state of India, Ural Mountain in Siberia and in America at Superior region (Mustafa & AlSharif, 2018).

Table 1. Physicochemical properties of copper (Royal Society of Chemistry, 2023)

Physicochemical Characteristic	Value
Position on the periodic table	Group Ib (11), d-block, period 4
Atomic number	29
Relative atomic mass	63.546
State	Solid at 20 °C
Electronic configuration	$[Ar]3d^{10}4s^{1}$
Boiling point	2560 °C
Melting point	1084.6 °C
Density	8.96 g/cm^3
Isotopes	^{63}Cu with a natural abundance of 69.15% ^{64}Cu with a natural abundance of 30.85%
Electronegativity	1.9 on the Pauling scale
Electron affinity	119.16 KJ/mol
Energy of first and second ionization	743.5 and 1946 KJ/mol
Van der Waal radius	0.128 nm
Covalent radius	1.22 Å

The metal is reddish brown (Figure 1), reflects orange and red light in the visible spectrum in addition to reacting with moist air to form a green coating of basic copper sulfate {$CuSO_4.3Cu\ (OH)_2$} that is only dissolved in concentrated acids (Nierengarten, 2018). Cu is a good conductor of heat and electricity in addition to being ductile and malleable. It has low chemical reactivity, is corrosion resistant and is softer compared to Zn. Some of the common uses of Cu include the manufacture of pesticides, industrial machinery, transport equipment, electrical and electronic products (Raffa et al., 2021). Apart from iron and zinc, the metal is one of the most abundant trace metals in human bodies. At 80,000 to 100,000 μg, Cu is deemed to be in excess in living systems and manifests toxic effects such as inhibition of the function of antioxidant enzymes, oxidative modification of proteins, lipids and DNA, interference of Fe transport, suppression of Zn use in the bodies and activation of genes that are redox sensitive (Mustafa & AlSharif, 2018; Haynes, 2015). Other physicochemical characteristics of Cu are summarized in Table 1.

Figure 1. The physical appearance of copper and zinc
(National Center for Biotechnology Information, 2023; Royal Society of Chemistry, 2023)

Copper

Zinc

Zinc

Pure Zn is a ductile, bluish-silver (Figure 1) metal of low boiling and melting point. The metal has an average mass of 65.38 as a result of five stable isotopes (Zn 64, 66, 67, 68 and 70) and an atomic number of 30. Zn has four oxidation states of -2, 0, +1 and +2 although the +2-oxidation state is the commonest. Zn (II) ions have an unfilled d subshell with an electronic configuration of $1S^2$, $2S^2$, $2P^6$, $3d^{10}$. Ores containing the metal include silicates such as hemimorphite {$Zn_4Si2O_7(OH)_2.H_2O$} and willemite (Zn_2SiO_4), phosphates such as hopeite {$Zn_3(PO_4)_2.4H_2O$}, carbonates such as smithsonite ($ZnCO_3$), sulfides including sphalerite (ZnS), sulfates such as goslarite ($ZnSO_4.2H_2O$) and zincosite ($ZnSO_4$) and oxides including gahnite ($ZnAl_2O_4$), franklinite ($ZnFe_2O_4$) and zincite (ZnO) (Barak & Helmke, 1993). Hemimorphite, smithsonite and sphalerite are the principal commercial ores of Zn. Some of the physicochemical properties of Zn are as summarized in Table 2.

Table 2. Physicochemical characteristics of Zn (National Center for Biotechnology Information, 2023)

Physicochemical Characteristic	Value
Atomic weight	65.38
Atomic number	30
Position on the periodic table	Group IIb (12), d-block, period 4
Electronic configuration	$[Ar]4s^23d^{10}$
Van der Waal atomic radius	1.39 pm
Oxidation states	-2, 0, +1, +2
Boiling point	907 °C
Melting point	419.53 °C
Density	7.134 g/cm^3
Electronegativity	1.65 on a Pauling scale
Ionization energy	9.394 eV
Electron affinity	0 eV
State	Solid
Atomic radius	2.01Å
Covalent radius	1.20Å
Isotopes	^{64}Zn with a natural abundance of 49.17% ^{66}Zn with a natural abundance of 27.73% ^{67}Zn with a natural abundance of 4.04% ^{68}Zn with a natural abundance of 18.45% ^{70}Zn with a natural abundance of 0.61%

Zn can be readily molded or cast in addition being used in galvanizing due to its corrosion resistant properties. The metal reacts with oxygen to form zinc oxide that occurs as a green flame. The metal has high affinity to sulfur though and reacts explosively with the non-metal to produce zinc sulfide (Haynes, 2015). ZnS is used to make fluorescent light builds and television screens while ZnO is used in the manufacture of plastics, cosmetics and paints. Zn reacts slowly with weak acids but violently with halogens such as fluorine and chlorine of high electronegativity (National Center for Biotechnology Information, 2023). Zn is also used in the textile, batteries, soap, ink, plastic, pharmaceutical, cosmetics, rubber, and paints industries in addition to manufacture of electrical equipment (Raffa et al., 2021).

TOXICITY OF COPPER AND ZINC IN SOILS

The mining, processing and use of Cu and Zn is growing due to the metals' resistance to corrosion, high electrical and thermal conductivity. Cu is used in minting coins, manufacture of machines, making of furniture, in the electronics industry and in generation of power (Wyszkowska et al., 2022). The metal's nanoparticles (CuONP) are also important in the making of photovoltaic installations, solar energy conversion devices and gas sensors among other medical and agricultural uses. The manufacturing processes produce large quantities of Cu-containing wastes that end up in soils (Li et al., 2017). Zn mining and processing is also key in various industrial activities. The metal is used in the manufacture of semiconductors, biosensors, aeronautics, in car fuels, explosives, car batteries and manufacture of phosphate fertilizers (Uchimiya et al., 2020; Ahmed et al., 2021). Industrial processes using both Cu and Zn introduce the metals to the soils either in the form of solid wastes, effluents, particulate matter or emissions that are later deposited on soil surfaces. The concentrations of the metals, their distribution and bioavailability are a function of the soil type, topography, geology and erosion processes (Dos Santos et al., 2013). In another study, Mansour et al. (2020) noted that the biological, physical and chemical processes in soils determine the mobility, bioavailability, redistribution, speciation, sorption and toxicity of heavy metals in such environments.

The mobility and toxicity of the metals in soil is influenced by a number of factors. These include the clay fractions in the soil, the concentrations of Fe and Mn oxides and hydroxides, soil organic matter, concentrations of basic cations, redox potential, pH and phosphates (Mansour et al., 2020). Additionally, the surface complexation, adsorption and ion exchange capacity of soils also influences the mobility and ultimate toxicity of heavy metals in soils (Abhijith & Varghese, 2021). The factors control the sorption-desorption properties of heavy metals in soils regulating their bioavailability, toxicity and distribution. The texture of soils is a determinant of the levels of Cu in soils and ultimately, the metal's toxicity. In loamy soils, Cu content is higher compared to light sandy soils (Borah et al., 2020). The tendency is attributable to Cu adsorption by clay minerals, soil organic matter, Mn, Fe, oxyhydroxides and carbonates, which are found in loamy soils (Kabata-Pendias, 2011). In another study, Fan et al. (2016) noted that organic matter content of soils, pH, and geogenic sources of Cu regulate its bioavailability and ultimately, toxicity. Datsenko et al., (2021) reported a decrease in vertical migration and toxicity of Cu and Zn ions in soils with accumulated humus and minerals in addition to a high contrast of redox and acid-base conditions. Fe and Mg occurring in most soils as ferromagnesian silicates are exchangeable cations and parts of organic matter and as such, promote the bioavailability and toxicity of Cu in soils (Borah et al., 2020). With high cation exchange capacity, the levels of Cu in soils reduced due to increased sorption of the metal (Singo, 2013).

Zn is a mobile heavy metal in its free and/or complex ionic form in soils. As such, bioavailability and toxicity of the heavy metal is regulated by the presence of soil organic matter, clay fractions, phosphorous levels, calcium levels in soils and pH. In the presence of clay fractions, phosphorous, organic matter and neutral to alkaline pH in soils, Zn is strongly bound and hence immobilized leading to its reduced toxicity (De Souza et al., 2015). The dissolution of the metal is promoted by the presence of aluminum, amorphous silicon and high cation exchange capacity in soils since the factors enhance Zn interaction of soil moisture (Shaheen et al., 2017). At a pH of 6.3, the concentration of Zn ions in soils increased while at lower pH, the availability of Cu ions decreased (Datsenko et al., 2021). According to Angelaki et al. (2022), Zn ions are more soluble and available at acidic rather than alkaline pH.

In both Cu and Zn, their bioavailability increased as pH decreased, which was a result of decreased sorption of the soil as well as reduced organic complexation of the ionic forms of the metal. Zn is complexed in basic soils while Cu is complexed in both alkaline and acidic soils and in the presence of dissolved organic carbon, which reduces their bioavailability (Singo, 2013). The presence of clay minerals such as kaolinite, hematite and goethite was associated with decreased concentrations of Cu and Zn ions due to the sorption effect. The presence of calcium ions increased Zn concentrations in soils due to competitive adsorption while in Cu, the effect was contrary due to increases in dissolved organic carbon and especially in alkaline soils (Singo, 2013).

COPPER AND ZINC POLLUTION IN SOILS OF SUB-SAHARAN AFRICA

The increase in industrial activities in SSA region has resulted to rising levels of both Cu and Zn in soils of the region. In this section, the pollution of SSA region soils by the metals is discussed under three subheadings based on the contamination sources: 1) mining, 2) waste and wastewater management and 3) agricultural processes.

Mining

The mining of Cu and Zn ores involves processes such as crushing, extraction, burning, smelting, desulfurization and ore transportation, which result to production of tailings, mine dump, ore stockpiles and slags laden with the metals (Singo, 2013). In Zambia for instance, mining activities have resulted to production of 40 million tons of slug covering 279 ha of land, 77 million tons of waste rock occupying 388 ha of land space, 1899 million tons of overburden materials taking up 20, 646 ha of land space and 791 million tons of tailings taking up 9125 ha of land space (Sikaundi, 2013). Although the spatial undertaking of the activities is small, their environmental impact is extensive due to the dispersion of the wastes along with their heavy metal content by water and wind (Mileusnic et al., 2014). Such mobilization of heavy metal-laden wastes results to acid mine drainage that can further contaminate soils. In many SSA regions, past and on-going mining and smelting processes have led to production of such wastes and the prolonged exposure to their hazardous heavy metals leading to pollution of soils that spreads to water resources, plants and ends up in animals and human beings (Salem et al., 2020). The countries in the region depend on mining as a significant economic activity second from agricultural activities and so the burden of heavy metals is expected to grow unless remedial and reclamation measures are taken up.

Soils of Kitwe and Mufulira regions of Zambia, where Cu mining activities were carried out and mine tailings disposed in soils had elevated Cu and Zn levels (Dusengemungu et al., 2022). In another study in Chingola district of Zambia, soils have elevated concentrations of the two metals due to accumulation of Cu mine wastes including overburden waste rock and tailings (Mutale et al., 2020). Similar soil pollution tendencies by Cu and Zn among other heavy metals were reported in soils around abandoned mines of Limpopo (Singo, 2013) Gauteng (Olobatoke & Mathuthu, 2016; Oyuorou et al., 2019) province in South Africa where mining and smelting of chalcopyrite, pyrite, sphalerite and galena ores to extract Au, Zn, Cu, Fe and Pb had occurred. The mining activities in the areas exposed soils to heavy metal-laden tailings, overburden materials and gangue. In a related study in Kombat, Grootfontein district of Namibia, Cu mining resulted to the production of

300 million tons of mine tailings whose disposal to soils of the surrounding was associated with elevated levels of Cu (Mileusnic et al., 2014).

In Katanga copper belt of the Democratic Republic of Congo (DRC), the mining and smelting Cu, which was done in open pits and underground channels, developed large quantities of mine tailings. The mine tailings had elevated levels of Cu that were found in association with Zn, As, Cd, Pb and U and polluted soils of the area (Pourret et al., 2016). Mining of Ag, Cd, Co, Zn, Pb, Cu, gallium (Ga) and germanium (Ge) in Kipushi city of DRC was associated with overburden materials during extraction, tailings during the transport of the minerals and production of particulates that had high concentrations of Cu and Zn and polluted soils of the Lubumbashi-Kipushi Road (Muhune et al., 2020). In Kilembe, Uganda, Cu and Zn concentrations in soils were beyond the allowable limits due to the production of cobaltiferous and cupriferous pyrite tailings during the mining and smelting of Cu in mines of the vicinity (Mwesigye et al., 2016).

Solid Waste and Wastewater Management

With the expansion of the mining industry in most SSA countries, rapid urbanization and the rise of occidental lifestyles have emerged leading to increased generation of solid wastes and wastewater in large quantities (Ngole & Ekosse, 2012). Generation of such wastes especially in urban areas is unregulated due to the concentration of cottage industries. The manner of disposing such wastes pose as a pollution threat to soils. In Botswana for instance, lubricants, pigments, paints, shampoos, cosmetics among other chemical products used contain Cu and Zn and once disposed off at landfills and open dumpsites, they lead to soil contamination through leachate generation. It is because of this reason that soils around Gaborone landfill site were rated as moderate to heavily contaminated by Cu and Zn using the geo-accumulation index (Ngole & Ekosse, 2012). Soils of the vicinity of Roundhill (Nyika et al., 2019) and Onderstepoort (Tshibalo, 2017) landfills of South Africa, in two dumpsites of Kumasi, Ghana (Akanchise et al., 2020) and Tepi dumpsite of Ethiopia (Mekonnen et al., 2020) had elevated Cu and Zn concentrations among other heavy metals. Elevated levels of the metals were associated with unsystematic disposal of co-mixtures of municipal, industrial and agricultural solid wastes containing the metals such as electronic wastes in the landfills and dumpsites leading to the generation of leachate and its migration to soils. A similar trend was reported in soils near the Kingtom dumpsite of Sierra Leone and Agbogloshie landfill of Ghana, where elevated concentrations of Cu and Zn among other potentially toxic elements were attributable to the disposal of electrical and electronic wastes, which were laden with the metals (Moeckel et al., 2020).

Untreated municipal and industrial wastewater from Durban, South Africa was attributable to soil pollution with Cu and Zn and the eventual transfer of the metals to cabbage and lettuce after root uptake, which posed as a human health risk (Nkosi & Msimango, 2022). In another study in the suburbs of Nairobi, Kenya, the use of sewage for irrigation was associated with elevated levels of Cu and Zn among other trace elements (Nyika & Dinka, 2022). The wastewater was from domestic sources as well as chemical, plastic, textiles, and soap manufacturing industries located in the vicinity. Similar trends were also reported in Embu (Sayo et al., 2020) and Machakos (Tomno et al., 2020) towns of Kenya where sewage was used for irrigation in urban and suburban areas due to water scarcity and led to heavy metal introduction in soils.

Agricultural Activities

Agriculture is the background of most economies in the SSA region. To continue sustaining the growing population in the region and provide adequate food, agricultural activities are expanding and the demand for soil micro- and macro-nutrients also growing following continued mining (Wuana & Okieimen, 2011). Consequently, fertilizers are being used in the intensive farming systems to supplement already naturally occurring nutrients. Some of these fertilizers including the phosphatic fertilizers have compounds complexed with Cu and Zn impurities, which with overuse and/or misuse could contaminate soils with such heavy metals (Azzi et al., 2017). Governments in countries such as Nigeria, Ethiopia, Zambia, Senegal, Tanzania, Malawi, Mali, Ghana and Kenya in the SSA region use fertilizer subsidies and encourage the use of Zn-containing fertilizers in soils used to grow maize and wheat despite the potential to promote the buildup of the metal in affected farmlands (Joy et al., 2016). In tea estates of Malawi, continued application of zinc oxide and copper sulfate as foliar sprays was found to be detrimental to the soil quality since it led to localized bioaccumulation of the metals (Njoloma, 2012)

Some of the chemicals used in the manufacture of fungicides and insecticides are based on heavy metal containing compounds. Examples include copper oxychloride and copper sulfate (Bordeaux mixture), used as fungicidal sprays to control late blight in potato crops and contain Cu (Ferreira et al., 2014). Copper zinc chromate is also a Cu and Zn containing fungicide used against diseases in citrus, peanuts, cucurbits, tomatoes and potatoes (Van-Zwieten et al., 2004). Some Zn containing compounds are used in making formulations to preserve wood and contain additional trace metals such as As, Cr and Cu (Wuana & Okieimen, 2011). Once the chemicals are applied on plants and wood, they are introduced to soils and can bioaccumulate with time in areas where application is done on a regular basis. In cocoa farms of southwestern Nigeria, the application of copper fungicides was associated with elevated levels of

Cu and Zn in soils compared to reference farms (Azeez et al., 2021). In East Akyem region of Ghana, application of phosphatic fertilizers to improve soil fertility and Cu-based fungicides to control the black pod disease in cocoa plants was reported to enhance the accumulation of both Cu and Zn metals in soils to levels beyond the allowable limits (Fei-baffoe et al., 2017).

Application of biosolids such as municipal sewage sludge, composts and livestock (pig, cattle and poultry) manures is a common practice in SSA region to supplement or substitute organic fertilizers. However, the manures add to the natural occurring Cu and Zn in soils because they are wastes of livestock diets with the heavy metals where they are used as growth promoters. The continued application of the manures on agricultural soils could lead to the bioaccumulation of the metals. Urban garden soils of Kaduna region, Nigeria were laden with Cu and Zn concentrations following the use of cow dung manure to grow vegetables (Akpa & Agbenin, 2012). The use of sewage sludge and wastewater in the agricultural practices especially in urban and peri-urban areas is a growing practice that can lead to the buildup of Cu and Zn in soils as reported in several studies (Sayo et al., 2020; Tomno et al., 2020; Nyika & Dinka, 2022).

CONCLUSION

Potentially toxic metals such as Cu and Zn in this chapter were found to be essential to plants in low concentrations. However, at elevated levels the metals interfere with soil processes such as respiration, microbial activity, enzyme activity and the balance of essential nutrients, which play essential roles in determining its fertility and quality. The toxicity inflicted by Cu and Zn is a function of their bioavailability and mobility, which is regulated by the soil type, geological forms of the metals in soils, Fe and Mg oxides and hydroxides, redox potential and pH among other factors that influence organic complexation of metals, sorption and desorption processes in soils. Mining of Cu and Zn from ores such as copper pyrite, sphalerite, smithsonite and hemimorphite and their pyrometallurgical processing in countries of SSA introduces the metals to soils through resultant wastes where they can bioaccumulate. Industrial wastes and effluents from chemical cottage industries and their introduction to soils enhanced metal contamination especially in urban areas due to the non-regulation of such industries in SSA region and their non-adherence to stipulated environmental laws. The use of fungicides as foliar sprays to prevent plant diseases, biosolids and phosphatic fertilizers to enhance soil fertility was associated with Cu and Zn pollution in SSA soils. Therefore, regulated use of heavy metal containing raw materials and products in the environment is key in preventing their associated toxicity once they bioaccumulate in soils.

REFERENCES

Abhijith, L., & Varghese, E. (2021). Removal of zinc and copper from contaminated soil by using adsorbents and mulches. *AIP Conference Proceedings, 2396*(030005), 1–9. doi:10.1063/5.0066408

Ahmed, T., Noman, M., Ijaz, M., Ali, S., Rizwan, M., Ijaz, U., Hameed, A., Ahmad, U., Wang, Y., Sun, G., & Li, B. (2021). Current trends and future prospective in nanoremediation of heavy metals contaminated soils: A way forward towards sustainable agriculture. *Ecotoxicology and Environmental Safety, 227*, 112888. doi:10.1016/j.ecoenv.2021.112888 PMID:34649136

Akachise, T., Boakye, S., Borquaye, L., Dodd, M., & Darko, G. (2020). Distribution of heavy metals in soils from abandoned dumpsites in Kumasi, Ghana. *Scientific African, 10*, e00614. doi:10.1016/j.sciaf.2020.e00614

Akpa, S., & Agbenin, J. (2012). Impact of cow dung manure on the solubility of copper, lead and zinc in urban garden soils from Northern Nigeria. *Communications in Soil Science and Plant Analysis, 43*(21), 2789–2800. doi:10.1080/00103624.2012.719976

Angelaki, A., Dionysidis, A., Sihag, P., & Golia, E. (2022). Assessment of contamination management caused by copper and zinc cations leaching and their impact on the hydraulic properties of a sandy and a loamy clay soil. *Land (Basel), 11*(2), 290. doi:10.3390/land11020290

Azeez, M., Adesanwo, O., & Adepetu, J. (2021). Effects of copper fungicides spray on nutrient content in soils of cocoa growing areas of Southwestern Nigeria. *Tanzania Journal of Science, 47*(5), 1546–1559. doi:10.4314/tjs.v47i5.5

Azzi, V., Samrani, A., Lartiges, B., Kobeissi, A., & Kanso, A. (2017). Trace metals in phosphate fertilizers used in Eastern Mediterranean countries. *Clean (Weinheim), 45*(1), 1–30. doi:10.1002/clen.201500988

Barak, P., & Helmke, P. (1993). The chemistry of zinc. In A. Robson (Ed.), *Zinc in soils and plants*. Kluewer Academic Press. doi:10.1007/978-94-011-0878-2_1

Borah, P., Gujre, N., Rene, E., Rangan, L., & Kumar, R. (2020). Assessment of mobility and environmental risks associated with copper, manganese and zinc in soils of a dumping site around a Ramsar site. *Chemosphere, 254*, 126852. doi:10.1016/j.chemosphere.2020.126852 PMID:32957277

Datsenko, V., Borzenko, O., & Kaliuzhna, L. (2021). Vertical migration of copper and zinc ions in soils polluted by electroplating sludge. *Clean (Weinheim)*, *49*(12), 2100087. doi:10.1002/clen.202100087

De Souza, J., Abrahao, L., de Mello, W., da Silva, J., da Costa, M., & de Oliveira, T. S. (2015). Geochemistry and spatial variability of metal (loid) concentrations in soils of the state of Minas Gerais. Brazil. *The Science of the Total Environment*, *505*, 338–349. doi:10.1016/j.scitotenv.2014.09.098 PMID:25461035

Dos Santos, G., Valladares, G., Abreu, C., Camargo, O., & Grego, C. (2013). Assessment of copper and zinc in soils of a vineyard region in the state of Sao Paulo, Brazil. *Applied and Environmental Soil Science*, *790795*, 1–10. doi:10.1155/2013/790795

Dusengemungu, L., Mubemba, B., & Gwanama, C. (2022). Evaluation of heavy metal contamination in copper mine tailing soils of Kitwe and Mufulira, Zambia for reclamation prospects. *Scientific Reports*, *12*(1), 11283. doi:10.103841598-022-15458-2 PMID:35787645

Fan, T., Wang, J., Li, B., He, Z., Gao, J., Zhou, D.-M., Friedman, S. P., & Sparks, D. L. (2016). Effect of organic matter on sorption of Zn on soil: Elucidation by Wien effect measurements and EXAFS spectroscopy. *Environmental Science & Technology*, *50*(6), 2931–2937. doi:10.1021/acs.est.5b05281 PMID:26894796

Fei-baffoe, B., Kani, A., Afrifa, A., & Yeboah, S. (2017). Copper and zinc pollution in cocoa growing areas in a low-income country. *Journal of Natural Sciences Research*, *7*(14), 28–34.

Ferreira, L., Scacroni, J., Da Silva, J., Cataneo, A., & Martins, D. (2014). Copper oxychloride fungicide and its effect on growth and oxidative stress of potato plants. *Pesticide Biochemistry and Physiology*, *112*, 63–69. doi:10.1016/j.pestbp.2014.04.010 PMID:24974119

Garcia-Gomez, C., Garcia-Gutierrez, S., Obrador, A., & Fernandez, D. (2020). Study of Zn availability, uptake, and effects on earthworms of zinc oxide nanoparticle versus bulk applied to two agricultural soils: Acidic and calcareous. *Chemosphere*, *239*, 124814. doi:10.1016/j.chemosphere.2019.124814 PMID:31527003

Genova, G., Chiesa, D., Mimmo, T., Borruso, L., & Cesco, S. (2022). Copper and zinc as a window to past agricultural land-use. *Journal of Hazardous Materials*, *424*(Part C), 126631. doi:10.1016/j.jhazmat.2021.126631 PMID:34334215

Gong, B., He, E., Qiu, H., Van Gestel, C., & Freire, A. (2020). Interactions of arsenic, copper and zinc in soil-plant system: Partition, uptake and phytotoxicity. *The Science of the Total Environment*, *745*, 140926. doi:10.1016/j.scitotenv.2020.140926 PMID:32712499

Haynes, W. (2015). *CRC handbook of chemistry and physics*. CRC Press/ Taylor and Francis.

Joy, E., Stein, A., Young, S., Ander, E., Watts, M., & Broadley, M. R. (2015). Zinc-enriched fertilizers as a potential public health intervention in Africa. *Plant and Soil*, *389*(1-2), 1–24. doi:10.100711104-015-2430-8

Kabata-Pendias, A. (2011). *Trace metals in soils and plants*. CRC Press.

Karimi, B., Masson, V., Guilland, C., Leroy, E., Pellegrinelli, S., Giboulot, E., Maron, P.-A., & Ranjard, L. (2021). Ecotoxicity of copper input and accumulation for soil biodiversity in vineyards. *Environmental Chemistry Letters*, *19*(3), 2013–2030. doi:10.100710311-020-01155-x

Li, L., Pan, D., Li, B., Wu, Y., Wang, H., Gu, Y., & Zuo, T. (2017). Patterns and challenges in the copper industry in China. *Resources, Conservation and Recycling*, *127*, 1–7. doi:10.1016/j.resconrec.2017.07.046

Mansour, H., Awad, F., Saber, M., & Zaghloul, A. (2020). Effect of contamination sources on the rate of zinc, copper and nickel release from various soil ecosystems. *Bulletin of the National Research Center*, *44*(1), 178. doi:10.118642269-020-00431-8

Marastoni, L., Sandri, M., Pii, Y., Valentinuzzi, F., Brunetto, G., Cesco, S., & Mimmo, T. (2019). Synergism and antagonisms between nutrients induced by copper toxicity in grapevine rootstocks: Monocropping vs. intercropping. *Chemosphere*, *214*, 563–578. doi:10.1016/j.chemosphere.2018.09.127 PMID:30286423

Mekonnen, B., Haddism, A., & Zeine, W. (2020). Assessment of the effect of solid waste dumpsite on surrounding soil and river water quality in Tepi town, Southwest Ethiopia. *Journal of Environmental and Public Health*, *5157046*, 1–9. doi:10.1155/2020/5157046 PMID:32587623

Mileusnic, M., Mapani, S., Kamona, F., Ruzicic, S., Mapaure, I., & Chimwamurombe, P. M. (2014). Assessment of agricultural soil contamination by potentially toxic metals dispersed from improperly disposed tailings, Kombat mine, Namibia. *Journal of Geochemical Exploration*, *144*, 409–420. doi:10.1016/j.gexplo.2014.01.009

Moeckel, C., Breivik, K., Nost, T., Sankoh, A., Jones, K., & Sweetman, A. (2020). Soil pollution at a major West African e-site recycling site: Contamination pathways and implications for potential mitigation strategies. *Environment International, 137*, 105563. doi:10.1016/j.envint.2020.105563 PMID:32106045

Muhune, S., Marsi, M., Tshisand, P., Nonga, W., & Kayembe, O. (2020). evaluation of soil contamination by metallic trace elements to the roadside on the Lubumbashi-Kipushi section (DRC). *International Journal of Advanced Research, 8*(9), 1187–1195. doi:10.21474/IJAR01/11775

Mustafa, K., & AlSharif, M. (2018). Copper (Cu) an essential redox-active transition metal in living system-a review article. *American Journal of Analytical Chemistry, 9*(1), 15–26. doi:10.4236/ajac.2018.91002

Mutale, C., Syampungani, S., Festin, E., Tigabu, M., & Daneshvar, A. (2020). Physicochemical characteristics and heavy metal concentrations of copper mine wastes in Zambia: Implications for pollution risk and restoration. *Journal of Forestry Research, 31*(4), 1283–1293. doi:10.100711676-019-00921-0

Mwesigye, A., Young, S., Bailey, E., & Tumwebaze, S. (2016). Population exposure to trace metals in the Kilembe copper mine area, Western Uganda: A pilot study. *The Science of the Total Environment, 15*(573), 366–375. doi:10.1016/j.scitotenv.2016.08.125 PMID:27572529

National Center for Biotechnology Information. (2023). *PubChem Compound Summary for CID 935, Nickel. [Online]* NCBI. https://pubchem.ncbi.nlm.nih.gov/#query=zinc; https://pubchem.ncbi.nlm.nih.gov/#query=copper

Ngole, V., & Ekosse, G. (2012). Copper, nickel and zinc contamination in soils within the precincts of mining and landfilling environments. *International Journal of Environmental Science and Technology, 9*(3), 485–494. doi:10.100713762-012-0055-5

Nierengarten, J. (2018). In my element: Copper. *Chemistry (Weinheim an der Bergstrasse, Germany), 25*(1), 16–18. doi:10.1002/chem.201805277

Njoloma, C. (2012). *Application of foliar sprays containing copper, zinc and boron to mature clonal tea (Camellia sinensis): effect on yield and quality*. [MSc Thesis, University of Pretoria, South Africa].

Nkosi, S., & Msimango, N. (2022). Screening of zinc, copper and iron in Chinese cabbage cultivated in Durban, South Africa, towards human health risk assessment. *South African Journal of Science, 118*(11/12), 12099. doi:10.17159ajs.2022/12099

Nyika, J., Onyari, E., Dinka, M., & Shivani, M. (2019). Heavy metal pollution and mobility in soils within a landfill vicinity: A South African case study. *Oriental Journal of Chemistry*, *35*(4), 1286–1296. doi:10.13005/ojc/350406

Olobatoke, R., & Mathuthu, M. (2016). Heavy metal concentration in soil in the tailing dam vicinity of an old gold mine in Johannesburg, South Africa. *Canadian Journal of Soil Science*, *96*(3), 299–304. doi:10.1139/cjss-2015-0081

Oyourou, J., McCrindle, R., Combrinck, S., & Fourie, C. (2019). Investigation of zinc and lead contamination of soil at the abandoned Edendale mine, Mamelodi (Pretoria, South Africa) using a field portable spectrometer. *Journal of the Southern African Institute of Mining and Metallurgy*, *119*(1), 55–62. doi:10.17159/2411-9717/2019/v119n1a7

Pourret, O., Lange, B., Bonhoure, J., Colinet, G., Decree, S., Mahy, G., Séleck, M., Shutcha, M., & Faucon, M.-P. (2016). Assessment of soil metal distribution and environmental impact of mining in Katanga (Democratic Republic of Congo). *Applied Geochemistry*, *64*, 43–55. doi:10.1016/j.apgeochem.2015.07.012

Raffa, M., Chiampo, F., & Shanthakumar, S. (2021). Remediation of metal/metalloid-polluted soils: A short review. *Applied Sciences (Basel, Switzerland)*, *11*(9), 4134. doi:10.3390/app11094134

Rajput, D., Minkina, M., Behal, A., Sushkova, N., Mandzhieva, S., Singh, R., Gorovtsov, A., Tsitsuashvili, V. S., Purvis, W. O., Ghazaryan, K. A., & Movsesyan, H. S. (2018). Effects of zinc-oxide nanoparticles on soil, plants, animals and soil organisms: A review. *Environmental Nanotechnology, Monitoring & Management*, *9*, 76–84. doi:10.1016/j.enmm.2017.12.006

Royal Society of Chemistry. (2023). *Copper*. [Online] RSC. https://www.rsc.org/periodic-table/element/29/copper

Salem, M., Bedade, D., Al-Ethawi, L., & Al-Waleed, S. (2020). Assessment of physiochemical properties and concentration of heavy metals in agricultural soils fertilized with chemical fertilizers. *Heliyon*, *6*(10), e05224. doi:10.1016/j.heliyon.2020.e05224 PMID:33102850

Sayo, S., Kiratu, J., & Nyamato, G. (2020). Heavy metal concentrations in soil and vegetables irrigated with sewage effluent: A case study of Embu sewage treatment plant, Kenya. *Scientific African*, *8*, e00337. doi:10.1016/j.sciaf.2020.e00337

Shaheen, M., Kwon, E., Biswas, K., Tack, M., Ok, S., & Rinklebe, J. (2017). Arsenic, chromium, molybdenum, and selenium: Geochemical fractions and potential mobilization in riverine soil profiles originating from Germany and Egypt. *Chemosphere*, *180*, 553–563. doi:10.1016/j.chemosphere.2017.04.054 PMID:28432892

Sikaundi, G. (2013). *Copper mining industry in Zambia: environmental challenges.* [Online] UN. https://unstats.un.org/unsd/environment/envpdf/UNSD_UNEP_ECA%20Workshop/Session%2008-5%20Mining%20in%20Zambia%20(Zambia).pdf [Accessed on 31st January 2023].

Singo, N. (2013). *An assessment of heavy metal pollution near an old copper mine dump in Musina, South Africa.* [MSc Thesis, University of South Africa, Pretoria, South Africa].

Tomno, R., Nzeve, J., Mailu, S., Shitanda, D., & Waswa, F. (2020). Heavy metal contamination of water, soil and vegetables in urban streams in Machakos municipality, Kenya. *Scientific African*, *9*, e00539. doi:10.1016/j.sciaf.2020.e00539

Tshibalo, R. (2017). *Assessment of municipal solid waste leachate pollution on soil and groundwater system at Onderstepoort landfill site in Pretoria.* [MSc Thesis, University of South Africa].

Uchimiya, M., Bannon, D., Nakanishi, H., McBride, M., Williams, M., & Yoshihara, T. (2020). Chemical speciation, plant uptake, and toxicity of heavy metals in agricultural soils. *Journal of Agricultural and Food Chemistry*, *68*(46), 12856–12869. doi:10.1021/acs.jafc.0c00183 PMID:32155055

Van- Zwieten, L., Merrington, G., & Van-Zwieten, M. (2004). Review of impacts of soil biota caused by copper residues from fungicide application. *3rd Australian New Zealand Soils Conference*. University of Sydney, Australia.

Wuana, R., & Okieimen, F. (2011). Heavy metals in contaminated soils: A review of sources, chemistry, risks and best available strategies for remediation. *ISRN Ecology*, *402647*, 1–20. doi:10.5402/2011/402647

Wyszkowska, J., Borowik, A., Zaborowska, M., & Kucharski, J. (2022). Mitigation of the adverse impact of copper, nickel, and zinc on soil microorganisms and enzymes by mineral sorbents. *Materials (Basel)*, *15*(15), 5198. doi:10.3390/ma15155198 PMID:35955133

Chapter 14
Approaches to the Management of Heavy Metals in Polluted Soils

ABSTRACT

The need to manage the increasing concentrations of heavy metals in soils of modern day is indispensable. Remediation aims to curtail the expanding pollution in such environments and the subsequent entry of heavy metals to food chains. In this chapter, techniques used to manage soil pollution by heavy metals and their merits and demerits were discussed. Findings showed that the methods are categorized as physical, chemical, and biological, depending on the mechanisms used. Evidently, physical methods were found to be unsustainable and only efficient for small-scale remediation; chemical methods were simple to execute but resulted to secondary pollution; while bioremediation was easy, natural, and green, but slow and only applicable for some metal types. Finding the best fit remediation technique based on the targeted metal, soil type, and level of pollution is key in enhancing the efficacy of the discussed techniques.

INTRODUCTION

Soils are fundamental but fragile non-renewable resources that are essential for sustenance of life on earth (Hou, 2020; Wang et al., 2021). However, in the contemporary society characterized with the rise of anthropic activities, soils are under threat of degradation, contamination and pollution, which are trends that reduce soil's capacity to provide ecological good and services (FAO, 2018). Pollution by

DOI: 10.4018/978-1-6684-7116-6.ch014

heavy metals including As, Cd, Cr, Cu, Hg, Ni, Pb and Zn in soil environments is growing leading to widespread and persistent biological toxicity and a public health concern globally (Zhao et al., 2022). The persistent nature of the metals enhances their accumulation and toxic effects in soils and other natural resources as well as uptake by plants and animals. For instance, the semi-removal duration through erosion, deflation, plant uptake and/or leaching for Pb, Cu, Zn and Cd was estimated at a range of 740-5900, 310-1500, 70-510, 13-110 years, respectively (Koptsik, 2014). During the periods, the metals accumulate, spread to natural resources and enter in the trophic chains in order to exert their noxious effects on living things including human beings.

The effects of heavy metals to the environment and living things are nutritional, evolutionary and ecological (Alsafran et al., 2022). The metals are poisonous/ toxic at low concentrations and have densities that are 4-5 g/cm^3 more than the density of water. The characteristics of heavy metals induce disturbances in the soils' physicochemical aspects and threaten the survival and growth of plants, reduces the availability of macro- and micronutrients for uptake by plants and interferes with plant metabolic processes (Vardhan et al., 2019; Ubando et al., 2020; Saini et al., 2021). Due to these effects, there is a growing environmental and public health concern over the pollution effects resulting from heavy metals in recent years. In addition, the exposure of the trace metals to humans and natural environments has grown to correspond to their exponential technological, domestic, agricultural and industrial applications (Cristaldi et al., 2017). Apart from lithologic sources, heavy metals are introduced to soils from pharmaceutical, agricultural, mining and manufacturing industrial activities and their resultant solid wastes and effluents in addition to wastes from domestic activities (Cristaldi et al., 2017; Hasanpour & Hatami, 2020). Foundry, smelting, oil refining, production of chemicals, pesticides and petrochemicals, coal combustion and its resultant by-products, traffic and metal piping among other diffuse metal uses also introduce the contaminants to soils either as emissions leading to atmospheric depositions and precipitation, particulate matter or solid wastes and wastewater (Vardhan et al., 2019; Saleem et al., 2020; Zaheer et al., 2020). Figure 1 summarizes some of the geogenic and anthropogenic sources of heavy metal pollution in soils (Rajendran et al., 2022).

Unlike organic compounds, heavy metals in soils resist chemical and biological degradation in addition to persisting in the environments for many years. Eventually, they accumulate and destroy soil ecosystems. The accumulation of heavy metals in soils interferes with their physicochemical aspects including cation exchange capacity, electrical conductivity, pH and mineralogy (Shahid et al., 2012; Khalid et al., 2017). Additionally, the metals alter the functioning of soil organic and inorganic ligands, which control macro- and micro-nutrient bioavailability for uptake by plants as well as interfere with the function of soil microorganisms (Minnikova et al., 2017). The

Figure 1. Sources of heavy metals that pollute soils
(Alsafran et al., 2022; Rajendran et al., 2022)

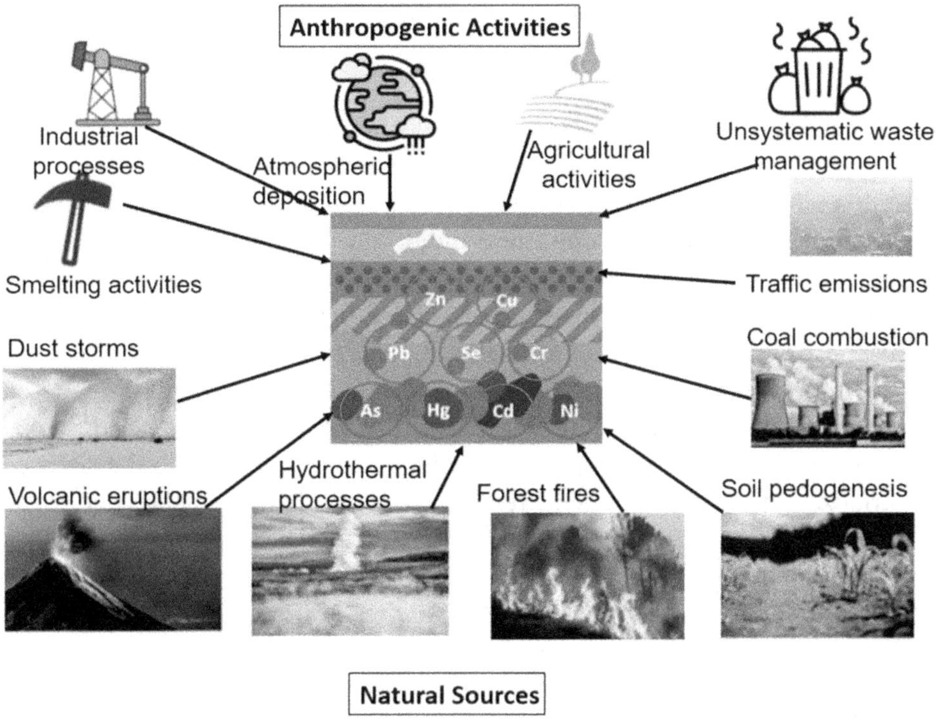

presence of Co, Cr, Cu, Mn, Ni and Zn in soils has been associated with inhibition of enzymatic activities of alkaline phosphatase, β-glucosidase, dehydrogenase, acid phosphatase and urease among other soil enzymes (Pattnaik & Equeenuddin, 2016; Minnikova et al., 2017). Overall, heavy metals reduce the biochemical functions of soils by interfering with the physiology of their microorganisms and enzymes and deteriorating the quality and fertility in such environments, which has an overall effect on ecological balance (Khalid et al., 2017).

With knowledge on the negative effects of heavy metals to soils and the ecosystem, it is essential to deploy effective, site-specific and innovative remedial strategies to control their toxicity in affected environments. In the last 20 years, many remediation strategies have been and are still being developed to minimize the bioavailable and total fractions of heavy metals in soils and control their bioaccumulation in food chains (Sabir et al., 2015). The remediation strategies also prevent the infiltration of heavy metals from soils to aquatic, atmospheric and terrestrial ecosystems in addition to reclaiming polluted land (Yan et al., 2020). Alsafran et al. (2022) noted that effective heavy metal remediation strategies should be stable and reliable to be deemed as effective in cleansing heavy metals from soils. This chapter discusses the remedial methods to manage heavy metals in polluted soils and compares their effectiveness based on their advantages and disadvantages using existent literature.

HEAVY METALS IN SOILS

Soils are essential sinks of heavy metals introduced to the environment through various human activities or from lithogenic origin. Various regions of the globe are confronted with heavy metal pollution in soils although the distribution of the contaminants differ spatially and temporally. The differences are a result of the level of awareness of the problem and measures in place to remove the heavy metals in such environs (Baldantoni et al., 2016). World around, more than 5 million sites are reported to be polluted by heavy metals with most of the locations being in developed nations of the USA, Germany, Australia, China and Sweden due to the advanced industrialization technologies in the regions (He et al., 2015a; Khalid et al., 2017).

In the USA, more than 50,000 sites covering 600,000 ha of land space have been identified as polluted with heavy metals and are awaiting remediation (Khalid et al., 2017). In Europe, France has more than 11,000 sites while Greece, Poland, Portugal and Ireland have >20,000 such areas in total, already polluted with heavy metals and awaiting their remediation (Perez, 2012; Agnello et al., 2015). In Germany, pollution by heavy metals has led to 10,000 ha of land losses used in crop production and financial losses translating to more than 17 billion euros for the country as a result of the soil pollution trend (Lewandowski et al., 2006). The reported numbers could even be higher as reported by Wang et al. (2021) who stated that more than 350, 000 sites in the USA and at least 2.5 million sites in Europe are potentially contaminated by heavy metals and require urgent remediation.

In China, 25% of arable farmland covering more than 20 million acres of land is already polluted with heavy metals, which leads to crop losses amounting to more than 10 million tons (Hongbo et al., 2011). In another study, the Ministry of Environmental Protection, MEP (2014) reported that 16.1% of China's land is laden with heavy metals whose concentrations are beyond the national predefined soil quality standards. In developing countries of Pakistan and India, most soils were laden with heavy metals as a result of the use of untreated municipal and industrial effluents for irrigation (Khan et al., 2015). The situation is even more dire in poor developing countries of sub-Saharan Africa (SSA) where industrial production, agricultural, artisanal and small-scale mining activities are growing for economic development with low precedence on pollution prevention initiatives and technologies. Tindwa and Singh (2023) noted that out of the top 80 most polluted countries in the world, 36 of them are from SSA region. With heavy metal pollution in soils, realizing sustainable development globally will be impossible. This observation is based on the fact that sustainable development goals on poverty reduction (SDG1), zero hunger (SDG2), good health and wellbeing (SDG3), universal water access (SDG6), building sustainable communities and cities (SDG11), responsible production and consumption (SDG12), climate action (SDG13) and life on land (SDG15) cannot be realized when soils are polluted (Arora & Khosla, 2021). Heavy metal pollution in soils is a global dilemma and environmental concern that requires urgent intervention to remediate the pollutants and rehabilitate the polluted sites.

INFORMATION TECHNOLOGY MANAGEMENT OF HEAVY METAL POLLUTION

To manage heavy metal polluted environments including soils, data administration and monitoring is a key component. Several basic analytical approaches are used to manage such data. For instance, studies employing high-precision remote sensing data to device heavy metal pollution monitoring in the soil spectra of mining areas and achieve rapid data collection, monitoring and evaluation have been employed (Dkhala et al., 2020; Liu et al., 2020; Tan et al., 2021). Other studies have utilized existing spatial data operation models to come up with integrated evaluation frameworks (Du et al., 2017; Wang et al., 2020). Such information technology (IT) management systems provide solutions to monitor heavy metal pollution and information on the best remediation approaches that can be adopted in SSA region. However, they rely on a single data source, which could lead to inaccurate deductions from the data. According to Cheng et al. (2022), using a single data source limits the availability of spatial differences in the minerals causing soil pollution by heavy metals and the complex geological environments of their location.

To diversify data sources and facilitate accurate monitoring and evaluation of heavy metal pollution in soils, diversified collection approaches are utilized nowadays. These include remote geological, geophysical, geochemical and groundwater data in addition to remote sensing data (Cheng et al., 2022). Collected data is managed and analyzed through a number of approaches. These include cloud computing, pollution risk assessment and Hadoop big data technology to share and processed the data without any transfer uncertainties (Zhou et al., 2016; Yan et al., 2021). The IT management systems on heavy metal pollution in soils enable the mining of knowledge from gathered data (Song et al., 2018). Additionally, it informs of the best management practices for heavy metal polluted soils. The techniques used are described in subsequent subtitles.

MANAGEMENT OF SOILS POLLUTED BY HEAVY METALS

Depending on the soil type and metal physicochemical characteristics, contaminants can persist in soils for a longtime. Remediation techniques are aimed at either complete removal of heavy metals from soils or their transformation to less harmful products (Ashraf et al., 2019). The technologies are categorized as 1) physical, 2) chemical and 3) biological based mechanisms applied to remove the trace metals. Physical remediation uses a number of technologies that involve hot desorption and replacement to swap polluted soils with safe soils and reduce heavy metal concentrations of a given area (Rajendran et al., 2022). Chemical techniques modify the pH of soils to precipitable and make heavy metals insoluble and hence less toxic. Biological remediation is a natural biotechnological process, which applies plant and microbial species among other organisms to treat and cure heavy metal polluted soils (Leong et al., 2019; Khoo et al., 2021). Bioremediation technologies can also be categorized as 1) in-situ, 2) on-site and 3) ex-situ based on the location where they are applied. In-situ techniques are applied at the pollution point and do not require soil excavation, on-site remediation requires removal of soil and its subsequent processing on a site near the polluted site while in ex-situ removal, soil is carried off the polluted site for processing before its return to the original site (Raffa et al., 2021). In some cases, the techniques can be combined for better outcomes. The following subtopics provide a detailed discussion of the techniques.

Physical Remediation Techniques

Soil Washing

Soil washing technique uses extraction solutions to remove heavy metals from the soil in-situ (Xu et al., 2022). In the technique, the extract solutions desorb and extract contaminants from soils. The process can be enhanced through mechanical or ultrasonic mixing of the extract and soil for better physical contact (Raffa et al., 2021). The process was initially developed for mineral processing and can be combined with high gradient magnetic segregation, leaching and flotation to enhance its efficiency (Rajendran et al., 2022). Washing is not specific to heavy metals but eliminates soil fractions with high contaminant levels using its wet release and categorization phases (Dhaliwal et al., 2020). The method is technically simple though capital installation of collection wells and procurement of extracting solutions could make it costly.

Surface Covering

Surface covering involves the use of an impermeable coating to cover contaminated soils. The technique is not restorative since it does not attempt to extract heavy metals from contaminated sites but reduces the exposure risk to such environs (Dhaliwal et al., 2020). By preventing the entry of water, the technique deters contaminant spread to groundwater resources by slowing heavy metal mobility. Some of the impermeable coatings used include high-density polyethylene (HDPE), asphalt, concrete and/or cement. Affected soils in this technique lose their natural environmental function. The technique can be expensive if the polluted area covers an expansive land space.

Soil Separation and Replacement

Soil separation involves isolation of soils polluted with trace metals as they await further remediation processes. The method confines pollutants to a specific area until advanced remediation methods are ready and prevents groundwater pollution through installation of subsurface boundaries (Tiwari et al., 2013).

In soil replacement, contaminated soils are partially of fully replaced with non-contaminated ones. The replaced soil is disposed off-site as waste or used for heavy metal extraction. The method is restorative since it enhances soil durability by reducing the trace metal concentrations through spading and putting cleaner soil imports (Kim et al., 2015). In spading, polluted soil is dug out deeply and replaced with clean soil while in importing, clean soil is mixed with some polluted soil to reduce the concentrations of the contaminants. The method is laborious and costly and is only suitable for small-scale polluted areas due to the bulk excavation and transport activities involved. In the event that the technique is extended to land-based activities, soil fertility can be lost (Lakkireddy & Kues, 2017). The approach is expensive and burdensome for large-scale areas with severe heavy metal contamination (Raffa et al., 2021).

Soil Encapsulation

Encapsulation technique does not focus on destroying pollutants unlike soil washing but, it neutralizes it to prevent contaminant spread and migration to other environmental compartments such as soil and water resources. Using an impervious material such as synthetic polymers, concrete, clay and lime, heavy metal cations in polluted soils are encircled to prevent their mobility and leaching. Encapsulation of heavy metal contaminated soils is done by installing watertight encryptions that are impermeable and do not allow entry of gases or liquids in such environs. The aim is to slow the movement and bioavailability of the pollutants. Encapsulation can be done through frozen barriers and construction of bored-pile-cut off walls, injection wells, jet grouting curtains, sheet pile walls, cement-bentonite-water slurries and slurry walls over such contaminated soils (Anekwe & Isa, 2021). Encapsulation also occurs using stabilization and solidification technologies. In the processes, geopolymers combined with metakaolin and aluminosilicates are used to stabilize heavy metals so that they are no longer mobile in soils (Carrilo et al., 2021). Encapsulation combines both physical and chemical processes. In a study aimed at encapsulating soil polluted by Pb, a geopolymer with AlO_4 and SiO_2 tetrahedral cross linkages combined with calcined kaolin was used to bind the contaminated soil and metal cations and stabilize them (Carrillo et al., 2021).

Thermal Treatment

In thermal treatment, soils are heated under conditioned temperatures with the aim of eliminating volatile heavy metals such as As and Hg. Some processes used in this technique include radio-frequency, steam-based, electrical resistive and conductive heating (Song et al., 2017). After heating, the metals are collected via carrier gases or through vacuum-negative pressure. Depending on the boiling point of a heavy metal, temperatures between 90 and 560 °C can be applied in thermal treatment (Raffa et al., 2021). For instance, thermal desorption of Hg occurs in temperatures below 357 °C (which is the boiling point for Hg) (Chang & Yen, 2006). Thermal treatment has also been used to fix elevated concentrations of Cu and Zn in soils (Wang et al., 2018). The approach consumes less energy in addition to being able to safely eliminate contaminants without secondary pollution. However, its capital expenses are high and control measures are required to regulate gas emissions and prevent atmospheric pollution (Chang & Yen, 2006). Additionally, the technique destroys the physicochemical characteristics of soils and is only effective if targeted pollutants are in high concentrations.

Chemical Remediation Techniques

Vitrification

Vitrification as a remediation process is based on the principle of fixing heavy metals using vitreous (glass-based) materials and at temperatures above 1500 °C. Using an electrical discharge-induced gas plasma or high voltage electricity, soil is heated, melted and then cooled before being transformed to a vitrified structure (Rajendran et al., 2022). Heavy metals are encapsulated ex-situ and/or in-situ using this technique while other soil contaminants are destroyed (Dhaliwal et al., 2020). The method is not applicable to soils with flammable or volatile organic contaminants, high moisture content and organic matter content. Additionally, it is complex, expensive and can destroy soil physicochemical aspects due to the high temperatures applied (Raffa et al., 2021).

Chemical Stabilization/ Fixation

Chemical stabilization and fixation are aimed at reducing the bio-accessibility, bioavailability and mobility of heavy metals in soils. In the technique, immobilizing and chelating agents enhance adsorption, complexation and precipitation of heavy metal pollutants in soils. The agents work by donating their functional groups such as P, S, O and N to react with heavy metal ions (Liu et al., 2018). Examples of stabilizers used include low-cost phosphates and carbonates and inorganic materials such as iron oxide, zeolite, manganese and sulfur complexes (Ullah et al., 2020). A common chemical stabilizer is glass-like apatite bone feasting ($Ca10(PO_4).6H_2O$) used to immobilize metal phosphates in polluted soils (Bilgin & Tulun, 2016). Waste-sourced biomaterials such as biochar (Xing et al., 2019), cockle- and egg-shells (Islam et al., 2017) have been used to immobilize Cd, Pb and Zn in contaminated soils. To be effective, the stabilizers are mixed with contaminated soils and/or sprayed along with aqueous solutions over polluted sites if they are soluble. The technique modifies the chemical characteristics in applied soils and hence, they require frequent monitoring.

Treatment With Nanoparticles

The synthesis of nanoparticles with a diameter of <100 nm and their varied applications is growing in modern day even in soil remediation. Through their large and specific surface areas, the particles remediate heavy metals through processes such as precipitation, co-precipitation, redox reactions and adsorption (Raffa et al., 2021). The use of zero-valent iron nanoparticles for removal of heavy metals in soils has been reported where starch, carboxymethylcellulose, rhamnolipid, vinegar residues, bentonite, biochar and zeolite can be used to support the iron (Sumiahadi & Acar, 2018; Rajendran et al., 2022). Although the nanoparticles are effective in heavy metal bioremediation, their toxicity and interactions in soils as well as their effect on ecosystems, natural resources and biodiversity in direct contact with polluted soils is largely speculative. Therefore, the use of the particles for soil cleanup requires more research prior to their largescale application for remediation of heavy metals.

Chemical Soil Washing

Chemical soil washing entails the use of surfactants, chelators, water, organic and inorganic acids to extract and leach heavy metals through formation of phosphates, carbonates, sulfides and hydroxides. The resultant solid particles are removed via filtration and/or sedimentation. The technique immobilizes the mobile metals but poses a risk of destabilizing the pollutant fractions that are strongly bound. Ethylenediaminetetraacetic acid (ETDA) is a common agent used in chemical soil washing though it is poorly biodegradable, can persist in soils and modify their functions (Raffa et al., 2021). Other friendly chemicals used in soil washing include ethylenediamine tetra (methylene phosphonic acid) (EDTMP), polyacrylic acid (PAA), polyaspartic acid (PASP), iminodisuccininc acid (ISA) and glutamate-N-N-diacetic acid (Wang et al., 2020). The chemicals are more biodegradable but less effective in heavy metal remediation compared to EDTA. Carbon black, lime and biochar have also been used to immobilize heavy metals in soils and are more effective since they do not alter the chemistry of soils and are sourced from biomaterials (Zhai et al. 2018). Although the use of chemical soil washers is common, their ability to modify soil characteristics and form secondary pollutants is a deterring factor.

Electrochemical/Electrokinetic Remediation

In electrochemical remediation, processes such as electromigration, diffusion, electrophoresis and electroosmosis are applied to enhance migration of heavy metals to electrodes of opposite charge under direct current (He et al., 2015b). The success of electrokinetic remediation is pegged on factors such as metal concentration, ion mobility, specific metal species and soil characteristics such as its moisture content, chemistry and structure (Rajendran et al., 2022). Environmental factors such as dissolution, precipitation and sorption also influence electrokinetic remediation. In the technique, electrodes are mounted in polluted regions and electrical current is passed to enable a response to the electrode surface. Figure 2 shows a schematic representation of electrochemical remediation and the processes involved (Raff et al., 2021). The technique is effective in remediation of fine-grained soils, which are polluted with As, Cd, Cr, Hg and Pb in in-situ conditions (Dhaliwal et al., 2020). The technique is applied in high depths, can monitor soils conditions and the electrical current rate does not interfere with soil characteristics especially in soils of low permeability (Fasani et al., 2018). Although the technique is simple to set up and dismantle, it requires constant buffering to prevent pH changes (Rad et al., 2014).

Figure 2. Schematic representation of electrokinetic remediation of polluted soils (Raffa et al., 2021)

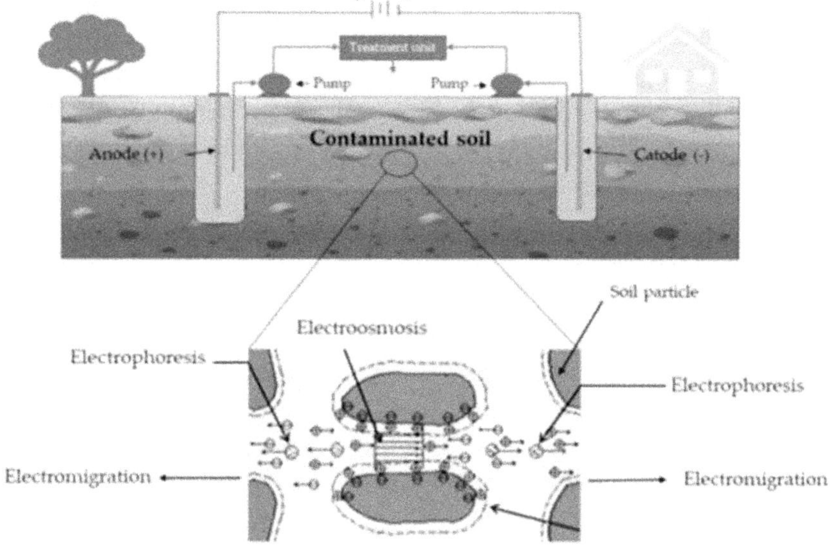

Chemical Leaching, Precipitation, and Adsorption

In chemical leaching, soils are discharged artificially using reagents, solvents and gases suitable in dirt removal (Rajendran et al., 2022). The technique applies inorganic compounds, surfactants and chelating operators in processes such as adsorption, chelation, particle trading and precipitation in soils to transfer toxic metals to leachate (Fasani et al., 2018). In precipitation, metal hydroxides from soils are mixed with precipitants such as Ca and Mg compounds to form insoluble complexes that sediment from the solution containing them (Tahoon et al., 2020). Some of the insoluble compounds formed include carbonates, sulfides and hydroxides. The process is dependent on pH levels so that once the levels increase, metal precipitation in soils is eased. The availability of a wide variety of adsorbents makes adsorption a common method to remediate heavy metals from contaminated soils (Wadhawan et al., 2020). Examples of adsorbents include activated carbon (Nyika & Dinka, 2021) and agricultural wastes. Nano-filters that are created at low temperatures and have a high surface area are currently being tested for use as adsorbents of soil heavy metals (Wadhawan et al., 2020).

Bioremediation

Bioremediation involves the use of plants and microorganisms that respond to heavy metal contaminants in soils using their defense mechanisms such as changes in cell morphology or through the production of enzymes in-situ and/or on-site (Nyika & Dinka, 2022). With bioremediation, heavy metals are not metabolized by plants and microbes but instead, they are bonded, immobilized and accumulated to become less toxic that in their original state in soils.

Microbial Bioremediation

The process involves the growth of specific microorganisms in polluted soils through bioaugmentation and biostimulation processes. Biostimulation involves the addition of nutrient and/or substrates in soils to enhance the activity of autochthonous microorganisms while bioaugmentation adds new microorganisms to supplement the activity of natural soil microbes in biometabolizing heavy metal contaminants (Nyika & Dinka, 2022). Some of the microbial processes used in bioremediation of polluted soils include biosorption, bioleaching, biosparging and bioinventing.

In biosorption, metal cations and cell surfaces of microbes that are anionic in nature bind to immobilize heavy metals on microbial cell structures. The technique involves processes such as precipitation, reduction, complex formation, ion exchange and physical adsorption. Factors such as the type of metal ion, the immobilizing agent and the density of sorption centers involved influence biosorption outcome (Velkova et al., 2018). Bioadsorbents used include algae, fungi and bacteria (Raffa et al., 2021). Algae have functional groups such as sulfonate, sulfhydryl, amino and carboxyl functional groups in their cell wall, which react with heavy metal cations to enable adsorption. Some of the algal species used in cleaning up contaminated soils include *Fucus serratus*, *Enteromorpha* sp., *Chaetomorpha* sp., *Cladophora* sp. and *Ulva* sp. (Alsafran et al., 2022). Fungi have polysaccharides, phosphate groups, glucuronic acid and chitin-chitosan complexes on their cells, which heavy metals from soils bind on. Fungal species used to bioremediate heavy metals in soils include *Coprinopsis atramentaria*, *Aspergillus* sp., and *Phanerochate chrysosporium* (Srivastava et al., 2015). Bacteria species use their sulfate, phosphate, amino and carboxyl groups to bind on heavy metals. Bacterial species such as *Bacillus* sp., *Desulfuromonas palmitatis*, *Pesudomonas* sp. and *Torulopsis bombicola* have been successfully used to bioremediate soils polluted by heavy metals (Dhaliwal et al., 2020).

In bioleaching, microbial secretions including organic acids of low molecular weight are used to dissolve, stabilize and reduce mobility of heavy metals in soils. Using microbial metabolites or during microbial metabolism, heavy metals can solubilize directly or indirectly (Wen et al., 2012). In bioleaching, biosurfactants such as lipopeptides, lipids and polysaccharides produced by microorganisms enhance faster activity due to their large surface area that allows the binding of metal ions to dissolve them (Yang et al., 2018).

Biosparging works on the principle of injecting air into polluted soils to enhance availability of oxygen and subsequently promote aerobic degradation and volatilization of heavy metal pollutants (Hussain et al., 2021). During the injection of oxygen, pressure is controlled to prevent the escape of pollutants to the atmosphere from polluted soils. The technique is effective in soils with high bioactivity and permeability and has been used in bioremediation of soils polluted by gasoline and jet fuel and their inherent metal content (Mosa et al., 2016). Bionventing is a form of biostimulation that works by using wells to add supplements and air to polluted soils to enhance development of microorganisms necessary for adsorbing heavy metals (Anekwe & Isa, 2021). Air flow is gradual and adequate oxygen promotes biodegradation while limiting the escape of pollutants to the air (Mosa et al., 2016).

Remediation by Plants

Some plants have the capacity to stabilize, adsorb and accumulate heavy metals from soils thereby, phyto-remediating them. In the technique, toxic metals are sequestered in plant roots and shoots to be transformed to less toxic substances (Ancona et al., 2019). The plants that take up such toxic elements from soils are known as hyperaccumulators and have been reports to take up and tolerate more than 1000 mg/kg of Cd, Co, Cu, Ni and Pb (Nwaichi & Dhankher, 2016). Depending on the size, large plants with large shoots and roots accumulate more toxic metals compared to smaller ones. Some of the plant species that are hyperaccumulators include smilo grass (*Piptatherum milliaceum*), sorghum (*Sorghum bicolor*), wheat (*Triticumaestivum* L.), poplar (*Populus* spp.) and willow (*Salix* spp.) (Raffa et al., 2021). Plants in the *Asteraceae, Caryophyllaceae, Euphorbiaceae, Fabaceae, Flacourtiaceae, Brassicaceae, Violaceae, Lamiaceae*, and *Poaceae* families are also known hyperaccumulators (Rajendran et al., 2022). The process of phytoremediation can be enhanced by addition to natural chelators to soils to promote metal adsorption and bioavailability. Based on the heavy metal being removed, the plant being used and the removal mechanism applied phytoremediation occurs via phytostabilization, phytoevaporation and phytorextraction.

During phytostabilization, metal ions are complexed with cell walls and roots of plants where they bind with metallothioneins and phytochelatins and thereafter they are sequestered in the plant cell root vacuoles (Shackira & Puthur, 2016). In soils contaminated by Cd and Pb, *Helianthus petiolaris* (lesser sunflower) was used to phytostabilize and remediate the metals from such environments (Sarah et al.,

2020). In phytoevaporation, volatile organic compounds and heavy metals from soils bind and they are transpired by plant leaves to the atmosphere. The transpired forms of heavy metals are less toxic compared to their original forms. Arsenic from soils was phytoevaporated with *Pteris vittate* (ladder brake) and reduced to arsenate and arsenite (Sakakibara et al., 2006). Few plants are able to remediate heavy metals from soils via phytovolatilization although some species can be enabled to do so via genetic modification. For instance, the transgenic tobacco plant has been genetically modified to phytoevaporate Hg from contaminated soils (Raffa et al., 2021). Although the method is effective in heavy metal removal from soils, the products released to the atmosphere can cause secondary pollution once they interact with the constituents of air.

Phytoextraction refers to the absorption of heavy metals pollutants from soils via the roots and their transport and accumulation in the plant shoots. Plants such as marigold, calendula, pansy and antirrhinum have been used to phytoextract Cd, Cr, Ni and Pb from polluted soils (Mahmood et al., 2020). The process can further be enhanced by adding EDTA and bacterial inoculum to speed up the growth of biomass (Xu et al., 2022). Due to the slow nature of all phytoremediation processes, it is recommended to combine them with other removal techniques previously discussed for enhanced and speedier efficacy (Nyika & Dinka, 2022).

Ex-Situ Remediation Techniques

Ex-situ remediation processes involve the excavation of polluted soils and their subsequent transportation to treatment sites before disposal in approved sites. The laborious nature of the techniques and the need to refill excavated sites makes them expensive (Fasani et al., 2018). Some of the bioremediation techniques applied ex-situ include land farming, composting and the use of biopiles. In land farming, polluted land is cultivated to turn the soil and expose it on the ground surface for microbial biodegradation of the trace minerals. Biodegradation activity is a result of mesophilic and thermophilic microorganisms, which can breakdown and adsorb heavy metals (Dhaliwal et al., 2020). Biopilling is a large-scale technology that decontaminates soils through cycles of land cultivation and soil treatment in a designated area fitted with an aeration device and leachate collection system (Gogoi et al., 2021).

A summary of the merits and demerits of the techniques to manage heavy metal polluted soils discussed herein is as shown in Table 1 (Khalid et al., 2017; Dhaliwal et al., 2020; Anekwe & Isa, 2021; Raffa et al., 2021; Rajendran et al., 2022).

Emerging Trends and Green Sustainable Remediation (GSR) Techniques

Although the uptake of physical, chemical and biological remediation methods to clean up contaminated soils has grown, new tendencies are emerging. The trends are motivated by three reasons. First, regulators and practitioners have realized that it is unrealistic to remediate all heavy metals from contaminated sites. Secondly, it is impossible to regain the original pristine conditions of land following contamination by heavy metals. For this reason, there is a growing call for risk-based remediation where heavy metal concentrations should be reduced to certain levels and the land used for designated purposes (Hou et al., 2020; Awasthi et al., 2022). Thirdly, researchers have reported that some remediation approaches can result to adverse environmental effects such as air pollution through greenhouse gas emissions, eutrophication and groundwater pollution (Wang et al., 2021). Owing to these challenges, there is need for sustainable remediation.

Table 1. A comparison of different methods used in the cleanup of heavy metals from contaminated soils (Khalid et al., 2017; Dhaliwal et al., 2020; Anekwe & Isa, 2021; Raffa et al., 2021; Rajendran et al., 2022)

Remediation Type	Technique	Advantages	Disadvantages
Physical Methods	All	-Most methods involve removal of the top soil layer that holds many of the contaminants -Techniques are rapid and easy	-Methods are time-consuming, laborious and commercially unsustainable
	Soil replacement	-Isolates heavy metal from polluted soils	-Only applicable in small scale and is not sustainable
	Soil isolation	-Reduces off-site transport of trace metals in highly contaminated soils	

continues on following page

Table 1. Continued

Remediation Type	Technique	Advantages	Disadvantages
Chemical techniques	All	-Simple, rapid and highly efficient	-Techniques are expensive -They alter the physicochemical properties of soils
	Vitrification	-An effective technique to decontaminate soils off heavy metals by precipitating them at high temperatures	-The technique has restricted execution, is expensive and high energy consuming
	Chemical stabilization	-Through adsorption, heavy metals in soils become less available for plant uptake -Fast and easy to apply	-The method is not sustainable if pollutants are discharged in soils when environmental factors are suitable
	Chemical leaching	-Allows for the growth of plants once toxic metals are removed from top layers	-It is not effective for plants that are deep-rooted
	Electrochemical remediation	-It is fast and easy to install -The technique is effective in soils of low permeability -The method does not modify soil physicochemical characteristics	-The technique cannot use ion exchange membranes and causes changes in pH of soils, which require constant adjustments -Not sustainable and effective for large-sale decontamination
	Chemical soil washing	-It is simple and effective	-The effectiveness depends on the extractant used and soil type -The extractants may cause secondary pollution
Bioremediation	All	-The techniques are eco-friendly, cheap and simple to execute	-Methods are slow and highly efficient only if pollutants are in low concentrations in soils
	Ex-situ methods	-Safe and allow application of exogenous and endogenous microbes to enhance efficiency -Can be tolerated in various environments	-Processes are slow and require adequate time
	In-situ methods	-Convert heavy metal contaminants to less harmful products	-Not effective for decontaminating all heavy metals
	Phytostabilization	-Less disruptive and economical	-Unsustainable and depends on the contaminant levels, the soil type and plant used
	Phytoextraction	-Economical, less disruptive and eco-friendly	-Depends on metal bioavailability, plant tolerance and its growing conditions -Hyperaccumulators are rare
	Use of microorganisms	-Effective in cleansing heavy metals from low-grade minerals, is efficient, easy and economical	-It is sometimes tedious and requires the provision of suitable conditions to enable growth and multiplication -Efficacy depends on the microorganism used and the metal being removed
	Use of plants	-Effective because roots, shoots and leaves of some plant can withstand high concentrations of heavy metals	- Difficulty in choosing the suitable plant to decontaminate a specific metal from soils with high efficacy

With sustainable remediation, greener and long-term approaches with minimal social, economic and environmental impacts will be taken up and hence the concept of green sustainable remediation (GSR) (Wang et al., 2021; Awasthi et al., 2022). GSR practices therefore look beyond the conventional cleaning practices to consider environmental risks of application and avoid secondary pollution. They also consider impacts of remediation on future generations while remaining focused on the economic and social sustainability and resilience to unprecedented changes. Lastly, the practices should be nature-based to enhance the sustainability aspect. In remediation of heavy metals from polluted soils and application of GSR, biochar, coal fly ash, industrial waste-based materials, natural clay minerals, red mud, metal (Fe and Mg) oxides and nanomaterials produced via green synthesis methods have been used to cleanse soils off heavy metals (Wang et al., 2021). The green restoration materials have been used in immobilizing heavy metal pollutants in soils successfully (Hou et al., 2020; Wang et al., 2021; Aswathi et al., 2022) though their stability in the long-term remains largely speculative. Additionally, some of the materials used in GSR such as industrial wastes and biochar could enhance heavy metal mobilization and dissolution in soils rathe than fixation, which could antagonize remediation efforts. It is therefore essential to conduct evidence-based assessment of the efficacy of GSR techniques for their sustainable use in decontaminating polluted soils without causing secondary contamination before any field-scale applications are done.

CONCLUSION

In this chapter, an evaluation of remediation techniques used to decontaminate soils off heavy metals was conducted. Findings showed that the efficacy of the techniques was not equivalent to the expanding extent of pollution. The trend was attributable to the economic and technical components required to set up and implement the techniques successfully. Remediation was found to be unsustainable, laborious, expensive and using some approaches, very slow. Although techniques such as bioremediation were found to be cost-effective and eco-friendly, chemical remediation methods were associated with secondary pollution and physical methods were deemed unsustainable and expensive. The chapter also highlighted the need to consider aspects of sustainability by adopting GSR techniques. Overall, the need for extensive research on the management of heavy metals in polluted soils was highlighted to enable field-scale applications of the techniques and enhance their efficacy for environmental sustainability without causing secondary pollution.

REFERENCES

Agnello, C., Bagard, M., van Hullebusch, D., Esposito, G., & Huguenot, D. (2015). Comparative bioremediation of heavy metals and petroleum hydrocarbons co-contaminated soil by natural attenuation, phytoremediation, bioaugmentation and bioaugmentation-assisted phytoremediation. *The Science of the Total Environment*, *564*, 693–703. doi:10.1016/j.scitotenv.2015.10.061 PMID:26524994

Alsafran, M., Saleem, M., Jabri, H., Rizwan, M., & Usman, K. (2022). Principles and applicability of integrated remediation strategies for heavy metal removal/recovery from contaminated environments. *Journal of Plant Growth Regulation*, 1–22. doi:10.100700344-022-10803-1

Ancona, V., Caracciolo, A., Campanale, C., Rascio, I., Grenni, P., Di Lenola, M., Bagnuolo, G., & Uricchio, V. F. (2019). Heavy metal phytoremediation of a poplar clone in a contaminated soil in southern Italy. *Journal of Chemical Technology and Biotechnology*, *95*, 940–949. doi:10.1002/jctb.6145

Anekwe, M., & Isa, Y. (2021). Wastewater and bioventing treatment systems for acid mine drainage–contaminated soil. *Soil & Sediment Contamination*, *30*(4), 1–14. do i:10.1080/15320383.2020.1863909

Arora, V., & Khosla, B. (2021). *Conventional and contemporary techniques for removal of heavy metals from soil.* IntechOpen. doi:10.5772/intechopen.98569

Ashraf, S., Ali, Q., Zahir, Z., Ashraf, S., & Asghar, H. (2019). Phytoremediation: Environmentally sustainable way for reclamation of heavy metal polluted soils. *Ecotoxicology and Environmental Safety*, *174*, 714–727. doi:10.1016/j. ecoenv.2019.02.068 PMID:30878808

Baldantoni, D., Morra, L., Zaccardelli, M., & Alfani, A. (2016). Cadmium accumulation in leaves of leafy vegetables. *Ecotoxicology and Environmental Safety*, *123*, 89–94. doi:10.1016/j.ecoenv.2015.05.019 PMID:26004982

Bilgin, M., & Tulun, S. (2016). Removal of heavy metals (Cu, Cd and Zn) from contaminated soils using EDTA and $FeCl_3$. *Global NEST Journal*, *18*(10), 98–107.

Carrillo, H., Tirado, J., Romero, J., Luis, A., Munoz, M., Pinon, J., Bueno, J., Montes, C., Naranjo, E., & Ramirez, A. (2021). Encapsulation of toxic heavy metals from waste CRT using calcined kaolin base-polymer. *Materials Chemistry and Physics*, *257*, 123745. doi:10.1016/j.matchemphys.2020.123745

Chang, T., & Yen, J. (2006). On-site mercury-contaminated soils remediation by using thermal desorption technology. *Journal of Hazardous Materials*, *128*(2-3), 208–217. doi:10.1016/j.jhazmat.2005.07.053 PMID:16144741

Cheng, Y., Zhou, K., Wang, J., Cui, S., Yan, J., Maeyer, P., & Voorde, T. (2022). Regional metal pollution risk assessment based on a big data framework: A case study of the eastern Tianshan mining area, China. *Ecological Indicators*, *145*, 109585. doi:10.1016/j.ecolind.2022.109585

Cristaldi, A., Conti, G., Jho, E., Zuccarello, P., Grasso, A., Copat, C., & Ferrante, M. (2017). Phytoremediation of contaminated soils by heavy metals and PAHs. A brief review. *Environmental Technology & Innovation*, *8*, 309–326. doi:10.1016/j.eti.2017.08.002

Dhaliwal, S., Singh, J., Taneja, K., & Mandal, A. (2020). Remediation techniques for removal of heavy metals from the soil contaminated through different sources: A review. *Environmental Science and Pollution Research International*, *27*(2), 1319–1333. doi:10.100711356-019-06967-1 PMID:31808078

Dkhala, B., Mezned, N., Gomez, C., & Abdeljaouad, S. (2020). Hyperspectral field spectroscopy and SENTINEL-2 multispectral data for minerals with high pollution potential content estimation and mapping. *The Science of the Total Environment*, *740*, 140160. doi:10.1016/j.scitotenv.2020.140160 PMID:32927579

Du, D., Ren, X., Yan, S., Shi, X., Liu, Y., & He, G. (2017). An integrated method for the quantitative evaluation of mineral resources of cobalt-rich crusts on seamounts. *Ore Geology Reviews*, *84*, 174–184. doi:10.1016/j.oregeorev.2017.01.011

FAO. (2018). *Soil Pollution: A Hidden Reality*. Food and Agricultural Organization of the United Nations.

Fasani, E., Manara, A., Martini, F., Furini, A., & DalCorso, G. (2018). The potential of genetic engineering of plants for the remediation of soils contaminated with heavy metals. *Plant, Cell & Environment*, *41*(5), 1201–1232. doi:10.1111/pce.12963 PMID:28386947

Gogoi, N., Baroowa, B., & Gogoi, N. (2021). *Ecological tools for remediation of soil pollutants. Bioremediation Science from Theory to Practice*. CRC Press.

Hasanpour, M., & Hatami, M. (2020). Application of three-dimensional porous aerogels as adsorbent for removal of heavy metal ions from water/wastewater: A review study. *Advances in Colloid and Interface Science*, *284*, 102247. doi:10.1016/j.cis.2020.102247 PMID:32916456

He, F., Gao, J., Pierce, E., Strong, P., Wang, H., & Liang, L. (2015b). In situ remediation technologies for mercury-contaminated soil. *Environmental Science and Pollution Research International, 22*(11), 8124–8147. doi:10.100711356-015-4316-y PMID:25850737

He, Z., Shen, J., Ni, Z., Tang, J., Song, S., Chen, J., & Zhao, L. (2015a). Electrochemically created roughened lead plate for electrochemical reduction of aqueous CO2. *Catalysis Communications, 72*, 38–42. doi:10.1016/j.catcom.2015.08.024

Hongbo, S., Liye, C., Gang, X., Kun, Y., & Lihua, Z. (2011). Progress in phytoremediating heavy-metal contaminated soils. In I. Sherameti & A. Varma (Eds.), *Detoxification of Heavy Metals. Soil Biology* (Vol. 30). Springer. doi:10.1007/978-3-642-21408-0_4

Hou, D. (2020). *Sustainable remediation of contaminated soil and groundwater: materials, processes, and assessment.* Butterworth-Heinemann Publishers. doi:10.1016/C2018-0-03003-0

Hou, D., O'Connor, D., Igalavithana, A., Alessi, S., Luo, J., Tsang, D. C. W., Sparks, D. L., Yamauchi, Y., Rinklebe, J., & Ok, Y. S. (2020). Metal contamination and bioremediation of agricultural soils for food safety and sustainability. *Nature Reviews. Earth & Environment, 1*(7), 366–381. doi:10.103843017-020-0061-y

Hussain, K., Haris, M., Qamar, H., Hussain, T., & Ahmad, G. (2021). Bioremediation of waste gases and polluted soils. In G. Panpatte & Y. Jhala (Eds.), *Microbial Rejuvenation of Polluted Environment.*, doi:10.1007/978-981-15-7455-9_5

Islam, M., Taki, G., Nguye, X., Jo, Y., & Kim, J. (2017). Heavy metal stabilization in contaminated soil by treatment with calcined cockle shell. *Environmental Science and Pollution Research International, 24*(8), 7177–7183. doi:10.100711356-016-8330-5 PMID:28097482

Khalid, S., Shahid, M., Niazi, N., Murtaza, B., & Bibi, I. (2017). A comparison of technologies for remediation of heavy metal contaminated soils. *Journal of Geochemical Exploration, 182* (part B), 247-268. doi:10.1016/j.gexplo.2016.11.021

Khan, A., Chattha, R., Farooq, K., Jawed, A., & Farooq, M.. (2015). Effect of farmyard manure levels and NPK applications on the pea plant growth, pod yield and quality. *Life Sciences International Journal, 9*, 3178–3181.

Khoo, S., Chia, Y., Chew, W., & Show, L. (2021). Microalgal-Bacterial consortia as future prospect in wastewater bioremediation, environmental management and bioenergy production. *Indian Journal of Microbiology*, *61*(3), 262–269. doi:10.100712088-021-00924-8 PMID:34294991

Kim, H., Choi, H., Joo, O., Kim, K., Choi, W., & Oh, B.-K. (2015). Development of a microbe-zeolite carrier for the effective elimination of heavy metals from seawater. *Journal of Microbiology and Biotechnology*, *25*(9), 1542–1546. doi:10.4014/jmb.1504.04067 PMID:26032363

Koptsik, G. (2014). Modern approaches to remediation of heavy metal polluted soils: A review. *Eurasian Soil Science*, *47*(7), 707–722. doi:10.1134/S1064229314070072

Lakkireddy, K., & Kues, U. (2017). Bulk isolation of basidiospores from wild mushrooms by electrostatic attraction with low risk of microbial contaminations. *AMB Express*, *7*(1), 1–22. doi:10.118613568-017-0326-0 PMID:28124290

Leong, H., Zaine, A., Ho, C., Uemura, Y., & Lam, K. (2019). Impact of various microalgal-bacterial populations on municipal wastewater bioremediation and its energy feasibility for lipid-based biofuel production. *Journal of Environmental Management*, *249*, 109384. doi:10.1016/j.jenvman.2019.109384 PMID:31419674

Lewandowski, I., Schmidt, U., Londo, M., & Faaij, A. (2006). The economic value of the phytoremediation function - assessed by the example of cadmium remediation by willow (Salix ssp). *Agricultural Systems*, *89*(1), 68–89. doi:10.1016/j.agsy.2005.08.004

Liu, G., Zhou, X., Li, Q., Shi, Y., Guo, G. L., Zhao, L., Wang, J., Su, Y., & Zhang, C. (2020). Spatial distribution prediction of soil as in a large-scale arsenic slag contaminated site based on an integrated model and multi-source environmental data. *Environmental Pollution*, *267*, 115631. doi:10.1016/j.envpol.2020.115631 PMID:33254608

Liu, S., Jiang, J., Wang, S., Guo, Y., & Ding, H. (2018). Assessment of water-soluble thiourea-formaldehyde (WTF) resin for stabilization/solidification (S/S) of heavy metal contaminated soils. *Journal of Hazardous Materials*, *346*, 167–173. doi:10.1016/j.jhazmat.2017.12.022 PMID:29274510

Mahmood, M., Yousra, M., Saman, L., & Ahmad, R. (2020). Floriculture: Alternate non-edible plants for phyto-remediation of heavy metal contaminated soils. *International Journal of Phytoremediation*, *22*(7), 725–732. doi:10.1080/15226514.2019.1707772 PMID:31916455

MEP, Ministry of Environmental Protection. (2014). *National soil contamination survey report*. MEP. https://www.pollutionsolutions-online.com/white-paper/soil remediation/18/xportreporter/soil-contamination-industry-report/9

Minnikova, V., Denisova, V., Mandzhieva, S., Kolesnikov, I., Minkina, M., Chaplygin, V. A., Burachevskaya, M. V., Sushkova, S. N., & Bauer, T. V. (2017). Assessing the effect of heavy metals from the Novocherkassk power station emissions on the biological activity of soils in the adjacent areas. *Journal of Geochemical Exploration*, *174*, 70–78. doi:10.1016/j.gexplo.2016.06.007

Mosa, K., Saadoun, I., Kumar, K., Helmy, M., & Dhankher, O. (2016). Potential biotechnological strategies for the cleanup of heavy metals and metalloids. *Frontiers in Plant Science*, *7*. doi:10.3389/fpls.2016.00303 PMID:27014323

Nwaichi, E., & Dhankher, O. (2016). Heavy metals contaminated environments and the road map with phytoremediation. *Journal of Environmental Protection*, *7*(1), 41–51. doi:10.4236/jep.2016.71004

Nyika, J., & Dinka, M. (2021). Activated bamboo charcoal in water treatment: A mini-review. *Materials Today: Proceedings*, *56*(4), 1904–1907. doi:10.1016/j.matpr.2021.11.167

Nyika, J., & Dinka, M. (2022). A min-review on wastewater treatment through bioremediation towards enhanced field applications of the technology. *AIMS Environmental Science*, *9*(4), 403–431. doi:10.3934/environsci.2022025

Pattnaik, K., & Equeenuddin, M. (2016). Potentially toxic metal contamination and enzyme activities in soil around chromite mines at Sukinda ultramafic complex, India. *Journal of Geochemical Exploration*, *168*, 127–136. doi:10.1016/j.gexplo.2016.06.011

Perez, J. (2012). *The soil remediation industry in Europe: The recent past and future perspectives*. Europa. https://ec.europa.eu/environment/archives/soil/pdf/may2012/08%20-%20Julien%20Perez%20-%20final.pdf [Accessed on 5 February 2023]

Rad, L., Momeni, A., Ghazani, B., Irani, M., Mahmoudi, M., & Noghreh, B. (2014). Removal of Ni2+ and Cd2+ ions from aqueous solutions using electrospun PVA/zeolite nanofibrous adsorbent. *Chemical Engineering Journal*, *256*, 119–127. doi:10.1016/j.cej.2014.06.066

Raffa, M., Chiampo, F., & Shanthakumar, S. (2021). Remediation of metal/metalloid-polluted soils: A short review. *Applied Sciences (Basel, Switzerland)*, *11*(9), 4134. doi:10.3390/app11094134

Rajendran, S., Priya, T., Khoo, K., Hoang, T., Ng, H., Munawaroh, H. S. H., Karaman, C., Orooji, Y., & Show, P. L. (2022). A critical review on various remediation approaches for heavy metal contaminants removal from contaminated soils. *Chemosphere*, *287*, 132369. doi:10.1016/j.chemosphere.2021.132369 PMID:34582930

Sabir, M., Waraich, A., Hakeem, R., Ozturk, M., & Ahmad, R. (2015). *Phytoremediation, soil remediation and plants*. Elsevier Inc., doi:10.1016/B978-0-12-799937-1.00004-8

Saini, S., Kaur, N., & Pati, P. (2021). Phytohormones: Key players in the modulation of heavy metal stress tolerance in plants. *Ecotoxicology and Environmental Safety*, *223*, 112578. doi:10.1016/j.ecoenv.2021.112578 PMID:34352573

Sakakibara, M., Watanabe, A., Sano, S., Inoue, M., & Kaise, T. (2006). *Phytoextraction and phytovolatilization of arsenic from As-contaminated soils by* Pteris vittata. In *Proceedings of the Annual International Conference on Soils, Sediments, Water and Energy*, Amherst, MA, USA.

Saleem, M., Ali, S., Irshad, S., Hussaan, M., & Rizwan, M. (2020). Copper uptake and accumulation, ultra-structural alteration, and bast fiber yield and quality of fibrous jute (*Corchorus capsularis* L.) plants grown under two different soils of China. *Plants*, *9*, 404. doi:10.3390/plants9030404 PMID:32213938

Saran, A., Fernandez, L., Cora, F., Savio, M., Thijs, S., Vangronsveld, J., & Merini, L. J. (2020). Phytostabilization of Pb and Cd polluted soils using *Helianthus petiolaris* as pioneer aromatic plant species. *International Journal of Phytoremediation*, *22*(5), 459–467. doi:10.1080/15226514.2019.1675140 PMID:31602996

Shackira, A., & Puthur, J. (2016). Phytostabilization of heavy metals: understanding of principles and practices. In S. Srivastava, A. Srivastava, & P. Suprasanna (Eds.), *Plant-Metal Interactions*. Springer. doi:10.1007/978-3-030-20732-8_13

Shahid, M., Pinelli, E., & Dumat, C. (2012). Review of Pb availability and toxicity to plants in relation with metal speciation; role of synthetic and natural organic ligands. *Journal of Hazardous Materials*, *219-220*, 1–12. doi:10.1016/j.jhazmat.2012.01.060 PMID:22502897

Song, B., Zeng, G., Gong, J., Liang, J., Xu, P., Liu, Z., Zhang, Y., Zhang, C., Cheng, M., Liu, Y., Ye, S., Yi, H., & Ren, X. (2017). Evaluation methods for assessing effectiveness of in situ remediation of soil and sediment contaminated with organic pollutants and heavy metals. *Environment International*, *105*, 43–55. doi:10.1016/j.envint.2017.05.001 PMID:28500873

Song, M., Fisher, R., Wang, J., & Cui, L. (2018). Environmental performance evaluation with big data: Theories and methods. *Annals of Operations Research, 270*(1–2), 459–472. doi:10.100710479-016-2158-8

Srivastava, S., Agrawal, S., & Mondal, K. (2015). A review on progress of heavy metal removal using adsorbents of microbial and plant origin. *Environmental Science and Pollution Research International, 22*(20), 15386–15415. doi:10.100711356-015-5278-9 PMID:26315592

Sumiahadi, A., & Acar, R. (2018). A review of phytoremediation technology: Heavy metals uptake by plants. *IOP Conference Series. Earth and Environmental Science, 012023*, 012023. doi:10.1088/1755-1315/142/1/012023

Tahoon, M., Siddeeg, S., Salem Alsaiari, N., Mnif, W., & Rebah, F. (2020). Effective heavy metals removal from water using nanomaterials: A review. *Processes (Basel, Switzerland), 8*(6), 645. doi:10.3390/pr8060645

Tan, K., Ma, W., Chen, L., Wang, H., Du, Q., Du, P., Yan, B., Liu, R., & Li, H. (2021). Estimating the distribution trend of soil heavy metals in mining area from HyMap airborne hyperspectral imagery based on ensemble learning. *Journal of Hazardous Materials, 401*, 123288. doi:10.1016/j.jhazmat.2020.123288 PMID:32645545

Tindwa, H., & Singh, B. (2023). Soil pollution and agriculture in sub-Saharan Africa: State of the knowledge and remediation technologies. *Frontiers in Soil Science, 2*, 1101944. doi:10.3389/fsoil.2022.1101944

Tiwari, S., Singh, S., & Garg, S. (2013). Microbially enhanced phytoextraction of heavy metal fly-ash amended soil. *Communications in Soil Science and Plant Analysis, 44*(21), 3161–3176. doi:10.1080/00103624.2013.832287

Ubando, A., Africa, A., Maniquiz-Redillas, M., Culaba, A., Chen, W., & Chang, J.-S. (2020). Microalgal biosorption of heavy metals: A comprehensive bibliometric review. *Journal of Hazardous Materials, 402*, 123431. doi:10.1016/j.jhazmat.2020.123431 PMID:32745872

Ullah, A., Ma, Y., Li, J., Tahir, N., & Hussain, B. (2020). Effective amendments on cadmium, arsenic, chromium and lead contaminated paddy soil for rice safety. *Agronomy (Basel), 10*(3), 359. doi:10.3390/agronomy10030359

Vardhan, K., Kumar, P., & Panda, R. (2019). A review on heavy metal pollution, toxicity and remedial measures: Current trends and future perspectives. *Journal of Molecular Liquids, 290*, 111197. doi:10.1016/j.molliq.2019.111197

Velkova, Z., Kirova, G., Stoytcheva, M., Kastadinova, S., & Todorova, K. (2018). Immobilized microbial biosorbents for heavy metals removal. *Engineering in Life Sciences, 18*(12), 871–881. doi:10.1002/elsc.201800017 PMID:32624881

Wadhawan, S., Jain, A., Nayyar, J., & Mehta, S. (2020). Role of nanomaterials as adsorbents in heavy metal ion removal from waste water: A review. *Journal of Water Process Engineering, 33*, 101038. doi:10.1016/j.jwpe.2019.101038

Wang, G., Pan, X., Zhang, S., Zhong, Q., Zhou, W., Zhang, X., Wu, J., Vijver, M. G., & Peijnenburg, W. J. G. M. (2020). Remediation of heavy metal contaminated soil by biodegradable chelator–induced washing: Efficiencies and mechanisms. *Environmental Research, 186*, 109554. doi:10.1016/j.envres.2020.109554 PMID:32344210

Wang, L., Rinklebe, J., Tack, F., & Hou, D. (2021). A review of green remediation strategies for heavy metal contaminated soil. *Soil Use and Management, 27*, 936-963. 9. doi:10.1111/sum.12717

Wang, P., Hu, X., He, Q., Waigi, M., Wang, J., & Ling, W. (2018). Using calcination remediation to stabilize heavy metals and simultaneously remove polycyclic aromatic hydrocarbons in soil. *International Journal of Environmental Research and Public Health, 15*(8), 1731. doi:10.3390/ijerph15081731 PMID:30104500

Wen, Y., Wang, Q., Tang, C., & Chen, Z. (2012). Bioleaching of heavy metals from sewage sludge by Acidithiobacillus thiooxidans—A comparative study. *Journal of Soils and Sediments, 12*(6), 900–908. doi:10.100711368-012-0520-2

Xing, J., Li, L., Li, G., & Xu, G. (2019). Feasibility of sludge-based biochar for soil remediation: Characteristics and safety performance of heavy metals influenced by pyrolysis temperatures. *Ecotoxicology and Environmental Safety, 180*, 457–465. doi:10.1016/j.ecoenv.2019.05.034 PMID:31121552

Xu, D., Shen, Z., Dou, C., Dou, Z., Li, Y., Gao, Y., & Sun, Q. (2022). Effects of soil properties on heavy metal bioavailability and accumulation in crop grains under different farmland use patterns. *Scientific Reports, 12*(1), 9211. doi:10.103841598-022-13140-1 PMID:35654920

Yan, A., Wang, Y., Tan, S., Yusof, M., & Ghosh, S. (2020). Phytoremediation: A promising approach for revegetation of heavy metal-polluted land. *Frontiers in Plant Science, 11*, 359. doi:10.3389/fpls.2020.00359 PMID:32425957

Yan, J., Liu, Y., Wang, L., Wang, Z., & Liu, H. (2021). An efficient organization method for large-scale and long time-series remote sensing data in a cloud computing environment. *IEEE Journal of Selected Topics in Applied Earth Observations and Remote Sensing,* (99), 1–1. . doi:10.1109/JSTARS.2021.3110900

Yang, Z., Shi, W., Yang, W., Liang, L., Yao, W., Chai, L., Gao, S., & Liao, Q. (2018). Combination of bioleaching by gross bacterial biosurfactants and flocculation: A potential remediation for the heavy metal contaminated soils. *Chemosphere, 206,* 83–91. doi:10.1016/j.chemosphere.2018.04.166 PMID:29730568

Zaheer, I., Ali, S., Saleem, M., Ali, M., Riaz, M., Javed, S., Sehar, A., Abbas, Z., Rizwan, M., El-Sheikh, M. A., & Alyemeni, M. N. (2020). Interactive role of zinc and iron lysine on *Spinacia oleracea* L. growth, photosynthesis and antioxidant capacity irrigated with tannery wastewater. *Physiology and Molecular Biology of Plants, 26*(12), 2435–2452. doi:10.100712298-020-00912-0 PMID:33424157

Zhai, X., Li, Z., Huang, B., Luo, N., Huang, M., Zhang, Q., & Zeng, G. (2018). Remediation of multiple heavy metal-contaminated soil through the combination of soil washing and in situ immobilization. *The Science of the Total Environment, 635,* 92–99. doi:10.1016/j.scitotenv.2018.04.119 PMID:29660731

Zhao, H., Wu, Y., Lan, X., Yang, Y., Wu, X., & Du, L. (2022). Comprehensive assessment of harmful heavy metals in contaminated soil in order to score pollution level. *Scientific Reports, 12*(1), 3552. doi:10.103841598-022-07602-9 PMID:35241759

Zhou, L., Chen, N., Chen, Z., & Xing, C. (2016). ROSCC: An efficient remote sensing observation-sharing method based on cloud computing for soil moisture mapping in precision agriculture. *IEEE Journal of Selected Topics in Applied Earth Observations and Remote Sensing, 9*(12), 5588–5598. doi:10.1109/JSTARS.2016.2574810

Chapter 15
Measures to Control and Prevent Heavy Metal Pollution in Soils of Sub-Saharan Africa

ABSTRACT

The exponential growth in industrial activities in sub-Saharan Africa (SSA) has resulted in serious environmental problems including heavy metal pollution in soils. Without effective measures to avert the growing trend, the situation is going to escalate resulting in unproductive soils, transfer of toxic metals to food chains, which will negatively affect humans' and animals' health. In this chapter, specific measures to control and prevent heavy metal pollution in soils of SSA are discussed. Findings show the need to avoid pollution sources, use advanced technology to assay and predict soil pollution, use stringent legislations to curtail unsystematic industrial emissions, wastewater, solid waste, agrochemical, and fertilizer uses in agricultural activities. Such initiatives must be supported by geoscientific and technical capacity building to be effective.

INTRODUCTION

Land use changes is a key environmental factor that causes variations in the functions of ecosystems and biodiversity (Zhen et al., 2019). The changes are a result of growing urbanization trends and intensive human activities in advances to achieve economic growth. As a result of the changes, environmental problems such as heavy metal pollution have emerged (Huang et al., 2018). Previous chapters of this book have shown that the accumulation of these metals in soils especially from industrial

DOI: 10.4018/978-1-6684-7116-6.ch015

activities has the capacity to alter soil physicochemical features and ultimately, its quality and functioning. The trend is attributable to the non-biodegradable, persistent and toxic nature of the metals. The toxic effects of the heavy metals can be transferred from soils to plants, animals and human beings causing negative effects to the environment and public health concerns (Okereafor et al., 2020).

In sub-Saharan Africa (SSA) region, where the priority has been economic growth with little attention on environmental sustainability, industrial, agricultural, mining and smelting activities have grown exponential and so are the resultant wastes. Consequently, heavy metal pollution in soils is becoming common and so are its effects (Tindwa & Singh, 2023). In this book, pollution of soils of SSA region by As, Cd, Cr, Cu, Hg, Ni, Pb and Zn has been discussed and varied sources of the metals related to lithogenic and anthropogenic origin detailed. Of concern is the fact that if no remedial measures are taken, the pollution trend and its associated effects will worsen. According to Lee et al. (2021), SSA region has seen increases in human activities, which are attributable to the growing prevalence of environmental hazards and modifications of natural landscapes. This trend is despite the fact that most countries in the region do not have formal governance or regulations to cleanup resultant pollution or ensure that it does not escalate. Tindwa and Singh (2023) also noted the laxity in taking care of the environment in SSA region, noting that waste generation from anthropogenic activities and its unsystematic disposal leading to soil contamination by heavy metals is growing and set to peak at 2100 if corrective measures are not taken. Similar trends have been reported in other studies of the region necessitating the need to take action to control and prevent heavy metal contamination in soils of the SSA (Tully et al., 2015; Nyika et al., 2019; Okereafor et al., 2020; Nyika & Dinka, 2022). Based on knowledge about the lack of precedence on environmental conservation and the effects of heavy metals on soils of SSA region discussed in previous chapters of this book, this chapter focuses on the control and preventative measures that the region can take to avert the worsening trend in soil contamination.

SPECIFIC MEASURES TO CONTROL AND PREVENT HEAVY METAL POLLUTION

1. Develop soil heavy metal detection techniques that are low-cost, have high precision, and use integrated, automated and intelligent technology.

In order to take action to control and prevent heavy metal pollution in SSA region, their assessment and prediction is a pre-requisite. Intelligent techniques combined with spectra data enable accurate estimation of heavy metals in soils and overcome costs associated with using laboratory procedures (Discussed in Chapter 4) to assay the contaminants. Example of such a technique is the use of artificial neural network (ANN), which is an artificial intelligent approach to determine Fe, Mn and Zn concentrations in polluted soils based on their Ca, K and Mg levels (Sari et al., 2022). Obtained results had errors of low significance compared to conventional methods of heavy metal assay in soils. Using an intelligent system combining satellite sensing edge cloud server (ECS), the federated machine learning technique and a deep learning predictive model, heavy metals in soils can be detected and their pollution extent predicted accurately (Yaseen, 2022). Additionally, such automated models also substitute spatial interpolation models that are used in predicting the extent of heavy metal pollution in soils by collecting and storing information on the pollutants to enable prediction via cloud computing (Yaseen, 2022). Other machine learning algorithms for modelling of heavy metals in contaminated soils include ensemble models, evolutionary computing, kernel methods (Li et al., 2021) genetic algorithms, error back propagation neural network (BPNN) and least absolute shrinkage and selection operator (LASSO) (Shi et al., 2022). The methods can be adopted in SSA countries where laboratory assay of soils for heavy metals is limited due to unavailability of advanced laboratory facilities.

2. Formulate, enact and improve legislations and laws aimed at regulating discharge from mining and smelting industries.

Mining and smelting industries are key economic activities in SSA region in addition to their role in introducing heavy metals to soils leading to their contamination. As such, existent regulation in the mining and smelting sector should be better implemented while the quality of existent environmental impact management plans of the sector should be enhanced (Lindahl et al., 2014). According to Agboola et al. (2020), an overhaul in the management of metal processing activities is key to reduce its resultant environmental impacts. Through proper regulation of metal processing industries, heavy metal containing tailings, acid mine drainage, emissions and particulate matter content that end up contaminating soils could be reduced. The spread of heavy metal pollution in soils and other environments can thereby be slowed to allow the implementation of reclamation plans in already polluted areas (Fashola et al., 2016). In mining countries of SSA region such as South Africa, Zambia, Democratic Republic of Congo (DRC), Nigeria and Zimbabwe, such improvements could control and prevent heavy metal pollution of soils. In Nigeria and South Africa, amendments meant to prevent heavy metal pollution of soils from the mining and smelting industries have been made under the State Environmental Protection Agency and the Department of Mineral Resources and Energy, respectively (Agboola et al., 2020) However, such initiatives must be supported by geoscientific and technical capacity building that incorporates education and training of parties involved, institutional and public policy revisions to prioritize reduced pollution of soils and other environmental resources (Agboola et al., 2020).

3. Raise community awareness on the need to conserve natural (land and water) resources as well as the environment to prevent pollution, restore human health and promote sustainable ecological development.

In many developing countries including those of SSA, inadequate awareness of the causatives of heavy metal pollution in soils by communities is attributable to the widespread contamination. Additionally, community members do not know the importance of taking up eco-friendly techniques of heavy metal remediation (Discussed in Chapter 14) (Wuana & Okieimen, 2011). Greater awareness creation is the key to more interest in the uptake of remediation measures since governments and the public in SSA will be more sensitive to the toxic effects of heavy metal pollution on food and animal production and quality as well as human health. With greater public awareness of the industrial sources of heavy metal pollution in soils, pressure can be put on regulatory agencies and the governments of SSA to enact and implement effective legislative frameworks to manage resultant wastes (Okereafor et al., 2020). Local governments in nations of SSA region should intensify education and publicity activities to enhance community protection and intensify law enforcement and supervision against heavy metal pollution of soils and other resources (Li & Liu et al., 2017).

4. Regulate the use of fertilizers, agrochemicals, biosolids and sewage for irrigation purposes.

Depending on the origin and raw materials used in manufacturing them, agrochemicals, sewage sludge and fertilizers could introduce heavy metals in soils (Adnan et al., 2022). The reutilization and overexploitation of biosolids such as manure, wastewater and sewage sludge could also lead to contaminant bioaccumulation in soils (Nunes et al., 2021). In SSA countries of Kenya, Tanzania and Senegal, untreated wastewater was reported to introduce metal contaminants in soils resulting to their pollution and phyto-transfer (Nyika, 2022). Although they are all potential biofertilizers, the use of such additives should be regulated to prevent and control the introduction of metal contaminants in soils. Such advances demand thorough screening of sewage sludge for heavy metals prior to use, strict adherence of regulations on the use agrochemicals and fertilizers and implementation, oversight and enforcement of guidelines to control soil contamination (Nunes et al., 2020). Quality assessment and treatment of wastewater prior to its use for irrigation purposes is also key in controlling metal introduction and subsequent pollution in soils (Nyika, 2022). In a country such as Qatar, strict regulations on the use and non-harmful application of biosolids are attributable to the minimal contamination risk associated with the use of such soil additives in the region (Ali et al., 2021). Such guidelines can be formulated in SSA and enforced stringently to control the pollution associated with such additives to soils.

5. Taking up soil dressing

Soil dressing entails covering heavy metal polluted soils with unpolluted soils and using them to grow shallow-rooted crops (Arao et al., 2010). The measure was used to remediate 7,327 ha of land that was polluted by Cd before being used to grow paddy rice in China (Arao et al., 2010). Although the method was effective in controlling heavy metal bioavailability, it was limited to shallow-rooted plants and was costly due to the difficulty in procuring uncontaminated soil. Additionally, the method is unsustainable since it does not prevent heavy metal mobility from soils to other resources such as groundwater.

6. Structural adjustments of economies and industries

Most of the heavy metals polluting soils globally are a result of industrial activities. It is therefore imperative to restructure industrial activities and transform their processes to be economically and technological improved rather than labor-intensive. Such measures enhance novelty, innovation and cleaner production (Li & Liu, 2017). With cleaner production, wastewater and solid wastes will be treated, reused or recycled to prolong their life cycle and reduce their potential to cause heavy metal pollution in soils. Using clean and green production approaches and bio-products in the automobile, pharmaceutical, manufacturing, agricultural and electronic industries can be adopted and used to promote environmental sustainability that is cautious on avoiding soil pollution.

7. Control and prevention of atmospheric emissions and precipitation pollutants

Industrial emissions from electroplating factories, printing workshops, chemical plants, transport emissions, coal combustion in various manufacturing and energy production processes introduce heavy metals such as As, Cu, Ni and Pb to the atmosphere before their deposition on soils to contaminate them (Zwolak et al., 2019; Su et al., 2022). Therefore, controlling such emissions is key to reducing potential soil pollution by heavy metals. This is especially so in most SSA countries whose ecological environments are vulnerable to pollution owing to the presence of arid land, loose soils and short wet seasons, which allow the mixing and transportation of soils along with heavy metal containing particulate matter. Some of the specific measures that should be taken include vegetating polluted areas to control dust pollution and dispersion of heavy metals, reduce point (industrial) emissions through treatment prior to release to the atmosphere, control linear emissions by avoiding Pb-based fuels, through car polling, walking and using biofuels in the transport sector (Ali & Liu, 2017).

8. Effective and stringent control of pollution sources

The key to preventing and controlling heavy metal pollution in soils is in identifying and avoiding their sources. In SSA region, industrial activities, the transport sector (dust and vehicular exhausts), atmospheric precipitation and depositions and untreated solid wastes and wastewater were found to be the main sources of heavy metal pollution in soils. However, discerning the specific source of heavy metal pollution in a given soil is difficult due to the multi-occurrence of the sources (Long et al., 2021). Such sources can be avoided through a number of measures including the use of clean energy (biofuels), systematic management of scrap cars, regularly servicing the vehicles, enhanced solid waste/wastewater treatment, reuse, recycling and recovery to prevent exposure of soils to heavy metals (Ali & Liu, 2017). Strict regulations requiring industries to treat wastes should be implemented, enforced and punitive measures should be taken to non-adherers (Long et al., 2021). The use of fertilizers and other agrochemicals should adhere to the manufacturers' instructions to prevent misuse or overuse. According to Li and Xing (2020), the abuse of soil additives can be avoided by strengthening agricultural soil monitoring (Tully et al., 2015), using rational portions of the soil amendments as recommended by manufacturers, using eco-friendly additives and enhancing the awareness and publicity to farmers on the need to control and avoid soil and environmental pollution.

9. Adoption of remediation technologies at Field- Scale Levels

The use of physical, chemical and biological techniques of soil remediation (Discussed in Chapter 14) is key to controlling the spread of heavy metals in soils and other environments. Remediation aims at removal of the contaminants or their chemical transformation to less- toxic substances, which reduces the concentrations in soils (Ashraf et al., 2019). Some bioremediation techniques are natural, easy to implement, eco-friendly, use sustainable materials and are less disruptive to soil functions and hence effective in removal of heavy metals from polluted soils (Wang et al., 2021). However, in using remediation techniques, a best-fit removal method based on the soil type, soil characteristics and the metal targeted for removal must be identified for better efficacy in contamination reduction and control.

10. Increased funding to overcome heavy metal pollution

Government and non-governmental organizations (NGOs) of SSA region must increase their funding to enable fight heavy metal pollution in the respective countries. Such advances could enhance compliance to environmental standards, promote the adoption of circular production processes in the region's industries, train personnel on the handling, treatment and remediation of heavy metal polluted soils and adopt greener production systems that promote less heavy metal pollution. Such funding was provided by organizations such as the Cleaner Production Program (CPP) and the United Nations Development Program (UNDP) in developing counties of SSA and Asia (Hira et al., 2022). Developed nations such as United Kingdom also fund SSA countries to promote reduced heavy metal pollution. For instance, through the Sustainable Manufacturing and Environmental Pollution (SMEP) program, the UK government provided £24.6 million to help SSA and Asian countries transit to circular industrial processes that reduced production of wastes containing toxic heavy metals rather than investing money to remediate already polluted areas (Hira et al., 2022). However, the success of such increased investments remains uncertain, although more funding is required as polluted areas are expansive and industrial processes used in SSA favor heavy metal introduction to the environment due to their unsustainability.

CONCLUSION

Pollution of soils by heavy metals owing to the rise of anthropogenic activities is growing in SSA region. The situation is dire because the region prioritizes on economic development rather than environmental conservation. Additionally, existent policies to prevent soil pollution are ineffective, non-adhered to and non-enforced. In this chapter, a number of approaches to control and prevent such pollution tendencies are discussed. These include accurate identification and prediction of polluted soils using automated techniques, avoidance of sources of heavy metals, establishment of workable regulatory frameworks on the use of soil additives and release of industrial emissions to the atmosphere as well as community awareness and action plans on environmental conservation. The uptake of these measures must be supported by good governance, effective legislations, geoscientific and technical capacity building even at the grassroot level. With control and preventative measures of soil pollution by heavy metals, soils can be more productive and pollution of water resources can be avoided towards environmental sustainability.

REFERENCES

Adnan, M., Xiao, B., Xiao, P., Zhao, P., Li, R., & Bibi, S. (2022). Research progress on heavy metals pollution in the soil of smelting sites in China. *Toxics*, *10*(5), 231. doi:10.3390/toxics10050231 PMID:35622644

Agboola, O., Babatunde, D., Fayomi, O., Sadiku, E., & Popoola, P. (2020). A review on the impact of mining operation: Monitoring, assessment and management. *Results in Engineering*, *8*, 100181. doi:10.1016/j.rineng.2020.100181

Ali, M., Ahmed, T., Dieyeh, M., & Al-Ghouti, M. (2021). Environmental impacts of municipal biosolids on soil, plant and groundwater qualities. *Sustainability (Basel)*, *13*(15), 8368. doi:10.3390u13158368

Arao, T., Ishikawa, S., Murakami, M., Abe, K., Maejima, Y., & Makino, T. (2010). Heavy metal contamination of agricultural soil and counter measures in Japan. *Paddy and Water Environment*, *8*(3), 247–257. doi:10.100710333-010-0205-7

Ashraf, S., Ali, Q., Zahir, Z., Ashraf, S., & Asghar, H. (2019). Phytoremediation: Environmentally sustainable way for reclamation of heavy metal polluted soils. *Ecotoxicology and Environmental Safety*, *174*, 714–727. doi:10.1016/j.ecoenv.2019.02.068 PMID:30878808

Fashola, M., Jeme, V., & Babalola, O. (2016). Heavy metal pollution from gold mines: Environmental effects and bacterial strategies for resistance. *International Journal of Environmental Research and Public Health*, *13*(11), 1047. doi:10.3390/ijerph13111047 PMID:27792205

Hira, A., Pacini, H., Attafuah-Wadee, K., Sikander, M., Oruko, R., & Dinan, A. (2022). Mitigating tannery pollution in Sub-Saharan Africa and South Asia. *Journal of Developing Societies*, *38*(3), 360–383. doi:10.1177/0169796X221104856

Huang, Y., Chen, Q., Deng, M., Japenga, J., Li, T., Yang, X., & He, Z. (2018). Heavy metal pollution and health risk assessment of agricultural soils in a typical peri-urban area in southeast China. *Journal of Environmental Management*, *207*, 159–168. doi:10.1016/j.jenvman.2017.10.072 PMID:29174991

Lee, J., Kaunda, R., Sinkala, T., Workman, C., Bazilian, M., & Clough, G. (2021). Phytoremediation and phytoextraction in sub-Saharan Africa: Addressing economic and social challenges. *Ecotoxicology and Environmental Safety*, *226*, 112864. doi:10.1016/j.ecoenv.2021.112864 PMID:34627045

Li, G., & Xing, J. (2020). The present situation of soil pollution in agricultural production and the countermeasures. *IOP Conference Series. Earth and Environmental Science*, *512*(1), 012032. doi:10.1088/1755-1315/512/1/012032

Li, Y., Yang, K., Gao, W., Han, Q., & Zhang, J. (2021). A spectral characteristic analysis method for distinguishing heavy metal pollution in crops: VMD-PCA-SVM. *Spectrochimica Acta. Part A: Molecular and Biomolecular Spectroscopy*, *255*, 119649. doi:10.1016/j.saa.2021.119649 PMID:33744840

Li, Z., & Liu, Z. (2017). Hazards, sources and control measures of heavy metal pollution of forest soil: Taking Jin-Jing-Ji region of China as an example. *Nature Environment and Pollution Technology*, *16*(4), 1141–1147.

Lindahl, J. (2014). *Environmental Impacts of Mining in Zambia: towards Better Environmental Management and Sustainable Exploitation of Mineral Resources, Geological Survey of Sweden.* Semantic Scholar. https://pdfs.semanticscholar.org/6ff8/34376b87410e8f618ebbf80271e6d23eb937.pdF

Long, Z., Zhu, H., Bing, H., Tian, X., Wang, Z., Wang, X., & Wu, Y. (2021). Contamination, sources and health risk of heavy metals in soil and dust from different functional areas in an industrial city of Panzhihua city, Southwest China. *Journal of Hazardous Materials*, *420*, 126638. doi:10.1016/j.jhazmat.2021.126638 PMID:34280716

Nunes, N., Ragonezi, C., Gouveia, C., & Carvalho, M. (2020). Review of sewage sludge as a soil amendment in relation to current international guidelines: A heavy metal perspective. *Sustainability (Basel)*, *13*(4), 2317. doi:10.3390u13042317

Nyika, J. (2022). Wastewater for Agricultural Production, Benefits, Risks, and Limitations. In H. Chatoui, M. Merzouki, H. Moummou, M. Tilaoui, & N. Saadaoui, (Eds.), *Nutrition and Human Health*. Springer. doi:10.1007/978-3-030-93971-7_6

Nyika, J., & Dinka, M. (2022). Heavy metal pollution in soils and vegetable from suburban regions of Nairobi, Kenya and their community health implications. *Pollution*, *8*(4), 1434–1447. doi:10.22059/POLL.2022.341522.1440

Nyika, J., Onyari, E., Dinka, M., & Shivani, B. (2019). Pollution and mobility in soils within a landfill vicinity; a South African case study. *Oriental Journal of Chemistry*, *35*(4), 1286–1296. doi:10.13005/ojc/350406

Okereafor, U., Makhatha, M., Mekuto, L., Okereafor, N., & Sebola, T. (2020). Toxic metal implications on agricultural soils, plants, animals, aquatic life and human health. *International Journal of Environmental Research and Public Health, 17*(7), 2204. doi:10.3390/ijerph17072204 PMID:32218329

Sari, M., Cosgun, T., Yalcin, I., Taner, M., & Ozyigit, I. (2022). Deciding heavy metal levels in soil based on various ecological information through artificial intelligence modelling. *Applied Artificial Intelligence, 36*(1), e2014189. doi:10.1080/08839514.2021.2014189

Shi, S., Hou, M., Gu, Z., Jiang, C., Zhang, W., Hou, M., Li, C., & Xi, Z. (2022). Estimation of heavy metal content in soil based on machine learning models. *Land (Basel), 11*(7), 1037. doi:10.3390/land11071037

Su, C., Meng, J., Zhou, Y., Bi, R., Chen, Z., Diao, J., Huang, Z., Kan, Z., & Wang, T. (2022). Heavy metals in soils from intense industrial areas in south China: Spatial distribution, source apportionment, and risk assessment. *Frontiers in Environmental Science, 10*, 820536. doi:10.3389/fenvs.2022.820536

Tindwa, H., & Singh, B. (2023). Soil pollution and agriculture in sub-Saharan Africa: State of the knowledge and remediation technologies. *Frontiers in Soil Science, 2*, 1101944. doi:10.3389/fsoil.2022.1101944

Tully, K., Sullivan, C., Weil, R., & Sanchez, P. (2015). The state of soil degradation in sub-Saharan Africa: Baselines, trajectories and solutions. *Sustainability (Basel), 7*(6), 6523–6552. doi:10.3390u7066523

Wang, L., Rinklebe, J., Tack, F., & Hou, D. (2021). A review of green remediation strategies for heavy metal contaminated soil. *Soil Use and Management, 27*, 936-963. doi:10.1111/sum.12717

Wuana, R., & Okieimen, F. (2011). Heavy metal in contaminated soils: A review of sources, chemistry, risks and best available strategies for remediation. *ISRN Ecology, 402647*, 1–20. doi:10.5402/2011/402647

Yaseen, Z. (2022). The next generation of soil and water bodies heavy metal prediction and detection: New expert system-based edge cloud server and federated learning technology. *Environmental Pollution, 315*, 120081. doi:10.1016/j.envpol.2022.120081 PMID:36075340

Zhen, Z., Wang, S., Luo, S., Ren, L., Liang, Y., Yang, R., Li, Y., Zhang, Y., Deng, S., Zou, L., Lin, Z., & Zhang, D. (2019). Significant impacts of both total amount and availability of heavy metals on the functions and assembly of soil microbial communities in different land use patterns. *Frontiers in Microbiology*, *10*, 2293. doi:10.3389/fmicb.2019.02293 PMID:31636621

Zwolak, A., Sarzynska, M., Szpyrka, E., & Stawarczyk, K. (2019). Sources of soil pollution by heavy metals and their accumulation in vegetables: A review. *Water, Air, and Soil Pollution*, *230*(7), 164. doi:10.100711270-019-4221-y

Compilation of References

Abdi, N. (2010). *Availability, transfer and balances of heavy metals in urban agriculture of West Africa.* [PhD Dissertation, University of Kassel, Germany].

Abeysinghe, S., Qiu, G., Goodale, E., Anderson, N., Bishop, K., Evers, C., Goodale, M. W., Hintelmann, H., Liu, S., Mammides, C., Quan, R.-C., Wang, J., Wu, P., Xu, X.-H., Yang, X.-D., & Feng, X. (2017). Mercury flow through an Asian rice-based food web. *Environmental Pollution, 229*, 219–228. doi:10.1016/j.envpol.2017.05.067 PMID:28599206

Abhijith, L., & Varghese, E. (2021). Removal of zinc and copper from contaminated soil by using adsorbents and mulches. *AIP Conference Proceedings, 2396*(030005), 1–9. doi:10.1063/5.0066408

Abiye, A., & Bhattacharya, P. (2019). Arsenic concentration in groundwater: Archetypal study from South Africa. *Groundwater for Sustainable Development, 9*, 100246. doi:10.1016/j.gsd.2019.100246

Aboh, K., Sampson, M., Nyaab, L., Caravanos, J., Ofosu, F., & Kuranchie-Mensah, H. (2013). Assessing levels of lead contamination in soil and predicting pediatric blood lead levels in Tema, Ghana. *Journal of Health & Pollution, 3*(5), 7–12. doi:10.5696/2156-9614-3.5.7

Achterberg, E., Browning, T., Gledhill, M., & Schlosser, C. (2019). Transition metals and heavy metal speciation. In: Cochran, J., Bokuniewicz, H, Yager, P. (eds) Encyclopedia of ocean sciences, 3rd edn. Academic, Oxford. . 11394-6 doi:10.1016/B978-0-12-409548-9.11394-6

Adams, W., Blust, R., Dwyer, R., Mount, D., Nordheim, E., Rodriguez, P., & Spry, D. (2020). Bioavailability assessment of metals in freshwater environments: A historical review. *Environmental Toxicology and Chemistry, 39*(1), 48–59. doi:10.1002/etc.4558 PMID:31880839

Addo-Bediako, A., Nukeri, S., & Kekana, M. (2021). Heavy metal and metalloid contamination in the sediments of the Spekboom River, South Africa. *Applied Water Science, 11*(7), 133. doi:10.100713201-021-01464-8

Adekola, F., & Eletta, O. (2007). A study of heavy metal pollution of ASA River, Ilorin. Nigeria; trace metal monitoring and geochemistry. *Environmental Monitoring and Assessment, 125*(1-3), 157–163. doi:10.100710661-006-9248-z PMID:17058013

Adelekan, B., & Abegunde, K. (2011). Heavy metals contamination of soil and groundwater at automobile mechanic villages in Ibadan, Nigeria. *International Journal of Physical Sciences*, *6*(5), 1045–1058. doi:10.5897/IJPS10.495

Adeyanju, E., & Okeke, C. (2019). Exposure effect to cement dust pollution: A minireview. *SN Applied Sciences*, *1*(12), 1572. doi:10.100742452-019-1583-0

Adeyi, A., & Babalola, B. (2017). Lead and cadmium levels in residential soils of Lagos and Ibadan, Nigeria. *Journal of Health & Pollution*, *7*(13), 42–55. doi:10.5696/2156-9614-7-13.42 PMID:30524813

Adhikari, S., Marcelo-Silva, J., Beukes, J., Van Zyl, P., Coetsee, Y., Boneschans, R., & Siebert, S. J. (2022). Contamination of useful plant leaves with chromium and other potentially toxic elements and associated health risks in a polluted mining-smelting region of South Africa. *Environmental Advances*, *9*, 100301. doi:10.1016/j.envadv.2022.100301

Adnan, M., Xiao, B., Xiao, P., Zhao, P., & Bibi, S. (2022). Heavy metal, waste, COVID-19, and rapid industrialization in this modern era—Fit for sustainable future. *Sustainability (Basel)*, *14*(8), 4746. doi:10.3390u14084746

Adnan, M., Xiao, B., Xiao, P., Zhao, P., Li, R., & Bibi, S. (2022). Research progress on heavy metals pollution in the soil of smelting sites in China. *Toxics*, *10*(5), 231. doi:10.3390/toxics10050231 PMID:35622644

Advanced Refractory Metals. (2023). *Five uses of chromium/ uses of chromium in industry and everyday life*. Refractory Metals. https://www.refractorymetal.org/uses-of-chromium/

Agboola, O., Babatunde, D., Fayomi, O., Sadiku, E., & Popoola, P. (2020). A review on the impact of mining operation: Monitoring, assessment and management. *Results in Engineering*, *8*, 100181. doi:10.1016/j.rineng.2020.100181

Aggarwal, V., Tuli, H., Varol, A., Thakral, F., Yerer, M., Sak, K., Varol, M., Jain, A., Khan, M., & Sethi, G. (2019). Role of reactive oxygen species in cancer progression: Molecular mechanisms and recent advancements. *Biomolecules*, *9*(11), 735. doi:10.3390/biom9110735 PMID:31766246

Agnan, Y., Le Dantec, T., Moore, C., Edwards, G., & Obrist, D. (2016). New constraints on terrestrial surface atmosphere fluxes of gaseous elemental mercury using a global database. *Environmental Science & Technology*, *50*(2), 507–524. doi:10.1021/acs.est.5b04013 PMID:26599393

Agnello, C., Bagard, M., van Hullebusch, D., Esposito, G., & Huguenot, D. (2015). Comparative bioremediation of heavy metals and petroleum hydrocarbons co-contaminated soil by natural attenuation, phytoremediation, bioaugmentation and bioaugmentation-assisted phytoremediation. *The Science of the Total Environment*, *564*, 693–703. doi:10.1016/j.scitotenv.2015.10.061 PMID:26524994

Agoro, M., Adeniji, A., Adefisoye, M., & Okoh, O. (2020). Heavy metals in wastewater and sewage sludge from selected municipal treatment plants in Eastern Cape province, South Africa. *Water (Basel)*, *12*(10), 2746. doi:10.3390/w12102746

Ahivar, B., Das, P., Srivastava, V., & Kumar, M. (2023). Perspectives of heavy metal pollution indices for soil, sediment and water pollution evaluation: An insight. *Total Environment Research Themes*, *6*, 100039. doi:10.1016/j.totert.2023.100039

Ahmed, S., Yesmin, M., Jeba, F., Hoque, S., Jamee, R., & Salam, A. (2020). Risk assessment and evaluation of heavy metals concentrations in blood samples of plastic industry workers in Dhaka, Bangladesh. *Toxicology Reports*, *7*, 1373–1380. doi:10.1016/j.toxrep.2020.10.003 PMID:33102140

Ahmed, T., Noman, M., Ijaz, M., Ali, S., Rizwan, M., Ijaz, U., Hameed, A., Ahmad, U., Wang, Y., Sun, G., & Li, B. (2021). Current trends and future prospective in nanoremediation of heavy metals contaminated soils: A way forward towards sustainable agriculture. *Ecotoxicology and Environmental Safety*, *227*, 112888. doi:10.1016/j.ecoenv.2021.112888 PMID:34649136

Ahoule, G., Lalanne, F., Mendret, J., Brosillon, S., & Maiga, H. (2015). Arsenic in African waters: A review. *Water, Air, and Soil Pollution*, *226*(9), 302. Advance online publication. doi:10.100711270-015-2558-4

Akachise, T., Boakye, S., Borquaye, L., Dodd, M., & Darko, G. (2020). Distribution of heavy metals in soils from abandoned dumpsites in Kumasi, Ghana. *Scientific African*, *10*, e00614. doi:10.1016/j.sciaf.2020.e00614

Akoto, O., Sam, N., Ikenaka, Y., Nakayama, S., Baidoo, E., & Yohannes, B. (2017). Contaminated levels and sources of heavy metals and a metalloid in surface soils in the Kumasi metropolis, Ghana. *Journal of Health & Pollution*, *7*(5), 28–39. doi:10.5696/2156-9614-7.15.28 PMID:30524828

Akpambang, V., Ebuzeme, G., & Oluwatobi, J. (2022). Heavy metal contamination of topsoil around a cement factory-a case study of Obajana cement plc. *Environmental Pollutants and Bioavailability*, *34*(1), 12–20. doi:10.1080/26395940.2021.2024090

Akpan, V., & Olukanni, D. (2020). Hazardous waste management: An African overview. *Recycling*, *5*(3), 15. doi:10.3390/recycling5030015

Akpanyung, E., Akanemesang, U., Akpakpan, E., & Anodoze, N. (2014). Levels of heavy metals in fish obtained from two fishing sites in Akwa Ibom State, Nigeria. *African Journal of Environmental Science and Technology*, *8*(7), 416–421. doi:10.5897/AJEST2014.1730

Akpa, S., & Agbenin, J. (2012). Impact of cow dung manure on the solubility of copper, lead and zinc in urban garden soils from Northern Nigeria. *Communications in Soil Science and Plant Analysis*, *43*(21), 2789–2800. doi:10.1080/00103624.2012.719976

Akporido, S., & Asagba, S. (2013). Quality characteristics of soil close to the Benin River in the vicinity of a lubricating oil producing factory, Koko, Nigeria. *International Journal of Soil Science*, *8*(1), 1–16. doi:10.3923/ijss.2013.1.16

Akporido, S., & Onianwa, P. (2015). Heavy Metals and Total Petroleum Hydrocarbon Concentrations in Surface Water of ESI River, Western Niger Delta. *Research Journal of Environmental Sciences*, *9*(2), 88–100. doi:10.3923/rjes.2015.88.100

Aksoy, A. (2016). Geleneksel devletten modern devlete: Sanayi devrimi ve kamu yönetimi düşüncesinde değişim. *Uluslararası Politik Araştırmalar Dergisi*, *2*(3), 31–37. doi:10.25272/j.2149-8539.2016.2.3.04

Akundi, A., Euresti, D., Luna, S., Ankobiah, W., Lopes, A., & Edinbarough, I. (2022). State of Industry 5.0—Analysis and Identification of Current Research Trends. *Applied System Innovation*, *5*(1), 27. doi:10.3390/asi5010027

Al-Anbari, R., Al Obaidy, H., & Ali, F. (2015). Pollution loads and ecological risk assessment of heavy metals in the urban soil affected by various anthropogenic activities. *International Journal of Advanced Research*, *3*(2), 104–110.

Albanese, S., Sadeghi, M., Lima, A., Cicchella, D., Dinelli, E., Valera, P., Falconi, M., Demetriades, A., & De Vivo, B. (2015). GEMAS: Cobalt, Cr, Cu and Ni distribution in agricultural and grazing land soil of Europe. *Journal of Geochemical Exploration*, *154*, 81–93. doi:10.1016/j.gexplo.2015.01.004

Alemu, A., & Tegegne, A. (2022). Assessment of chromium contamination in the soil and khat leaves (Catha edulis Forsk) and its health risks located in the vicinity of tannery industries; a case study of Bahir Dar city, Ethiopia. *Heliyon*, *8*(12), e11914. doi:10.1016/j.heliyon.2022.e11914 PMID:36506399

Alghobar, M., & Suresha, S. (2017). Evaluation of metal accumulation in soil and tomatoes irrigated with sewage water from Mysore city, Karnataka, India. *Journal of the Saudi Society of Agricultural Sciences*, *16*(1), 49–59. doi:10.1016/j.jssas.2015.02.002

Ali, H., & Khan, E. (2018). Assessment of potentially toxic heavy metals and health risk in water, sediments, and different fish species of River Kabul, Pakistan. *Human and Ecological Risk Assessment*, *24*(8), 2101–2218. doi:10.1080/10807039.2018.1438175

Ali, H., & Khan, E. (2018). What are heavy metals? Long standing controversy over scientific use of the term 'heavy metal'- proposal of a comprehensive definition. *Toxicological and Environmental Chemistry*, *100*(1), 619. doi:10.1080/02772248.2017.1413652

Ali, H., Khan, E., & Ilahi, I. (2019). Environmental chemistry and ecotoxicology of hazardous heavy metals: Environmental persistence, toxicity and bioaccumulation. *Journal of Chemistry*, *6730305*, 1–14. doi:10.1155/2019/6730305

Ali, M., Ahmed, T., Dieyeh, M., & Al-Ghouti, M. (2021). Environmental impacts of municipal biosolids on soil, plant and groundwater qualities. *Sustainability (Basel)*, *13*(15), 8368. doi:10.3390u13158368

Ali, M., Hossain, D., Al-Imran, M., Khan, S., Begum, M., & Osman, M. (2021). *Environmental pollution with heavy metals: a public health concern*. Intech Open., doi:10.5772/intechopen.96805

Ali, S., Awan, Z., Mumtaz, S., Shakir, H., Ahmad, F., Ulhaq, M., Tahir, H. M., Awan, M. S., Sharif, S., Irfan, M., & Khan, M. A. (2020). Cardiac toxicity of heavy metals (cadmium and mercury) and pharmacological intervention by vitamin C in rabbits. *Environmental Science and Pollution Research International*, *27*(23), 29266–29279. doi:10.100711356-020-09011-9 PMID:32436095

Alsafran, M., Saleem, M., Jabri, H., Rizwan, M., & Usman, K. (2022). Principles and applicability of integrated remediation strategies for heavy metal removal/recovery from contaminated environments. *Journal of Plant Growth Regulation*, 1–22. doi:10.100700344-022-10803-1

Ameen, N., Amjad, M., Murtaza, B., Abbas, G., Shahid, M., Imran, M., Naeem, M. A., & Niazi, N. K. (2019). Biogeochemical behavior of nickel under different abiotic stresses: Toxicity and detoxification mechanisms in plants. *Environmental Science and Pollution Research International*, *26*(11), 10496–10514. doi:10.100711356-019-04540-4 PMID:30835069

Amos, H., Sonke, J., Obrist, D., Robins, N., Hagan, N., Horowitz, H., Mason, R., et al (2015). Observational and modeling constraints on global anthropogenic enrichment of mercury. *Environmental Science and Technology, 49*, 4036– 4047. . doi:10.1021/es5058665

Amusan, A., Bada, S., & Salami, A. (2003). Nigeria. *West African Journal of Applied Ecology*, *4*, 107–114.

Ancona, V., Caracciolo, A., Campanale, C., Rascio, I., Grenni, P., Di Lenola, M., Bagnuolo, G., & Uricchio, V. F. (2019). Heavy metal phytoremediation of a poplar clone in a contaminated soil in southern Italy. *Journal of Chemical Technology and Biotechnology*, *95*, 940–949. doi:10.1002/jctb.6145

Anekwe, M., & Isa, Y. (2021). Wastewater and bioventing treatment systems for acid mine drainage–contaminated soil. *Soil & Sediment Contamination*, *30*(4), 1–14. doi:10.1080/15320383.2020.1863909

Angelaki, A., Dionysidis, A., Sihag, P., & Golia, E. (2022). Assessment of contamination management caused by copper and zinc cations leaching and their impact on the hydraulic properties of a sandy and a loamy clay soil. *Land (Basel)*, *11*(2), 290. doi:10.3390/land11020290

Antoniadis, V., Shaheen, M., Levizou, E., Shahid, M., Niazi, K., Vithanage, M., Ok, Y. S., Bolan, N., & Rinklebe, J. (2019). A critical prospective analysis of the potential toxicity of trace element regulation limits in soils worldwide: Are they protective concerning health risk assessment? – a review. *Environment International*, *127*, 819–847. doi:10.1016/j.envint.2019.03.039 PMID:31051325

Anyanwu, B., Ezejiofor, A., Igweze, Z., & Orisakwe, O. (2018). Heavy metal mixture and effects in developing nations: An update. *Toxics*, *6*(4), 65. doi:10.3390/toxics6040065 PMID:30400192

Aparicio, J., Garcia-Velasco, N., Urionabarrenetxea, E., Soto, M., Alvarez, A., & Polti, M. (2019). Evaluation of the effectiveness of a bioremediation process in experimental soils polluted with chromium and lindane. *Ecotoxicology and Environmental Safety*, *181*, 255–263. doi:10.1016/j.ecoenv.2019.06.019 PMID:31200198

Aparicio, J., Raimondo, E., Gil, R., Benimeli, C., & Polti, M. (2018). Actinobacteria consortium as an efficient biotechnological tool for mixed polluted soil reclamation: Experimental factorial design for bioremediation process optimization. *Journal of Hazardous Materials*, *342*, 408–417. doi:10.1016/j.jhazmat.2017.08.041 PMID:28854393

Apeti, D., & Hartwell, S. (2015). Baseline assessment of heavy metal concentrations in surficial sediment from Kachemak Bay, Alaska. *Environmental Monitoring and Assessment*, *187*(1), 1–11. doi:10.100710661-014-4106-x PMID:25394770

Apostoli, P., & Catalani, S. (2011). *Metal ions in toxicology: effects, interactions, interdependencies*. De Gruyter. doi:10.1515/9783110436624-016

Appiah-Brempong, M., Essandoh, H., Asiedu, N., Dadzie, S., & Momade, F. (2022). Artisanal tannery wastewater: Quantity and characteristics. *Heliyon*, *8*(1), e08680. doi:10.1016/j.heliyon.2021.e08680 PMID:35024490

Apriliyanti, M., & Ilham, M. (2022). Challenges of the industrial revolution era 1.0 to 5.0: university digital library in Indonesia. *Library Philosophy and Practice (e-journal),* 6994.

Arao, T., Ishikawa, S., Murakami, M., Abe, K., Maejima, Y., & Makino, T. (2010). Heavy metal contamination of agricultural soil and counter measures in Japan. *Paddy and Water Environment*, *8*(3), 247–257. doi:10.100710333-010-0205-7

Arctic Monitoring and Assessment Program/ United Nations Environment (AMAP/UN Environment) (2019). *Technical Background Report for the Global Mercury Assessment 2018*. Arctic Monitoring and Assessment Program, Oslo, Norway/UN Environment Program, Chemicals and Health Branch, Geneva, Switzerland

Ariya, P., Amyot, M., Dastoor, A., Deeds, D., Feinberg, A., Kos, G., Poulain, A., Ryjkov, A., Semeniuk, K., Subir, M., & Toyota, K. (2015). Mercury physicochemical and biogeochemical transformation in the atmosphere and at atmospheric interfaces: A review and future directions. *Chemical Reviews*, *115*(10), 3760–3802. doi:10.1021/cr500667e PMID:25928690

Arora, V., & Khosla, B. (2021). *Conventional and contemporary techniques for removal of heavy metals from soil*. IntechOpen. doi:10.5772/intechopen.98569

Arshad, H., Mehmood, M., Shah, M., & Abbasi, A. (2020). Evaluation of heavy metals in cosmetic products and their health risk assessment. *Saudi Pharmaceutical Journal*, *28*(7), 779–790. doi:10.1016/j.jsps.2020.05.006 PMID:32647479

Asamoah, B., Asare, A., Okpati, S., & Aidoo, P. (2021). Heavy metal levels and their ecological risks in surface soils at Sunyani magazine in the Bono region of Ghana. *Scientific African*, *13*, e00937. doi:10.1016/j.sciaf.2021.e00937

Ashraf, S., Ali, Q., Zahir, Z., Ashraf, S., & Asghar, H. (2019). Phytoremediation: Environmentally sustainable way for reclamation of heavy metal polluted soils. *Ecotoxicology and Environmental Safety*, *174*, 714–727. doi:10.1016/j.ecoenv.2019.02.068 PMID:30878808

Atobatele, O., & Olutona, G. (2015). Distribution of arsenic (As) in water, sediment and fish from a shallow tropical reservoir (Aiba reservoir, Iwo, Nigeria). *Journal of Applied Science & Environmental Management, 19*(1), 95. doi:10.4314/jasem.v19i1.13

Attard, T., & Attard, E. (2022). *Heavy metals in cosmetics.* IntechOpen., doi:10.5772/intechopen.102406

Audi, G., Bersillon, O., Blachot, J., & Wapstra, A. (2003). The NUBASE evaluation of nuclear and decay properties. *Nuclear Physics. A., 729*(1), 3–128. doi:10.1016/j.nuclphysa.2003.11.001

Avkopashvili, M., Avkopashvili, G., Avkopashvili, I., Asanidze, L., Matchavariani, L., Gongadze, A., & Gakhokidze, R. (2022). Mining-related metal pollution and ecological risk factors in South Eastern Georgia. *Sustainability (Basel), 14*(9), 5621. doi:10.3390u14095621

Azeez, M., Adesanwo, O., & Adepetu, J. (2021). Effects of copper fungicides spray on nutrient content in soils of cocoa growing areas of Southwestern Nigeria. *Tanzania Journal of Science, 47*(5), 1546–1559. doi:10.4314/tjs.v47i5.5

Azzi, V., Kazpard, V., Lartiges, B., Kobeissi, A., Kanso, A., & El Samrani, A. (2017). Trace metals in phosphate fertilizers used in eastern Mediterranean countries. *Clean (Weinheim), 45*(1), 1–10. doi:10.1002/clen.201500988

Baba, H., Tsuneyama, K., Yazaki, M., Nagata, K., Minamisaka, T., Tsuda, T., Nomoto, K., Hayashi, S., Miwa, S., Nakajima, T., Nakanishi, Y., Aoshima, K., & Imura, J. (2013). The liver in itai-itai disease (chronic cadmium poisoning): Pathological features and metallothionein expression. *Modern Pathology, 26*(9), 1228–1234. doi:10.1038/modpathol.2013.62 PMID:23558578

Babula, P., Adam, V., Opatrilova, R., Zehnalek, J., Havel, L., & Kizek, R. (2008). Uncommon heavy metals, metalloids and their plant toxicity: A review. *Environmental Chemistry Letters, 6*(4), 189–213. doi:10.100710311-008-0159-9

Bacon, R., & Davidson, M. (2008). Is there a future for sequential chemical extraction? *Analyst (London), 133*(1), 25–46. doi:10.1039/B711896A PMID:18087610

Bailey, T., Robinson, N., Farrell, M., MacDonald, B., Weaver, T., Antille, D., Chin, A., & Brackin, R. (2022). Storage of soil samples leads to over-representation of the contribution of nitrate to plant available nitrogen. *Soil Research (Collingwood, Vic.), 60*(1), 22–32. doi:10.1071/SR21013

Bakirdere, S. (2013). *Speciation studies in soil, sediment and environmental samples.* CRC Press., doi:10.1201/b15501

Bakshi, A. (2016). *Analysis of anthropogenic disturbances and impact of pollution on fish fauna of River Churni with special reference to chromium pollution.* [PhD Thesis, University of Kalyani, Shodhganga]. http://hdl. handle.net/10603/241694

Bakshi, A., & Panigrahi, A. (2022). Chromium contamination in soil and its bioremediation: An Overview. In J. A. Malik (Ed.), *Advances in bioremediation and phytoremediation for sustainable soil management.* Springer. doi:10.1007/978-3-030-89984-4_15

Balali-Mood, M., Naseri, K., Tahergorabi, Z., Khazdair, M., & Sadeghi, M. (2021). Toxic mechanisms of five heavy metals: Mercury, lead, chromium, cadmium, and arsenic. *Frontiers in Pharmacology*, *12*, 643972. doi:10.3389/fphar.2021.643972 PMID:33927623

Baldantoni, D., Morra, L., Zaccardelli, M., & Alfani, A. (2016). Cadmium accumulation in leaves of leafy vegetables. *Ecotoxicology and Environmental Safety*, *123*, 89–94. doi:10.1016/j.ecoenv.2015.05.019 PMID:26004982

Bali, S., & Sidhu, S. (2021). Heavy metal contamination indices and ecological risk assessment index to assess metal pollution status in different soils. In V. Kumar, A. Sharma, & A. Cerda (Eds.), *Heavy Metals in the Environment*. Elsevier. doi:10.1016/B978-0-12-821656-9.00005-5

Banfalvi, G. (2011). *Cellular effects of heavy metals*. Springer. doi:10.1007/978-94-007-0428-2

Banza, L., Casas, L., Haufroid, V., De Putter, T., & Saenen, N. (2018). Sustainability of artisanal mining of cobalt in DR Congo. *Nature Sustainability*, *1*(9), 495–504. doi:10.103841893-018-0139-4 PMID:30288453

Barbour, K., Sabra, H., Bianu, G., Jaber, L., & Shaib, H. (2009). Oppositional dynamics of organic versus inorganic contaminants in oysters following an oil spill. *Journal of Coastal Research*, *25*(4), 864–869. doi:10.2112/08-1059.1

Bartrem, C., Tirima, S., Lindern, I., Braun, M., Worrell, M., & Anka, S. (2014). Unknown risk: Co-exposure to lead and other heavy metals among children living in small-scale mining communities in Zamfara State, Nigeria. *International Journal of Environmental Health Research*, *24*(4), 304–319. doi:10.1080/09603123.2013.835028 PMID:24044870

Bar-Zeev, Y., Greenberg, D., Ling, G., & Lifshitz, M. (2002). Acrodynia: A case report of two siblings. *Archives of Disease in Childhood*, *86*(6), 453. doi:10.1136/adc.86.6.453 PMID:12023189

Basu, N., Goodrich, M., & Head, J. (2014). Ecogenetics of mercury: From genetic polymorphisms and epigenetics to risk assessment and decision-making. *Environmental Toxicology and Chemistry*, *33*(6), 1248–1258. doi:10.1002/etc.2375 PMID:24038486

Beckers, F., & Rinklebe, J. (2017). Cycling of mercury in the environment: sources, fate and human health implications: a review. *Critical Reviews in Environmental Science and Technology*, *47*(9), 693–794. doi:10.1080/10643389.2017.1326277

Bekabil, U. (2020). Industrialization and environmental pollution in Africa: An empirical review. *Journal of Resource Development and Management*, *69*, 1–4. doi:10.7176/JRDM/69-03

Belles, M., Albina, M., Sanchez, D., Corbella, J., & Domingo, L. (2002). Interactions in developmental toxicology: Effects of concurrent exposure to lead, organic mercury, and arsenic in pregnant mice. *Archives of Environmental Contamination and Toxicology*, *42*(1), 93–98. doi:10.1007002440010296 PMID:11706373

Benvenga, S., Marini, H., Micali, A., Freni, J., Pallio, G., Irrera, N., Squadrito, F., Altavilla, D., Antonelli, A., Ferrari, S. M., Fallahi, P., Puzzolo, D., & Minutoli, L. (2020). Protective effects of myo-inositol and selenium on cadmium-induced thyroid toxicity in mice. *Nutrients*, *12*(5), 1222. doi:10.3390/nu12051222 PMID:32357526

Bezerra, P., Takiyama, L., & Bezerra, C. (2009). Complexation of metal ions by dissolved organic matter: Modeling and application to real systems. *Acta Amazonica*, *39*, 639–648. doi:10.1590/S0044-59672009000300019

Bhattacharya, P., Jacks, G., Frisbie, S., Smith, E., Naidu, R., & Sarkar, B. (2002). Arsenic in the environment: a global perspective. In B. Sarkar (Ed.), *Heavy Metals in the Environment* (pp. 147–215). Marcel Dekker. Inc. doi:10.1201/9780203909300.ch6

Bidar, G., Pelfrene, A., Schwartz, C., Waterlot, C., Sahmer, K., Marot, F., & Douay, F. (2020). Urban kitchen gardens: Effect of the soil contamination and parameters on the trace element accumulation in vegetables-a review. *The Science of the Total Environment*, *738*, 139569. doi:10.1016/j.scitotenv.2020.139569 PMID:32516675

Bigalke, M., Ulrich, A., Rehmus, A., & Keller, A. (2017). Accumulation of cadmium and uranium in arable soils in Switzerland. *Environmental Pollution*, *221*, 85–93. doi:10.1016/j.envpol.2016.11.035 PMID:27908488

Bigham, G., Murray, K., Slowey, Y., & Henry, E. (2017). Biogeochemical controls on methylmercury in soils and sediments: Implications for site management. *Integrated Environmental Assessment and Management*, *13*(2), 249–263. doi:10.1002/ieam.1822 PMID:27427265

Bildirici, M. (2022). The impacts of governance on environmental pollution in some countries of Middle East and sub-Saharan Africa: The evidence from panel quantile regression and causality. *Environmental Science and Pollution Research International*, *29*(12), 17382–17393. doi:10.100711356-021-15716-2 PMID:34665419

Bilgin, M., & Tulun, S. (2016). Removal of heavy metals (Cu, Cd and Zn) from contaminated soils using EDTA and $FeCl_3$. *Global NEST Journal*, *18*(10), 98–107.

Birke, M., Reimann, C., Rauch, U., Ladenberger, A., Demetriades, A., Jähne-Klingberg, F., Oorts, K., Gosar, M., Dinelli, E., & Halamić, J. (2017). GEMAS: Cadmium distribution and its sources in agricultural and grazing land soil of Europe - original data versus clr-transformed data. *Journal of Geochemical Exploration*, *173*, 13–30. doi:10.1016/j.gexplo.2016.11.007

Blacksmith Institute. (2015). *Kabwe lead mines, 2015*. Blacksmith Institute. http://www.blacksmithinstitute.org/projects/display/3

Blois, L., & Aime, L. (2021). Environmental impacts from atmospheric emission of heavy metals: A case study of a cement plant. *Measurement: Sensors*, *18*, 100313. doi:10.1016/j.measen.2021.100313

Bodaghi-Namileh, V., Sepand, R., Omidi, A., Aghsami, M., Seyednejad, A., Kasirzadeh, S., et al. (2018). Acetyl-l-carnitine attenuates arsenic-induced liver injury by abrogation of mitochondrial dysfunction, inflammation, and apoptosis in rats. *Environmental Toxicology and Pharmacology, 58*, 11–20. https://doi.org/. 2017.12.005 doi:10.1016/j.etap

Bodeau-Livinec, F., Glorennec, P., Cot, M., Dumas, P., Durand, S., Massougbodji, A., Ayotte, P., & Le Bot, B. (2016). Elevated blood lead levels in infants and mothers in Benin and potential sources of exposure. *International Journal of Environmental Research and Public Health, 13*(3), 316. doi:10.3390/ijerph13030316 PMID:26978384

Bokar, H., Traore, Z., Mariko, A., Diallo, T., Traore, A., Sy, A., Soumare, O., Dolo, A., Bamba, F., Sacko, M., & Touré, O. (2020). Geogenic influence and impact of mining activities on water soil and plants in surrounding areas of Morila Mine, Mali. *Journal of Geochemical Exploration, 209*, 106429. Advance online publication. doi:10.1016/j.gexplo.2019.106429

Bolan, S., Adriano, D., & Naidu, R. (2003). Role of phosphorus in mobilization and bioavailability of heavy metals in the soil-plant system. *Reviews of Environmental Contamination and Toxicology, 177*, 1–44. doi:10.1007/0-387-21725-8_1 PMID:12666817

Boldyrev, M. (2018). Lead: Properties, history and applications. *WikiJournal of Science, 1*(2), 7. doi:10.15347/wjs/2018.007

Bonzongo, J., Donkor, A., Nartey, V., & Lacerda, L. (2004). Mercury pollution in Ghana: a case study of environmental impacts of artisanal gold mining in sub-Saharan Africa. In L. Drude, R. Santelli, E. Duursma, & J. Abrao (Eds.), *Environmental Geochemistry in the tropical and subtropical environments*. Springer. doi:10.1007/978-3-662-07060-4_12

Borah, P., Gujre, N., Rene, E., Rangan, L., & Kumar, R. (2020). Assessment of mobility and environmental risks associated with copper, manganese and zinc in soils of a dumping site around a Ramsar site. *Chemosphere, 254*, 126852. doi:10.1016/j.chemosphere.2020.126852 PMID:32957277

Borah, S., Bora, T., Baruah, S., & Dutta, J. (2015). Heavy metal ion sensing in water using surface plasmon resonance of metallic nanostructures. *Groundwater for Sustainable Development, 1*(1), 1–11. doi:10.1016/j.gsd.2015.12.004

Bouida, L., Rafatullah, M., Kerrouche, A., Qutob, M., Alosaimi, A., Alorfi, H., & Hussein, M. (2022). Review on cadmium and lead contamination: Sources, fate, mechanism, health effects and remediation methods. *Water (Basel), 14*(21), 3432. doi:10.3390/w14213432

Boyd, A., Seger, D., Vannucci, S., Langley, M., Abraham, L., & King, E. Jr. (2000). Mercury exposure and cutaneous disease. *Journal of the American Academy of Dermatology, 43*(1), 81–90. doi:10.1067/mjd.2000.106360 PMID:10863229

Branca, J., Morucci, G., & Pacini, A. (2018). Cadmium-induced neurotoxicity: Still much ado. *Neural Regeneration Research, 13*(11), 1879–1882. doi:10.4103/1673-5374.239434 PMID:30233056

Bret, E., Otieno, V., Nganga, C., Fort, J., & Taylor, M. (2019). Assessment of the presence of soil lead contamination near a former lead smelter in Mombasa, Kenya. *Journal of Health & Pollution*, *9*(21), 190307. doi:10.5696/2156-9614-9.21.190307 PMID:30931167

Brevik, E., Slaughter, L., Singh, B., Steffan, J., Collier, D., Barnhart, P., & Pereira, P. (2020). Soil and human health: Current status and future needs. *Air, Soil and Water Research*, *13*, 1–12. doi:10.1177/1178622120934441

Brezova, V., Valko, M., Breza, M., Morris, H., Telser, J., Dvoranova, D., Kaiserova, K., Varecka, L., Mazur, M., & Leibfritz, D. (2003). Role of radicals and singlet oxygen in photoactivated DNA cleavage by the anticancer drug camptothecin: An electron paramagnetic resonance study. *The Journal of Physical Chemistry B*, *107*(10), 2415–2425. doi:10.1021/jp027743m

Briffa, J., Sinagra, E., & Blundell, R. (2020). Heavy metal pollution in the environment and their toxicological effects on humans. *Heliyon*, *6*(9), e04691. doi:10.1016/j.heliyon.2020.e04691 PMID:32964150

Brown, A., Thomas, G., & Lindsley, C. W. (2017). Targeting phospholipase D in cancer, infection and neurodegenerative disorders. *Nature Reviews. Drug Discovery*, *16*(5), 351–367. doi:10.1038/nrd.2016.252 PMID:28209987

Brusseau, M., Gerba, C., & Pepper, I. (2019). *Environmental and pollution Science*. Academic Press Publisher.

Bu, Q., Li, Q., Zhang, H., Cao, H., Gong, W., Zhang, X., Ling, K., & Cao, Y. (2020). Concentrations, spatial distributions and sources of heavy metals in surface soils of the coal mining city Wuhai, China. *Journal of Chemistry*, *4705954*, 1–10. Advance online publication. doi:10.1155/2020/4705954

Burkitt, M. (2001). A critical overview of the chemistry of copper-dependent low density lipoprotein oxidation: Roles of lipid hydroperoxides, α-tocopherol, thiols, and ceruloplasmin. *Archives of Biochemistry and Biophysics*, *394*(1), 117–135. doi:10.1006/abbi.2001.2509 PMID:11566034

Buta, M., Hubeny, J., Zielinski, W., Harnisz, M., & Korzeniewska, E. (2021). Sewage sludge in agriculture - the effects of selected chemical pollutants and emerging genetic resistance determinants on the quality of soil and crops – a review. *Ecotoxicology and Environmental Safety*, *214*, 112070. doi:10.1016/j.ecoenv.2021.112070 PMID:33652361

Butler, T., Cook, J., Davidson, C., Harrington, C., & Miles, L. (2009). Atomic spectrometry update: Environmental analysis. *Journal of Analytical Atomic Spectrometry*, *24*(2), 131–177. doi:10.1039/b821579k

Buxton, S., Garman, E., Heim, K., Darden, T., Schlekat, C., & Taylor, M. (2019). Concise review of nickel human health toxicology and ecotoxicology. *Inorganics*, *7*(7), 89. doi:10.3390/inorganics7070089

Byers, H., McHenry, L., & Grundl, T. (2019). XRF techniques to quantify heavy metals in vegetables at low detection limits. *Food Chemistry: X*, *1*, 100001. doi:10.1016/j.fochx.2018.100001

Cabral, M., Dieme, D., Verdin, A., Garcon, G., Fall, M., Bouhsina, S., Dewaele, D., Cazier, F., Tall-Dia, A., Diouf, A., & Shirali, P. (2012). Low-level environmental exposure to lead and renal adverse effects: A cross-sectional study in the population of children bordering the Mbeubeuss landfill near Dakar, Senegal. *Human and Experimental Toxicology*, *31*(12), 1280–1291. doi:10.1177/0960327112446815 PMID:22837546

Cai, C., Xiong, B., Zhang, Y., Li, X., & Nunes, L. (2015). Critical comparison of soil pollution indices for assessing contamination with toxic metals. *Water, Air, and Soil Pollution*, *226*(10), 352. doi:10.100711270-015-2620-2

Candeias, C., Da Silva, F., Salgueiro, A., Pereira, H., Peis, A., & Patinha, C., &. (2011). Assessment of soil contamination by potentially toxic elements in the Aljustrel mining area in order to implement soil reclamation strategies. *Land Degradation & Development*, *22*(6), 565–585. doi:10.1002/ldr.1035

Carrillo, H., Tirado, J., Romero, J., Luis, A., Munoz, M., Pinon, J., Bueno, J., Montes, C., Naranjo, E., & Ramirez, A. (2021). Encapsulation of toxic heavy metals from waste CRT using calcined kaolin base-polymer. *Materials Chemistry and Physics*, *257*, 123745. doi:10.1016/j.matchemphys.2020.123745

Carr, R., Zhang, S., Moles, N., & Harder, M. (2008). Identification and mapping of heavy metal pollution in soils from a sports ground in Galway City, Ireland, using a portable XRF analyzer and GIS. *Environmental Geochemistry and Health*, *30*(1), 45–52. doi:10.100710653-007-9106-0 PMID:17610027

Cempel, M., & Nikel, G. (2005). Nickel: A review of its sources and environmental toxicology. *Polish Journal of Environmental Studies*, *15*(3), 375–382.

Cesarani, A., Minoia, C., Pigatto, D., & Guzzi, G. (2010). Mercury, dental amalgam, and hearing loss. *International Journal of Audiology*, *49*(1), 2. doi:10.3109/14992020902962439 PMID:20001448

Cetin, H. (2002). Liberalizmin tarihsel kökenleri. *Cumhuriyet Üniversitesi İktisadi ve İdari Bilimler Dergisi*, *3*(1), 79–96. doi:10.25272/j.2149-8539.2018.4.1.02

Cevik, D. (2017). *Sanayi Devrimlerinin Süreci ve 4.* Sanayi Devrimi. https://www.alomaliye.com/2017/05/29/sanayi-devrimlerinin-sureci-4-sanayi-devrimi/

Chang, T., & Yen, J. (2006). On-site mercury-contaminated soils remediation by using thermal desorption technology. *Journal of Hazardous Materials*, *128*(2-3), 208–217. doi:10.1016/j.jhazmat.2005.07.053 PMID:16144741

Charles, J. (1980). The coming copper and copper-base alloys and iron: A metallurgical sequence. In T. Wertime & J. Muhley (Eds.), *The coming of the age of iron*. Yale University Press.

Chemistry Explained. (2023). Mercury. *Chemistry Explained.* http://www.chemistryexplained.com/elements/L-P/Mercury.html

Cheng, M., Wu, L., Huang, Y., Luo, Y., & Christie, P. (2014). Total concentrations of heavy metals and occurrence of antibiotics in sewage sludges from cities throughout China. *Journal of Soils and Sediments*, *14*(6), 1123–1135. doi:10.100711368-014-0850-3

Cheng, Y., Zhou, K., Wang, J., Cui, S., Yan, J., Maeyer, P., & Voorde, T. (2022). Regional metal pollution risk assessment based on a big data framework: A case study of the eastern Tianshan mining area, China. *Ecological Indicators*, *145*, 109585. doi:10.1016/j.ecolind.2022.109585

Chen, M., & Ma, L. (2001). Comparison of three aqua regia digestion methods for twenty Florida soils. *Soil Science Society of America Journal*, *65*(2), 491–499. doi:10.2136ssaj2001.652491x

Chen, S., Huang, Y., Liu, L., Cai, P., Liang, W., & Li, M. (2010). Poultry manure compost alleviates the phytotoxicity of soil cadmium: Influence on growth of pak choi (*Brassica chinensis* L.). *Pedosphere*, *20*(1), 63–70. doi:10.1016/S1002-0160(09)60283-6

Chen, S., Kimirei, A., Yu, C., Shen, Q., & Gao, Q. (2022). Assessment of urban river water pollution with urbanization in East Africa. *Environmental Science and Pollution Research International*, *29*(27), 40812–40825. doi:10.100711356-021-18082-1 PMID:35083687

Chen, X., Kumari, D., Cao, C. J., Plaza, G., & Achal, V. (2020). A review on remediation technologies for nickel-contaminated soil. *Human and Ecological Risk Assessment*, *26*(3), 571–585. doi:10.1080/10807039.2018.1539639

Chindah, A., Braide, A., & Sibeudu, O. (2004). Distribution of hydrocarbons and heavy metals in sediment and a crustacean (shrimps- *Penaeus notialis*) from the Bonny/New Calabar River Estuary, Niger Delta. *African Journal of Environmental Assessment and Management*, *9*, 1–17.

Chinedu, E., & Chukwuemeka, C. (2018). Oil spillage and heavy metals toxicity risk in the Niger Delta, Nigeria. *Journal of Health & Pollution*, *19*, 180905. doi:10.5696/2156-9614-8.19.180905 PMID:30524864

Chinhanga, J. (2010). Impact of industrial effluent from an iron and steel company on the physicochemical quality of Kwekwe river water in Zimbabwe. *International Journal of Engineering Science and Technology*, *2*(7), 129–140. doi:10.4314/ijest.v2i7.63754

Choi, R., Kim, H., Yang, J., & Kim, J. (2020). Ecological impact of fast industrialization inferred from a sediment core in Seocheon, West Coast of Korean Peninsula. *Journal of Ecology and Environment*, *44*(24), 1–10. doi:10.118641610-019-0144-1

Chung, J., Yu, S., & Hong, Y. (2014). Environmental source of arsenic exposure. *Journal of Preventive Medicine and Public Health*, *47*(5), 253–257. doi:10.3961/jpmph.14.036 PMID:25284196

Clough, R., Harrington, C., Hill, S., Madrid, Y., & Tyson, J. (2018). Atomic spectrometry update: Review of advances in elemental speciation. *Journal of Analytical Atomic Spectrometry*, *33*(7), 1103–1149. doi:10.1039/C8JA90025F

Coetzee, J., Bansal, N., & Chirwa, E. (2020). Chromium in environment, its toxic effect from chromite-mining and ferrochrome industries, and its possible bioremediation. *Exposure and Health*, *12*(1), 51–62. doi:10.100712403-018-0284-z PMID:33748533

Crans, C., Smee, J., Gaidamauskas, E., & Yang, L. (2004). The chemistry and biochemistry of vanadium and the biological activities exerted by vanadium compounds. *Chemical Reviews*, *104*(2), 849–902. doi:10.1021/cr020607t PMID:14871144

Cristaldi, A., Conti, G., Jho, E., Zuccarello, P., Grasso, A., Copat, C., & Ferrante, M. (2017). Phytoremediation of contaminated soils by heavy metals and PAHs. A brief review. *Environmental Technology & Innovation*, *8*, 309–326. doi:10.1016/j.eti.2017.08.002

Crompton, T. (2015). Determination of metals in natural waters, sediments, and soils (1st ed.). Elsevier Science. 1016/B978-0-12-802654-0.00001-5.

Csavina, J., Taylor, M., Felix, O., Rine, K., Saez, A., & Betterton, E. (2014). Size-resolved dust and aerosol contaminants associated with copper and lead smelting emissions: Implication for emissions management and human health. *The Science of the Total Environment*, *493*, 750–756. doi:10.1016/j.scitotenv.2014.06.031 PMID:24995641

Dahri, N., Abdelfattah, A., Manel, E., & Habib, A. (2018). Assessment of streambed sediment contamination by heavy metals: The case of the Gabes Catchment, South-eastern Tunisia. *Journal of African Earth Sciences*, *140*, 29–41. doi:10.1016/j.jafrearsci.2017.12.033

Dankoub, Z., Ayoubi, S., Khademi, H., & Lu, S. (2012). Spatial distribution of magnetic properties and selected heavy metals in calcareous soils as affected by land use in the Isfahan region, Central Iran. *Pedosphere*, *22*(1), 33–47. doi:10.1016/S1002-0160(11)60189-6

Darma, M., Kankara, I., & Abdullahi, S. (2016). Effect of artisanal gold mining at Maiwayo environ northern Nigeria: implication for environmental risk. *Sustainable Economic Development Conference*, Abuja, Nigeria.

Das, K., & Chakraborty, R. (1997). Electrothermal atomic absorption spectrometry in the study of metal ion speciation. *Fresenius' Journal of Analytical Chemistry*, *357*(1), 1–17. doi:10.1007002160050102

Das, P., Das, B., & Dash, P. (2021). Chromite mining pollution, environmental impact, toxicity and phytoremediation: A review. *Environmental Chemistry Letters*, *19*(2), 1369–1381. doi:10.100710311-020-01102-w

Datsenko, V., Borzenko, O., & Kaliuzhna, L. (2021). Vertical migration of copper and zinc ions in soils polluted by electroplating sludge. *Clean (Weinheim)*, *49*(12), 2100087. doi:10.1002/clen.202100087

Davidson, C. (2013). Methods for the Determination of Heavy Metals and Metalloids in Soils. In B. Alloway (Ed.), *Heavy Metals in Soils. Environmental Pollution* (Vol. 22). Springer. doi:10.1007/978-94-007-4470-7_4

David, W., & Somsubhra, C. (2020). Portable X-ray fluorescence spectrometry analysis of soils. *Soil Science Society of America Journal*, *84*(5), 1384–1392. doi:10.1002aj2.20151

Davis, J. (2009). Nickel, cobalt, and their alloys. Cobalt market review. ASM International.

Davis, B., Price, H., O'Connor, R., Fernando, R., Rowland, A., & Morgan, D. (2001). Mercury vapor and female reproductive toxicity. *Toxicological Sciences*, *59*(2), 291–296. doi:10.1093/toxsci/59.2.291 PMID:11158722

De Souza, J., Abrahao, L., de Mello, W., da Silva, J., da Costa, M., & de Oliveira, T. S. (2015). Geochemistry and spatial variability of metal (loid) concentrations in soils of the state of Minas Gerais. Brazil. *The Science of the Total Environment*, *505*, 338–349. doi:10.1016/j.scitotenv.2014.09.098 PMID:25461035

De Villiers, S., Thiart, C., & Basson, N. (2010). Identification of sources of environmental lead in South Africa from sur - face soil geochemical maps. *Environmental Geochemistry and Health*, *32*(5), 451–459. doi:10.100710653-010-9288-8 PMID:20848346

Debrah, J., Teye, G., & Dinis, M. (2022). Barriers and challenges to waste management hindering the circular economy in Sub-Saharan Africa. *Urban Science (Basel, Switzerland)*, *6*(3), 57. doi:10.3390/urbansci6030057

Defarge, N., Vendomois, J., & Seralini, G. (2017). Toxicity of formulants and heavy metals in glyphosphate-based herbicides and other pesticides. *Toxicology Reports*, *5*, 156–163. doi:10.1016/j.toxrep.2017.12.025 PMID:29321978

Deng, W., Li, X., An, Z., & Yang, L. (2016). The occurrence and sources of heavy metal contamination in peri-urban and smelting contaminated sites in Baoji, China. *Environmental Monitoring and Assessment*, *188*(4), 251. doi:10.100710661-016-5246-y PMID:27021694

Department of Environmental Affairs (DEA) (2013). National Environmental Management, Waste Act, 2008. National norms and standards for the remediation of contaminated land and soil quality in the Republic of South Africa. *Government Gazette, No. 36447.*

Dessie, B., Tessema, B., Asegide, E., Tibebe, D., Alamirew, T., Walsh, C., & Zeleke, G. (2022). Physicochemical characterization and heavy metal analysis from industrial discharges in Upper Awash River basin, Ethiopia. *Toxicology Reports*, *9*, 1297–1307. doi:10.1016/j.toxrep.2022.06.002 PMID:36518430

Devanesan, E., Suresh-Gandhi, M., Selvapandiyan, M., Senthilkumar, G., & Ravisankar, R. (2017). Heavy metal and potential ecological risk assessment in sediments collected from Poombuhar to Karaikal Coast of Tamil Nadu using energy dispersive X-ray fluorescence (EDXRF) technique. *Beni-Suef University Journal of Basic and Applied Sciences*, *6*(3), 285–292. doi:10.1016/j.bjbas.2017.04.011

Dhal, B., Thatoi, N., Das, N., & Pandey, D. (2013). Chemical and microbial remediation of hexavalent chromium from contaminated soil and mining/metallurgical solid waste: A review. *Journal of Hazardous Materials*, *250*, 272–291. doi:10.1016/j.jhazmat.2013.01.048 PMID:23467183

Dhaliwal, S., Singh, J., Taneja, K., & Mandal, A. (2020). Remediation techniques for removal of heavy metals from the soil contaminated through different sources: A review. *Environmental Science and Pollution Research International*, *27*(2), 1319–1333. doi:10.100711356-019-06967-1 PMID:31808078

Dias-Ferreira, C., Kirkelund, G., & Ottosen, L. (2015). Ammonium citrate as enhancement for electrodialytic soil remediation and investigation of soil solution during the process. *Chemosphere*, *119*, 889–895. doi:10.1016/j.chemosphere.2014.08.064 PMID:25240953

Dietler, D., Babu, M., Cisse, G., Halage, A., Malambala, E., & Fuhrimann, S. (2019). Daily variation of heavy metal contamination and its potential sources along the major urban wastewater channel in Kampala, Uganda. *Environmental Monitoring and Assessment*, *191*(52), 1–13. doi:10.100710661-018-7175-4 PMID:30617634

Dignam, T., Kaufmann, R., LeStourgeon, L., & Brown, M. (2019). Control of Lead Sources in the United States, 1970-2017: Public Health Progress and Current Challenges to Eliminating Lead Exposure. *Journal of Public Health Management and Practice*, *25*(1), S13–S22. doi:10.1097/PHH.0000000000000889 PMID:30507765

Ding, W., Stewart, D., Humphreys, P., Rout, S., & Burke, I. (2016). Role of an organic carbon-rich soil and Fe (III) reduction in reducing the toxicity and environmental mobility of chromium (VI) at a COPR disposal site. *The Science of the Total Environment*, *541*, 1191–1199. doi:10.1016/j.scitotenv.2015.09.150 PMID:26476060

Ding, Y., Xia, G., Ji, H., & Xiong, X. (2019). Accurate quantitative determination of heavy metals in oily soil by laser induced breakdown spectroscopy (LIBS) combined with interval partial least squares (IPLS). *Analytical Methods*, *11*(29), 3657–3664. doi:10.1039/C9AY01030K

Diop, C., Dewaele, D., Cazier, F., Diouf, A., & Ouddane, B. (2015). Assessment of trace metals contamination level, bioavailability and toxicity in sediments from Dakar coast and Saint Louis estuary in Senegal, West Africa. *Chemosphere*, *138*, 980–987. doi:10.1016/j.chemosphere.2014.12.041 PMID:25592460

Djoko, K., Ong, C., Walker, M., & McEwan, A. (2015). The role of copper and zinc toxicity in innate immune defense against bacterial pathogens. *The Journal of Biological Chemistry*, *290*(31), 18954–18961. doi:10.1074/jbc.R115.647099 PMID:26055706

Dkhala, B., Mezned, N., Gomez, C., & Abdeljaouad, S. (2020). Hyperspectral field spectroscopy and SENTINEL-2 multispectral data for minerals with high pollution potential content estimation and mapping. *The Science of the Total Environment*, *740*, 140160. doi:10.1016/j.scitotenv.2020.140160 PMID:32927579

Doka, M., & Mohammed, M. (2020). Effects of sludge on some soil properties and heavy metals uptake by some vegetables grown on tannery sludge amended soils in Kano metropolis, Nigeria. *Dutse Journal of Pure and Applied Sciences*, *6*(1), 173–181.

Dos Santos, G., Valladares, G., Abreu, C., Camargo, O., & Grego, C. (2013). Assessment of copper and zinc in soils of a vineyard region in the state of Sao Paulo, Brazil. *Applied and Environmental Soil Science*, *790795*, 1–10. doi:10.1155/2013/790795

Dube, M., MacLatchy, D., Kieffer, J., Glozier, N., Culp, J., & Cash, K. (2005). Effects of metal mining effluent on Atlantic salmon (*Salmo salar*) and slimy sculpin (*Cottus cognatus*): Using artificial streams to assess existing effects and predict future consequences. *The Science of the Total Environment*, *343*(1-3), 135–154. doi:10.1016/j.scitotenv.2004.09.037 PMID:15862841

Dubiella, A., Wasik, A., Przyjazny, A., & Namiesnik, J. (2007). Preparation of soil and sediment samples for determination of organometallic compounds. *Polish Journal of Environmental Studies*, *16*, 159–176.

Du, D., Ren, X., Yan, S., Shi, X., Liu, Y., & He, G. (2017). An integrated method for the quantitative evaluation of mineral resources of cobalt-rich crusts on seamounts. *Ore Geology Reviews*, *84*, 174–184. doi:10.1016/j.oregeorev.2017.01.011

Duressa, T., & Leta, S. (2015). Determination of levels of As, Cd, Cr, Hg and Pb in soils and some vegetables taken from Rive Mojo water irrigated farmland at Koka village, Oromia state, East Ethiopia. *International Journal of Sciences: Basic Applied Sciences*, *21*(2), 352–372.

Dusengemungu, L., Mubemba, B., & Gwanama, C. (2022). Evaluation of heavy metal contamination in copper mine tailing soils of Kitwe and Mufulira, Zambia for reclamation prospects. *Scientific Reports*, *12*(1), 11283. doi:10.103841598-022-15458-2 PMID:35787645

Dutta, S., Joshi, K., Sengupta, P., & Bhattacharya, K. (2013). Unilateral and bilateral cryptorchidism and its effect on the testicular morphology, histology, accessory sex organs and sperm count in laboratory mice. *Journal of Human Reproductive Sciences*, *6*(2), 106–110. doi:10.4103/0974-1208.117172 PMID:24082651

Dzvinamurungu, T., Rose, D., Vijoen, K., & Bafubiandi, A. (2020). A process mineralogical evaluation of chromite at the Nkomati nickel mine, Uitkomst complex, South Africa. *Minerals (Basel)*, *10*(8), 709. doi:10.3390/min10080709

Eaton, J., & Qian, M. (2002). Molecular bases of cellular iron toxicity. *Free Radical Biology & Medicine*, *32*(9), 833–840. doi:10.1016/S0891-5849(02)00772-4 PMID:11978485

Ebenebe, P., Shale, K., Sedibe, M., Tikilili, P., & Achilonu, M. (2017). South African mine effluents: Heavy metal pollution and impact on the ecosystem. *International Journal of Chemical Science*, *15*(4), 1–13.

Ebenebe, P., Shale, K., Sedibe, M., Tikilili, P., & Achilonu, M. (2017). South African Mine Effluents: Heavy Metal Pollution and Impact on the Ecosystem. *International Journal of Chemical Science*, *15*(4), 198.

Ebo, A., De Vries, N., & Nyarko, K. (2018). Assessment of heavy metal pollution in the main Pra River and its tributaries in the Pra basin of Ghana. *Environmental Nanotechnology, Monitoring & Management*, *10*, 264–271. doi:10.1016/j.enmm.2018.06.003

Ebong, G., Dan, E., Inam, E., & Offiong, N. (2019). Total concentration, speciation, source identification and associated health implications of trace metals in Lemma dumpsite soil, Calabar, Nigeria. *Journal of King Saud University. Science*, *31*(4), 886–897. doi:10.1016/j.jksus.2018.01.005

Ediene, V., & Umoetok, S. (2017). Concentration of heavy metals in soils at the municipal dumpsite in Calabar metropolis. *Asian Journal of Environment and Ecology*, *3*(2), 1–11. doi:10.9734/AJEE/2017/34236

Edokpayi, J., Odiyo, J., Durowoju, O., & Adetoro, A. (2017). Household hazardous waste management in sub-Saharan Africa. In D. Mmereki (Ed.), *Household hazardous waste management*. IntechOpen. doi:10.5772/66292

EEPA. (2003). Guideline ambient environmental standards for Ethiopia. Ethiopian Environmental Protection Authority and United Nations Industrial Development Organization, UNIDO, Addis Ababa, Ethiopia.

Egwuonwu, G., Olabode, V., Bukar, P., Okolo, V., & Odunze, A. (2011). Characterization of topsoil and groundwater at leather industrial area, Challawa, Kano, Northern Nigeria. *Pacific Journal of Science and Technology*, *12*(1), 628.

Ekengele, N., Myung, C., Ombolo, A., Ngatcha, N., Georges, E., & Lape, M. (2008). Metal pollution in freshly deposited sediments from river Mingoa, main tributary to the Municipal Lake of Yaounde, Cameroon. *Geosciences Journal*, *12*(4), 337–347. doi:10.100712303-008-0034-5

Ekeocha, C., Nwoko, I., & Onyeke, L. (2017). Impact of automobile repair activities on physicochemical and microbial properties of soils in selected automobile repair sites in Abuja, central Nigeria. *Chemical Science International Journal*, *20*(2), 1–15. doi:10.9734/CSJI/2017/36065

Ekoa Bessa, Z., Ngueutchoua, G., Kwewouo Janpou, A., El-Amier, A., & Nguetnga, O. A. (2020). Heavy metal contamination and its ecological risks in the beach sediments along the Atlantic Ocean (Limbe coastal fringes, Cameroon). *Earth Systems and Environment*, *5*(2), 433–444. doi:10.100741748-020-00167-5

Elenge, M., & De Brouwer, C. (2011). Identification of hazards in the workplaces of artisanal mining in Katanga. *International Journal of Occupational Medicine and Environmental Health*, *245*(1), 7–66. doi:10.247813382-011-0012-4 PMID:21468903

Elliott, M. (2003). Biological pollutants and biological pollution—An increasing cause for concern. *Marine Pollution Bulletin*, *46*(3), 275–280. doi:10.1016/S0025-326X(02)00423-X PMID:12604060

El-Naggar, A., Ahmed, N., Mosa, A., Niazi, N., Yousaf, B., Sharma, A., Sarkar, B., Cai, Y., & Chang, S. X. (2021). Nickel in soil and water: Sources, biogeochemistry and remediation using biochar. *Journal of Hazardous Materials*, *419*, 126421. doi:10.1016/j.jhazmat.2021.126421 PMID:34171670

El-Naggar, A., Rajapaksha, U., Shaheen, M., Rinklebe, J., & Ok, S. (2018b). Potential of biochar to immobilize nickel in contaminated soils. In *Nickel in Soils and Plants* (pp. 293–318). CRC Press. doi:10.1201/9781315154664-13

El-Naggar, A., Shaheen, M., Ok, S., & Rinklebe, J. (2018a). Biochar affects the dissolved and colloidal concentrations of Cd, Cu, Ni, and Zn and their phytoavailability and potential mobility in a mining soil under dynamic redox-conditions. *The Science of the Total Environment*, *624*, 1059–1071. doi:10.1016/j.scitotenv.2017.12.190 PMID:29929223

El-Shahawi, M., Hamza, A., Bashammakhb, A., & Al-Saggaf, W. (2010). An overview on the accumulation, distribution, transformations, toxicity and analytical methods for the monitoring of persistent organic pollutants. *Talanta*, *80*(5), 1587–159. doi:10.1016/j.talanta.2009.09.055 PMID:20152382

Eludoyin, A., Ojo, A., Ojo, T., & Awotoye, O. (2017). Effects of artisanal gold mining activities on soil properties in a part of southwestern Nigeria. *Cogent Environmental Science*, *3*(1), 1305650. doi:10.1080/23311843.2017.1305650

Emilie, E., Harue, M., Takahiro, S., Aki, N., Yusuke, S., Yusuke, M., & Hitoshi, C. (2017). Geochemical distribution and fate of arsenic in water and sediments of rivers from the Hokusetsu area. *Journal of Hydrology. Regional Studies*, *9*, 34–47. doi:10.1016/j.ejrh.2016.09.008

Eneh, O. (2021). Health effects of selected trace elements in hairdressing cosmetics on hairdressers in Enugu, Nigeria. *Scientific Reports*, *11*(1), 20352. doi:10.103841598-021-00022-1 PMID:34645821

Engwa, G., Udoka, P., Nwalo, F., & Unachukwu, M. (2019). *Mechanism and health effects of heavy metal toxicity in humans*. Intech Open. doi:10.5772/intechopen.82511

Erboz, G. (2017). *How to define industry 4.0: main pillars of industry 4.0. Managerial trends in the development of enterprises in globalization era conference*. Slovak University of Agriculture in Nitra.

Ertani, A., Mietto, A., Borin, M., & Nardi, S. (2017). Chromium in agricultural soils and crops: A review. *Water, Air, and Soil Pollution*, *228*(5), 190. doi:10.100711270-017-3356-y

Essien, J., Ikpe, D., Inam, E., Okon, A., Ebong, G., & Benson, N. (2022). Occurrence and spatial distribution of heavy metals in landfill leachates and impacted freshwater ecosystem: An environmental and human health threat. *PLoS ONE, 17*(2): e0263279. https://doi.org/. pone.0263279 doi:10.1371/journal

Essumang, D. (2009). Analysis and human health risk assessment of arsenic, cadmium, and mercury in Manta birostris (manta ray) caught along the Ghanaian coastline. *Human and Ecological Risk Assessment*, *15*(5), 985–998. doi:10.1080/10807030903153451

Ettler, V., Kribek, B., Majer, V., Knesl, I., & Mihaljevic, M. (2012). Differences in the bioaccessibility of metals/metalloids in soils from mining and smelting areas (Copperbelt, Zambia). *Journal of Geochemical Exploration*, *113*, 68–75. doi:10.1016/j.gexplo.2011.08.001

European Commission (EC) (2022). *Industry 5.0 roundtable, Brussels, 27 April 2022 meeting report.* EC, Brussels, Belgium. doi:10.2777/982391

Ezeh, B., & Chukwe, G. (2011). Small scale mining and heavy metals pollution of agricultural soils: The case of Ishiagu Mining District, South Eastern Nigeria. *Journal of Geology and Mining Research*, *3*, 87–104.

Fakayode, S., & Olu-Owolabi, B. (2003). Heavy metal contamination of roadside topsoil in Osogbo, Nigeria: Its relationship to traffic density and proximity to highways. *Environmental Geology (Berlin)*, *44*(2), 150–157. doi:10.100700254-002-0739-0

Fang, Z., Zhao, M., Zhen, H., Chen, L., Shi, P., & Huang, Z. (2014). Genotoxicity of tri- and hexavalent chromium compounds in vivo and their modes of action on DNA damage in vitro. *PLoS One*, *9*(8), e103194. doi:10.1371/journal.pone.0103194 PMID:25111056

Fan, T., Wang, J., Li, B., He, Z., Gao, J., Zhou, D.-M., Friedman, S. P., & Sparks, D. L. (2016). Effect of organic matter on sorption of Zn on soil: Elucidation by Wien effect measurements and EXAFS spectroscopy. *Environmental Science & Technology*, *50*(6), 2931–2937. doi:10.1021/acs.est.5b05281 PMID:26894796

FAO. (2018). *Soil Pollution: A Hidden Reality*. Food and Agricultural Organization of the United Nations.

Farias, P., Alamo-Hernandez, U., Mancilla-Sanchez, L., Texcalac-Sangrador, J., & Carrizales-Yanez, L. (2014). Lead in school children from Morelos, Mexico: Levels, sources and feasible interventions. *International Journal of Environmental Research and Public Health*, *12*(11), 12668–12682. doi:10.3390/ijerph111212668 PMID:25493390

Fasani, E., Manara, A., Martini, F., Furini, A., & DalCorso, G. (2018). The potential of genetic engineering of plants for the remediation of soils contaminated with heavy metals. *Plant, Cell & Environment*, *41*(5), 1201–1232. doi:10.1111/pce.12963 PMID:28386947

Fashola, M., Ngolo-Jeme, V., & Babalola, O. (2016). Heavy metal pollution from gold mines: Environmental effects and bacterial strategies for resistance. *International Journal of Environmental Research and Public Health*, *13*(11), 1047. doi:10.3390/ijerph13111047 PMID:27792205

Fasinu, P., & Orisakwe, O. (2013). Heavy metal pollution in sub-Saharan Africa and possible implications in cancer epidemiology. *Asian Pacific Journal of Cancer Prevention*, *14*(6), 3393–3402. doi:10.7314/APJCP.2013.14.6.3393 PMID:23886118

Fayiga, A., Ipinmoroti, M., & Chirenje, T. (2018). Environmental pollution in Africa. *Environment, Development and Sustainability*, *20*(1), 41–73. doi:10.100710668-016-9894-4

Fei-baffoe, B., Kani, A., Afrifa, A., & Yeboah, S. (2017). Copper and zinc pollution in cocoa growing areas in a low-income country. *Journal of Natural Sciences Research*, *7*(14), 28–34.

Fei, C., Min, X., Wang, Z., Pang, Z., Liang, Y., & Ke, Y. (2017). Health and ecological risk assessment of heavy metals pollution in an antimony mining region: A case study from South China. *Environmental Science and Pollution Research International, 24*(35), 27573–27586. doi:10.100711356-017-0310-x PMID:28980103

Ferreira, L., Scacroni, J., Da Silva, J., Cataneo, A., & Martins, D. (2014). Copper oxychloride fungicide and its effect on growth and oxidative stress of potato plants. *Pesticide Biochemistry and Physiology, 112*, 63–69. doi:10.1016/j.pestbp.2014.04.010 PMID:24974119

Fianko, J., Osae, S., Adomako, D., Adotey, D., & Serfor-Armah, Y. (2007). Assessment of heavy metal pollution of the Iture Estuary in the central region of Ghana. *Environmental Monitoring and Assessment, 131*(1-3), 467–473. doi:10.100710661-006-9492-2 PMID:17171259

Figueiredo, E., Soares, E., Baptista, P., Castro, M., & Bastos, M. L. (2007). Validation of an electrothermal atomization atomic absorption spectrometry method for quantification of total chromium and chromium (VI) in wild mushrooms and underlying soils. *Journal of Agricultural and Food Chemistry, 55*(17), 7192–7198. doi:10.1021/jf0710027 PMID:17661487

Fillion, M., Mergler, D., Carlos, J., Larribe, F., Melanie, L., & Jean, G. (2006). A preliminary study of mercury exposure and blood pressure in the Brazilian Amazon. *The Science of the Total Environment, 5*(1), 29. doi:10.1186/1476-069X-5-29 PMID:17032453

Food and Agricultural Organization (FAO) & United Nations Environmental Program (UNEP) (2021). *Global assessment of soil pollution: report*. FAO & UNEP. doi:10.4060/cb4894en

Foso-Mensah, B., Addae, E., Tawiah, E., & Nyame, F. (2017). Heavy metals concentration and distribution in soils and vegetation at Korle Lagoon area in Accra, Ghana. *Cogent Environmental Science, 3*(1), 1405887. doi:10.1080/23311843.2017.1405887

Freije, A. (2014). Heavy metal, trace element and petroleum hydrocarbon pollution in the Arabian Gulf: A review. *Journal of the Association of Arab Universities for Basic and Applied Sciences, 17*(1), 90–100. doi:10.1016/j.jaubas.2014.02.001

Friberg, L., Kjellström, T., Elinder, G., & Nordberg, F. (2019). Cadmium and health: a toxicological and epidemiological appraisal. *Cadmium Health: A Toxicological and Epidemiological Appraisal.* . doi:10.1201/9780429260599

Fu, J., Wang, Q., Wang, H., Yu, H., & Zhang, X. (2014). Monitoring of non-destructive sampling strategies to assess the exposure of avian species in Jiangsu Province, China to heavy metals. *Environmental Science and Pollution Research International, 21*(4), 2898–2906. doi:10.100711356-013-2242-4 PMID:24154854

Fuller, R., Landrigan, P., Balakrishnan, K., Bose, S., & Braver, M. (2022). Pollution and health: A progress update. *The Lancet. Planetary Health, 6*(6), 6537–6547. doi:10.1016/S2542-5196(22)00090-0 PMID:35594895

Fu, X., Li, G., Tian, H., & Dong, D. (2018). Detection of cadmium in soils using laser-induced breakdown spectroscopy combined with spatial confinement and resin enrichment. *RSC Advances*, *8*(69), 39635–39640. doi:10.1039/C8RA07799A PMID:35558063

Gao, X., & Chen, A. (2012). Heavy metal pollution status in surface sediments of the coastal Bohai Bay. *Water Research*, *46*(6), 1901–1911. doi:10.1016/j.watres.2012.01.007 PMID:22285040

Garcia, E., & Boluda, V. (1996). Heavy metals incidence in the application of inorganic fertilizers and pesticides to rice farming soils. *Environmental Pollution*, *92*(1), 19–25. doi:10.1016/0269-7491(95)00090-9 PMID:15091407

Garcia-Gomez, C., Garcia-Gutierrez, S., Obrador, A., & Fernandez, D. (2020). Study of Zn availability, uptake, and effects on earthworms of zinc oxide nanoparticle versus bulk applied to two agricultural soils: Acidic and calcareous. *Chemosphere*, *239*, 124814. doi:10.1016/j.chemosphere.2019.124814 PMID:31527003

Garg, U., Kaur, M., Garg, V., & Sud, D. (2007). Removal of hexavalent chromium from aqueous solution by agricultural waste biomass. *Journal of Hazardous Materials*, *140*(1-2), 60–68. doi:10.1016/j.jhazmat.2006.06.056 PMID:16879918

Garza-Lombó, C., Pappa, A., Panayiotidis, M., Gonsebatt, E., & Franco, R. (2019). Arsenic-induced neurotoxicity: A mechanistic appraisal. *Journal of Biological Inorganic Chemistry*, *24*(8), 1305–1316. doi:10.100700775-019-01740-8 PMID:31748979

Gautam, P., Gautam, R., Chattopadhyaya, M., Banerjee, S., Chattopadhyaya, M., & Pandey, M. (2016). *Heavy metals in the environment: fate, transport, toxicity and remediation technologies Thermodynamic profiling of pollutants View project Materials for Solid oxide fuel cells View project Heavy Metals in the Environment: Fate*. Transport, Toxicity and Rem.

Gbogbo, F., Otoo, D., Asomaning, O., & Huago, Q. (2017). Contamination status of arsenic in fish and shellfish from three river basins in Ghana. *Environmental Monitoring and Assessment*, *189*(8), 400. doi:10.100710661-017-6118-9 PMID:28718096

Gebeyehu, H., & Bayissa, L. (2020). Levels of heavy metals in soil and vegetables and associated health risks in Mojo area, Ethiopia. *PLoS One*, *15*(1), e0227883. doi:10.1371/journal.pone.0227883 PMID:31999756

Genchi, G., Sinicropi, M., Lauria, G., Carocci, A., & Catalano, A. (2020). The effects of cadmium toxicity. *International Journal of Environmental Research and Public Health*, *17*(11), 3782. doi:10.3390/ijerph17113782 PMID:32466586

Genova, G., Chiesa, D., Mimmo, T., Borruso, L., & Cesco, S. (2022). Copper and zinc as a window to past agricultural land-use. *Journal of Hazardous Materials*, *424*(Part C), 126631. doi:10.1016/j.jhazmat.2021.126631 PMID:34334215

Gezahegn, A., Feyessa, F., Tekeste, E., & Beyene, M. (2021). Chromium laden soil, water and vegetables nearby tanning industries: Speciation and spatial distribution. *Journal of Chemistry*, *5531349*, 1–10. doi:10.1155/2021/5531349

Ghane, E., Khanverdiluo, S., & Mehri, F. (2022). The concentration and health risk of potentially toxic elements (PTEs) in the breast milk of mothers: A systematic revies and metal analysis. *Journal of Trace Elements in Medicine and Biology*, *73*, 126998. doi:10.1016/j.jtemb.2022.126998 PMID:35617722

Ghosh, I., Chatterjee, S., & Mukherjea, K. (2012). Chromium (VI) in tannery effluents: Assessment, biodistribution, and environmental health impact. *Journal of the Indian Chemical Society*, *89*(4), 479–483.

Gnandi, K., & Tobschall, H. (2002). Heavy metals distribution of soils around mining sites of cadmium-rich marine sedimentary phosphorites of Kpogamé and Hahotoé (southern Togo). *Environmental Geology (Berlin)*, *41*(5), 593–600. doi:10.1007002540100425

Gogoi, N., Baroowa, B., & Gogoi, N. (2021). *Ecological tools for remediation of soil pollutants. Bioremediation Science from Theory to Practice*. CRC Press.

Goldhaber, S. (2003). Trace element risk assessment: Essentiality vs. toxicity. *Regulatory Toxicology and Pharmacology*, *38*(2), 232–242. doi:10.1016/S0273-2300(02)00020-X PMID:14550763

Golia, E., Tsiropoulos, G., Dimirkou, A., & Mitsios, I. (2007). Distribution of heavy metals of agricultural sols of central Greece using the modified BCR sequential extraction method. *International Journal of Environmental Analytical Chemistry*, *87*(13-14), 1053–1063. doi:10.1080/03067310701451012

Gomezulu, E., Mwakaje, A., & Katima, J. (2018). Heavy metals and cyanide distribution in the villages surrounding Buzwagi gold mine in Tanzania. *Tanzania Journal of Science*, *44*(1), 107–122.

Gong, B., He, E., Qiu, H., Van Gestel, C., & Freire, A. (2020). Interactions of arsenic, copper and zinc in soil-plant system: Partition, uptake and phytotoxicity. *The Science of the Total Environment*, *745*, 140926. doi:10.1016/j.scitotenv.2020.140926 PMID:32712499

Gottesfeld, P., & Pokhrel, K. (2011). Lead exposure in battery manufacturing and recycling in developing countries and among children in nearby communities. *Journal of Occupational and Environmental Hygiene*, *8*(9), 520–532. doi:10.1080/15459624.2011.601710 PMID:21793732

Gottesfeld, P., Were, F., Adogame, L., Gharbi, S., San, D., Nota, M., & Kuepouo, G. (2018). Soil contamination from lead battery manufacturing and recycling in seven Africa countries. *Environmental Research*, *161*, 609–614. doi:10.1016/j.envres.2017.11.055 PMID:29248873

Grant, C. (2011). Influence of phosphate fertilizer on cadmium in agricultural soils and crops. *Pedologist*, *3*, 143–155. doi:10.18920/pedologist.54.3_143 PMID:26557096

Green, C. (2017). *A comparison of factors affecting the small-scale distribution of mercury contamination in a Zimbabwean stream system*. [MSc Thesis, Sam Houston State University, Texas, USA].

Guedron, S., & Acha, D. (2021). Mercury and methylmercury contamination of terrestrial and aquatic ecosystems. *Applied Sciences (Basel, Switzerland)*, *11*(11), 4807. doi:10.3390/app11114807

Gunay, D. (2002). Sanayi ve sanayi tarihi. *Mimar ve Mühendis Dergisi*, *31*, 8–14.

Gupta, D., Chatterjee, S., Datta, S., Veer, V., & Walther, C. (2014). Role of phosphate fertilizers in heavy metal uptake and detoxification of toxic metals. *Chemosphere*, *108*, 134–144. doi:10.1016/j.chemosphere.2014.01.030 PMID:24560283

Gupta, N., Yadav, K., Kumar, V., Krishnan, S., Kumar, S., & Nejad, D. (2020). Evaluating heavy metals contamination in soil and vegetables in the region of North India: Levels, transfer and potential human health risk analysis. *Environmental Toxicology and Pharmacology*, *103563*. doi:10.1016/j.etap.2020.103563 PMID:33310081

Gupta, V., Jatav, K., Verma, R., Kothari, L., & Kachhwaha, S. (2017). Nickel accumulation and its effect on growth, physiological and biochemical parameters in millets and oats. *Environmental Science and Pollution Research International*, *24*(30), 23915–23925. doi:10.100711356-017-0057-4 PMID:28875293

Gustavsson, B., Luthbom, K., & Lagerkvist, A. (2006). Comparison of analytical error and sampling error for contaminated soil. *Journal of Hazardous Materials*, *138*(2), 252–260. doi:10.1016/j.jhazmat.2006.01.082 PMID:17030410

Gworek, B., Dmuchowski, W., & Dabrowska, A. (2020). Mercury in the terrestrial environment: A review. *Environmental Sciences Europe*, *32*(1), 128. doi:10.118612302-020-00401-x

Habashi, F. (2013). Nickel, Physical and Chemical Properties. In R. H. Kretsinger, V. N. Uversky, & E. A. Permyakov (Eds.), *Encyclopedia of Metalloproteins*. Springer. doi:10.1007/978-1-4614-1533-6_338

Habib, E. (2017). Heavy metals pollution of soil; toxicity and phytoremediation techniques. *International Journal of Advanced Research and Publications*, *1*(1), 29–41.

Hakanson, L. (1980). An ecological risk index for aquatic pollution control: A sedimentological approach. *Water Research*, *14*(8), 975–1001. doi:10.1016/0043-1354(80)90143-8

Hamidu, H., Halilu, F., Yerima, K., Garba, L., Suleiman, A., Kankara, A. I., & Abdullahi, I. M. (2021). Heavy metals pollution indexing, geospatial and statistical approaches of groundwater within Challawa and Sharada industrial areas, Kano city, North-western Nigeria. *SN Applied Sciences*, *3*(7), 690. doi:10.100742452-021-04662-w

Hamilton, E., Lark, R., Young, S., Bailey, E., Sakala, G., Maseka, K., & Watts, M. J. (2020). Reconnaissance sampling and determination of hexavalent chromium in potentially-contaminated agricultural soils in Copperbelt province, Zambia. *Chemosphere*, *247*, 125984. doi:10.1016/j.chemosphere.2020.125984 PMID:32079057

Hansson, S., Grusson, Y., Chimienti, M., Claustres, A., Jean, S., & Le Roux, G. (2019). Legacy Pb pollution in the contemporary environment and its potential bioavailability in three mountain catchments. *The Science of the Total Environment*, *671*, 1227–1236. doi:10.1016/j.scitotenv.2019.03.403

Han, Y., Park, J., Kim, S., Jeong, H., & Ahn, J. (2019). Redox transformation of soil minerals and arsenic in arsenic-contaminated soil under cycling redox conditions. *Journal of Hazardous Materials, 378*, 120745. doi:10.1016/j.jhazmat.2019.120745 PMID:31203129

Hapke, M., & Gerhard, H. (2020). Introduction to cobalt chemistry and catalysis. In M. Hapke & H. Gerhard (Eds.), *Cobalt catalysis in organic synthesis: methods and reactions*. Wiley Publishers., doi:10.1002/9783527814855.ch1

Hararuk, O., Obrist, D., & Luo, Y. (2013). Modelling the sensitivity of soil mercury storage to climate-induced changes in soil carbon pools. *Biogeosciences, 10*(4), 2393–2407. doi:10.5194/bg-10-2393-2013

Harischandra, S., Ghaisas, S., Zenitsky, G., Jin, H., Kanthasamy, A., Anantharam, V., & Kanthasamy, A. G. (2019). Manganese-induced neurotoxicity: New insights into the triad of protein misfolding, mitochondrial impairment, and neuroinflammation. *Frontiers in Neuroscience, 13*, 654. doi:10.3389/fnins.2019.00654 PMID:31293375

Hartwig, A., & Schwerdtle, T. (2002). Interactions by carcinogenic metal compounds with DNA repair processes: Toxicological implications. *Toxicology Letters, 127*(1-3), 47–54. doi:10.1016/S0378-4274(01)00482-9 PMID:12052640

Hasanpour, M., & Hatami, M. (2020). Application of three-dimensional porous aerogels as adsorbent for removal of heavy metal ions from water/wastewater: A review study. *Advances in Colloid and Interface Science, 284*, 102247. doi:10.1016/j.cis.2020.102247 PMID:32916456

Hasimuna, O., Maulu, S., & Chibesa, M. (2022). Assessment of heavy metal contamination in water and largescale yellowfish (*Labeobarbus marequensis*, smith 1841) from Solwezi river, north-western Zambia. *Cogent Food & Agriculture, 8*(1), 1–17. doi:10.1080/23311932.2022.2121198

Haynes, M., Bruno, T., & Lide, D. (2014). *CRC Handbook of Chemistry and Physics*. CRC Press. doi:10.1201/b17118

Haynes, W. (2015). *CRC handbook of chemistry and physics*. CRC Press/ Taylor and Francis.

Hazardous Substances Research Center. (2003). Environmental impact of the petroleum industry. CF Pub. https://cfpub.epa.gov/ncer_abstracts/index.cfm/fuseaction/display.files/fileID/14522

Hazelhoff, H., & Torres, A. (2018). Gender differences in mercury-induced hepatotoxicity: Potential mechanisms. *Chemosphere, 202*, 330–338. doi:10.1016/j.chemosphere.2018.03.106 PMID:29574386

He, F., Gao, J., Pierce, E., Strong, P., Wang, H., & Liang, L. (2015b). In situ remediation technologies for mercury-contaminated soil. *Environmental Science and Pollution Research International, 22*(11), 8124–8147. doi:10.100711356-015-4316-y PMID:25850737

Helmenstine, A. (2019). *Nickel element facts and properties*. ThoughtCo. https://www.thoughtco.com/nickel-facts-606565

Hemmaphan, S., & Bordeerat, N. (2022). Genotoxic Effects of Lead and Their Impact on the Expression of DNA Repair Genes. *International Journal of Environmental Research and Public Health*, *19*(7), 4307. doi:10.3390/ijerph19074307 PMID:35409986

Hemminki, K., Niemi, M., Kostinen, K., & Vainio, H. (1980). Spontaneous abortions among women employed in the metal industry in Finland. *International Archives of Occupational and Environmental Health*, *47*(1), 53–60. doi:10.1007/BF00378328 PMID:7429646

Heyden, C., & New, G. (2004). Sediment chemistry: A history of mine contaminant remediation and an assessment of processes and pollution potential. *Journal of Geochemical Exploration*, *82*(1-3), 35–57. doi:10.1016/j.gexplo.2003.11.001

He, Z., Shen, J., Ni, Z., Tang, J., Song, S., Chen, J., & Zhao, L. (2015a). Electrochemically created roughened lead plate for electrochemical reduction of aqueous CO2. *Catalysis Communications*, *72*, 38–42. doi:10.1016/j.catcom.2015.08.024

Hilson, G. (2016). Artisanal and small-scale mining and agriculture: Exploring their links in rural sub-Saharan Africa. London: IIED. https://pubs.iied.org/sites/ default/files/pdfs/migrate/16617IIED.pdf

Hira, A., Pacini, H., Wadee, K., Sikander, M., Oruko, R., & Dinan, A. (2022). Mitigating tannery pollution in sub-Saharan Africa and South Asia. *Journal of Developing Societies*, *38*(3), 360–383. doi:10.1177/0169796X221104856

Hirwa, H., Nshimiyimana, F., Ngendahayo, E., Akimpaye, B., Nahayo, L., & Ngamata, O. (2019). Evaluation of soil contamination in mining areas of Rwanda. *American Journal of Water Science and Engineering*, *5*(1), 9–15. doi:10.11648/j.ajwse.20190501.12

Hocaoglu-Ozyigit, A., & Genc, B. (2020). Cadmium in plants, humans and the environment. *Frontiers in Life Sciences and Related Technologies*, *1*(1), 12–21.

Hoenig, M. (2001). Preparation steps in environmental trace element analysis – facts and traps. *Talanta*, *54*(6), 1021–1038. doi:10.1016/S0039-9140(01)00329-0 PMID:18968324

Hogarh, J., Gyamfi, E., Nukpezah, D., Akoto, O., & Adu-Kumi, S. (2016). Contamination from mercury and other heavy metals in a mining district in Ghana: Discerning recent trends from sediment core analysis. *Environmental Systems Research*, *5*(15), 1–9. doi:10.118640068-016-0067-0

Hongbo, S., Liye, C., Gang, X., Kun, Y., & Lihua, Z. (2011). Progress in phytoremediating heavy-metal contaminated soils. In I. Sherameti & A. Varma (Eds.), *Detoxification of Heavy Metals. Soil Biology* (Vol. 30). Springer. doi:10.1007/978-3-642-21408-0_4

Hong, Y., Kim, Y., & Lee, K. (2012). Methylmercury exposure and health effects. *Journal of Preventive Medicine and Public Health*, *45*(6), 353–363. doi:10.3961/jpmph.2012.45.6.353 PMID:23230465

Hooda, S. (2010). *Trace elements in soils. trace elements in soils*. Blackwell Publishing Ltd. doi:10.1002/9781444319477

Hoornweg, D., Bhada-Tata, P., & Kennedy, C. (2015). Peak waste: When is it likely to occur? *Journal of Industrial Ecology*, *19*(1), 117–128. doi:10.1111/jiec.12165

Horn, C., & Ramudzuli, M. (2020). Arsenic Contamination of soil in relation to water in Northeastern South Africa. In A. Fares & S. Singh (Eds.), *Arsenic water resources contamination. Advances in water security*. Springer. doi:10.1007/978-3-030-21258-2_7

Hou, D. (2020). *Sustainable remediation of contaminated soil and groundwater: materials, processes, and assessment*. Butterworth-Heinemann Publishers. doi:10.1016/C2018-0-03003-0

Hou, D., O'Connor, D., Igalavithana, A., Alessi, S., Luo, J., Tsang, D. C. W., Sparks, D. L., Yamauchi, Y., Rinklebe, J., & Ok, Y. S. (2020). Metal contamination and bioremediation of agricultural soils for food safety and sustainability. *Nature Reviews. Earth & Environment*, *1*(7), 366–381. doi:10.103843017-020-0061-y

Hou, D., & Ok, S. (2019). Soil pollution — Speed up global mapping. *Nature*, *566*(7745), 455. doi:10.1038/d41586-019-00669-x PMID:30809065

Hsiao, C., Wu, H., & Wan, S. (2011). Effects of environmental lead exposure on T-helper cell-specific cytokines in children. *Journal of Immunotoxicology*, *8*(4), 284–287. doi:10.3109/1547691X.2011.592162 PMID:21726182

Hsu, L., Liu, Y., & Tzou, Y. (2015). Comparison of the spectroscopic speciation and chemical fractionation of chromium in contaminated paddy soils. *Journal of Hazardous Materials*, *296*, 230–238. doi:10.1016/j.jhazmat.2015.03.044 PMID:25935296

Huang, F. (2011). *Study on the toxic effects mercury on crops and the toxicity threshold value to crops of soil mercury*. Fujian Agriculture and Forestry University. doi:10.7666/d.y1878345

Huang, H., Lee, C., & Yu, H. (2019). Arsenic-induced carcinogenesis and immune dysregulation. *International Journal of Environmental Research and Public Health*, *16*(15), 2746. doi:10.3390/ijerph16152746 PMID:31374811

Huang, Y., Chen, Q., Deng, M., Japenga, J., Li, T., Yang, X., & He, Z. (2018). Heavy metal pollution and health risk assessment of agricultural soils in a typical peri-urban area in southeast China. *Journal of Environmental Management*, *207*, 159–168. doi:10.1016/j.jenvman.2017.10.072 PMID:29174991

Hubeny, J., Harnisz, M., Korzeniewska, E., Buta, M., Zieliński, W., Rolbiecki, D., Giebułtowicz, J., Nałęcz-Jawecki, G., & Płaza, G. (2021). Industrialization as a source of heavy metals and antibiotics which can enhance the antibiotic resistance in wastewater, sewage sludge and river water. *PLoS One*, *16*(6), e0252691. doi:10.1371/journal.pone.0252691 PMID:34086804

Hussain, K., Haris, M., Qamar, H., Hussain, T., & Ahmad, G. (2021). Bioremediation of waste gases and polluted soils. In G. Panpatte & Y. Jhala (Eds.), *Microbial Rejuvenation of Polluted Environment*., doi:10.1007/978-981-15-7455-9_5

Hussain, T., & Gondal, A. (2008). Monitoring and assessment of toxic metals in Gulf War oil spill contaminated soil using laser-induced breakdown spectroscopy. *Environmental Monitoring and Assessment*, *136*(1-3), 391–399. doi:10.100710661-007-9694-2 PMID:17406995

IARC. (2012). Monographs on the evaluation of carcinogenic risk to human. International Agency for Research on Cancer.

Ibrahim, Z., Dan-Badjo, T., Guero, Y., Idi, M., & Feidt, C. (2019). Distribution spatiale des éléments traces métalliques dans les sols de la zone aurifère de Komabangou au Niger. *International Journal of Biological and Chemical Sciences*, *13*(1), 557–573. doi:10.4314/ijbcs.v13i1.43

Idris, M., Khalid, D., & Abdullahi, Z. (2015). Comparative assessment of heavy metals concentration in the soil in the vicinity of tannery industries, Kumbotso old dump site and River Challawa, conference at Challawa industrial estate, Kano State, Nigeria. *International Journal of Innovative Research and Development*, *4*(6), 122–128.

IITA. (1982). *Automated and semi-automated methods for soil and plant analysis*. International Institute of Tropical Agriculture.

Ikenaka, Y., Nakayama, S., Muroya, T., Yabe, J., Konnai, S., Darwish, W., Muzandu, K., Choongo, K., Mainda, G., Teraoka, H., Umemura, T., & Ishizuka, M. (2012). Effects of environmental lead contamination on cattle in a lead/zinc mining area: Changes in cattle immune systems on exposure to lead in vivo and in vitro. *Environmental Toxicology and Chemistry*, *31*(10), 2300–2305. doi:10.1002/etc.1951 PMID:22821446

Iloms, E., Ololade, O., Ogola, H., & Selvarajan, R. (2020). Investigating industrial effluent impact on municipal wastewater treatment plant in Vaal, South Africa. *International Journal of Environmental Research and Public Health*, *17*(3), 1096. doi:10.3390/ijerph17031096 PMID:32050467

Imran, A., & Hassan, A. (2002). Determination of metal ions in water, soil and sediment by capillary electrophoresis. *Analytical Letters*, *35*(13), 2053–2076. doi:10.1081/AL-120015519

Inengite, K., Abasi, Y., & Walter, C. (2015). Application of pollution indices for the assessment of heavy metal pollution in flood impacted soil. *International Research Journal of Pure and Applied Chemistry*, *8*(3), 175–189. doi:10.9734/IRJPAC/2015/17859

Irina, R., & John, R. (2020). Does manganese contribute to methamphetamine-induced psychosis? *Current Emergency and Hospital Medicine Reports*, *8*(4), 133–141. doi:10.100740138-020-00221-6

Irunde, R., Ijumulana, J., Ligate, F., Maity, J., Ahmad, A., Mtamba, J., Mtalo, F., & Bhattacharya, P. (2022). Arsenic in Africa: Potential sources, spatial variability and the state-of-the-art arsenic removal using locally available materials. *Groundwater for Sustainable Development*, *18*, 100746. doi:10.1016/j.gsd.2022.100746

Islam, M., Taki, G., Nguye, X., Jo, Y., & Kim, J. (2017). Heavy metal stabilization in contaminated soil by treatment with calcined cockle shell. *Environmental Science and Pollution Research International, 24*(8), 7177–7183. doi:10.100711356-016-8330-5 PMID:28097482

Israr, M., & Shivendra, V. (2006). Antioxidative responses to mercury in the cell cultures of *Sesbania drummondii. Plant Physiology and Biochemistry, 44*(10), 590–595. doi:10.1016/j.plaphy.2006.09.021 PMID:17070690

Itai, T., Otsuka, M., Asante, K., Muto, M., Ankomah, Y., Asare, O., & Tanabe, S. (2014). Variation and distribution of metals and metalloids in soil/ash mixtures from Agbogbloshie e-waste recycling site in Accra, Ghana. *The Science of the Total Environment, 470-471*, 707–716. doi:10.1016/j.scitotenv.2013.10.037 PMID:24184547

Izah, S., Richard, G., Aigberua, A., & Ekakitie, O. (2021). Variations in reference values utilized for evaluation of complex pollution indices of potentially toxic elements: A critical review. *Environmental Challenges, 5*, 100322. doi:10.1016/j.envc.2021.100322

JaccobA. (2020). Evaluation of Lead and Copper content in hair of workers from oil product distribution companies in Iraq. *Brazilian Journal of Pharmaceutical Sciences*, 56. https://doi.org/ doi:10.1590/s2175-97902019000318061

Jaishankar, M., Tseten, T., Anbalagan, N., Mathew, B., & Beeregowda, N. (2014). Toxicity, mechanism and health effects of some heavy metals. *Interdisciplinary Toxicology, 7*(2), 60–72. doi:10.2478/intox-2014-0009 PMID:26109881

Jakubus, M., & Graczyk, M. (2020). Availability of nickel in soil evaluated by various chemical extractants and plant accumulation. *Agronomy (Basel), 10*(11), 1805. doi:10.3390/agronomy10111805

Jan, A., Azam, M., Siddiqui, K., Ali, A., Choi, I., & Haq, Q. (2015). Heavy metals and human health: Mechanistic insight into toxicity and counter defense system of antioxidants. *International Journal of Molecular Sciences, 16*(12), 29592–29630. doi:10.3390/ijms161226183 PMID:26690422

Jang, Y., Somanna, Y., & Kim, H. (2016). Source, distribution, toxicity and remediation of arsenic in the environment-a review. *International Journal of Applied Environmental Sciences, 11*(2), 559–581.

Jaworska, H., & Klimek, J. (2021). Assessment of the impact of a motorway on content and spatial distribution of mercury in adjacent agricultural soils. *Minerals (Basel), 11*(11), 1221. doi:10.3390/min11111221

Ji, W, Yang, T., Ma, S., & Ni, W. (2012). Heavy metal pollution of soils in the site of a retired paint and ink factory. *Energy Procedia, 16*(Part A), 21-26. doi:10.1016/j.egypro.2012.01.005

Jiang, B., Gong, Y., Gao, J., Sun, T., Liu, Y., Oturan, N., & Oturan, A. (2019). The reduction of Cr (VI) to Cr (III) mediated by environmentally relevant carboxylic acids: State-of-the-art and perspectives. *Journal of Hazardous Materials, 365*, 205–226. doi:10.1016/j.jhazmat.2018.10.070 PMID:30445352

Jiang, J., Bauer, I., Paul, A., & Kappler, A. (2009). Arsenic redox changes by microbially and chemically formed semiquinone radicals and hydroquinones in a humic substance model quinone. *Environmental Science & Technology*, *15*(10), 3639–3645. doi:10.1021/es803112a PMID:19544866

Jiang, Y., & Zhao, F. (2001). Mechanism of heavy metal injury and resistance of plants. *Chinese Journal of Applied and Environmental Biology*, *7*(1), 92–99. doi:10.3321/j.issn:1006-687X.2001.01.022

Jibrin, M., Abdulhameed, A., Nayaya, A., & Ezra, G. (2021). Health risk effect of heavy metals from pesticides in vegetables and soils: A review. *Dutse Journal of Pure and Applied Sciences*, *7*(3b), 24–33. doi:10.4314/dujopas.v7i3b.3

Jimoh, A., Agbaji, E., Ajibola, V., & Funtua, M. (2020). Application of pollution load indices, enrichment factors, contamination factor and health risk assessment of heavy metals pollution of soils of welding workshops at old Panteka market, Kaduna-Nigeria. *Open Journal of Analytical and Bioanalytical Chemistry, 4*(1), 011–019. doi:10.17352/ojabc.000019

Joy, E., Stein, A., Young, S., Ander, E., Watts, M., & Broadley, M. R. (2015). Zinc-enriched fertilizers as a potential public health intervention in Africa. *Plant and Soil*, *389*(1-2), 1–24. doi:10.100711104-015-2430-8

Jozsef, P., David, N., Sandor, K., & Aron, B. (2019). A comparison study of analytical performance of chromium speciation methods. *Microchemical Journal, 149*, 103958. https://doi.org/. microc.2019.05.058 doi:10.1016/j

Junaid, M., Hashmi, Z., Tang, M., Malik, R., & Pei, D. (2017). Potential health risk of heavy metals in the leather manufacturing industries in Sialkot, Pakistan. *Scientific Reports*, *7*(1), 8848. doi:10.103841598-017-09075-7 PMID:28821790

Jyothi, N. (2020). *Heavy metal sources and their effects on human health*. Intech Open.

Kabata-Pendias, A. (2010). *Trace elements in soils and plants*. CRC Press. doi:10.1201/b10158

Kabata-Pendias, A. (2011). *Trace metals in soils and plants*. CRC Press.

Kabir, E., Ray, S., Kim, K., Yoon, H., Jeon, E., Kim, Y., Cho, Y.-S., Yun, S.-T., & Brown, R. J. C. (2012). Current status of trace metal pollution in soils affected by industrial activities. *TheScientificWorldJournal*, *916705*, 1–18. doi:10.1100/2012/916705 PMID:22645468

Kagambega, N., Sawadogo, S., & Gordio, A. (2014). High arsenic enrichment in water and soils from Sambayourou watershed-Burkina Faso (West Africa). *International Journal of Environmental Monitoring and Analysis, 2*(6-1), 6-12. doi:10.11648/j.ijema.s.2014020601.12

Kagambega, N., Sawadogo, S., Bamba, O., Zombre, P., & Galvez, R. (2014). Acid mine drainage and heavy metals contamination of surface water and soil in southwest Burkina Faso–West Africa. *International Journal of Multidisciplinary Academic Research*, *2*, 9–19.

Kalenga, J. (2019). Assessment of heavy metal concentrations in streams and economic effects in Haut-Katanga province of the Democratic Republic of Congo. *The Hosei University Economic Review*, *86*(3-4), 1–21. doi:10.15002/00021806

Kamunda, C., Mathuthu, M., & Madhuku, M. (2016). Health risk assessment of heavy metals in soils from Witwatersrand gold mining basin, South Africa. *International Journal of Environmental Research and Public Health*, *13*(7), 663. doi:10.3390/ijerph13070663 PMID:27376316

Kang, Z., Wang, S., Qin, J., Wu, R., & Li, H. (2020). Pollution characteristics and ecological risk assessment of heavy metals in paddy fields of Fujian province, China. *Scientific Reports*, *10*(1), 12244. doi:10.103841598-020-69165-x PMID:32699372

Kapusta, P., & Sobczyk, L. (2015). Effects of heavy metal pollution from mining and smelting on enchytraeid communities under different land management and soil conditions. *The Science of the Total Environment*, *536*, 517–526. doi:10.1016/j.scitotenv.2015.07.086 PMID:26233783

Karimi, B., Masson, V., Guilland, C., Leroy, E., Pellegrinelli, S., Giboulot, E., Maron, P.-A., & Ranjard, L. (2021). Ecotoxicity of copper input and accumulation for soil biodiversity in vineyards. *Environmental Chemistry Letters*, *19*(3), 2013–2030. doi:10.100710311-020-01155-x

Karim, Z., Qureshi, A., & Mumtaz, M. (2015). Geochemical baseline determination and pollution assessment of heavy metals in urban soils of Karachi, Pakistan. *Ecological Indicators*, *48*, 358–364. doi:10.1016/j.ecolind.2014.08.032

Karn, R., Ojha, N., Abbas, S., & Bhugra, S. (2021). A review on heavy metal contamination at mining sites and remedial techniques. *IOP Conference Series. Earth and Environmental Science*, *796*(1), 012013. doi:10.1088/1755-1315/796/1/012013

Kartal, S., Aydin, Z., & Tokalioglu, S. (2006). Fractionation of metals in street sediment samples by using the BCR sequential extraction procedures and multivariate statistical elucidation of the data. *Journal of Hazardous Materials*, *132*(1), 80–89. doi:10.1016/j.jhazmat.2005.11.091 PMID:16466857

Kasten-Jolly, J., Heo, Y., & Lawrence, D. (2010). Impact of developmental lead exposure on splenic factors. *Toxicology and Applied Pharmacology*, *247*(2), 105–115. doi:10.1016/j.taap.2010.06.003 PMID:20542052

Katana, C., Murungi, J., & Mbuvi, H. (2013). Speciation of chromium and nickel in open air automobile mechanic workshop soils in Ngara, Nairobi, Kenya. *World Environment*, *3*(5), 143–154. doi:10.5923/j.env.20130305.01

Katarzyna, S., Grazyna, D., Lukasz, Z., Grazyna, P., Katarzyna, G., & Barbara, A. (2016). Determination of mercury in selected environmental components using cold vapor atomic absorption spectrometry. *Bulletin off the Maritime Institute in Gdansk*, *31*(1), 73-79. https://doi.org/ doi:10.5604/12307424.1201262

Kayode, O., Aizebeokhai, A., & Odukoya, A. (2021). Arsenic in agricultural soils and implications for sustainable agriculture. *IOP Conference Series. Earth and Environmental Science, 655*(1), 012081. doi:10.1088/1755-1315/655/1/012081

Kayode, O., Ogunyemi, E., Odukoya, A., & Aizebeokhai, A. (2022). Assessment of chromium and nickel in agricultural soils: Implications for sustainable agriculture. *IOP Conference Series. Earth and Environmental Science, 993*(1), 012014. doi:10.1088/1755-1315/993/1/012014

Kaza, S., Yao, L., Bhada-Tata, P., & Woerden, V. (2018). *What is waste 2. a global snapshot of solid waste management to 2050*. World Bank Publications, The World Bank Group. doi:10.1596/978-1-4648-1329-0

Kelishadi, R. (2012). Environmental pollution: Health effects and operational implications for pollutants removal. *Journal of Environmental and Public Health, 341637*, 1–2. doi:10.1155/2012/341637 PMID:22619687

Kepa, P., & Zaborska, A. (2021). Sources, fate and distribution of inorganic contaminants in the Svalbard area, representative of a typical arctic critical environment-a review. *Environmental Monitoring and Assessment, 193*(11), 724. doi:10.100710661-021-09305-6 PMID:34648070

Khalid, S., Shahid, M., Niazi, N., Murtaza, B., & Bibi, I. (2017). A comparison of technologies for remediation of heavy metal contaminated soils. *Journal of Geochemical Exploration, 182* (part B), 247-268. doi:10.1016/j.gexplo.2016.11.021

Khan, A., Chattha, R., Farooq, K., Jawed, A., & Farooq, M.. (2015). Effect of farmyard manure levels and NPK applications on the pea plant growth, pod yield and quality. *Life Sciences International Journal, 9*, 3178–3181.

Khan, I., Chowdhury, S., & Techato, K. (2022). Waste to energy in developing countries–a rapid review: Opportunities, challenges, and policies in selected countries of Sub-Saharan Africa and south Asia towards sustainability. *Sustainability (Basel), 7*(7), 3740. doi:10.3390u14073740

Khan, M., Khan, S., Khan, A., & Alam, M. (2017). Soil contamination with cadmium, consequences and remediation using organic amendments. *The Science of the Total Environment, 601*, 1591–1605. doi:10.1016/j.scitotenv.2017.06.030 PMID:28609847

Khan, N., Mobin, M., Abbas, K., & Alamri, A. (2017). Fertilizers and their contaminants in soils, surface and groundwater. In A. Dominick, S. Della, & M. Goldstein (Eds.), *The encyclopedia of the Anthropocene* (pp. 225–240). Elsevier. doi:10.1016/B978-0-12-809665-9.09888-8

Khan, S., Naushad, M., Lima, E., Zhang, S., Shaheen, S., & Rinklebe, J. (2021). Global soil pollution by toxic elements: Current status and future perspectives on the risk assessment and remediation strategies-a review. *Journal of Hazardous Materials, 417*, 126039. doi:10.1016/j.jhazmat.2021.126039 PMID:34015708

Kharazi, A., Leili, M., Khazaei, M., Alikhani, M., & Shokoohi, R. (2021). Human health risk assessment of heavy metals in agricultural soil and food crops in Hamadan, Iran. *Journal of Food Composition and Analysis, 100*, 103890. doi:10.1016/j.jfca.2021.103890

Kharb, A. (2018). Industrial revolution–from industry 1.0 to industry 4.0. *Journal of Advances in Computational Intelligence and Communication Technologies*, *2*(1), 1–3.

Khoo, S., Chia, Y., Chew, W., & Show, L. (2021). Microalgal-Bacterial consortia as future prospect in wastewater bioremediation, environmental management and bioenergy production. *Indian Journal of Microbiology*, *61*(3), 262–269. doi:10.100712088-021-00924-8 PMID:34294991

Kihampa, C., Kaisi, G., & Kihampa, H. (2016). Assessing the contribution of industrial wastewater to toxic metals contamination in receiving urban rivers, Dar es Salaam city, Tanzania. *Elixir Pollution*, *93*, 39532–39541.

Kim, D., Ock, J., Moon, K., & Park, C. (2021). Associations between Pb, Cd and Hg exposure and liver injury among Korean adults. *International Journal of Environmental Research and Public Health*, *18*(13), 6783. doi:10.3390/ijerph18136783 PMID:34202682

Kim, H., Choi, H., Joo, O., Kim, K., Choi, W., & Oh, B.-K. (2015). Development of a microbe-zeolite carrier for the effective elimination of heavy metals from seawater. *Journal of Microbiology and Biotechnology*, *25*(9), 1542–1546. doi:10.4014/jmb.1504.04067 PMID:26032363

Kim, H., Kim, Y., & Seo, Y. (2015). An overview of carcinogenic heavy metal: Molecular toxicity mechanism and prevention. *Journal of Cancer Prevention*, *20*(4), 232–241. doi:10.15430/JCP.2015.20.4.232 PMID:26734585

Kim, J., Park, J., Choi, J., & Kim, J. (2020). Determination of metal concentration in road-side trees from an industrial area using laser ablation inductively coupled plasma mass spectrometry. *Minerals (Basel)*, *10*(2), 175. doi:10.3390/min10020175

Kim, Y., & Kim, M. (2015). Arsenic toxicity in male reproduction and development. *Development & Reproduction*, *19*(4), 167–180. doi:10.12717/DR.2015.19.4.167 PMID:26973968

Kinimo, C., Yao, M., Marcotte, S., Kouassi, B., & Trokourey, A. (2021). Trace metal(loid)s contamination in paddy rice (Oryza sativa L.) from wetlands near two goldmines in Cote d'Ivoire and health risk assessment. *Environmental Science and Pollution Research International*, *28*(18), 22779–22788. Advance online publication. doi:10.100711356-021-12360-8 PMID:33423204

Kinuthia, G., Ngure, V., Lugalia, R., Wangila, A., & Kamau, L. (2020). Levels of heavy metals in wastewater and soil samples from open drainage channels in Nairobi, Kenya: Community health implication. *Scientific Reports*, *10*(1), 8434. doi:10.103841598-020-65359-5 PMID:32439896

Kirsti, L., Muller, I., Reichel, S., Jones, C., Brunet, F., & Guedard, M. (2022). Risk management for arsenic in agricultural soil-water systems: Lessons learned from case studies in Europe. *Journal of Hazardous Materials*, *424*, 127677. doi:10.1016/j.jhazmat.2021.127677 PMID:34774350

Kishe, M., & Machiwa, J. (2003). Distribution of heavy metals in sediments of Mwanza Gulf of Lake Victoria, Tanzania. *Environment International*, *28*(7), 619–625. doi:10.1016/S0160-4120(02)00099-5 PMID:12504158

Kocadal, K., Alkas, F., Battal, D., & Saygi, S. (2020). Cellular pathologies and genotoxic effects arising secondary to heavy metal exposure: A review. *Human and Experimental Toxicology*, *39*(1), 3–13. doi:10.1177/0960327119874439 PMID:31496299

Koch, J., & Gunther, D. (2017). Laser ablation inductively coupled plasma mass spectrometry. In J. Lindon, G. Tranter, & D. Koppenaal (Eds.), *Encyclopedia of spectroscopy and spectrometry* (pp. 526–532). Academic Press., doi:10.1016/B978-0-12-803224-4.00024-8

Kocot, K., Pytlakowska, K., Zawisza, B., & Sitko, R. (2016). How to detect metal species preconcentrated by microextraction techniques? *Trends in Analytical Chemistry*, *82*, 412–424. doi:10.1016/j.trac.2016.07.003

Kong, F., Chen, Y., Huang, L., Yang, Z., & Zhu, K. (2021). Human health risk visualization of potentially toxic elements in farmland soil: A combined method of source and probability. *Ecotoxicology and Environmental Safety*, *211*, 111922. doi:10.1016/j.ecoenv.2021.111922 PMID:33472110

Koptsik, G. (2014). Modern approaches to remediation of heavy metal polluted soils: A review. *Eurasian Soil Science*, *47*(7), 707–722. doi:10.1134/S1064229314070072

Kortei, K., Koryo-Dabrah, A., Akonor, P., Manaphraim, N., Akonor, M., & Boadi, N. (2020). Potential health risk assessment of toxic metals contamination in clay eaten as pica (geophagia) among pregnant women of Ho in the Volta Region of Ghana. *BMC Pregnancy and Childbirth*, *20*(1), 160. doi:10.118612884-020-02857-4 PMID:32169034

Kouadio, L., Tillous, K., Coulibaly, V., Sei, J., & Martinez, H. (2023). Assessment of metallic pollution of water and soil from illegal gold mining sites in Kong 2, Hire and Degbezere (Ivory Coast). *Journal of Materials & Environmental Sciences*, *14*(1), 41–61.

Koumolou, L., Edorh, P., Montcho, S., Aklikokou, K., Loko, F., Boko, M., & Creppy, E. E. (2013). Health-risk market garden production linked to heavy metals in irrigation water in Benin. *Comptes Rendus Biologies*, *336*(5-6), 278–283. doi:10.1016/j.crvi.2013.04.002 PMID:23916203

Kowalska, J., Mazurek, R., Gasiorek, M., Setlak, M., Zaleski, T., & Waroszewski, J. (2016). Soil pollution indices conditioned by medieval metallurgical activity: A case study from Krakow (Poland). *Environmental Pollution*, *218*, 1023–1036. doi:10.1016/j.envpol.2016.08.053 PMID:27574802

Kowalska, J., Mazurek, R., Gasiorek, M., & Zaleski, T. (2018). Pollution indices as useful tools for the comprehensive evaluation of the degree of soil contamination-a review. *Environmental Geochemistry and Health*, *40*(6), 2395–2420. doi:10.100710653-018-0106-z PMID:29623514

Kozlova, T., Wood, C., & McGeer, J. (2009). The effect of water chemistry on the acute toxicity of nickel to the cladoceran Daphnia pulex and the development of a biotic ligand model. *Aquatic Toxicology (Amsterdam, Netherlands)*, *91*(3), 221–228. doi:10.1016/j.aquatox.2008.11.005 PMID:19111357

Kribek, B., De Vivo, B., & Davies, T. (2014). Impacts of mining and mineral processing on the environment and human health in Africa. *Journal of Geochemical Exploration*, *144*(part C), 387–390. doi:10.1016/j.gexplo.2014.07.018

Kribek, B., Majer, V., Pasava, J., Kamona, F., Mapani, B., Keder, J., & Ettler, V. (2014). Contamination of soils with dust fallout from the tailings dam at the Rosh Pinah area, Namibia: Regional assessment, dust dispersion modelling and environmental consequences. *Journal of Geochemical Exploration*, *144*, 391–408. doi:10.1016/j.gexplo.2014.01.010

Kubier, A., Wilkin, R., & Pichler, T. (2019). Cadmium in soils and groundwater: A review. *Applied Geochemistry*, *108*, 1–16. doi:10.1016/j.apgeochem.2019.104388 PMID:32280158

Kuffour, R., Tiimub, B., & Agyapong, D. (2018). Impacts of illegal mining (Galamsey) on the environment (water and soil) at Bontefufuo area in the Amansie west district. *Journal of Environment and Earth Science*, *8*(7), 98–107.

Kulikova, T., Hiller, E., Jurkovič, L., Filová, L., Šottník, P., & Lacina, P. (2019). Total mercury, chromium, nickel and other trace chemical element contents in soils at an old cinnabar mine site (Merník, Slovakia): Anthropogenic versus natural sources of soil contamination. *Environmental Monitoring and Assessment*, *191*(5), 263. doi:10.100710661-019-7391-6 PMID:30953219

Kulka, M. (2016). A review of paraoxonase 1 properties and diagnostic applications. *Polish Journal of Veterinary Sciences*, *19*(1), 225–232. doi:10.1515/pjvs-2016-0028 PMID:27096809

Kumar, A., Bawge, G., & Kumar, V. (2021). An overview of industrial revolution and technology of industrial 4.0. *International Journal of Research in Engineering and Science*, *9*(1), 64–71.

Kumar, M., Sawhney, N., & Lal, R. (2021). Chemistry of heavy metals in the environment. In V. Kumar, A. Sharma, & A. Cerdia (Eds.), *Heavy metals in the environment*. Elsevier. doi:10.1016/B978-0-12-821656-9.00002-X

Kumar, N., Bauddh, K., Kumar, S., Dwivedi, N., Singh, D., & Barman, S. (2012). Extractability and phytotoxicity of heavy metals present in petrochemical industry sludge. *Clean Technologies and Environmental Policy*, *15*(6), 1033–1039. doi:10.100710098-012-0559-1

Kumi-Boateng, B. (2007). *Assessing the spatial distribution of arsenic concentration from goldmine for environmental management at Obuasi, Ghana*. [Msc Thesis, International Institute for Geo-information Science and Earth Observation, The Netherlands].

Kupper, M., Weinbruch, S., Skaug, V., Skogstad, A., Thornér, E., Benker, N., Ebert, M., Chashchin, V., Odland, J. Ø., & Thomassen, Y. (2015). Electron microscopy of particles deposited in the lungs of nickel refinery workers. *Analytical and Bioanalytical Chemistry*, *407*(21), 6435–6445. doi:10.100700216-015-8806-z PMID:26077746

Kurfurst, U., Desaules, A., Rehnert, A., & Muntau, H. (2004). Estimation of measurement uncertainty by the budget approach for heavy metal content in soils under different land use. *Accreditation and Quality Assurance*, *9*, 64–75. doi:10.100700769-003-0697-6

Kurt, M. (2018). Comparison of trace element and heavy metal concentrations of top and bottom soils in a complex land use area. *Carpathian Journal of Earth and Environmental Sciences*, *13*(1), 47–56. doi:10.26471/cjees/2018/013/005

Kylander, M., Rauch, S., Morrison, G., & Andam, K. (2003). Impact of automobile emissions on the levels of platinum and lead in Accra, Ghana. *Journal of Environmental Monitoring*, *5*(1), 91–95. doi:10.1039/b211736c PMID:12619761

Lakkireddy, K., & Kues, U. (2017). Bulk isolation of basidiospores from wild mushrooms by electrostatic attraction with low risk of microbial contaminations. *AMB Express*, *7*(1), 1–22. doi:10.118613568-017-0326-0 PMID:28124290

Lamas, G., Goertz, C., Boineau, R., Mark, D., Rozema, T., Nahin, R. L., Drisko, J. A., & Lee, K. L. (2012). Design of the trial to assess chelation therapy (TACT). *American Heart Journal*, *163*(1), 7–12. doi:10.1016/j.ahj.2011.10.002 PMID:22172430

Laribi, A., Shand, C., Wendler, R., Mouhouche, B., & Colinet, G. (2019). Concentrations and sources of Cd, Cr, Cu, Fe, Ni, Pb and Zn in soil of the Mitidja plain, Algeria. *Toxicological and Environmental Chemistry*, *101*(1-2), 59–74. doi:10.1080/02772248.2019.1619744

Lee, J., Kaunda, R., Sinkala, T., Workman, C., Bazilian, M., & Clough, G. (2021). Phytoremediation and phytoextraction in sub-Saharan Africa: Addressing economic and social challenges. *Ecotoxicology and Environmental Safety*, *226*, 112864. doi:10.1016/j.ecoenv.2021.112864 PMID:34627045

Lee, P., Kang, M., Yu, S., & Kwon, Y. (2020). Assessment of trace metal pollution in roof dusts and soils near a large Zn smelter. *The Science of the Total Environment*, *713*, 136536. doi:10.1016/j.scitotenv.2020.136536 PMID:31955082

Lentini, P., Zanoli, L., Granata, A., Signorelli, S., Castellino, P., & Aquila, R. (2017). Kidney and heavy metals-the role of environmental exposure [review]. *Molecular Medicine Reports*, *15*(5), 3413–3419. doi:10.3892/mmr.2017.6389 PMID:28339049

Leong, H., Zaine, A., Ho, C., Uemura, Y., & Lam, K. (2019). Impact of various microalgal-bacterial populations on municipal wastewater bioremediation and its energy feasibility for lipid-based biofuel production. *Journal of Environmental Management*, *249*, 109384. doi:10.1016/j.jenvman.2019.109384 PMID:31419674

Lepp, N. (2012). *Effect of heavy metal pollution on plants. metals in the environment, pollution monitoring series, applied science publishers. department of biology*. Liverpool Polytechnic.

Lewandowski, I., Schmidt, U., Londo, M., & Faaij, A. (2006). The economic value of the phytoremediation function - assessed by the example of cadmium remediation by willow (Salix ssp). *Agricultural Systems*, *89*(1), 68–89. doi:10.1016/j.agsy.2005.08.004

Lewicki, S., Zdanowski, R., Krzyzowska, M., Lewicka, A., Debski, B., Niemcewicz, M., & Goniewicz, M. (2014). The role of chromium III in the organism and its possible use in diabetes and obesity treatment. *Annals of Agricultural and Environmental Medicine*, *21*(2), 331–335. doi:10.5604/1232-1966.1108599 PMID:24959784

Lian, G., Wang, B., Lee, X., Li, L., Liu, T., & Lyu, W. (2019). Enhanced removal of hexavalent chromium by engineered biochar composite fabricated from phosphogypsum and distillers' grains. *The Science of the Total Environment*, *697*, 134119. doi:10.1016/j.scitotenv.2019.134119 PMID:32380611

Liaw, J., Marshall, G., Yuan, Y., Ferreccio, C., Steinmaus, C., & Smith, H. (2008). Increased childhood liver cancer mortality and arsenic in drinking water in northern Chile. *Cancer Epidemiology, Biomarkers & Prevention*, *17*(8), 1982–1987. doi:10.1158/1055-9965.EPI-07-2816 PMID:18708388

Li, F., Li, X., Hou, L., & Shao, A. (2018). Impact of the coal mining on the spatial distribution of potentially toxic metals in farmland tillage soil. *Scientific Reports*, *8*(1), 14925. doi:10.103841598-018-33132-4 PMID:30297728

Li, F., Ma, C., & Zhang, P. (2020). Mercury deposition, climate change and anthropogenic activities: A review. *Frontiers in Earth Science (Lausanne)*, *8*, 1–17. doi:10.3389/feart.2020.00316

Li, G., & Xing, J. (2020). The present situation of soil pollution in agricultural production and the countermeasures. *IOP Conference Series. Earth and Environmental Science*, *512*(1), 012032. doi:10.1088/1755-1315/512/1/012032

Li, L., Pan, D., Li, B., Wu, Y., Wang, H., Gu, Y., & Zuo, T. (2017). Patterns and challenges in the copper industry in China. *Resources, Conservation and Recycling*, *127*, 1–7. doi:10.1016/j.resconrec.2017.07.046

Li, M., Yang, W., Sun, T., & Jin, Y. (2016). Potential ecological risk of heavy metal contamination in sediments and macrobenthos in coastal wetlands induced by freshwater releases: A case study in the Yellow River Delta, China. *Marine Pollution Bulletin*, *103*, 227–239. doi:10.1016/j.marpolbul.2015.12.014 PMID:26719069

Lim, D., Roh, T., Kim, M., Kwon, Y., Choi, S., Kwack, S., Kim, K. B., Yoon, S., Kim, H. S., & Lee, B.-M. (2018). Non-cancer, cancer, and dermal sensitization risk assessment of heavy metals in cosmetics. *Journal of Toxicology and Environmental Health. Part A.*, *20*(81), 432–452. doi:10.1080/15287394.2018.1451191 PMID:29589992

Lindahl, J. (2014). *Environmental Impacts of Mining in Zambia: towards Better Environmental Management and Sustainable Exploitation of Mineral Resources, Geological Survey of Sweden.* Semantic Scholar. https://pdfs.semanticscholar.org/6ff8/34376b87410e8f618ebbf80271e6d23eb937.pdF

Liu, G., Cai, Y., O'Driscoll, N., Feng, X., & Jiang, G. (2012). Overview of mercury in the environment. In G. Liu & Y. Cai (Eds.), *N. O'Driscoll. Environmental chemistry and toxicology of mercury*. John Wiley & Sons.

Liu, G., Zhou, X., Li, Q., Shi, Y., Guo, G. L., Zhao, L., Wang, J., Su, Y., & Zhang, C. (2020). Spatial distribution prediction of soil as in a large-scale arsenic slag contaminated site based on an integrated model and multi-source environmental data. *Environmental Pollution*, *267*, 115631. doi:10.1016/j.envpol.2020.115631 PMID:33254608

Liu, J., Liu, C., Fang, G., & Wang, S. (2015). Advanced analytical methods and sample preparation for ion chromatography techniques. *RSC Advances*, *5*(72), 58713–58726. doi:10.1039/C5RA10348G

Liu, K., & Shi, X. (2001). In vivo reduction of chromium (VI) and its related free radical generation. *Molecular and Cellular Biochemistry*, *222*(1/2), 41–47. doi:10.1023/A:1017994720562 PMID:11678610

Liu, L., Li, W., Song, W., & Guo, M. (2018). Remediation techniques for heavy metal-contaminated soils: Principles and applicability. *The Science of the Total Environment*, *633*, 206–219. doi:10.1016/j.scitotenv.2018.03.161 PMID:29573687

Liu, Q., Li, M., Duan, J., Wu, H., & Hong, X. (2013). Analysis on influence factors of soil Pb and Cd in agricultural soils of Changsha suburb based on geographically weighted regression model. *Nongye Gongcheng Xuebao (Beijing)*, *29*, 225–234.

Liu, S., Jiang, J., Wang, S., Guo, Y., & Ding, H. (2018). Assessment of water-soluble thiourea-formaldehyde (WTF) resin for stabilization/solidification (S/S) of heavy metal contaminated soils. *Journal of Hazardous Materials*, *346*, 167–173. doi:10.1016/j.jhazmat.2017.12.022 PMID:29274510

Liu, S., Wang, X., Guo, G., & Yan, Z. (2021). Status and environmental management of soil mercury pollution in China: A review. *Journal of Environmental Management*, *277*, 111442. doi:10.1016/j.jenvman.2020.111442 PMID:33069151

Liu, Z., Xiao, F., Perkins, B., Zhu, M., Zhu, J., Xiong, Y., & Ning, P. (2017). Geogenic cadmium pollution and potential health risks, with emphasis on black shale. *Journal of Geochemical Exploration*, *176*, 42–49. doi:10.1016/j.gexplo.2016.04.004

Li, Y., Yang, K., Gao, W., Han, Q., & Zhang, J. (2021). A spectral characteristic analysis method for distinguishing heavy metal pollution in crops: VMD-PCA-SVM. *Spectrochimica Acta. Part A: Molecular and Biomolecular Spectroscopy*, *255*, 119649. doi:10.1016/j.saa.2021.119649 PMID:33744840

Li, Z., & Liu, Z. (2017). Hazards, sources and control measures of heavy metal pollution of forest soil: Taking Jin-Jing-Ji region of China as an example. *Nature Environment and Pollution Technology*, *16*(4), 1141–1147.

Lloyd, R., Hanna, P., & Mason, R. (1997). The origin of the hydroxyl radical oxygen in the Fenton reaction. *Free Radical Biology & Medicine*, *22*(5), 885–888. doi:10.1016/S0891-5849(96)00432-7 PMID:9119257

Loha, K., Lamoree, M., Weiss, J., & Boer, J. (2018). Import, disposal and health impacts of pesticides in East Africa Rift (EAR) zone: A review on management and policy analysis. *Crop Protection (Guildford, Surrey)*, *112*, 322–331. doi:10.1016/j.cropro.2018.06.014

Lombi, E., & Susini, J. (2009). Synchrotron-based techniques for plant and soil science: Opportunities, challenges and future perspectives. *Plant and Soil*, *320*(1-2), 1–35. doi:10.100711104-008-9876-x

Long, Z., Zhu, H., Bing, H., Tian, X., Wang, Z., Wang, X., & Wu, Y. (2021). Contamination, sources and health risk of heavy metals in soil and dust from different functional areas in an industrial city of Panzhihua city, Southwest China. *Journal of Hazardous Materials*, *420*, 126638. doi:10.1016/j.jhazmat.2021.126638 PMID:34280716

Loska, K., Wiechula, D., Barska, B., Cebula, E., & Chojnecka, A. (2003). Assessment of arsenic enrichment of cultivated soils in southern Poland. *Polish Journal of Environmental Studies*, *12*(2), 187.

Loska, K., Wiechuła, D., & Korus, I. (2004). Metal contamination of farming soils affected by industry. *Environment International*, *30*(2), 159–165. doi:10.1016/S0160-4120(03)00157-0 PMID:14749104

Lu, G., & Bai, Q. (2010). Contamination and potential mobility assessment of heavy metals in urban soils of Hangzhou, China: Relationship with different land uses. *Environmental Earth Sciences*, *60*(7), 1481–1490. doi:10.100712665-009-0283-2

Luilo, G., & Othman, O. (2006). Lead pollution in urban roadside environment of Dar es Salaam city. *Tanzania Journal of Science*, *32*(20), 61–67.

Luo, L., Mei, K., Qu, Y., Zhang, C., Chen, H., et al. (2019). Assessment of the geographical detector method for investigating heavy metal source apportionment in an urban watershed of Eastern China. *Science of the Total Environment*, *653*, 714e722. doi:10.1016/j.scitotenv.2018.10.424

Lusilao-Makiese, J., Cukrowska, E., Tessier, E., Almouroux, D., & Weiersbye, I. (2013). The impact of post gold mining on mercury pollution in the West Rand region, Gauteng, South Africa. *Journal of Geochemical Exploration*, *134*, 111–119. doi:10.1016/j.gexplo.2013.08.010

Lu, X., Gu, A., Huang, C., Wei, Y., Xu, M., Yin, H., & Hu, X.-F. (2021). Assessments of heavy metal pollution of a farmland in an urban area based on the environmental geochemical baselines. *Journal of Soils and Sediments*, *21*(7), 2659–2671. doi:10.100711368-021-02945-8

Lu, X., Wang, L., Lei, K., Huang, J., & Zhai, Y. (2009). Contamination assessment of copper, lead, zinc, manganese and nickel in street dust of Baoji, NW China. *Journal of Hazardous Materials*, *161*(2–3), 1058–1062. doi:10.1016/j.jhazmat.2008.04.052 PMID:18502044

Lu, Y., Jenkins, A., Ferrier, R., Bailey, M., Gordon, I., Song, S., Huang, J., Jia, S., Zhang, F., Liu, X., Feng, Z., & Zhang, Z. (2015). Addressing China's grand challenge of achieving food security while ensuring environmental sustainability. *Science Advances*, *1*(1), e1400039. doi:10.1126ciadv.1400039 PMID:26601127

Madden, C., Pringle, K., Jeffery, A., Wisniewski, K., Heaton, V., Oliver, I., Glanville, H., Stimpson, I. G., Dick, H. C., Eeley, M., & Goodwin, J. (2022). Portable X-ray fluorescence (pXRF) analysis of heavy metal contamination in church graveyards with contrasting soil types. *Environmental Science and Pollution Research International*, *29*(36), 55278–55292. doi:10.100711356-022-19676-z PMID:35318600

Magalhaes, D., Marques, M., Baptista, D., & Buss, D. (2015). Metal bioavailability and toxicity in freshwaters. *Environmental Chemistry Letters*, *13*(1), 69–87. doi:10.100710311-015-0491-9

Magos, L., & Clarkson, T. (2006). Overview of the clinical toxicity of mercury. *Annals of Clinical Biochemistry*, *43*(4), 257–268. doi:10.1258/000456306777695654 PMID:16824275

Mahey, S., Kumar, R., Sharma, M., Kumar, V., & Bhardwaj, R. (2020). A critical review on toxicity of cobalt and its remediation strategies. *SN Applied Sciences*, *2*(7), 1279. doi:10.100742452-020-3020-9

Mahmood, M., Yousra, M., Saman, L., & Ahmad, R. (2020). Floriculture: Alternate non-edible plants for phyto-remediation of heavy metal contaminated soils. *International Journal of Phytoremediation*, *22*(7), 725–732. doi:10.1080/15226514.2019.1707772 PMID:31916455

Makinen, E., Korhonen, M., Viskari, L., Haapamaki, S., Jarvinen, M., & Lu, L. (2006). Comparison of XRF and FAAS methods in analyzing CCA contaminated soils. *Water, Air, and Soil Pollution*, *171*(1-4), 95–110. doi:10.100711270-005-9017-6

Makokha, O., Mghweno, L., Magoha, H., Nakajugo, A., & Wekesa, J. (2011). The effects of environmental lead pollution in Kisumu, Mwanza and Kampala. *The Open Environmental Engineering Journal*, *4*, 133–140. doi:10.2174/1874829501104010133

Makondo, C., Mundike, J., & Mwaanga, P. (2013). Lead deposition from mobile sources: A case study of Ndola-Kitwe dual carriage highway. *American Journal of Environmental Protection*, *2*(6), 128–133. doi:10.11648/j.ajep.20130206.12

Malan, M., Muller, F., Cyster, L., Raitt, L., & Aalbers, J. (2015). Heavy metals in the irrigation water, soils and vegetables in the Philippi horticultural area in the Western Cape Province of South Africa. *Environmental Monitoring and Assessment*, *187*(1), 4085. doi:10.100710661-014-4085-y PMID:25380711

Maleki, N., Safavi, A., & Doroodmand, M. (2005). Determination of selenium in water and soil by hydride generation atomic absorption spectrometry using solid reagents. *Talanta*, *66*(4), 858–862. doi:10.1016/j.talanta.2004.12.053 PMID:18970063

Malkoc, S., & Yazici, B. (2017). Multivariate analyses of heavy metals in surface soil around an organized industrial area in Eskisehir, Turkey. *Bulletin of Environmental Contamination and Toxicology*, *98*(2), 244–250. doi:10.100700128-016-1991-4 PMID:27942760

Malunguja, G., Thakur, B., & Devi, A. (2022). Heavy metal contamination of forest soils by vehicular emissions: Ecological risks and effects on tree productivity. *Environmental Processes*, *9*(1), 11. doi:10.100740710-022-00567-x

Ma, M., Dong, S., Jin, W., Zhang, C., & Zhou, W. (2019). Fate of the organo-phosphorous pesticide profenofos in cotton fiber. *Journal of Environmental Science and Health. Part B, Pesticides, Food Contaminants, and Agricultural Wastes*, *54*(1), 70–75. doi:10.1080/03601234.2018.1505036 PMID:30633718

Mandal, B., & Suzuki, K. (2002). Arsenic round the world: A review. *Talanta*, *58*(1), 201–235. doi:10.1016/S0039-9140(02)00268-0 PMID:18968746

Mandina, S., & Tawanda, M. (2013). Chromium, an essential nutrient and pollutant: A review. African. *Journal of Pure Applied Chemistry*, *7*, 310–317. doi:10.5897/AJPAC2013.0517

Mandina, S., & Tawanda, M. (2013). Speciation of chromium in soils, plants and wastewater at a ferrochrome slag dump in Gweru. *IOSR Journal of Environmental Science, Toxicology and Food Technology*, *7*(4), 43–49. doi:10.9790/2402-0744349

Mansour, H., Awad, F., Saber, M., & Zaghloul, A. (2020). Effect of contamination sources on the rate of zinc, copper and nickel release from various soil ecosystems. *Bulletin of the National Research Center*, *44*(1), 178. doi:10.118642269-020-00431-8

Mantey, J., Nyarko, K., Owusu-Nimo, F., Awua, K., Bempah, C., Amankwah, R., Akatu, W., & Effah, A. (2020). Mercury contamination of soil and water media from different illegal artisanal small-scale gold mining operations (galamsey). *Heliyon*, *6*(6), e04312. doi:10.1016/j.heliyon.2020.e04312 PMID:32637700

Manyiwa, T., Ultra, V. Jr, Rantong, G., Opaletswe, K., Gabankitse, G., Taupedi, S., & Gajaje, K. (2022). Heavy metals in soil, plants and associated risk on grazing ruminants in the vicinity of Cu-Ni mine in Selebi-Phikwe, Botswana. *Environmental Geochemistry and Health*, *44*(5), 1633–1648. doi:10.100710653-021-00918-x PMID:33855629

Marastoni, L., Sandri, M., Pii, Y., Valentinuzzi, F., Brunetto, G., Cesco, S., & Mimmo, T. (2019). Synergism and antagonisms between nutrients induced by copper toxicity in grapevine rootstocks: Monocropping vs. intercropping. *Chemosphere*, *214*, 563–578. doi:10.1016/j.chemosphere.2018.09.127 PMID:30286423

Marcelo, B., Isabel, D., & Mariela, P. (2017). A low-cost device for sample introduction and determination of mercury by cold vapor atomic absorption spectrometry-application for irrigation water and paddy soil. *Brazilian Journal of Analytical Chemistry*, *4*(14), 34.

Marchetti, C. (2014). Interaction of metal ions with neurotransmitter receptors and potential role in neurodiseases. *Biometals*, *27*(6), 1097–1113. doi:10.100710534-014-9791-y PMID:25224737

Marin, B., Chopin, B., Jupinet, B., & Gauthier, D. (2008). Comparison of microwave-assisted digestion procedures for total trace element determination in calcareous soils. *Talanta*, *77*(1), 282–288. doi:10.1016/j.talanta.2008.06.023 PMID:18804634

Markus, J., & McBratney, A. (2001). A review of the contamination of soil with lead II. Spatial distribution and risk assessment of soil lead. *Environment International*, *27*(5), 399–411. doi:10.1016/S0160-4120(01)00049-6 PMID:11757854

Marove, C., Sotozono, R., Tangviroon, P., Baltazar, C., & Igarashi, T. (2022). Assessment of soil, sediment and water contaminations around open pit coal mines in Moatize, Tete Province, Mozambique. *Environmental Advances*, *8*, 100215. doi:10.1016/j.envadv.2022.100215

Mar, S., & Okazaki, M. (2012). Investigation of Cd contents in several phosphate rocks used for the production of fertilizer. *Microchemical Journal*, *104*, 17–21. doi:10.1016/j.microc.2012.03.020

Martinez, D., Vucic, A., Becker-Santos, D., Gil, L., & Lam, L. (2011). Arsenic exposure and the induction of human cancers. *Journal of Toxicology*, *2011*, 1–13. doi:10.1155/2011/431287 PMID:22174709

Martínez-Guijarro, R., Paches, M., Romero, I., & Aguado, D. (2021). Sources, mobility, reactivity, and remediation of heavy metal(loid) pollution: a review. *Advances in Environmental and Engineering Research, 2*(4), 033. doi:10.21926/aeer.2104033

Masindi, V., & Muedi, K. (2019). *Environmental contamination by heavy metals*. Intech Open Publishers.

Mason, R., Choi, A., Fitzgerald, W., Hammerschmidt, C., Lamborg, C., Soerensen, A., & Sunderland, E. M. (2012). Mercury biogeochemical cycling in the ocean and policy implications. *Environmental Research*, *119*, 101–117. doi:10.1016/j.envres.2012.03.013 PMID:22559948

Mathebula, M., Panichev, N., & Mandiwana, K. (2020). Determination of mercury thermospecies in South African coals in the enhancement of mercury removal by pre-combustion technologies. *Scientific Reports*, *10*(1), 19282. doi:10.103841598-020-76453-z PMID:33159166

Mathee, A. (2014). Towards the prevention of lead exposure in South Africa: Contemporary and emerging challenges. *Neurotoxicology*, *45*, 220–223. doi:10.1016/j.neuro.2014.07.007 PMID:25086205

Mathee, A., Kootbodien, T., Kapwata, T., & Naicker, N. (2018). Concentrations of arsenic and lead in residential garden soil from four Johannesburg neighborhoods. *Environmental Research*, *167*, 524–527. doi:10.1016/j.envres.2018.08.012 PMID:30142628

Mathee, A., Street, R., Teare, J., & Naicker, N. (2020). Lead exposure in the home environment: An overview of risks from cottage industries in Africa. *Neurotoxicology*, *81*, 34–39. doi:10.1016/j.neuro.2020.08.003 PMID:32835764

Mazurek, R., Kowalska, J., Gasiorek, M., Zadrozny, P., Jozefowska, A., Zaleski, T., Kępka, W., Tymczuk, M., & Orłowska, K. (2017). Assessment of heavy metals contamination in surface layers of Roztocze National Park Forest soils (SE Poland) by indices of pollution. *Chemosphere*, *168*, 839–850. doi:10.1016/j.chemosphere.2016.10.126 PMID:27829506

Mbongwe, B., Barnes, B., Tshabang, J., Zhai, M., Rajoram, S., & Mpuchane, S. (2010). Exposure to lead among children aged 1-6 years in the city of Gaborone, Botswana. *Journal of Environmental Health Research*, *10*, 17–26. doi:10.13140/RG.2.2.17136.20484

Mekki, A., & Sayadi, S. (2017). Study of heavy metal accumulation and residual toxicity in soil saturated with phosphate processing wastewater. *Water, Air, and Soil Pollution*, *228*(6), 215. doi:10.100711270-017-3399-0 PMID:28603317

Mekonnen, B., Haddus, A., & Zeine, W. (2020). Assessment of the effect of solid waste dump site on surrounding soil and river water quality in Tepi town, Southwest Ethiopia. *Journal of Environmental and Public Health*, *5157046*, 1–9. doi:10.1155/2020/5157046 PMID:32587623

Mekuria, D., Kassegne, A., & Asfaw, S. (2021). Assessing pollution profiles among little Akaki river receiving municipal and industrial wastewaters, central Ethiopia: Implications for emvironmental and public health safety. *Heliyon*, *7*(7), e07526. doi:10.1016/j.heliyon.2021.e07526 PMID:34337176

Meng, D., Zhao, N., Ma, M., Fang, L., Gu, Y., Jia, Y., Liu, J., & Liu, W. (2017, June 20). Application of a mobile laser- induced breakdown spectroscopy system to detect heavy metal elements in soil. *Applied Optics*, *56*(18), 5204. doi:10.1364/AO.56.005204 PMID:29047571

Mensah, K., Marschner, B., Shaheen, M., Wang, J., Wang, L., & Rinklebe, J. (2020). Arsenic contamination in abandoned and active gold mine spoils in Ghana: Geochemical fractionation, speciation, and assessment of the potential human health risk. *Environmental Pollution*, *261*, 114116. doi:10.1016/j.envpol.2020.114116 PMID:32220748

MEP, Ministry of Environmental Protection. (2014). *National soil contamination survey report*. MEP. https://www.pollutionsolutions-online.com/white-paper/soil remediation/18/xportreporter/soil-contamination-industry-report/9

Merian, E., & Clarkson, T. (1991). *Metals and their compounds in the environment: occurrence, analysis, and biological relevance*. VCH.

Messner, B., Knoflach, M., Seubert, A., Ritsch, A., Pfaller, K., Henderson, B., Shen, Y. H., Zeller, I., Willeit, J., Laufer, G., Wick, G., Kiechl, S., & Bernhard, D. (2010). Cadmium is a novel and independent risk factor for early atherosclerosis mechanisms and in vivo relevance. *Arteriosclerosis, Thrombosis, and Vascular Biology*, *29*(9), 1392–1398. doi:10.1161/ATVBAHA.109.190082 PMID:19556524

Metcalf, S., & Veiga, M. (2012). Using street theatre to increase awareness of and reduce mercury pollution in the artisanal gold mining sector: A case from Zimbabwe. *Journal of Cleaner Production*, *37*, 179–184. doi:10.1016/j.jclepro.2012.07.004

Michalski, R. (2018). Ion Chromatography applications in wastewater analysis. *Separations*, *5*(1), 16. doi:10.3390eparations5010016

Mico, C., Peris, M., Sanchez, J., & Recatala, L. (2008). Trace element analysis via open-vessel or microwave-assisted digestion in calcareous Mediterranean soils. *Communications in Soil Science and Plant Analysis*, *39*(5-6), 890–904. doi:10.1080/00103620701881246

Mico, C., Recatala, L., Peris, M., & Sanchez, J. (2007). A comparison of two digestion methods for the analysis of heavy metals by flame atomic absorption spectroscopy. *Spectroscopy Europe*, *19*(1), 23–27.

Mihaljevic, M., Ettler, V., Sebek, O., Sracek, O., Kribek, B., Kyncl, T., Majer, V., & Veselovský, F. (2011). Lead isotopic and metallic pollution record in tree rings from the Copperbelt mining-smelting area, Zambia. *Water, Air, and Soil Pollution*, *216*(1-4), 657–668. doi:10.100711270-010-0560-4

Mikoczy, Z., & Hagmar, L. (2005). Cancer incidence in the Swedish leather tanning industry: Updated findings 1958-99. *Occupational and Environmental Medicine*, *62*(7), 461–464. doi:10.1136/oem.2004.017038 PMID:15961622

Mileusnic, M., Mapani, S., Kamona, F., Ruzicic, S., Mapaure, I., & Chimwamurombe, P. M. (2014). Assessment of agricultural soil contamination by potentially toxic metals dispersed from improperly disposed tailings, Kombat mine, Namibia. *Journal of Geochemical Exploration*, *144*, 409–420. doi:10.1016/j.gexplo.2014.01.009

Miller, D., Collins, M., Omotayo, M., Martin, L., Dickin, L., & Young, L. (2018). Geophagic earths consumed by women in western Kenya contain dangerous levels of lead, arsenic, and iron. *American Journal of Human Biology*, *30*(4), e23130. doi:10.1002/ajhb.23130 PMID:29722093

Milton, A., Hussain, S., Akter, S., Rahman, M., Mouly, T., & Mitchell, K. (2017). A review of the effects of chronic arsenic exposure on adverse pregnancy outcomes. *International Journal of Environmental Research and Public Health*, *14*(6), 556. doi:10.3390/ijerph14060556 PMID:28545256

Minamata Convention on Mercury. (2019). *Home*. UN. https://www.mercuryconvention.org/

Minnesota Department of Health. (2021). Heavy metal in fertilizers. Minnesota Department of Health. https://www.health.state.mn.us/communities/environment/risk/studies/metals.html

Minnikova, V., Denisova, V., Mandzhieva, S., Kolesnikov, I., Minkina, M., Chaplygin, V. A., Burachevskaya, M. V., Sushkova, S. N., & Bauer, T. V. (2017). Assessing the effect of heavy metals from the Novocherkassk power station emissions on the biological activity of soils in the adjacent areas. *Journal of Geochemical Exploration*, *174*, 70–78. doi:10.1016/j.gexplo.2016.06.007

Miranda, L., Wijesiri, B., Ayoko, G., Egodawatta, P., & Goonetilleke, A. (2021). Water-sediment interactions and mobility of heavy metals in aquatic environments. *Water Research*, *202*(1), 117386. doi:10.1016/j.watres.2021.117386 PMID:34229194

Mireles, F., Davila, J., Pinedo, J., Reyes, E., Speakman, R., & Glascock, M. (2012). Assessing urban soil pollution in the cities of Zacatecas and Guadalupe, Mexico by instrumental neutron activation analysis. *Microchemical Journal*, *103*, 158–164. doi:10.1016/j.microc.2012.02.009

Mishra, K., & Singh, S. (2020). Heavy metals exposure and risk to autoimmune diseases: A review. *Archives of Immunology and Allergy*, *3*(2), 22–26.

Mitra, S., Chakraborty, A., Tareq, A., Emran, T., Nainu, F., Khusro, A., Idris, A. M., Khandaker, M. U., Osman, H., Alhumaydhi, F. A., & Simal-Gandara, J. (2022). Impact of heavy metals on the environment and human health: Novel therapeutic insights to counter toxicity. *Journal of King Saud University. Science*, *34*(3), 101865. doi:10.1016/j.jksus.2022.101865

Miura, S., Takahashi, K., Imagawa, T., Uchida, K., Saito, S., Tominaga, M., & Ohta, T. (2013). Involvement of TRPA1 activation in acute pain induced by cadmium in mice. *Molecular Pain*, *9*, 1744-8069-9-7. Advance online publication. doi:10.1186/1744-8069-9-7 PMID:23448290

Miyazawa, M., Pavan, M., Oliveira, E., Ionashiro, M., & Silva, A. (2000). Gravimetric determination of soil organic matter. Brazilian Archives of Biology and Biotechnology, 43(5), 475-478. https://doi.org/ doi:10.1590/S1516-89132000000500005

Mkhize, T. (2020). *Assessment of heavy metal contamination in soils around Krugersdorp mining area, Johannesburg, South Africa.* [Msc Thesis, University of Kwa-Zulu Natal, South Africa].

Mladenov, N., Wolski, P., Hettiarachchi, M., Murray-Hudson, M., Enriquez, H., & Damaraju, S. (2014). Abiotic and biotic factors influencing the mobility of arsenic in groundwater of a through-flow island in the Okavango Delta, Botswana. *Journal of Hydrology (Amsterdam)*, *518*, 326–341. doi:10.1016/j.jhydrol.2013.09.026

Mmbaga, T., & Semu, E. (1999). Contents of heavy metals in some soils of the Morogoro municipality, Tanzania, as a result of cottage-scale metal working operations. *The International Journal of Environmental Studies*, *56*(3), 373–383. doi:10.1080/00207239908711211

Mngadi, S., Sihlahla, M., Lekoadu, S., Moja, S., & Nomngongo, P. (2020). Evaluation of mobility, fractionation and potential risk of trace metals present in soils from Struibult gold mine dumps. *Journal of African Earth Sciences*, *127*, 104008. doi:10.1016/j.jafrearsci.2020.104008

Moeckel, C., Breivik, K., Nost, T., Sankoh, A., Jones, K., & Sweetman, A. (2020). Soil pollution at a major West African e-site recycling site: Contamination pathways and implications for potential mitigation strategies. *Environment International*, *137*, 105563. doi:10.1016/j.envint.2020.105563 PMID:32106045

Moeletsi, M., & Tongwane, M. (2020). Projected direct carbon dioxide emission reductions as a result of the adoption of electric vehicles in Gauteng province in South Africa. *Atmosphere (Basel)*, *11*(6), 591. doi:10.3390/atmos11060591

Mohajane, C., & Manjoro, M. (2022). Sediment-associated heavy metal contamination and potential ecological risk along an urban river in South Africa. *Heliyon*, *8*(12), e12499. doi:10.1016/j.heliyon.2022.e12499 PMID:36643299

Mohammed, S., & Folorunsho, J. (2015). Heavy metals concentration in soil and Amaranthus retroflexus grown on irrigated farmlands in the Makera Area, Kaduna, Nigeria. *Journal of Geography and Regional Planning*, *8*(8), 210–217. doi:10.5897/JGRP2015.0498

Mohanty, M., & Patra, H. (2011). Attenuation of chromium toxicity by bioremediation technology. In D. M. Whitacre (Ed.), *Reviews of environmental contamination and toxicology* (Vol. 210, pp. 1–34). Springer. doi:10.1007/978-1-4419-7615-4_1

Mondal, M., Begum, W., Nasrollahzadeh, M., Ghorbannezhad, F., & Antoniadis, V. (2021). A comprehensive review on chromium chemistry along with detection, speciation, extraction and remediation of hexavalent chromium in contemporary science and technology. *Vietnam Journal of Chemistry*, *59*(6), 711–732. doi:10.1002/vjch.202100048

Mongi, R., & Chove, L. (2020). Heavy metal contamination in cocoyam crops and soils in countries around the Lake Victoria basin (Tanzania, Uganda and Kenya). *Tanzanian Journal of Agricultural Sciences*, *19*(2), 148–160.

Morrison, M., Goldhaber, B., Lee, L., Holloway, M., Wanty, B., Wolf, E., & Ranville, J. F. (2009). A regional-scale study of chromium and nickel in soils of northern California, USA. *Applied Geochemistry*, *24*(8), 1500–1511. doi:10.1016/j.apgeochem.2009.04.027

Mosa, K., Saadoun, I., Kumar, K., Helmy, M., & Dhankher, O. (2016). Potential biotechnological strategies for the cleanup of heavy metals and metalloids. *Frontiers in Plant Science*, *7*. doi:10.3389/fpls.2016.00303 PMID:27014323

Moyo, B., Matodzi, V., Legodi, M., Pakade, V., & Tavengwa, N. (2020). Determination of Cd, Mn and Ni in fruits, vegetables and soil in the Thohayandou town area, South Africa. *Water S.A.*, *46*(2), 285–290. doi:10.17159/wsa/2020.v46.i2.8244

Muchuweti, M., Birkett, J., Chinyanga, E., Zvauya, R., Scrimshaw, M., & Lester, J. (2006). Heavy metal content of vegetables irrigated with mixtures of wastewater and sewage sludge in Zimbabwe: Implications for human health. *Agriculture, Ecosystems & Environment*, *112*(1), 41–48. doi:10.1016/j.agee.2005.04.028

Mudimbu, D., Davies, T., Tagwireyi, D., & Meck, M. (2022). Application of geochemical indices evaluating potentially harmful element contamination at mining centers in the Sanyati catchment, Zimbabwe. *Frontiers in Environmental Science*, *10*, 829900. doi:10.3389/fenvs.2022.829900

Mudzielwana, R., Gitari, W., Akinyemi, A., Talabi, O., & Ndungu, P. (2020). Hydrogeochemical characteristics of arsenic rich groundwater in Greater Giyani municipality, Limpopo province, South Africa. *Groundwater for Sustainable Development*, *10*, 100336. doi:10.1016/j.gsd.2020.100336

Muhammed, A., Hussen, A., Redi, M., & Kaneta, T. (2021). Remote investigation of total chromium determination in environmental samples of the Kombolcha industrial zone, Ethiopia, using microfluidic paper-based analytical devices. *Analytical Sciences*, *37*(4), 585–591. doi:10.2116/analsci.20P325 PMID:33041309

Muhune, S., Marsi, M., Tshisand, P., Nonga, W., & Kayembe, O. (2020). evaluation of soil contamination by metallic trace elements to the roadside on the Lubumbashi-Kipushi section (DRC). *International Journal of Advanced Research*, *8*(9), 1187–1195. doi:10.21474/IJAR01/11775

Muimba, K., Banza, L., Mwitwa, J., Kampemba, F., & Mulele, N. (2022). Impacts of trace metals pollution of water, food crops and ambient air on population health in Zambia and the DR Congo. *Journal of Environmental and Public Health, 4515115*, 1–14. doi:10.1155/2022/4515115

Mukhopadhyay, S., Masto, R., Tripathi, R., & Srivastava, N. (2019). Application of soil quality indicators for the phyto-restoration of mine spoil dumps. In V. Pandey & K. Bauddh (Eds.), *Phytomanagement of polluted sites* (pp. 361–388)., doi:10.1016/B978-0-12-813912-7.00014-4

Muller, G. (1981). The heavy metal pollution of the sediments of Neckars and its tributary: A stocktaking. *Chemiker-Zeitung, 105*, 157–164.

Muller, J. (2020). *Enabling Technologies for Industry 5.0: Results of a Workshop with Europe's Technology Leaders*. European Commission.

Mungai, T., Owino, A., Makokha, V., Gao, Y., Yan, X., & Wang, J. (2016). Occurrences and toxicological risk assessment of eight heavy metals in agricultural soils from Kenya, Eastern Africa. *Environmental Science and Pollution Research International, 23*(18), 18533–18541. doi:10.100711356-016-7042-1 PMID:27291978

Mungai, T., & Wang, J. (2020). Heavy metal pollution in suburban topsoil in Nyeri, Kapsabet, Voi, Ngong and Juja towns in Kenya. *SN Applied Sciences, 1*(9), 960. doi:10.100742452-019-0996-0

Musa, D., Sadiya, M., Yusuf, M., Garba, Y., & Gimba, E. (2018). Study of water and soil contamination by heavy metal from industrial effluents at Bompai industrial area Kano, Nigeria. *FUW Trends in Science and Technology Journal, 3*(1), 234–238.

Mustafa, S., & AlSharif, M. (2018). Copper (Cu) an Essential Redox-Active Transition Metal in Living System—A Review Article. *American Journal of Analytical Chemistry, 9*(1), 15–26. doi:10.4236/ajac.2018.91002

Mutale, C., Syampungani, S., Festin, E., Tigabu, M., & Daneshvar, A. (2020). Physicochemical characteristics and heavy metal concentrations of copper mine wastes in Zambia: Implications for pollution risk and restoration. *Journal of Forestry Research, 31*(4), 1283–1293. doi:10.100711676-019-00921-0

Mutune, A., Makobe, M., & Abukutsa-Onyango, M. (2014). Heavy metal content of selected African leafy vegetables planted in urban and peri-urban Nairobi, Kenya. *African Journal of Environmental Science and Technology, 8,* 66–74. http://dx.doi.org/. 1573 doi:10.5897/ajest2013

Mutune, A., Makobe, M., & Onyango, M. (2014). Heavy metal content of selected leafy vegetables planted in urban and peri-urban Nairobi, Kenya. *African Journal of Environmental Science and Technology, 8*(1), 66-74. https://doi.org/. 1573 doi:10.5897/AJEST2013

Muwanga, A., & Barifaijo, E. (2006). Impact of industrial activities on heavy metal loading and their physicochemical effects on wetlands of Lake Victoria basin (Uganda). *African Journal of Science and Technology, 7*(1). doi:10.4314/ajst.v7i1.55197

Mwaanga, P., Silondwa, M., Kasali, G., & Banda, P. (2019). Preliminary review of mine air pollution in Zambia. *Heliyon, 5*(9), e02485. doi:10.1016/j.heliyon.2019.e02485 PMID:31687579

Mwalongo, D., Haneklaus, N., Lisuma, J., Kivevele, T., & Mtei, K. (2022). Uranium in phosphate rocks and mineral fertilizers applied to agricultural soils in East Africa. *Environmental Science and Pollution Research International*, *30*(12), 1–9. doi:10.100711356-022-24574-5 PMID:36496520

Mwamburi, J. (2016). Chromium distribution and spatial variations in the finer sediment grain size fraction and unfractioned surficial sediments on Nyanza gulf, of Lake Victoria (East Africa). *Journal of Waste Management*, *7528263*, 1–15. doi:10.1155/2016/7528263

Mwesigye, R., Young, S., Bailey, E., & Tumwebaze, B. (2016). Population exposure to trace elements in the Kilembe copper mine area, Western Uganda: A pilot study. *The Science of the Total Environment*, *15*(573), 366–375. doi:10.1016/j.scitotenv.2016.08.125 PMID:27572529

Nabulo, G., Young, C., & Black, C. (2010). Assessing risk to human health from tropical leafy vegetables grown on contaminated urban soils. *The Science of the Total Environment*, *408*(22), 5338–5535. doi:10.1016/j.scitotenv.2010.06.034 PMID:20739044

Nael, M., Khademi, H., Jalalain, A., Schulin, R., Kalbasi, M., & Sotohian, F. (2009). Effect of geo-pedological conditions on the distribution and chemical speciation of selected trace elements in forest soils of Western Alborz, Iran. *Geoderma*, *152*(1-2), 157–170. doi:10.1016/j.geoderma.2009.06.001

Naicker, N., & Mathee, A. (2015). Trends in lead exposure in a rural mining town in South Africa, 1991-2008. *South African Medical Journal*, *105*(7), 515. doi:10.7196/SAMJnew.7809 PMID:26447246

Nakayama, S., Ikenaka, Y., Hamada, K., Muzandu, K., Choongo, K., Teraoka, H., Mizuno, N., & Ishizuka, M. (2011). Metal and metalloid contamination in roadside soil and wild rats around a Pb-Zn mine in Kabwe, Zambia. *Environmental Pollution*, *59*(1), 175–181. doi:10.1016/j.envpol.2010.09.007 PMID:20971538

Nakkeeran, E., Patra, C., Shahnaz, T., Rangabhashiyam, S., & Selvaraju, N. (2018). Continuous biosorption assessment for the removal of hexavalent chromium from aqueous solutions using *Strychnos nux vomica* fruit shell. *Bioresource Technology Reports*, *3*, 256–260. doi:10.1016/j.biteb.2018.09.001

Namuhani, N., & Kimumwe, C. (2015). Soil contamination with heavy metals around Jinja steel rolling mills in Jinja, Municipality, Uganda. *Journal of Health & Pollution*, *9*(9), 61–67. doi:10.5696/2156-9614-5-9.61 PMID:30524777

National Center for Biotechnology Information. (2023). *PubChem Compound Summary for CID 935, Nickel. [Online]* NCBI. https://pubchem.ncbi.nlm.nih.gov/#query=zinc; https://pubchem.ncbi.nlm.nih.gov/#query=copper

National Center for Biotechnology Information. (2023). PubChem Compound Summary for CID 935, Nickel. NCBI. https://pubchem.ncbi.nlm.nih.gov/compound/Nickel

Naz, B., & Batool, S. (2017). Infertility related issues and challenges: Perspectives of patients, spouses, and infertility experts. *Pakistan Journal of Clinical Sociology and Psychology*, *15*, 3–11.

Nduka, J., & Orisakwe, O. (2010). Assessment of environmental distribution of lead in some municipalities of South-Eastern Nigeria. *International Journal of Environmental Research and Public Health*, *7*(6), 2501–2513. doi:10.3390/ijerph7062501 PMID:20644686

Ndungu, A., Yan, X., Makokha, V., Githiga, K., & Wang, J. (2019). Occurrence and risk assessment of heavy metals and organochlorine pesticides in surface soils, central Kenya. *Journal of Environmental Health Science & Engineering*, *17*(1), 63–73. doi:10.100740201-018-00326-x PMID:31321038

Neeraj, A., Hiranmai, R., & Iqbal, K. (2022). Comprehensive assessment of pollution indices, sources apportionment and ecological risk mapping of heavy metals in agricultural soils of Raebareli district, Uttar Pradesh, India, employing a GIS approach. *Land Degradation & Development*, *34*(1), 173–195. doi:10.1002/ldr.4451

Ngole, V., & Ekosse, G. (2012). Copper, nickel and zinc contamination in soils within the precincts of mining and landfilling environments. *International Journal of Environmental Science and Technology*, *9*(3), 485–494. doi:10.100713762-012-0055-5

Ngure, V., & Kinuthia, G. (2020). Health risk implications of lead, cadmium, zinc and nickel for consumers of food items in Migori gold mines, Kenya. *Journal of Geochemical Exploration*, *209*, 106430. doi:10.1016/j.gexplo.2019.106430

Nierengarten, J. (2018). In my element: Copper. *Chemistry (Weinheim an der Bergstrasse, Germany)*, *25*(1), 16–18. doi:10.1002/chem.201805277

Nigam, H., Das, M., Chauhan, S., Pandey, P., Swati, P., & Yadav, M. (2015). Effect of chromium generated by solid waste of tannery and microbial degradation of chromium to reduce its toxicity: A review. *Advances in Applied Science Research*, *6*(3), 129–136.

Nipen, M., Jorgensen, S., Nizzetto, P., Borga, K., Breivik, K., Mmochi, A., Mwakalapa, E., & … . (2022). Mercury in air and soil on an urban-rural transect in East Africa. *Environmental Science. Processes & Impacts*, *24*(6), 921–931. doi:10.1039/D2EM00040G PMID:35583028

Njoku, C., & Agwu, O. (2017). Assessment of the solid waste disposal on ground water quality in selected Tube wells and Bore holes in Kano metropolis North West Nigeria. *Journal of Applied chemistry*, *10*(7), 56–60. doi:10.9790/5736-1007035660

Njoloma, C. (2012). *Application of foliar sprays containing copper, zinc and boron to mature clonal tea (Camellia sinensis): effect on yield and quality*. [MSc Thesis, University of Pretoria, South Africa].

Nkonya, E., Johnson, T., Kwon, H., & Kato, E. (2016). Economics of land degradation in Sub-Saharan Africa. In E. Nkonya, A. Mirzabaev, & J. von Braun (Eds.), *Economics of land degradation and improvement – a global assessment for sustainable development*. Springer. doi:10.1007/978-3-319-19168-3_9

Nkosi, S., & Msimango, N. (2022). Screening of zinc, copper and iron in Chinese cabbage cultivated in Durban, South Africa, towards human health risk assessment. *South African Journal of Science*, *118*(11/12), 12099. doi:10.17159ajs.2022/12099

Nordstrom, S., Beckman, L., & Nordenson, I. (1978). Occupational and environmental risks in and around a smelter in northern Sweden. *Hereditas*, *88*(1), 43–46. doi:10.1111/j.1601-5223.1978.tb01600.x PMID:649423

Nriagu, J. (1994). *Arsenic in the environment*. Wiley.

Nriagu, J. (2002). Arsenic poisoning through the ages. In W. T. Frankenberger (Ed.), *Environmental chemistry of arsenic* (pp. 1–26). Marcel Dekker.

Nunes, N., Ragonezi, C., Gouveia, C., & Carvalho, M. (2020). Review of sewage sludge as a soil amendment in relation to current international guidelines: A heavy metal perspective. *Sustainability (Basel)*, *13*(4), 2317. doi:10.3390u13042317

Nunes, P., Munita, S., Vasconcellos, A., Oliveira, S., Croci, A., & Faleiros, F. (2009). Characterization of soil samples according to their metal content. *Journal of Radioanalytical and Nuclear Chemistry*, *281*(3), 359–363. doi:10.100710967-009-0016-8

Nwaichi, E., & Dhankher, O. (2016). Heavy metals contaminated environments and the road map with phytoremediation. *Journal of Environmental Protection*, *7*(1), 41–51. doi:10.4236/jep.2016.71004

Nyanza, C., Bernier, P., Manyama, M., Hatfield, J., Martin, W., & Dewey, D. (2019). Maternal exposure to arsenic and mercury in small-scale gold mining areas of Northern Tanzania. *Environmental Research*, *173*, 432–442. doi:10.1016/j.envres.2019.03.031 PMID:30974369

Nyanza, C., Dewey, D., Thomas, K., Davey, M., & Ngallaba, E. (2014). Spatial distribution of mercury and arsenic levels in water, soil and cassava plants in a community with long history of gold mining in Tanzania. *Bulletin of Environmental Contamination and Toxicology*, *93*(6), 716–721. doi:10.100700128-014-1315-5 PMID:24923470

Nyanza, E., Joseph, M., Premji, S., Thomas, D., & Mannion, C. (2014). Geophagy practices and the content of chemical elements in the soil eaten by pregnant women in artisanal and small scale gold mining communities in Tanzania. *BMC Pregnancy and Childbirth*, *14*(1), 144. doi:10.1186/1471-2393-14-144 PMID:24731450

Nyika, J. (2021). Tolerance of microorganisms to heavy metals. In S. Dey & B. Acharya (Eds.), *Recent advancements in bioremediation of metal contaminants*. IGI Global. doi:10.4018/978-1-7998-4888-2.ch002

Nyika, J. (2022). Wastewater for agricultural production, benefits, risks, and limitations. In H. Chatoui, M. Merzouki, H. Moummou, M. Tilaoui, N. Saadaoui, & A. Brhich (Eds.), *Nutrition and human health*. Springer. doi:10.1007/978-3-030-93971-7_6

Nyika, J., & Dinka, M. (2021). Activated bamboo charcoal in water treatment: A mini-review. *Materials Today: Proceedings*, *56*(4), 1904–1907. doi:10.1016/j.matpr.2021.11.167

Nyika, J., & Dinka, M. (2022). A min-review on wastewater treatment through bioremediation towards enhanced field applications of the technology. *AIMS Environmental Science*, *9*(4), 403–431. doi:10.3934/environsci.2022025

Nyika, J., & Dinka, M. (2022). Heavy metal pollution in soils and vegetables from suburban regions of Nairobi, Kenya and their community health implications. *Pollution*, *8*(4), 1434–1447. doi:10.22059/POLL.2022.341522.1440

Nyika, J., & Onyari, E. (2019). Hydrogeochemical analysis and spatial distribution of groundwater quality in Roundhill landfill vicinity of South Africa. *Air, Soil and Water Research*, *12*, 1–8. doi:10.1177/1178622119872771

Nyika, J., Onyari, E., Dinka, M., & Mishra, S. (2019). Heavy metal pollution and mobility in soils within a landfill vicinity: A South African case study. *Oriental Journal of Chemistry*, *35*(4), 1286–1296. doi:10.13005/ojc/350406

Nyika, J., Onyari, E., Dinka, M., & Mishra, S. (2019a). A comparison of reproducibility of inductively coupled spectrometric techniques in soil metal analyses. *Air, Soil and Water Research*, *12*, 1–13. doi:10.1177/1178622119869002

Nyika, J., Onyari, E., Dinka, M., & Mishra, S. (2019a). Analysis of particle size distribution of landfill contaminated soils and their mineralogical composition. *Particulate Science and Technology*, *38*(7), 843–853. doi:10.1080/02726351.2019.1635238

Nyika, J., Onyari, E., Dinka, M., & Mishra, S. (2020a). Comparative assessment of trace metal concentrations and their eco-risk analysis in soils of the vicinity of Roundhill landfill, Southern Africa. *Nature Environment and Pollution Technology*, *19*(2), 539–548. doi:10.46488/NEPT.2020.v19i02.009

Nyika, J., Onyari, E., Dinka, M., & Mishra, S. (2020b). Assessment of trace metal contamination of soil in a landfill vicinity: A southern Africa case study. *Current Chemistry Letters*, *9*, 171–182. doi:10.5267/j.ccl.2020.2.003

Nyika, J., Onyari, E., Dinka, M., & Shivani, B. (2020). Assessment of trace metal contamination of soils in a landfill vicinity: A southern Africa case study. *Chemistry Letters*, *9*(4), 171–182. doi:10.46488/nept.2020v19i02.009

Nyika, J., Onyari, E., Dinka, M., & Shivani, B. (2021). A review on methods of assessing pollution levels from landfills in South Africa. *International Journal of Environment and Waste Management*, *28*(4), 436–455. doi:10.1504/IJEWM.2021.118859

O'Connor, D., Hou, D., Ok, Y., Mulder, J., Duan, L., Wu, Q., Wang, S., Tack, F., & Rinklebe, J. (2019). Mercury speciation, transformation and transportation in soils, atmospheric flux, and implications for risk management: A critical review. *Environment International*, *126*, 747–761. doi:10.1016/j.envint.2019.03.019 PMID:30878870

O'Day, P. (2006). Chemistry and mineralogy of arsenic. *Elements (Quebec)*, *2*(2), 77–83. doi:10.2113/gselements.2.2.77

O'Shea, M., Toupal, J., Caballero-Gómez, H., McKeon, T., Howarth, M., Pepino, R., & Gieré, R. (2021). Lead Pollution, Demographics, and Environmental Health Risks: The Case of Philadelphia, USA. *International Journal of Environmental Research and Public Health*, *18*(17), 9055. doi:10.3390/ijerph18179055 PMID:34501644

O'Sullivan, J., Lupakisyo, G., Purcell, J., Turner, N., & Mtalo, F. (2016). Assessing sediment and water quality issues in expanding African wetlands: The case of the Mara River, Tanzania. *The International Journal of Environmental Studies*, *73*(1), 95–107. doi:10.1080/00207233.2015.1116226

Obeng-Gyasi, E. (2022). Sources of lead exposure in West Africa. *Science*, *4*(3), 33. doi:10.3390ci4030033

Obone, E., Chakrabarti, S., Bai, M., Malick, M., Lamontagne, L., & Subramanian, K. (1999). Toxicity and bioaccumulation of nickel sulfate in Sprague-Dawley rats following 13 weeks of subchronic exposure. *Journal of Toxicology and Environmental Health. Part A.*, *57*(6), 379–401. doi:10.1080/009841099157593 PMID:10478821

Obrist, D., Pearson, C., Webster, J., Kane, T., Lin, C., Aiken, G., & Alpers, C. N. (2016). A synthesis of terrestrial mercury in the western United States: Spatial distribution defined by land cover and plant productivity. *The Science of the Total Environment*, *568*, 522–535. doi:10.1016/j.scitotenv.2015.11.104 PMID:26775833

Ocal, F., & Altintas, K. (2018). Dördüncü sanayi devriminin emek piyasaları üzerindeki olası etkilerinin incelenmesi ve çözüm önerileri. *OPUS Uluslararası Toplum Araştırmaları Dergisi*, *8*(15), 35–35. doi:10.26466/opus.439952

Odai, S., Mensah, E., Sipitey, D., Ryo, S., & Awuah, E. (2008). Heavy metals uptake by vegetables cultivated on urban waste dumpsites: Case study of Kumasi, Ghana. *Research Journal of Environmental Toxicology*, *2*(2), 92–99. doi:10.3923/rjet.2008.92.99

Odero, D., Semu, E., & Kamau, G. (2000). Assessment of cottage industry derived heavy metal pollution of soils within Ngara and Gikomba areas of Nairobi city, Kenya. *African Journal of science and Technology. Science and Engineering Series*, *1*, 52–62.

Odumo, B., Carbonell, G., Angeyo, H., Patel, J., Torrijos, M., & Martin, A. (2014). Impact of gold mining associated with mercury contamination in soil, biota sediments and tailings in Kenya. *Environmental Science and Pollution Research International*, *21*(21), 12426–12435. doi:10.100711356-014-3190-3 PMID:24943890

Ogiyama, S., Sakamoto, K., Suzuki, H., Ushio, S., Anzai, T., & Inubushi, K. (2006). Measurement of trace metals in arable soils with animal manure application using INAA and the concentrated acid digestion method. *Soil Science and Plant Nutrition*, *52*, 114–121. doi:10.1111/j.1747-0765.2006.00013.x

Ogola, S., Mitullah, V., & Omulo, A. (2002). Impact of gold mining on the environment and human health: A case study in the Migori gold belt, Kenya. *Environmental Geochemistry and Health*, *24*(2), 141–157. doi:10.1023/A:1014207832471

Ogunkunle, C., & Fatoba, P. (2014). Contamination and spatial distribution of heavy metals in topsoil surrounding a mega cement factory. *Atmospheric Pollution Research*, *5*(2), 270–282. doi:10.5094/APR.2014.033

Okereafor, G., Makhatha, M., Mekuto, L., & Mavumengwana, V. (2019). *Evaluation of trace elemental levels as pollution indicators in an abandoned goldmine dump in Ekurhuleni area, South Africa*. Intech Open.

Okereafor, U., Makhatha, M., Mekuto, L., Okereafor, N., & Sebola, T. (2020). Toxic metal implications on agricultural soils, plants, animals, aquatic life and human health. *International Journal of Environmental Research and Public Health*, *17*(7), 2204. doi:10.3390/ijerph17072204 PMID:32218329

Okonkwo, J., & Maribe, F. (2004). Assessment of lead exposure in Thohoyandou, South Africa. *The Environmentalist*, *24*(3), 171–178. doi:10.100710669-005-6051-2

Okonkwo, O. (2007). Arsenic status and distribution in soils at disused cattle dip in South Africa. *Bulletin of Environmental Contamination and Toxicology*, *79*(4), 380–383. doi:10.100700128-007-9255-y PMID:17701088

Okunola, J., Uzairu, A., & Ndukwe, G. (2007). Levels of trace metals in soil and vegetation along major and minor roads in metropolitan city of Kaduna, Nigeria. *African Journal of Biotechnology*, *6*, 1703–1709.

Olafisoye, O., Adefioye, T., & Osibote, O. (2013). Heavy metals contamination of water, soil and plants around an electronic waste dumpsite. *Polish Journal of Environmental Studies*, *22*, 1431–1439.

Olatunji, A., Kolawole, T., Oloruntola, M., & Gunter, C. (2018). Evaluation of pollution of soils and particulate matter around metal recycling factories in Southwestern Nigeria. *Journal of Health & Pollution*, *8*(7), 20–30. doi:10.5696/2156-9614-8.17.20 PMID:30524846

Oliver, W., Graham, C., MacKenzie, B., Ellam, M., & Farmer, G. (2008). Distribution and partitioning of depleted uranium (DU) in soils at weapons test ranges – Investigations combining the BCR extraction scheme and isotopic analysis. *Chemosphere*, *72*(6), 932–939. doi:10.1016/j.chemosphere.2008.03.029 PMID:18457863

Olobatoke, R., & Mathuthu, M. (2016). Heavy metal concentration in soil in the tailing dam vicinity of an old gold mine in Johannesburg, South Africa. *Canadian Journal of Soil Science*, *96*(3), 299–304. doi:10.1139/cjss-2015-0081

Oloniran, A., Balgobind, A., & Pillay, B. (2013). Bioavailability of heavy metals in soil: Impact on microbial biodegradation of organic compounds and possible improvement strategies. *International Journal of Molecular Sciences*, *14*(5), 10197–10228. doi:10.3390/ijms140510197 PMID:23676353

Oloruntegbe, K., Akinsete, M., & Odutuyi, M. (2009). Fifty years of oil exploration in Nigeria: Physicochemical impacts and implication for environmental accounting and development. *Journal of Applied Sciences Research*, *5*, 2131–2137.

Olowoyo, J., Lion, N., Unathi, T., & Oladeji, O. (2022). Concentrations of Pb and other associated elements in soil dust 15 years after the introduction of unleaded fuel and the human health implications in Pretoria, South Africa. *International Journal of Environmental Research and Public Health*, *19*(16), 10238. doi:10.3390/ijerph191610238 PMID:36011873

Olowoyo, J., Okedeyi, O., Mkolo, N., Lion, G., & Mdakane, S. (2012). Uptake and translocation of heavy metals by medicinal plants growing around a waste dump site in Pretoria, South Africa. *South African Journal of Botany*, *78*, 116–121. doi:10.1016/j.sajb.2011.05.010

Olowoyo, J., van Heerden, E., Fischer, J., & Baker, C. (2010). Trace metals in soil and leaves of Jacaranda mimosifolia in Tshwane area, South Africa. *Atmospheric Environment*, *44*(14), 1826–1830. doi:10.1016/j.atmosenv.2010.01.048

Omatoso, A., & Ojo, J. (2015). Assessment of some heavy metals contamination in the soil of river Niger floodplain at Jebba, central Nigeria. *Water Utility Journal*, *9*, 71–80.

Omobowale, O., Oyagbemi, A., Akinrinde, S., Saba, B., Daramola, T., Ogunpolu, S., & Olopade, J. O. (2014). Failure of recovery from lead induced hepatoxicity and disruption of erythrocyte antioxidant defense system in Wistar rats. *Environmental Toxicology and Pharmacology*, *37*(3), 1202–1211. doi:10.1016/j.etap.2014.03.002 PMID:24814264

Omotehinse, A., & Ako, B. (2019). The environmental implications of the exploration and exploitation of solid minerals in Nigeria with a special focus on tin in Jos and coal in Enugu. *Journal of Sustainable Mining*, *18*(1), 18–24. doi:10.1016/j.jsm.2018.12.001

Ondayo, M., Simiyu, G., Raburu, P., & Were, F. (2016). Child exposure to lead in the vicinities of informal used lead-acid battery recycling operations in Nairobi slums, Kenya. *Journal of Health & Pollution*, *12*(12), 15–25. doi:10.5696/2156-9614-6.12.15 PMID:30524801

Onyia, P., Ozoko, D., & Ifediegwu, S. (2020). Phytoremediation of arsenic-contaminated soils by arsenic hyperaccumulating plants in selected areas of Enugu state, Southeastern, Nigeria. *Geology. Ecology and Landscapes*, *5*(4), 308–319. doi:10.1080/24749508.2020.1809058

Opiyo, R., Osano, P., Mbandi, A., Apondo, W., & Muhoza, C. (2020). Commentary using citizen science to assess the cumulative risk from air and other pollution sources in informal settlements. *Clean Air Journal*, *30*(1), 1–4. doi:10.17159/caj/2020/30/1.8374

Orr, S., & Bridges, C. (2017). Chronic kidney disease and exposure to nephrotoxic metals. *International Journal of Molecular Sciences*, *18*(5), 1039. doi:10.3390/ijms18051039 PMID:28498320

Oruko, R., Edokpayi, J., Msagati, T., Tavengwa, N., Ogola, H., Ijoma, G., & Odiyo, J. (2021). Investigating the chromium status, heavy metal contamination and ecological risk assessment via tannery waste disposal in sub-Saharan Africa (Kenya and South Africa). *Environmental Science and Pollution Research International*, *28*(31), 42135–42149. doi:10.100711356-021-13703-1 PMID:33797722

Oruko, R., Odiyo, J., & Edokpayi, J. (2018). Chromium tanning, management challenges and environmental legislation in sub–Saharan African tanneries. *Proceedings of ICSMNR2018*, Polokwane, South Africa.

Oruko, R., Selvarajan, R., Ogola, H., Edokpayi, J., & Odiyo, J. (2020). Contemporary and future direction of chromium tanning and management in sub-Saharan Africa tanneries. *Process Safety and Environmental Protection*, *133*, 369–386. doi:10.1016/j.psep.2019.11.013

Owino, J., Ragama, P., Maghanga, C., & Njeru, N. (2015). Lead metal exposure to residents residing in informal settlement: A case study of residents in Nakuru municipality, Kenya. *Kabarak Journal of Research and Innovation*, *3*(1), 41–47.

Oyourou, J., McCrindle, R., Combrinck, S., & Fourie, C. (2019). Investigation of zinc and lead contamination of soil at the abandoned Edendale mine, Mamelodi (Pretoria, South Africa) using a field portable spectrometer. *Journal of the Southern African Institute of Mining and Metallurgy*, *119*(1), 55–62. doi:10.17159/2411-9717/2019/v119n1a7

Ozyigit, I., Yilmaz, S., Dogan, I., Sakcali, S., Tombuloglu, G., & Demir, G. (2016). Detection of physiological and genotoxic damages reflecting toxicity in kalanchoe clones. *Global NEST Journal*, *18*(1), 223–232. doi:10.30955/gnj.001349

Packer, M. (2016). Cobalt cardiomyopathy: a critical reappraisal in light of a recent resurgence. *Circulation: Heart Failure 9*(12). . doi:10.1161/CIRCHEARTFAILURE.116.003604

Palansooriya, N., Shaheen, M., Chen, S., Tsang, W., Hashimoto, Y., Hou, D., Bolan, N. S., Rinklebe, J., & Ok, Y. S. (2020). Soil amendments for immobilization of potentially toxic elements in contaminated soils: A critical review. *Environment International*, *134*, 105046. doi:10.1016/j.envint.2019.105046 PMID:31731004

Panagopoulos, I., Karayannis, A., Kollias, K., Xenidis, A., & Papassiopi, N. (2015). Investigation of potential soil contamination with Cr and Ni in four metal finishing facilities at Asopos industrial area. *Journal of Hazardous Materials*, *281*, 20–26. doi:10.1016/j.jhazmat.2014.07.040 PMID:25112552

Parades-Aguilar, J., Reyes-Martínez, V., Bustamante, G., Almendariz, J., Martínez-Meza, G. et al. (202)1. Removal of nickel (II) from wastewater using a zeolite-packed anaerobic bioreactor: bacterial diversity and community structure shifts. *Journal of Environmental Management, 279*, 111558 doi:10.1016/j.jenvman.2020.111558

Park, Y.-H., Kim, D., Dai, J., & Zhang, Z. (2015). Human bronchial epithelial BEAS-2B cells, an appropriate in vitro model to study heavy metals induced carcinogenesis. *Toxicology and Applied Pharmacology*, *287*(3), 240–245. doi:10.1016/j.taap.2015.06.008 PMID:26091798

Pattnaik, K., & Equeenuddin, M. (2016). Potentially toxic metal contamination and enzyme activities in soil around chromite mines at Sukinda ultramafic complex, India. *Journal of Geochemical Exploration*, *168*, 127–136. doi:10.1016/j.gexplo.2016.06.011

Perez, J. (2012). *The soil remediation industry in Europe: The recent past and future perspectives*. Europa. https://ec.europa.eu/environment/archives/soil/pdf/may2012/08%20-%20Julien%20Perez%20-%20final.pdf [Accessed on 5 February 2023]

Petelka, J., Abraham, J., Bockreis, A., Deikumah, J., & Zerbe, S. (2019). Soil heavy metal(loid) pollution and phytoremediation potential of native plants on a former gold mine in Ghana. *Water, Air, and Soil Pollution*, *230*(11), 267. doi:10.100711270-019-4317-4

Petit, D., & Rucandio, I. (1999). Sequential extractions for determination of cadmium distribution in coal fly ash, soil and sediment samples. *Analytica Chimica Acta*, *401*(1-2), 283–291. doi:10.1016/S0003-2670(99)00487-0

Pi, H., Xu, S., Reiter, J., Guo, P., Zhang, L., & Li, Y. (2015). SIRT3-SOD2- mROS-dependent autophagy in cadmium-induced hepatotoxicity and salvage by melatonin. *Autophagy*, *11*(7), 1037–1051. https://doi.org/. 1052208 doi:10.1080/15548627.2015

Pierce, L., Kibriya, G., Tong, L., Jasmine, F., Argos, M., Roy, S., Paul-Brutus, R., Rahaman, R., Rakibuz-Zaman, M., Parvez, F., Ahmed, A., Quasem, I., Hore, S. K., Alam, S., Islam, T., Slavkovich, V., Gamble, M. V., Yunus, M., Rahman, M., & Ahsan, H. (2012). Genome-wide association study identifies chromosome 10q24.32 variants associated with arsenic metabolism and toxicity phenotypes in Bangladesh. *PLOS Genetics*, *8*(2), e1002522. doi:10.1371/journal.pgen.1002522 PMID:22383894

Pi, J., Horiguchi, S., Sun, Y., Nikaido, M., Shimojo, N., Hayashi, T., Yamauchi, H., Itoh, K., Yamamoto, M., Sun, G., Waalkes, M. P., & Kumagai, Y. (2003). A potential mechanism for the impairment of nitric oxide formation caused by prolonged oral exposure to arsenate in rabbits. *Free Radical Biology & Medicine*, *35*(1), 102–113. doi:10.1016/S0891-5849(03)00269-7 PMID:12826260

Pikula, D., & Stepien, W. (2021). Effect of the degree of soil contamination with heavy metals on their mobility in the soil profile in a microplot experiment. *Agronomy (Basel)*, *11*(5), 878. doi:10.3390/agronomy11050878

Pitiya, R., Jacob, L., & Emilinot, R. (2022). A pilot study on the concentration of heavy metals in sediments from the lower orange river, Karas region, Namibia. *Journal of Materials Science and Chemical Engineering*, *10*(3), 1–14. doi:10.4236/msce.2022.103001

Plumlee, G., Durant, J., Morman, S., Neri, A., Wolf, R., Dooyema, C., Hageman, P. L., Lowers, H. A., Fernette, G. L., Meeker, G. P., Benzel, W. M., Driscoll, R. L., Berry, C. J., Crock, J. G., Goldstein, H. L., Adams, M., Bartrem, C. L., Tirima, S., Behbod, B., & Brown, M. J. (2013). Linking geological and health sciences to assess childhood lead poisoning from artisanal gold mining in Nigeria. *Environmental Health Perspectives*, *121*(6), 744–750. doi:10.1289/ehp.1206051 PMID:23524139

Plunkett, M., & Castle, J. (2010). *Soil organic matter and soil nutrient analysis soil organic matter*. Teagasc, Agriculture and Food Development Authority.

Podolsky, F., Ettler, V., Sebek, O., Jezek, J., Mihaljevic, M., Kribek, B., Sracek, O., Vaněk, A., Penížek, V., Majer, V., Mapani, B., Kamona, F., & Nyambe, I. (2015). Mercury in soil profiles from metal mining and smelting areas in Namibia and Zambia: Distribution and potential sources. *Journal of Soils and Sediments*, *15*(3), 648–658. doi:10.100711368-014-1035-9

Postigo, C., Martinez, D., Grondona, S., & Miglioranza, K. (2018). Groundwater pollution: sources, mechanisms and prevention. In A. Dominick & M. Goldstein (Eds.), *Encyclopedia of the Anthropocene*. Elsevier Publishers., doi:10.1016/B978-0-12-809665-9.09880-3

Pourret, O., Lange, B., Bonhoure, J., Colinet, G., Decree, S., Mahy, G., Séleck, M., Shutcha, M., & Faucon, M.-P. (2016). Assessment of soil metal distribution and environmental impact of mining in Katanga (Democratic Republic of Congo). *Applied Geochemistry*, *64*, 43–55. doi:10.1016/j.apgeochem.2015.07.012

Prasad, S., Yadav, K., Kumar, S., Gupta, N., Cabral-Pinto, M., Rezania, S., Radwan, N., & Alam, J. (2021). Chromium contamination and the effect on environmental health and its remediation: A sustainable approach. *Journal of Environmental Management*, *285*, 112174. doi:10.1016/j.jenvman.2021.112174 PMID:33607566

Prematuri, R., Turjaman, M., Sato, T., & Tawaraya, K. (2020). The impact of nickel mining on soil properties and growth of two fast-growing tropical tree species. *International Journal of Forestry Research*, *8837590*, 1–9. doi:10.1155/2020/8837590

Prohaska, J. (2000). Long-term functional consequences of malnutrition during brain development: Copper. *Nutrition (Burbank, Los Angeles County, Calif.)*, *16*(7-8), 502–504. doi:10.1016/S0899-9007(00)00308-7 PMID:10906536

Proshad, R., Kormoker, T., Mursheed, N., Islam, M., Bhuyan, I., & Islam, S. (2018). Heavy metal toxicity in agricultural soil due to rapid industrialization in Bangladesh: A review. *International Journal of Advanced Geosciences*, *6*(1), 83–88. doi:10.14419/ijag.v6i1.9174

Pueyo, M., Mateu, J., Rigol, A., Vidal, M., Lopez-Sanchez, F., & Rauret, G. (2008). Use of the modified BCR three-step sequential extraction procedure for the study of trace element dynamics in contaminated soils. *Environmental Pollution*, *152*(2), 330–341. doi:10.1016/j.envpol.2007.06.020 PMID:17655986

Puga, P., Abreu, A., Melo, A., Paz-Ferreiro, P., & Beesley, L. (2015). Cadmium, lead and zinc mobility and plant uptake in a mine soil amended with sugarcane straw biochar. *Environmental Science and Pollution Research International*, *22*(22), 17606–17614. doi:10.100711356-015-4977-6 PMID:26146374

Punshon, T., Jackson, P., Meharg, A., Warczack, T., Scheckel, K., & Guerinot, M. (2017). Understanding arsenic dynamics in agronomic systems to predict and prevent uptake by crop plants. *Science of the Total Environment,* (581 – 582), 209 - 220. doi:10.1016/j.scitotenv.2016.12.111

Qian, J., Wei, L., Liu, R., Jiang, F., Hao, X., & Chen, G. (2016). An exploratory study on the pathways of Cr (VI) reduction in sulfate-reducing up-flow anaerobic sludge bed (UASB) reactor. *Scientific Reports*, *6*(1), 23694. doi:10.1038rep23694 PMID:27021522

Qin, G., Niu, Z., Yu, J., Li, Z., Ma, J., & Xiang, P. (2021). Soil heavy metal pollution and food safety in China: Effects, sources and removing technology. *Chemosphere*, *267*, 129205. doi:10.1016/j.chemosphere.2020.129205 PMID:33338709

Qingjie, G., Jun, D., Yunchuan, X., Qingfei, W., & Liqiang, Y. (2008). Calculating pollution indices by heavy metals in ecological geochemistry assessment and a case study in parks of Beijing. *Journal of China University of Geosciences*, *19*(3), 230–241. doi:10.1016/S1002-0705(08)60042-4

Qu, C., Ma, Z., Yang, J., Liu, Y., Bi, J., & Huang, L. (2012). Human exposure pathways of heavy metals in a lead-zinc mining area, Jiangsu Province, China. *PLoS One*, *7*(11), e46793. doi:10.1371/journal.pone.0046793 PMID:23152752

Quenea, K., Larny, I., Winterton, P., Bermond, A., & Dumat, C. (2009). Interactions between metals and soil organic matter in various particle size fractions of soil contaminated with waste water. *Geoderma*, *149*(3-4), 217–223. doi:10.1016/j.geoderma.2008.11.037

Qu, M., Li, W., Zhang, C., Huang, B., & Zhao, Y. (2015). Assessing the pollution risk of soil chromium based on loading capacity of paddy soil at a regional scale. *Scientific Reports*, *5*(1), 18451. doi:10.1038rep18451 PMID:26675587

Qu, R., Han, G., Liu, M., & Li, X. (2019). The mercury behavior and contamination in soil profiles of Mun River basin, Northeast Thailand. *International Journal of Environmental Research and Public Health*, *16*(21), 4131. doi:10.3390/ijerph16214131 PMID:31717757

Rada, M. (2018). *Industry 5.0-from Virtual to Physical*. LinkedIn. https://www.linkedin.com/pulse/industry-50-from-virtual-physical-michael-rada

Rad, L., Momeni, A., Ghazani, B., Irani, M., Mahmoudi, M., & Noghreh, B. (2014). Removal of Ni2+ and Cd2+ ions from aqueous solutions using electrospun PVA/ zeolite nanofibrous adsorbent. *Chemical Engineering Journal*, *256*, 119–127. doi:10.1016/j.cej.2014.06.066

Radomirovic, M., Cirovic, Z., Maksin, D., Bakic, T., Lukic, J., Stankovic, S., & Onjia, A. (2020). Ecological Risk Assessment of Heavy Metals in the Soil at a Former Painting Industry Facility. *Frontiers in Environmental Science*, *8*, 560415. doi:10.3389/fenvs.2020.560415

Raffa, M., Chiampo, F., & Shanthakumar, S. (2021). Remediation of metal/metalloid-polluted soils: A short review. *Applied Sciences (Basel, Switzerland)*, *11*(9), 4134. doi:10.3390/app11094134

Rafiq, M., Shahid, M., Abbas, G., Shamshad, S., Khalid, S., Niazi, N., & Dumat, C. (2017). Comparative effect of calcium and EDTA on arsenic uptake and physiological attributes of Pisum sativum. International *Journal of Phytoremediation, 19*(7), 662–669. https://doi.org/. 2016.1278426 doi:10.1080/15226514

Rahaman, S., Sinha, A., Pati, R., & Mukhopadhyay, D. (2013). Arsenic contamination: A potential hazard to the affected areas of West Bengal, India. *Environmental Geochemistry and Health*, *35*(1), 119–132. doi:10.100710653-012-9460-4 PMID:22618763

Rahimzadeh, M., Rahimzadeh, M., Kazemi, S., & Moghadamnia, A. (2017). Cadmium toxicity and treatment: An update. *Caspian Journal of Internal Medicine*, *8*, 135–145. doi:10.22088/acadpub.bums.8.2.67 PMID:28932363

Rahmanian, M., & Safari, Y. (2022). Contamination factor and pollution load index to estimate source apportionment of selected heavy metals in soils around a cement factory, SW Iran. *Archives of Agronomy and Soil Science*, *68*(7), 903–913. doi:10.1080/03650340.2020.1861252

Rai, P., Lee, S., Zhang, M., Tsang, Y., & Kim, K. (2019). Heavy metals in food crops: Health risks, fate, mechanisms and management. *Environment International*, *125*, 365–385. doi:10.1016/j.envint.2019.01.067 PMID:30743144

Rajaei, G., Mansouri, B., Jahantigh, H., & Hamidian, A. (2012). Metal concentrations in the water of Chah Nimeh reservoirs in Zabol. *Bulletin of Environmental Contamination and Toxicology*, *89*(3), 495–500. doi:10.100700128-012-0738-0 PMID:22885539

Rajapaksha, U., Alam, S., Chen, N., Alessi, S., Igalavithana, D., Tsang, D. C. W., & Ok, Y. S. (2018). Removal of hexavalent chromium in aqueous solutions using biochar: Chemical and spectroscopic investigations. *The Science of the Total Environment*, *625*, 1567–1573. doi:10.1016/j.scitotenv.2017.12.195 PMID:29996453

Rajendran, S., Priya, T., Khoo, K., Hoang, T., Ng, H., Munawaroh, H. S. H., Karaman, C., Orooji, Y., & Show, P. L. (2022). A critical review on various remediation approaches for heavy metal contaminants removal from contaminated soils. *Chemosphere*, *287*, 132369. doi:10.1016/j.chemosphere.2021.132369 PMID:34582930

Rajput, D., Minkina, M., Behal, A., Sushkova, N., Mandzhieva, S., Singh, R., Gorovtsov, A., Tsitsuashvili, V. S., Purvis, W. O., Ghazaryan, K. A., & Movsesyan, H. S. (2018). Effects of zinc-oxide nanoparticles on soil, plants, animals and soil organisms: A review. *Environmental Nanotechnology, Monitoring & Management*, *9*, 76–84. doi:10.1016/j.enmm.2017.12.006

Ramudzuli, R., & Horn, C. (2014). Arsenic residues in soil at cattle dip tanks in the Vhembe district, Limpopo Province, South Africa. *South African Journal of Science*, *110*(7/8), 1–7. doi:10.1590ajs.2014/20130393

Rani, A., Kumar, A., Lal, A., & Pant, M. (2014). Cellular mechanisms of cadmium-induced toxicity: A review. *International Journal of Environmental Health Research*, *24*(4), 378–399. doi:10.1080/09603123.2013.835032 PMID:24117228

Rastogi, K., Pandey, A., & Tripathi, S. (2008). Occupational health risks among the workers employed in leather tanneries at Kanpur. *Indian Journal of Occupational and Environmental Medicine*, *12*(3), 132. doi:10.4103/0019-5278.44695 PMID:20040972

Rattan, S., Zhou, C., Chiang, C., Mahalingam, S., Brehm, E., & Flaws, J. (2017). Exposure to endocrine disruptors during adulthood: Consequences for female fertility. *The Journal of Endocrinology*, *233*(3), R109–R129. doi:10.1530/JOE-17-0023 PMID:28356401

Rauret, G., Lopez-Sanchez, F., Sahuquillo, A., Rubio, R., Davidson, C., Ure, A., & Quevauviller, P. (1999). Improvement of the BCR three step sequential extraction procedure prior to the certification of new sediment and soil reference materials. *Journal of Environmental Monitoring*, *1*(1), 57–61. doi:10.1039/a807854h PMID:11529080

Ray, P., & Datta, S. (2016). Solid phase speciation of Zn and Cd in zinc smelter effluent irrigated soils. *Chemical Speciation and Bioavailability*, *29*(1), 6–14. doi:10.1080/09542299.2016.1247656

Reczek, R., & Chandel, N. (2017). The two faces of reactive oxygen species in cancer. *Annual Review of Cancer Biology*, *1*(1), 79–98. doi:10.1146/annurev-cancerbio-041916-065808

Reimann, C., Bjorvatn, K., Frengstad, B., Melaku, Z., Tekle-Haimanot, R., & Siewers, U. (2003). Drinking water quality in the Ethiopian section of the east African Rift valley I- data and health aspects. *The Science of the Total Environment*, *311*(1-3), 65–80. doi:10.1016/S0048-9697(03)00137-2 PMID:12826384

Renu, K., Chakraborty, R., Myakala, H., Koti, R., Famurewa, A., Madhyastha, H., Vellingiri, B., George, A., & Valsala Gopalakrishnan, A. (2021). Molecular mechanism of heavy metals (lead, chromium, arsenic, mercury, nickel and cadmium)-induced hepatoxicity-a review. *Chemosphere*, *271*, 129735. doi:10.1016/j.chemosphere.2021.129735 PMID:33736223

Rerkmitr, P., Kantikosum, K., Chottawornsak, N., Tangkijngamvong, N., Kerr, S., & Prueksapanich, P. (2019). Chronic occupational exposure to lead leads to significant mucocutanerous changes in lead factory workers. *Journal of the European Academy of Dermatology and Venereology*, *33*(10), 1993–2000. doi:10.1111/jdv.15678 PMID:31087433

Reyes, J., Molina-Jijón, E., Rodríguez-Muñoz, R., Bautista-García, P., Debray-García, Y., & Namorado, M. (2013). Tight junction proteins and oxidative stress in heavy metals-induced nephrotoxicity. *BioMed Research International*, *2013*, 1–14. doi:10.1155/2013/730789 PMID:23710457

Rinklebe, J., Shaheen, M., & Frohne, T. (2016). Amendment of biochar reduces the release of toxic elements under dynamic redox conditions in a contaminated floodplain soil. *Chemosphere*, *142*, 41–47. doi:10.1016/j.chemosphere.2015.03.067 PMID:25900116

Rinklebe, J., & Shaheen, S. (2017). Redox chemistry of nickel in soils and sediments: A review. *Chemosphere*, *179*, 265–278. doi:10.1016/j.chemosphere.2017.02.153 PMID:28371710

Robson, A. (1993). The chemistry of zinc. In P. Barak & P. Helmke (Eds.), *Zinc in soils and plants. Developments in plant and soil sciences*. Springer. doi:10.1007/978-94-011-0878-2_1

Rojko, A. (2017). Industry 4.0 concept: Background and overview. *International Journal of Interactive Mobile Technologies*, *11*(5), 77–90. doi:10.3991/ijim.v11i5.7072

Rollin, H., Olutola, B., Channa, K., & Odland, J. (2017). Reduction of in utero lead exposures in South Africans populations: Positive impact of unleaded petrol. *PLoS One*, *12*(10), e0186445. doi:10.1371/journal.pone.0186445 PMID:29036215

Rosaaen, D. (2017). *Hexavalent Chromium Exposure to Military Aircraft Painters*. Airverter. http://www.airverter.com/controlling-exposure-hexavalent-chromium-aerospaceair transport-painting/

Rotterdam Convention. (2010). *Overview.* UN. https://www.pic.int/TheConvention/Overview/tabid/1044/language/en-US/Default.aspx

Royal Society of Chemistry. (2023). *Copper.* [Online] RSC. https://www.rsc.org/periodic-table/element/29/copper

Roy, J., Chatterjee, D., Das, N., & Giri, K. (2018). Substantial evidences indicate that inorganic arsenic is a genotoxic carcinogen: A review. *Toxicological Research*, *34*(4), 311–324. doi:10.5487/TR.2018.34.4.311 PMID:30370006

Rytuba, J. (2003). Mercury from mineral deposits and potential environmental impact. *Environmental Geology (Berlin)*, *43*(3), 326–338. doi:10.100700254-002-0629-5

Ryzhenko, N., Zhavryda, D., Bokhonov, Y., & Ryzhenko, D. (2021). Mercury contamination in soil, water, plants and hydrobionts in Kyiv and the Kyiv region. *Polish Journal of Soil Science*, *2*(2), 185. doi:10.17951/pjss.2021.54.2.185

Sabir, M., Waraich, A., Hakeem, R., Ozturk, M., & Ahmad, R. (2015). *Phytoremediation, soil remediation and plants.* Elsevier Inc., doi:10.1016/B978-0-12-799937-1.00004-8

Saha, A., Gupta, B., Patidar, S., & Martínez-Villegas, N. (2022). Evaluation of Potential Ecological Risk Index of Toxic Metals Contamination in the Soils. Chemistry Proceedings, 10, 59. doi:10.3390/IOCAG2022-12214

Sahin, D. (2019). Atomic spectroscopy. In *M, Khan, G, Nascimento & M, El-Azazy. Modern spectroscopic techniques and applications.* Intech Open. doi:10.5772/intechopen.77480

Saini, S., Kaur, N., & Pati, P. (2021). Phytohormones: Key players in the modulation of heavy metal stress tolerance in plants. *Ecotoxicology and Environmental Safety*, *223*, 112578. doi:10.1016/j.ecoenv.2021.112578 PMID:34352573

Sakakibara, M., Watanabe, A., Sano, S., Inoue, M., & Kaise, T. (2006). *Phytoextraction and phytovolatilization of arsenic from As-contaminated soils by* Pteris vittata. In *Proceedings of the Annual International Conference on Soils, Sediments, Water and Energy*, Amherst, MA, USA.

Saleem, M., Ali, S., Irshad, S., Hussaan, M., & Rizwan, M. (2020). Copper uptake and accumulation, ultra-structural alteration, and bast fiber yield and quality of fibrous jute (*Corchorus capsularis* L.) plants grown under two different soils of China. *Plants*, *9*, 404. doi:10.3390/plants9030404 PMID:32213938

Saleh, M., Hamad, Z., & Hama, J. (2021). Assessment of heavy metals in crude oil workers from Kurdistan region, northern Iraq. *Environmental Monitoring and Assessment*, *193*(1), 49. doi:10.100710661-020-08818-w PMID:33415539

Salem, M., Bedade, D., Al-Ethawi, L., & Al-Waleed, S. (2020). Assessment of physiochemical properties and concentration of heavy metals in agricultural soils fertilized with chemical fertilizers. *Heliyon*, *6*(10), e05224. doi:10.1016/j.heliyon.2020.e05224 PMID:33102850

Salnikow, K., & Zhitkovich, A. (2008). Genetic and epigenetic mechanisms in metal carcinogenesis and cocarcinogenesis: Nickel, arsenic, and chromium. *Chemical Research in Toxicology*, *21*(1), 28–44. doi:10.1021/tx700198a PMID:17970581

Sanga, V., Fabian, C., & Kimbokota, F. (2022). Heavy metal pollution in leachates and its impacts on the quality of groundwater resources around Iringa municipal solid waste dumpsite. *Environmental Science and Pollution Research International*, 1–13. doi:10.100711356-022-22760-z PMID:36053421

Santos, S., Santos, P., Conti, M., Santos, N., & Oliveira, E. (2009). Evaluation of some metals in Brazilian coffees cultivated during the process of conversion from conventional to organic agriculture. *Food Chemistry*, *115*(4), 1405–1410. doi:10.1016/j.foodchem.2009.01.069

Saran, A., Fernandez, L., Cora, F., Savio, M., Thijs, S., Vangronsveld, J., & Merini, L. J. (2020). Phytostabilization of Pb and Cd polluted soils using *Helianthus petiolaris* as pioneer aromatic plant species. *International Journal of Phytoremediation*, *22*(5), 459–467. doi:10.1080/15226514.2019.1675140 PMID:31602996

Sari, M., Cosgun, T., Yalcin, I., Taner, M., & Ozyigit, I. (2022). Deciding heavy metal levels in soil based on various ecological information through artificial intelligence modelling. *Applied Artificial Intelligence*, *36*(1), e2014189. doi:10.1080/08839514.2021.2014189

Sastre, J., Sahuquillo, A., Vidal, M., & Rauret, G. (2002). Determination of Cd, Cu, Pb and Zn in environmental samples: Microwave-assisted total digestion versus aqua regia and nitric acid extraction. *Analytica Chimica Acta*, *462*(1), 59–72. doi:10.1016/S0003-2670(02)00307-0

Satarug, S. (2019). Cadmium sources and toxicity. *Toxics*, *7*(2), 25. doi:10.3390/toxics7020025 PMID:31064047

Saturday, A. (2018). Mercury and its associated impacts on environment and human health: A review. *Journal of Environment and Health Sciences*, *4*(2), 37–43. doi:10.15436/2378-6841.18.1906

Sayadi, H., Shabani, M., & Ahmadpour, N. (2015). Pollution index and ecological risk of heavy metals in the surface soils of Amir-Abad Area in Birjand City, Iran. *Health Scope*, *4*(1), 121–137. doi:10.17795/jhealthscope-21137

Saygili, S. (2013). Sanayi toplumundan bilgi toplumuna geçiş sürecinde eğitimde dönüştürücü bir entelektüel olarak öğretmenler. *Uşak Üniversitesi Sosyal Bilimler Dergisi, 6*(ÖYGE Özel Sayısı), 270-281.

Sayo, S., Kiratu, J., & Nyamato, G. (2020). Heavy metal concentrations in soil and vegetables irrigated with sewage effluent: A case study of Embu sewage treatment plant, Kenya. *Scientific African*, *8*, e00337. doi:10.1016/j.sciaf.2020.e00337

Scerri, E. (2020). Recent attempts to change the periodic table. *Philosophical Transactions - Royal Society. Mathematical, Physical, and Engineering Sciences*, *378*(2180), 20190300. doi:10.1098/rsta.2019.0300 PMID:32811365

Scholz, W., Nothbaum, N., & May, T. (1994). Fixed and hypothesis-guided soil sampling methods – Principles, strategies and examples. In B. Markert (Ed.), *Environmental sampling for trace analysis* (pp. 335–345). VCH. doi:10.1002/9783527615872.ch17

Seltenrich, N. (2021). A fix for fixtures: Addressing lead contamination in West African drinking water. *Environmental Health Perspectives*, *129*(8), 084003. doi:10.1289/EHP9610 PMID:34402631

Semu, E., Tindwa, H., & Singh, B. (2019). Heavy metals and organopesticides: Ecotoxicology, health effects and mitigation options with emphasis on sub-Saharan Africa. *HSOA Journal of Toxicology : Current Research*, *2*(3), 1–10. doi:10.24966/TCR-3735/100010

Şengel, U. (2019). *Türkiye'nin Turizm Talebini Etkileyen Faktörlerin Sosyo-Ekonomik Açıdan Ampirik Olarak Değerlendirilmesi*. Basılmamış Doktora Tezi, Sakarya Uygulamalı Bilimler Üniversitesi Lisansüstü Eğitim Enstitüsü, Sakarya.

Sengel, U. (2021). Chronology of the interaction between the industrial revolution and modern tourism flows. *Journal of Tourism Intelligence and Smartness*, *4*(1), 19–30.

Sengupta, P., Banerjee, R., Nath, S., Das, S., & Banerjee, S. (2015). Metals and female reproductive toxicity. *Human and Experimental Toxicology*, *34*(7), 679–697. doi:10.1177/0960327114559611 PMID:25425549

Sevim, C., Dogan, E., & Comakli, S. (2020). Cardiovascular disease and toxic metals. *Current Opinion in Toxicology*, *19*, 88–92. doi:10.1016/j.cotox.2020.01.004

Shaban, M., & Galaly, A. (2016). Highly sensitive and selective in-situ SERS detection of Pb^{2+}, Hg^{2+}, and Cd^{2+} using Nanoporous membrane functionalized with CNTs. *Scientific Reports*, *6*(1), 25307. doi:10.1038rep25307 PMID:27143512

Shackira, A., & Puthur, J. (2016). Phytostabilization of heavy metals: understanding of principles and practices. In S. Srivastava, A. Srivastava, & P. Suprasanna (Eds.), *Plant-Metal Interactions*. Springer. doi:10.1007/978-3-030-20732-8_13

Shaheen, M., Antoniadis, V., Kwon, E., Song, H., Wang, L., Hseu, Z.-Y., & Rinklebe, J. (2020). Soil contamination by potentially toxic elements and the associated human health risk in geo-and anthropogenic contaminated soils: A case study from the temperate region (Germany) and the arid region (Egypt). *Environmental Pollution*, *262*, 114312. doi:10.1016/j.envpol.2020.114312 PMID:32193081

Shaheen, M., El-Naggar, A., Antoniadis, V., Moghanm, S., Zhang, Z., Tsang, D. C. W., Ok, Y. S., & Rinklebe, J. (2020). Release of toxic elements in fishpond sediments under dynamic redox conditions: Assessing the potential environmental risk for a safe management of fisheries systems and degraded waterlogged sediments. *Journal of Environmental Management*, *255*, 109778. doi:10.1016/j.jenvman.2019.109778 PMID:32063315

Shaheen, M., El-Naggar, A., Wang, J., Hassan, E., & Niazi, K. (2018). Biochar as an (im)mobilizing agent for the potentially toxic elements in contaminated soils. In *Biochar from Biomass and Waste* (pp. 255–274). Elsevier. doi:10.1016/B978-0-12-811729-3.00014-5

Shaheen, M., Kwon, E., Biswas, K., Tack, M., Ok, S., & Rinklebe, J. (2017). Arsenic, chromium, molybdenum, and selenium: Geochemical fractions and potential mobilization in riverine soil profiles originating from Germany and Egypt. *Chemosphere*, *180*, 553–563. doi:10.1016/j.chemosphere.2017.04.054 PMID:28432892

Shaheen, M., Tawfik, W., Mankoula, A., Gagnon, J., Fryer, B., & El-Mekawy, F. (2021). Determination of heavy metal content and pollution indices in the agricultural soils using laser ablation inductively coupled plasma mass spectrometry. *Environmental Science and Pollution Research International*, *28*(27), 36039–36052. doi:10.100711356-021-13215-y PMID:33686601

Shahid, M., Shamshad, S., Rafiq, M., Khalid, S., Bibi, I., Niazi, N., Dumat, C. & Rashid, M. (2017). Chromium speciation, bioavailability, uptake, toxicity and detoxification in soil-plant system: a review. *Chemosphere, 178*, 513–533. https://doi.org/. 2017.03.074 doi:10.1016/j.chemosphere

Shahid, M., Austruy, A., Echevarria, G., Arshad, M., Sanaullah, M., Aslam, M., Nadeem, M., Nasim, W., & Dumat, C. (2014). EDTA-enhanced phytoremediation of heavy metals: A review. *Soil & Sediment Contamination*, *23*(4), 389–416. doi:10.1080/15320383.2014.831029

Shahid, M., Pinelli, E., & Dumat, C. (2012). Review of Pb availability and toxicity to plants in relation with metal speciation; role of synthetic and natural organic ligands. *Journal of Hazardous Materials*, *219-220*, 1–12. doi:10.1016/j.jhazmat.2012.01.060 PMID:22502897

Shankar, K., & Venkateswarlu, B. (2011). *Chromium: environmental pollution, health effects, and mode of action. Encyclopedia of Environmental Health.* Elsevier., doi:10.1016/B978-0-444-52272-6.00390-1

Shapi, M., Jordaan, A., Nadasan, S., Davies, C., Chirenje, E., Dube, M., & Lekoa, M. R. (2020). Analysis of the distribution of some potentially harmful elements (PHEs) in the Krugersdorp game reserve, Gauteng, South Africa. *Minerals (Basel)*, *10*(2), 1–18. doi:10.3390/min10020151

Sharifi, A., Darabi, R., Akbarloo, N., Larijani, B., & Khoshbaten, A. (2004). Investigation of circulatory and tissue ACE activity during development of lead-induced hypertension. *Toxicology Letters*, *153*(2), 233–238. doi:10.1016/j.toxlet.2004.04.013 PMID:15451554

Sharma, A., & Singh, B. (2020). Evolution of industrial revolutions: A review. *International Journal of Innovative Technology and Exploring Engineering*, *9*(11), 66–73. doi:10.35940/ijitee.I7144.0991120

Sharma, H., Rawal, N., & Mathew, B. (2015). The characteristics, toxicity and effects of cadmium. *International Journal of Nanotechnology and Nanoscience*, *3*, 1–9.

Sharma, R., & Agrawal, M. (2005). Biological effects of heavy metals: An overview. *Journal of Environmental Biology*, *26*, 301–313. PMID:16334259

Sharma, S. (2014). *Heavy metals in water: presence, removal and safety*. Royal Society of Chemistry. doi:10.1039/9781782620174

Sharma, S., Dietz, J., & Mimura, T. (2017). Vacuolar compartmentalization as indispensable component of heavy metal detoxification in plants. *Plant, Cell & Environment*, *39*(5), 1112–1126. doi:10.1111/pce.12706 PMID:26729300

Sharma, S., Nagpal, A., & Kaur, I. (2018). Heavy metal contamination in soil, food crops and associated health risks for residents of Ropar wetland, Punjab, India and its environs. *Food Chemistry*, *255*, 15–22. doi:10.1016/j.foodchem.2018.02.037 PMID:29571461

Shen, F., Mao, L., Sun, R., Du, J., Tan, Z., & Ding, M. (2019). Contamination evaluation and source identification of heavy metals in the sediments from the Lishui river watershed, Southern China. *International Journal of Environmental Research and Public Health*, *16*(3), 336. doi:10.3390/ijerph16030336 PMID:30691076

Shen, T., Kong, W., Liu, F., Chen, Z., Yao, J., Wang, W., Peng, J., Chen, H., & He, Y. (2018). Rapid determination of cadmium contamination in lettuce using laser induced breakdown spectroscopy. *Molecules (Basel, Switzerland)*, *23*(11), 2930. doi:10.3390/molecules23112930 PMID:30424009

Shezi, B., Street, R., Webster, C., Kunene, Z., & Mathee, A. (2022). Heavy metal contamination of soil in preschool facilities around industrial operations, Kuils river, Cape Town (South Africa). *International Journal of Environmental Research and Public Health*, *19*(7), 4380. doi:10.3390/ijerph19074380 PMID:35410061

Shiomi, N. (2015). An assessment of the causes of lead pollution and the efficiency of bioremediation by plants and microorganisms. In N. Shiomi (Ed.), *Advances in bioremediation of wastewater and polluted soil*. IntechOpen. doi:10.5772/60802

Shi, S., Hou, M., Gu, Z., Jiang, C., Zhang, W., Hou, M., Li, C., & Xi, Z. (2022). Estimation of heavy metal content in soil based on machine learning models. *Land (Basel)*, *11*(7), 1037. doi:10.3390/land11071037

Shrivastava, A., Barla, A., Singh, S., Mandraha, S., & Bose, S. (2017). Arsenic contamination in agricultural soils of Bengal deltaic region of West Bengal and its higher assimilation in monsoon rice. *Journal of Hazardous Materials*, *324*(Part B), 526–534. doi:10.1016/j.jhazmat.2016.11.022

Shrivastava, A., Ghosh, D., Dash, A., & Bose, S. (2015). Arsenic Contamination in Soil and Sediment in India: Sources, Effects, and Remediation. *Current Pollution Reports*, *1*(1), 35–46. doi:10.100740726-015-0004-2

Shu, Y., & Zhai, S. (2014). Study on soil heavy metals contamination of a lead refinery. *Chinese Journal of Geochemistry*, *33*(4), 393–397. doi:10.100711631-014-0703-1

Sikaundi, G. (2013). *Copper mining industry in Zambia: environmental challenges.* [Online] UN. https://unstats.un.org/unsd/environment/envpdf/UNSD_UNEP_ECA%20Workshop/Session%2008-5%20Mining%20in%20Zambia%20(Zambia).pdf [Accessed on 31st January 2023].

Singh, D., Iqbal, J., Bhat, A., Bhat, R., Dervash, M., & Ganei, S. (2018). Vehicular stress a cause for heavy metal accumulation and change in physico-chemical characteristics of roadside soils in Pahalgam. *Environmental Monitoring and Assessment, 190*(6), 353. doi:10.100710661-018-6731-2 PMID:29785575

Singh, H., Mahajan, P., Kaur, S., Batish, D., & Kohli, R. (2013). Chromium toxicity and tolerance in plants. *Environmental Chemistry Letters, 11*(3), 229–254. doi:10.100710311-013-0407-5

Singh, S., & Srivastava, P. (2020). Bioavailability of arsenic in agricultural soils under the influence of different soil properties. *SN Applied Sciences, 2*(2), 153. doi:10.100742452-019-1932-z

Singh, V., Agrawal, H., Joshi, G., Sudershan, M., & Sinha, A. (2011). Elemental profile of agricultural soil by the EDXRF technique and use of the Principal Component Analysis (PCA) method to interpret the complex data. *Applied Radiation and Isotopes, 69*(7), 969–974. doi:10.1016/j.apradiso.2011.01.025 PMID:21377884

Singo, N. (2013). *An assessment of heavy metal pollution near an old copper mine dump in Musina, South Africa.* [MSc Thesis, University of South Africa, Pretoria, South Africa].

Sisay, B., Debebe, E., Meresa, A., & Abera, T. (2019). Analysis of cadmium and lead using atomic absorption spectrophotometer in roadside soils of Jimma town. *Journal of Analytical & Pharmaceutical Research, 8*(4), 144–147. doi:10.15406/japlr.2019.08.00329

Six, L., & Smolders, E. (2014). Future trends in soil cadmium concentration under current cadmium fluxes to European agricultural soils. *The Science of the Total Environment, 485*, 319–328. doi:10.1016/j.scitotenv.2014.03.109 PMID:24727598

Skalnaya, M., & Skalny, V. (2018). *Essential trace elements in human health: A physician's view.* Publishing House of Tomsk State University.

Skobelev, P., & Borovik, S. (2017). On the way from Industry 4.0 to Industry 5.0: From digital manufacturing to digital society. *Industry 4.0, 2*, 307–311.

Slobodian, M., Petahtegoose, J., Wallis, A., Levesque, D., & Merritt, T. (2021). The effects of essential and non-essential metal toxicity in the drosophila melanogaster insect model: A review. *Toxics, 9*(10), 269. doi:10.3390/toxics9100269 PMID:34678965

Smith-Downey, N., Sunderland, E., & Jacob, D. (2010). Anthropogenic impacts on global storage and emissions of mercury from terrestrial soils: Insights from a new global model. *Journal of Geophysical Research, 115*(G3), G03008. doi:10.1029/2009JG001124

Soleimani, M., Amini, N., Sadeghian, B., Wang, D., & Fang, L. (2018). Heavy metals and their source identification in particulate matter (PM2.5) in Isfahan City, Iran. *Journal of Environmental Sciences (China), 72*, 166–175. doi:10.1016/j.jes.2018.01.002 PMID:30244743

Song, B., Zeng, G., Gong, J., Liang, J., Xu, P., Liu, Z., Zhang, Y., Zhang, C., Cheng, M., Liu, Y., Ye, S., Yi, H., & Ren, X. (2017). Evaluation methods for assessing effectiveness of in situ remediation of soil and sediment contaminated with organic pollutants and heavy metals. *Environment International*, *105*, 43–55. doi:10.1016/j.envint.2017.05.001 PMID:28500873

Song, M., Fisher, R., Wang, J., & Cui, L. (2018). Environmental performance evaluation with big data: Theories and methods. *Annals of Operations Research*, *270*(1–2), 459–472. doi:10.100710479-016-2158-8

Spahic, M., Sakan, S., Trbic, G., Tancic, P., Skrivanj, S., Kovacevic, J., & Manojlovic, D. (2018). Natural and anthropogenic sources of chromium, nickel and cobalt in soils impacted by agricultural and industrial activity (Vojvodina, Serbia). *Journal of Environmental Science and Health. Part A, Toxic/Hazardous Substances & Environmental Engineering*, *54*(3), 219–230. doi:10.1080/10934529.2018.1544802 PMID:30587075

Sracek, O., Kribek, B., Mihaljevic, M., Ettler, V., Vanek, A., Penizek, V., Veselovský, F., Bagai, Z., Kapusta, J., & Sulovský, P. (2021). Mobility of Mn and other trace elements in Mn-rich mine tailings and adjacent creek at Kanye, southeast Botswana. *Journal of Geochemical Exploration*, *220*, 106–658. doi:10.1016/j.gexplo.2020.106658

Srivastava, S., Agrawal, S., & Mondal, K. (2015). A review on progress of heavy metal removal using adsorbents of microbial and plant origin. *Environmental Science and Pollution Research International*, *22*(20), 15386–15415. doi:10.100711356-015-5278-9 PMID:26315592

Srivastava, V., Sarkar, A., Singh, S., Singh, P., Araujo, A., & Singh, R. (2017). Agroecological Responses of Heavy Metal Pollution with Special Emphasis on Soil Health and Plant Performances. *Frontiers in Environmental Science*, *5*, 64. doi:10.3389/fenvs.2017.00064

Steinnes, E. (2013). Mercury. In B. Alloway (Ed.), *Heavy Metals in Soils. Environmental Pollution* (Vol. 22). Springer., doi:10.1007/978-94-007-4470-7_15

Stockholm Convention. (2019). *Home.* UN. http://chm.pops.int/

Stofejova, L., Fazekas, J., & Fazekasova, D. (2021). Analysis of Heavy Metal Content in Soil and Plants in the Dumping Ground of Magnesite Mining Factory Jelšava-Lubeník (Slovakia). *Sustainability (Basel)*, *13*(8), 4508. doi:10.3390u13084508

Streets, D., Horowitz, H., Lu, Z., Levin, L., Thackray, C., & Sunderland, E. (2019). Global and regional trends in mercury emissions and concentrations. 2010–2015. *Atmospheric Environment*, *201*, 417–427. doi:10.1016/j.atmosenv.2018.12.031

Suanon, F., Tometin, L., Dimon, B., Agani, I., Mama, D., & Azandegbe, E. (2016). Utilization of sewage sludge in agricultural soil as fertilizer in the Republic of Benin (West Africa): What are the risks of heavy metals contamination and spreading? *American Journal of Environmental Sciences*, *12*(1), 5–15. doi:10.3844/ajessp.2016.8.15

Subasic, M., Samec, D., Selovic, A., & Karalija, E. (2022). Phytoremediation of cadmium polluted soils: Current status and approaches for enhancing. *Soil Systems*, *6*(3), 1–21. doi:10.3390oilsystems6010003

Su, C., Meng, J., Zhou, Y., Bi, R., Chen, Z., Diao, J., Huang, Z., Kan, Z., & Wang, T. (2022). Heavy metals in soils from intense industrial areas in South China: Spatial distribution, source apportionment, and risk assessment. *Frontiers in Environmental Science*, *10*, 820536. doi:10.3389/fenvs.2022.820536

Suciu, N., De Vivo, R., Rizzati, N., & Capri, E. (2022). Cd content in phosphate fertilizer: Which potential risk for the environment and human health? *Current Opinion in Environmental Science & Health*, *30*, 100392. doi:10.1016/j.coesh.2022.100392

Sumiahadi, A., & Acar, R. (2018). A review of phytoremediation technology: Heavy metals uptake by plants. *IOP Conference Series. Earth and Environmental Science*, *012023*, 012023. doi:10.1088/1755-1315/142/1/012023

Sun, C., Liu, J., Wang, Y., Sun, L., & Yu, H. (2013). Multivariate and geostatistical analyses of the spatial distribution and sources of heavy metals in agricultural soil in Dehui, Northeast China. *Chemosphere*, *92*(5), 517–523. doi:10.1016/j.chemosphere.2013.02.063 PMID:23608467

Sundseth, K., Pacyna, J., Pacyna, E., Pirrone, N., & Thorne, R. (2017). Global sources and pathways of mercury in the context of human health. *International Journal of Environmental Research and Public Health*, *14*(1), 105. doi:10.3390/ijerph14010105 PMID:28117743

Sun, H., Qi, Y., Zhang, D., Li, Q., & Wang, J. (2016). Concentrations, distribution, sources and risk assessment of organohalogenated contaminants in soils from Kenya, Eastern Africa. *Environmental Pollution*, *209*, 177–185. doi:10.1016/j.envpol.2015.11.040 PMID:26686059

Sun, X., Sun, M., Chao, Y., Shang, X., Wang, H., & Pan, H. (2022). Effects of lead pollution on soil microbial community diversity and biomass and on invertase activity. *Soil Ecology Letters*, 1–10. doi:10.100742832-022-0134-6

Sun, Z., Xie, X., Wang, P., Hu, Y., & Cheng, H. (2018). Heavy metal pollution caused by small-scale metal ore mining activities: A case study from a polymetallic mine in South China. *The Science of the Total Environment*, *639*, 217–227. doi:10.1016/j.scitotenv.2018.05.176 PMID:29787905

Sutherland, A. (2000). Bed sediment-associated trace metals in an urban stream, Oahu, Hawaii. *Environmental Geology (Berlin)*, *39*(6), 611–627. doi:10.1007002540050473

Swaran, F. (2015). Arsenic: chemistry, occurrence and exposure. In S. Flora (Ed.), *Handbook of arsenic toxicology*. Academic Press, Elsevier. doi:10.1016/B978-0-12-418688-0.00001-0

Tabelin, C., Igarashi, T., Villacorte-Tabelin, M., Park, I., Opiso, E., Ito, M., & Hiroyoshi, N. (2018). Arsenic, selenium, boron, lead, cadmium, copper, and zinc in naturally contaminated rocks: A review of their sources, modes of enrichment, mechanisms of release, and mitigation strategies. *The Science of the Total Environment*, *645*, 1522–1553. doi:10.1016/j.scitotenv.2018.07.103 PMID:30248873

Taghipour, M., & Jalali, M. (2016). Influence of organic acids on kinetic release of chromium in soil contaminated with leather factory waste in the presence of some adsorbents. *Chemosphere*, *155*, 395–404. doi:10.1016/j.chemosphere.2016.04.063 PMID:27139119

Tahmasebi, P., Taheri, M., & Gharaie, M. (2020). Heavy metal pollution associated with mining activity in the Kouh-e Zar region, NE Iran. *Bulletin of Engineering Geology and the Environment*, *79*(2), 1113–1123. doi:10.100710064-019-01574-3

Tahoon, M., Siddeeg, S., Salem Alsaiari, N., Mnif, W., & Rebah, F. (2020). Effective heavy metals removal from water using nanomaterials: A review. *Processes (Basel, Switzerland)*, *8*(6), 645. doi:10.3390/pr8060645

Taj, I., & Jhanjhi, N. (2022). Towards industrial revolution 5.0 and explainable artificial intelligence. Challenges and opportunities. *International Journal of Computing and Digital Systems*, *12*(1), 1–26. doi:10.12785/ijcds/120124

Tang, H., Liu, Z., Li, S., Sha, Y., & Qiu, P. (2017). Effects of mercury pollution stress on activity of several kinds of antioxidant enzymes in earthworms. *Journal of Shanghai Jiaotong University*, *35*(3), 17–23. doi:10.3969/J.ISSN.1671-9964.2017.03.003

Tan, K., Ma, W., Chen, L., Wang, H., Du, Q., Du, P., Yan, B., Liu, R., & Li, H. (2021). Estimating the distribution trend of soil heavy metals in mining area from HyMap airborne hyperspectral imagery based on ensemble learning. *Journal of Hazardous Materials*, *401*, 123288. doi:10.1016/j.jhazmat.2020.123288 PMID:32645545

Taylor, M., & Kesterton, R. (2002). Heavy metal contamination of an arid river environment: Gruben River, Namibia. *Geomorphology*, *42*(3-4), 311–327. doi:10.1016/S0169-555X(01)00093-9

Tchounwou, P., Yedjou, C., Patlolla, A., & Sutton, D. (2012). Heavy metals toxicity and the environment. *EXS*, *101*, 133–164. doi:10.1007/978-3-7643-8340-4_6 PMID:22945569

Teare, J., Kootbodien, T., Naicker, N., & Mathee, A. (2015). The extent, nature and environmental health implications of cottage industries in Johannesburg, south Africa. *International Journal of Environmental Research and Public Health*, *12*(2), 1894–1901. doi:10.3390/ijerph120201894 PMID:25664698

Teju, E., Megersa, N., Chandravanshi, B., & Zewge, F. (2014). Lead accumulation in the roadside soils from heavy metal density motor way towns of Eastern Ethiopia. *Bulletin of the Chemical Society of Ethiopia*, *28*(2), 161–176. doi:10.4314/bcse.v28i2.1

Tellez-Plaza, M., Guallar, E., Howard, V., Umans, G., Francesconi, A., Goessler, W., Silbergeld, E. K., Devereux, R. B., & Navas-Acien, A. (2013). Cadmium exposure and incident cardiovascular disease. *Epidemiology (Cambridge, Mass.)*, *24*(3), 421–429. doi:10.1097/EDE.0b013e31828b0631 PMID:23514838

Tembo, B., Sichilongo, K., & Cernak, J. (2006). Distribution of copper, lead, cadmium and zinc concentrations in soils around Kabwe town in Zambia. *Chemosphere*, *63*(3), 497–501. doi:10.1016/j.chemosphere.2005.08.002 PMID:16337989

Templeton, D., & Fujishiro, H. (2017). Terminology of elemental speciation – an IUPAC perspective. *Coordination Chemistry Reviews*, *352*, 424–431. doi:10.1016/j.ccr.2017.02.002

Teta, C., Ncube, M., & Naik, Y. (2017). Heavy metal contamination of water and fish in peri-urban dams around Bulawayo, Zimbabwe. *African Journal of Aquatic Science*, *42*(4), 351–358. doi:10.2989/16085914.2017.1392925

Thestorf, K., & Makki, M. (2021). Pseudo-total antimony content in topsoils of the Berlin Metropolitan Area. *Journal of Soils and Sediments*, *21*(5), 2102–2117. doi:10.100711368-020-02742-9

Tindwa, H., & Singh, B. (2023). Soil pollution and agriculture in sub-Saharan Africa: State of the knowledge and remediation technologies. *Frontiers in Soil Science*, *2*, 1101944. doi:10.3389/fsoil.2022.1101944

Tiruneh, A., Fadiran, A., & Mtshali, J. (2014). Evaluation of the risk of heavy metals in sewage sludge intended for agricultural application in Swaziland. *International Journal of Environmental Sciences*, *5*(1), 197–216.

Tiwari, S., Singh, S., & Garg, S. (2013). Microbially enhanced phytoextraction of heavy metal fly-ash amended soil. *Communications in Soil Science and Plant Analysis*, *44*(21), 3161–3176. doi:10.1080/00103624.2013.832287

Tomlinson, D., Wilson, J., Harris, C., & Jefrey, D. (1980). Problems in the assessment of heavy-metal levels in estuaries and the formation of a pollution index. *Helgoländer Wissenschaftliche Meeresuntersuchungen*, *33*(1), 566–575. doi:10.1007/BF02414780

Tomno, R., Nzeve, J., Mailu, S., Shitanda, D., & Waswa, F. (2020). Heavy metal contamination of water, soils and vegetables in urban streams in Machakos municipality, Kenya. *Scientific African*, *9*, e00539. doi:10.1016/j.sciaf.2020.e00539

Tong, D., Zhang, W., Deng, Y., & Wang, J. (2016). Pollution characteristics analysis and risk assessment of total mercury and methylmercury in aquatic. *Journal of Integrative Environmental Research*, *37*(3), 942–949. doi:10.13227/j.hjkx.2016.03.019 PMID:27337885

Toth, G., Hermann, T., Da Silva, M., & Montanarella, L. (2016). Heavy metals in agricultural soils of the European Union with implications for food safety. *Environmental Pollution*, *88*, 299–309. doi:10.1016/j.envint.2015.12.017 PMID:26851498

Tsadilas, C., & Rinklebe, J. (2018). Nickel in soils and plants, first edition. CRC Press, Taylor & Francis Group, New York, USA. doi:10.1201/9781315154664

Tshibalo, R. (2017). *Assessment of municipal solid waste leachate pollution on soil and groundwater system at Onderstepoort landfill site in Pretoria*. [MSc Thesis, University of South Africa].

Tsuma, J., Wandiga, S., & Abong'o, D. (2016). Methane and heavy metal levels from leachates at Dandora dumpsite, Nairobi County, Kenya. *IOSR Journal of Applied Chemistry*, *9*(9), 39–46. doi:10.9790/5736-0909023946

Tuakuila, J., Lison, D., Mbuyi, F., Haufroid, V., & Hoet, P. (2013). Elevated blood lead levels and sources of exposure in the population of Kinshasa, the capital of the Democratic Republic of Congo. *Journal of Exposure Science & Environmental Epidemiology*, *23*(1), 81–87. doi:10.1038/jes.2012.49 PMID:22617721

Tully, K., Sullivan, C., Weil, R., & Sanchez, P. (2015). The state of soil degradation in sub-Saharan Africa: Baselines, trajectories and solutions. *Sustainability (Basel)*, *7*(6), 6523–6552. doi:10.3390u7066523

Tumolo, M., Ancona, V., De Paola, D., Losacco, D., Campanale, C., Massarelli, C., & Uricchio, V. (2020). Chromium pollution in European water sources, health risk, and remediation strategies: An overview. *International Journal of Environmental Research and Public Health*, *17*(15), 5438. doi:10.3390/ijerph17155438 PMID:32731582

Ubando, A., Africa, A., Maniquiz-Redillas, M., Culaba, A., Chen, W., & Chang, J.-S. (2020). Microalgal biosorption of heavy metals: A comprehensive bibliometric review. *Journal of Hazardous Materials*, *402*, 123431. doi:10.1016/j.jhazmat.2020.123431 PMID:32745872

Uchimiya, M., Bannon, D., Nakanishi, H., McBride, M., Williams, M., & Yoshihara, T. (2020). Chemical speciation, plant uptake, and toxicity of heavy metals in agricultural soils. *Journal of Agricultural and Food Chemistry*, *68*(46), 12856–12869. doi:10.1021/acs.jafc.0c00183 PMID:32155055

Ullah, A., Ma, Y., Li, J., Tahir, N., & Hussain, B. (2020). Effective amendments on cadmium, arsenic, chromium and lead contaminated paddy soil for rice safety. *Agronomy (Basel)*, *10*(3), 359. doi:10.3390/agronomy10030359

Umvoto Africa. (2014). *Role of fertilizers in trace metal (specifically cadmium) contamination of groundwater*. Water Research Commission.

UNEP. (2019). Global mercury assessment 2018. Environment UN, Chemical and health branch, Geneva, Switzerland.

UNEP. (n.d). *Basel Convention on the control of transboundary movements of hazardous wastes*. UNEP. https://www.unep.org/resources/report/basel-convention-control-transboundary-movements-hazardous-wastes (Accessed July 13th, 2023).

UN-Habitat. (2022). *Africa's waste problem*. UN. https://unhabitat.org/ african-clean-cities-africas-waste-problems

United Nations Environmental Program. (2022). *The Lead Campaign*. UN. https://www.unep.org/explore-topics/transport/what-we-do/partnership-clean-fuels-and-vehicles/lead-campaign

United States-Environmental Protection Agency. US-EPA. (2014). Method 6020B: Inductively coupled plasma-mass spectrometry. Environmental Protection Agency, Washington DC, USA.

USEPA. (2016). Outdoor air-industry, business and home: paint and coating manufacturing. EPA. https://archive.epa.gov/airquality/community/web/html/paint_manuf_addl_info.html

USEPA. (2019). *Mercury emissions: the global context (2019).* United States Environmental Protection Agency. https://www.epa.gov/international-cooperation/mercury-emissions-global-context

Uugwanga, M., & Kgabi, N. (2020). Assessment of metals pollution in sediments and tailings of Kelin Aub and Oamites mine sites, Namibia. *Environmental Advances*, *2*, 100006. doi:10.1016/j.envadv.2020.100006

Vacha, R. (2021). Heavy Metal Pollution and Its Effects on Agriculture. *Agronomy (Basel)*, *11*(9), 1719. doi:10.3390/agronomy11091719

Vaidya, S., Ambad, P., & Bhosle, S. (2018). Industry 4.0–A glimpse. *Procedia Manufacturing*, *20*, 233–238. doi:10.1016/j.promfg.2018.02.034

Valko, M., Izakovic, M., Mazur, M., Rhodes, C., & Telser, J. (2004). Role of oxygen radicals in DNA damage and cancer incidence. *Molecular and Cellular Biochemistry*, *266*(1/2), 37–56. doi:10.1023/B:MCBI.0000049134.69131.89 PMID:15646026

Valsecchi, M., & Polesello, S. (1999). Analysis of inorganic species in environmental samples by capillary electrophoresis. *Journal of Chromatography. A*, *834*(1-2), 363–385. doi:10.1016/S0021-9673(98)00914-5 PMID:10189695

Van Brusselen, D., Kitenge, T., Musanzayi, S., Kasole, T., & Ngombe, L. (2020). Metal mining and birth defects: A case-control study in Lubumbashi, Democratic Republic of the Congo. *The Lancet. Planetary Health*, *4*(4), e158–e167. doi:10.1016/S2542-5196(20)30059-0 PMID:32353296

Van- Zwieten, L., Merrington, G., & Van-Zwieten, M. (2004). Review of impacts of soil biota caused by copper residues from fungicide application. *3rd Australian New Zealand Soils Conference*. University of Sydney, Australia.

Vardhan, K., Kumar, P., & Panda, R. (2019). A review on heavy metal pollution, toxicity and remedial measures: Current trends and future perspectives. *Journal of Molecular Liquids*, *290*, 111197. doi:10.1016/j.molliq.2019.111197

Varol, M. (2011). Assessment of heavy metal contamination in sediments of the Tigris River (Turkey) using pollution indices and multivariate statistical techniques. *Journal of Hazardous Materials*, *195*, 355–364. doi:10.1016/j.jhazmat.2011.08.051 PMID:21890271

Vaziri, D. (2008). Mechanisms of lead-induced hypertension and cardiovascular disease. *American Journal of Physiology. Heart and Circulatory Physiology*, *295*(2), H454–H465. doi:10.1152/ajpheart.00158.2008 PMID:18567711

Velkova, Z., Kirova, G., Stoytcheva, M., Kastadinova, S., & Todorova, K. (2018). Immobilized microbial biosorbents for heavy metals removal. *Engineering in Life Sciences*, *18*(12), 871–881. doi:10.1002/elsc.201800017 PMID:32624881

Venegas, A., Rigol, A., & Vidal, M. (2015). Viability of organic wastes and biochar as amendments for remediation of heavy metal contaminated soil. *Chemosphere*, *119*, 190–198. doi:10.1016/j.chemosphere.2014.06.009 PMID:24995385

Vinitha, K., Prabhu, R., Bhaskar, R., & Hariharan, R. (2020). Review on industrial mathematics and materials at Industry 1.0 to Industry 4.0. *Materials Today: Proceedings*, *33*, 3956–3960. doi:10.1016/j.matpr.2020.06.331

Vischetti, C., Marini, E., Casucci, C., & De Bernardi, A. (2022). Nickel in the environment: Bioremediation techniques for soils with low or moderate contamination in European Union. *Environments (Basel, Switzerland)*, *9*(10), 133. doi:10.3390/environments9100133

Wadhawan, S., Jain, A., Nayyar, J., & Mehta, S. (2020). Role of nanomaterials as adsorbents in heavy metal ion removal from waste water: A review. *Journal of Water Process Engineering*, *33*, 101038. doi:10.1016/j.jwpe.2019.101038

Wael, M., Khaled, A., Hussein, S., Marina, F., Svetlana, G., & Octavian, D. (2015). Instrumental neutron activation analysis of soil and sediment samples from Siwa Oasis, Egypt. *Physics of Particles and Nuclei Letters*, *12*(4), 637–644. doi:10.1134/S154747711504007X

Wagner, G., Mohr, M. E., Sprengart, J., Desaules, A., Muntau, H., Theocharopoulos, S., & Quevauviller, P. (2001). Objectives, concept and design of the CEEM soil project. *The Science of the Total Environment*, *264*(1-2), 3–15. doi:10.1016/S0048-9697(00)00608-2 PMID:11213187

Walker, C., Sibly, R., & Hopkin, D. (2012). *Principles of Ecotoxicology; Group, Taylor and Francis* (4th ed.). CRC Press.

Walters, C., Somerset, V., Leaner, J., & Nel, J. (2011). A review of mercury pollution in South Africa: Current status. *Journal of Environmental Science and Health. Part A, Toxic/Hazardous Substances & Environmental Engineering*, *46*(10), 1129–1137. doi:10.1080/10934529.2011.590729 PMID:21806457

Wandiga, S. (2001). Use and distribution of organochlorine pesticides. The future in Africa. *Pure and Applied Chemistry*, *73*(7), 1147–1156. doi:10.1351/pac200173071147

Wang, L., Rinklebe, J., Tack, F., & Hou, D. (2021). A review of green remediation strategies for heavy metal contaminated soil. *Soil Use and Management, 27*, 936-963. 9. doi:10.1111/sum.12717

Wang, B., Chiu, H., Lee, Y., Li, C., Wang, Y., & Lee, Y. (2018). Pterostilbene attenuates hexavalent chromium-induced allergic contact dermatitis by preventing cell apoptosis and inhibiting IL-1b-related NLRP3 inflammasome activation. *Journal of Clinical Medicine*, *7*(12), 489. doi:10.3390/jcm7120489 PMID:30486377

Wang, B., & Du, Y. (2013). Cadmium and its neurotoxic effects. *Oxidative Medicine and Cellular Longevity*, *898034*, 1–12. Advance online publication. doi:10.1155/2013/898034 PMID:23997854

Wang, G., Pan, X., Zhang, S., Zhong, Q., Zhou, W., Zhang, X., Wu, J., Vijver, M. G., & Peijnenburg, W. J. G. M. (2020). Remediation of heavy metal contaminated soil by biodegradable chelator–induced washing: Efficiencies and mechanisms. *Environmental Research*, *186*, 109554. doi:10.1016/j.envres.2020.109554 PMID:32344210

Wang, J., Zhu, H., Yang, Z., & Liu, Z. (2013). Antioxidative effects of hesperetin against lead acetate-induced oxidative stress in rats. *Indian Journal of Pharmacology*, *45*(4), 395–398. doi:10.4103/0253-7613.115015 PMID:24014918

Wang, L., Cho, W., Tsang, C., Cao, X., Hou, D., Shen, Z., Alessi, D. S., Ok, Y. S., & Poon, C. S. (2019). Green remediation of As and Pb contaminated soil using cement-free clay-based stabilization/solidification. *Environment International*, *126*, 336–345. doi:10.1016/j.envint.2019.02.057 PMID:30826612

Wang, L., Hou, D., Cao, Y., Ok, Y., Tack, F., Rinklebe, J., & O'Connor, D. (2020). Remediation of mercury contaminated soil, water and air: A review of emerging materials and innovative technologies. *Environment International*, *134*, 105281. doi:10.1016/j.envint.2019.105281 PMID:31726360

Wang, L., Tao, Y., Su, B., Wang, L., & Liu, P. (2022). Environmental and health risks posed by heavy metal contamination of groundwater in the Sunan Coal Mine, China. *Toxics*, *10*(7), 390. doi:10.3390/toxics10070390 PMID:35878294

Wang, P., Hua, C., Chen, Z., & Dai, H. (2008). Effects of water pollution caused by Hg^{2+} on the chlorophyll content and antioxidase activity in spinach seedlings. *Anhui Nongye Kexue*, *36*(14), 5738–5739. doi:10.3969/j.issn.0517-6611.2008.14.023

Wang, P., Hu, X., He, Q., Waigi, M., Wang, J., & Ling, W. (2018). Using calcination remediation to stabilize heavy metals and simultaneously remove polycyclic aromatic hydrocarbons in soil. *International Journal of Environmental Research and Public Health*, *15*(8), 1731. doi:10.3390/ijerph15081731 PMID:30104500

Wang, Q., Shaheen, M., Jiang, Y., Li, R., Slany, M., Abdelrahman, H., Kwon, E., Bolan, N., Rinklebe, J., & Zhang, Z. (2021). Fe/Mn- and P-modified drinking water treatment residuals reduced Cu and Pb phytoavailability and uptake in a mining soil. *Journal of Hazardous Materials*, *403*, 123628. doi:10.1016/j.jhazmat.2020.123628 PMID:32814241

Wang, Z., Sun, Y., Yao, W., Ba, Q., & Wang, H. (2021). Effects of cadmium exposure on the immune system and immunoregulation. *Frontiers in Immunology*, *12*, 695484. doi:10.3389/fimmu.2021.695484 PMID:34354707

Wang, Z., Wang, Y., Chen, L., Yan, C., Yan, Y., & Chi, Q. (2015). Assessment of metal contamination in coastal sediments of the Maluan Bay (China) using geochemical indices and multivariate statistical approaches. *Marine Pollution Bulletin*, *99*(1-2), 43–53. doi:10.1016/j.marpolbul.2015.07.064 PMID:26233304

Wedepohl, K. (1995). The composition of the continental crust. *Geochimica et Cosmochimica Acta*, *59*(7), 1217–1232. doi:10.1016/0016-7037(95)00038-2

Weiss, J. (2016). *Handbook of ion chromatography* (4th ed., Vol. 1). Wiley. org/ doi:10.1002/9783527651610

Weissmannova, H., & Pavlovsky, J. (2017). Indices of soil contamination by heavy metals-methodology of calculation for pollution assessment (minireview). *Environmental Monitoring and Assessment*, *189*(12), 616. doi:10.100710661-017-6340-5 PMID:29116419

Wei, W., Ma, R., Sun, Z., Zhou, A., Bu, J., Long, X., & Liu, Y. (2018). Effects of mining activities on the release of heavy metals (HMs) in a typical mountain headwater region, the Qinghai-Tibet Plateau in China. *International Journal of Environmental Research and Public Health*, *15*(9), 1987. doi:10.3390/ijerph15091987 PMID:30213099

Wen, Y., Wang, Q., Tang, C., & Chen, Z. (2012). Bioleaching of heavy metals from sewage sludge by Acidithiobacillus thiooxidans—A comparative study. *Journal of Soils and Sediments*, *12*(6), 900–908. doi:10.100711368-012-0520-2

WHO. (2000). *Air quality guidelines for Europe (second edition).* WHO. https://wedocs.unep.org/20.500.11822/8681

WHO. (2014). Chemicals of public health concern in the African Region and their management – Regional Assessment Report. WHO, Geneva, Switzerland.

WHO. (2017). *Mercury and health.* WHO. https://www.who.int/news-room/fact-sheets/ detail/mercury-and-health

Wiafe, S., Yeboah, E., Boakye, E., & Ofosu, S. (2022). Environmental risk assessment of heavy metals contamination in the catchment of small-scale mining enclave in Prestea Huni-Valley district, Ghana. *Sustainable Environment*, *8*(1), 1–15. doi:10.1080/27658511.2022.2062825

Wieczorek, J., & Baran, A. (2022). Pollution indices and biotests as useful tools for the evaluation of the degree of soil contamination by trace elements. *Journal of Soils and Sediments*, *22*(2), 559–576. doi:10.100711368-021-03091-x

Wieczorek, J., Baran, A., Urbanski, K., Mazurek, R., & Pawlas, A. (2018). Assessment of the pollution and ecological risk of lead and cadmium in soils. *Environmental Geochemistry and Health*, *40*(6), 2325–2342. doi:10.100710653-018-0100-5 PMID:29589150

Woldetsadik, D., Drechsel, P., Keraita, B., Itanna, F., & Gebrekidan, H. (2017). Heavy metal accumulation and health risk assessment in wastewater-irrigated urban vegetable farming sites of Addis Ababa, Ethiopia. *International Journal of Food Contamination*, *4*(1), 9. doi:10.118640550-017-0053-y

World Bank. (2011). Study of mercury-containing lamp waste in sub–Saharan Africa. World Bank.

World Health Organization. (2022). *Lead poisoning*. WHO. https://www.who.int/news-room/fact-sheets/detail/lead-poisoning-and-health

World Health Organization. WHO. (2015). Lead exposure in African children, contemporary sources and concerns. WHO Regional Office for Africa, Brazzaville, Republic of Congo. .

Wuana, R., & Okieimen, F. (2011). Heavy metals in contaminated soils: A review of sources, chemistry, risks and best available strategies for remediation. *ISRN Ecology*, *402647*, 1–20. doi:10.5402/2011/402647

Wu, S., Peng, S., Zhang, X., Wu, D., Luo, W., Zhang, T., Zhou, S., Yang, G., Wan, H., & Wu, L. (2015). Levels and health risk assessments of heavy metals in urban soils in Dongguan, China. *Journal of Geochemical Exploration*, *148*, 71–78. doi:10.1016/j.gexplo.2014.08.009

Wu, X., Cobbina, S. J., Mao, G., Xu, H., Zhang, Z., & Yang, L. (2016). A review of toxicity and mechanisms of individual and mixtures of heavy metals in the environment. *Environmental Science and Pollution Research International*, *23*(9), 8244–8259. doi:10.100711356-016-6333-x PMID:26965280

Wyszkowska, J., Borowik, A., Zaborowska, M., & Kucharski, J. (2022). Mitigation of the adverse impact of copper, nickel, and zinc on soil microorganisms and enzymes by mineral sorbents. *Materials (Basel)*, *15*(15), 5198. doi:10.3390/ma15155198 PMID:35955133

Xiao, W., Ye, X., Yang, X., Li, T., Zhao, S., & Zhang, Q. (2015). Effects of alternating wetting and drying versus continuous flooding on chromium fate in paddy soils. *Ecotoxicology and Environmental Safety*, *113*, 439–445. doi:10.1016/j.ecoenv.2014.12.030 PMID:25546832

Xiao, X., Zhang, J., Wang, H., Han, X., Ma, J., Ma, Y., & Luan, H. (2020). Distribution and health risk assessment of potentially toxic elements in soils around coal industrial areas: A global meta-analysis. *The Science of the Total Environment*, *713*, 135292. doi:10.1016/j.scitotenv.2019.135292 PMID:32019003

Xing, J., Li, L., Li, G., & Xu, G. (2019). Feasibility of sludge-based biochar for soil remediation: Characteristics and safety performance of heavy metals influenced by pyrolysis temperatures. *Ecotoxicology and Environmental Safety*, *180*, 457–465. doi:10.1016/j.ecoenv.2019.05.034 PMID:31121552

Xu, D., Shen, Z., Dou, C., Dou, Z., Li, Y., Gao, Y., & Sun, Q. (2022). Effects of soil properties on heavy metal bioavailability and accumulation in crop grains under different farmland use patterns. *Scientific Reports*, *12*(1), 9211. doi:10.103841598-022-13140-1 PMID:35654920

Xu, H., Jia, Y., Sun, Z., Su, J., Liu, Q., Zhou, Q., & Jiang, G. (2022a). Environmental pollution, a hidden culprit for health issues. *Eco-Environment and Health*, *1*(1), 31–45. doi:10.1016/j.eehl.2022.04.003

Xu, L., Xu, L., & Li, L. (2018). Industry 4.0: State of the art and future trends. *International Journal of Production Research*, *56*(8), 2941–2962. doi:10.1080/00207543.2018.1444806

Yabe, J., Ishizuka, M., & Umemura, T. (2010). Current levels of heavy metal pollution in Africa. *The Journal of Veterinary Medical Science*, *72*(10), 1257–1263. doi:10.1292/jvms.10-0058 PMID:20519853

Yabe, J., Nakayama, S., Ikenaka, Y., Muzandu, K., Choongo, K., Mainda, G., Kabeta, M., Ishizuka, M., & Umemura, T. (2013). Metal distribution in tissues of free-range chickens near a lead–zinc mine in Kabwe, Zambia. *Environmental Toxicology and Chemistry*, *32*(1), 189–192. doi:10.1002/etc.2029 PMID:23059509

Yabe, J., Nakayama, S., Ikenaka, Y., Yohannes, Y., Sam, N., & Oroszlany, B. (2015). Lead poisoning in children from townships in the vicinity of a lead-zinc mine in Kabwe, Zambia. *Chemosphere*, *119*, 941–947. doi:10.1016/j.chemosphere.2014.09.028 PMID:25303652

Yadav, K., Gupta, N., Kumar, V., & Singh, K. (2017). Bioremediation of heavy metals from contaminated sites using potential species: A review. *Indian Journal of Environmental Protection*, *37*, 65–84.

Yahaya, S., Abubakar, F., & Abdu, N. (2021). Ecological risk assessment of heavy metal-contaminated soils of selected villages in Zamfara state, Nigeria. *SN Applied Sciences*, *3*(2), 168. doi:10.100742452-021-04175-6

Yan, J., Liu, Y., Wang, L., Wang, Z., & Liu, H. (2021). An efficient organization method for large-scale and long time-series remote sensing data in a cloud computing environment. *IEEE Journal of Selected Topics in Applied Earth Observations and Remote Sensing,* (99), 1–1. . doi:10.1109/JSTARS.2021.3110900

Yan, A., Wang, Y., Tan, S., Yusof, M., & Ghosh, S. (2020). Phytoremediation: A promising approach for revegetation of heavy metal-polluted land. *Frontiers in Plant Science*, *11*, 359. doi:10.3389/fpls.2020.00359 PMID:32425957

Yan, B., & Zhang, X. (2021). Current status, causes and harm of soil arsenic pollution. *IOP Conference Series. Earth and Environmental Science*, *769*(2), 022034. doi:10.1088/1755-1315/769/2/022034

Yang, O., Kim, H., Weon, J., & Seo, Y. (2015). Endocrine-disrupting chemicals: Review of toxicological mechanisms using molecular pathway analysis. *Journal of Cancer Prevention*, *20*(1), 12–24. doi:10.15430/JCP.2015.20.1.12 PMID:25853100

Yang, Z., Shi, W., Yang, W., Liang, L., Yao, W., Chai, L., Gao, S., & Liao, Q. (2018). Combination of bioleaching by gross bacterial biosurfactants and flocculation: A potential remediation for the heavy metal contaminated soils. *Chemosphere*, *206*, 83–91. doi:10.1016/j.chemosphere.2018.04.166 PMID:29730568

Yaseen, Z. (2022). The next generation of soil and water bodies heavy metal prediction and detection: New expert system-based edge cloud server and federated learning technology. *Environmental Pollution*, *315*, 120081. doi:10.1016/j.envpol.2022.120081 PMID:36075340

Yehouenou, E., Adamou, R., Azehoun, P., Edorh, P., & Ahoyo, T. (2013). Monitoring of heavy metals in the complex "Nokoué lake-Cotonou and Porto-Novo lagoon" ecosystem during three years in the Republic of Benin. *Research Journal of Chemical Sciences*, *3*, 12–18.

Yiika, L., Tita, M., Suh, C., Mimba, M., & Lavenir, N. (2023). *Heavy metal speciation by Tessier sequential extraction applied to artisanal gold mine tailing sin eastern Cameroon*. Chemistry Africa. doi:10.100742250-023-00652-0

Yildiz, Y. (2017). *General aspects of the cobalt chemistry*. Intech Open. doi:10.5772/intechopen.71089

Yin, Y., Stecke, E., & Li, D. (2018). The evolution of production systems from Industry 2.0 through Industry 4.0. *International Journal of Production Research*, *56*(1-2), 848–861. doi:10.1080/00207543.2017.1403664

Yitagesu, Y., & Bekele, E. (2019). Impacts of cement dust deposition on heavy metal pollution in soil and barley crop grown around Abyssinia cement factory, Ethiopia. *Chemistry and Materials Research*, *11*(2), 1–11. doi:10.7176/cmr

Yiu, N., Zhu, F., Wei, F., Miao, J., Zhou, M., & Guan, J. (2016). Characteristics and evaluation of heavy metal pollution in different functional areas of Luoyang City, Henan Province. *Environmental Sciences (Ruse)*, *37*(6), 2322–2328. doi:10.13227/j.hjkx.2016.06.041

Yoshihisa, Y., & Shimizu, T. (2012). Metal allergy and systemic contact dermatitis: An overview. *Dermatology Research and Practice*, *1-5*, 1–5. Advance online publication. doi:10.1155/2012/749561 PMID:22693488

Yoshizawa, K., Rimm, B., Morris, S., Spate, L., Hsieh, C., Spiegelman, D., Stampfer, M. J., & Willett, W. C. (2002). Mercury and the risk of coronary heart disease in men. *The New England Journal of Medicine*, *347*(22), 1755–1760. doi:10.1056/NEJMoa021437 PMID:12456851

Young, M., Van Der Merwe, C., Linde, S., & Du Plessis, J. (2018). A retrospective analysis of nickel exposure data at a South African base metal refinery. *Journal of Occupational and Environmental Hygiene*, *15*(3), 204–213. doi:10.1080/15459624.2017.1411596 PMID:29194021

Yudovich, Y., & Ketris, M. (2005). Arsenic in coal: A review. *International Journal of Coal Geology*, *61*(3-4), 141–196. doi:10.1016/j.coal.2004.09.003

Yu, H., Liao, W., & Chai, C. (2006). Arsenic carcinogenesis in the skin. *Journal of Biomedical Science*, *13*(5), 657–666. doi:10.100711373-006-9092-8 PMID:16807664

Yu, L., Kaiyi, S., Jie, Y., & Qiyu, K. (2021). Evaluation of heavy metal pollutants from plateau mines in wetland surface deposits. *Frontiers in Environmental Science*, *8*, 5557302. doi:10.3389/fenvs.2020.557302

Yu, L., Zhang, F., Zang, K., He, L., Wan, F., Liu, H., Zhang, X., & Shi, Z. (2021). Potential ecological risk assessment of heavy metals in cultivated land based on soil geochemical zoning: Yishui County, North China Case Study. *Water (Basel)*, *13*(23), 3322. doi:10.3390/w13233322

Yumoto, T., Tsukahara, K., Naito, H., Iida, A., & Nakao, A. (2017). A successfully treated case of criminal thallium poisoning. *Journal of Clinical and Diagnostic Research : JCDR*, *11*, od01–od02. doi:10.7860/JCDR/2017/24286.9494 PMID:28571191

Zaheer, I., Ali, S., Saleem, M., Ali, M., Riaz, M., Javed, S., Sehar, A., Abbas, Z., Rizwan, M., El-Sheikh, M. A., & Alyemeni, M. N. (2020). Interactive role of zinc and iron lysine on *Spinacia oleracea* L. growth, photosynthesis and antioxidant capacity irrigated with tannery wastewater. *Physiology and Molecular Biology of Plants*, *26*(12), 2435–2452. doi:10.100712298-020-00912-0 PMID:33424157

Zahra, A., Hashmi, Z., Malik, N., & Ahmed, Z. (2014). Enrichment and geo-accumulation of heavy metals and risk assessment of sediments of the Kurang Nallah-Feeding tributary of the Rawal Lake Reservoir, Pakistan. *The Science of the Total Environment*, *470–471*, 925–933. doi:10.1016/j.scitotenv.2013.10.017 PMID:24239813

Zakoldaev, D., Shukalov, A., Zharinov, I., & Zharinov, O. (2019). Modernization stages of the industry 3.0 company and projection route for the industry 4.0 virtual factory. *IOP Conference Series. Materials Science and Engineering*, *537*(3), 032005. doi:10.1088/1757-899X/537/3/032005

Zambelli, B., Uversky, N., & Ciurli, S. (2016). Nickel impact on human health: An intrinsic disorder perspective. *Biochimica et Biophysica Acta (BBA)-. Biochimica et Biophysica Acta. Proteins and Proteomics*, *1864*(12), 1714–1731. doi:10.1016/j.bbapap.2016.09.008 PMID:27645710

Zare, B., Nami, M., & Shahverdi, A. (2017). Tracing tellurium and its nanostructures in biology. *Biological Trace Element Research*, *180*(2), 171–181. doi:10.100712011-017-1006-2 PMID:28378115

Zeiner, M., Rezic, I., & Steffan, I. (2007). Analytical methods for the determination of heavy metals in the textile industry. *Journal of Chemistry and Chemical Engineering*, *56*(11), 587–595.

Zeljka, Z., Branka, G., Aleksandra, P., Vlatka, J., Lola, G., & Nada, M. (2018). Comparison of two different CEC determination methods regarding the soil properties. *ACS. Agriculturae Conspectus Scientificus*, *84*(2), 151–158.

Zhai, X., Li, Z., Huang, B., Luo, N., Huang, M., Zhang, Q., & Zeng, G. (2018). Remediation of multiple heavy metal-contaminated soil through the combination of soil washing and in situ immobilization. *The Science of the Total Environment*, *635*, 92–99. doi:10.1016/j.scitotenv.2018.04.119 PMID:29660731

Zhang, J. (2019). Discussion on how to repair the soil environment of polluted sites. *China Resources Comprehensive Utilization*, *37*(2), 140–142. doi:10.1016/j.jcou.2018.12.002

Zhang, J., Gao, Y., Yang, N., Dai, E., Yang, M., Wang, Z., & Geng, Y. (2021). Ecological risk and source analysis of soil heavy metals pollution in the river irrigation area from Baoji, China. *PLoS One*, *16*(8), e0253294. doi:10.1371/journal.pone.0253294 PMID:34339446

Zhang, Q., Hu, J., Lee, D., Chang, Y., & Lee, Y. (2017). Sludge treatment: Current research trends. *Bioresource Technology*, *243*, 1159–1172. doi:10.1016/j.biortech.2017.07.070 PMID:28764130

Zhang, S., Song, J., Cheng, Y., & McBride, M. (2018). Derivation of regional risk screening values and intervention values for cadmium-contaminated agricultural land in the Guizhou plateau. *Land Degradation & Development*, *29*(8), 2366–2377. doi:10.1002/ldr.3034

Zhang, Y., Chen, M., Zhong, Y., Zhang, M., Cheng, M., & Li, X. (2015). Assessment of cadmium (Cd) concentration in arable soil in China. *Environmental Science and Pollution Research International*, *22*(7), 4932–4941. doi:10.100711356-014-3892-6 PMID:25483971

Zhao, H., Wu, Y., Lan, X., Yang, Y., Wu, X., & Du, L. (2022). Comprehensive assessment of harmful heavy metals in contaminated soil in order to score pollution level. *Scientific Reports*, *12*(1), 3552. doi:10.103841598-022-07602-9 PMID:35241759

Zhao, Y., Deng, Q., Lin, Q., Zeng, C., & Zhong, C. (2020). Cadmium source identification in soils and higher risk regions predicted by geographical detector method. *Environmental Pollution*, *263*, 114338. doi:10.1016/j.envpol.2020.114338 PMID:32304950

Zheng, W., Aschner, M., & Ghersi-Egea, J. (2003). Brain barrier systems: A new frontier in metal neurotoxicological. *Toxicology and Applied Pharmacology*, *192*(1), 1–11. doi:10.1016/S0041-008X(03)00251-5 PMID:14554098

Zhen, Z., Wang, S., Luo, S., Ren, L., Liang, Y., Yang, R., Li, Y., Zhang, Y., Deng, S., Zou, L., Lin, Z., & Zhang, D. (2019). Significant impacts of both total amount and availability of heavy metals on the functions and assembly of soil microbial communities in different land use patterns. *Frontiers in Microbiology*, *10*, 2293. doi:10.3389/fmicb.2019.02293 PMID:31636621

Zhong, T., Chen, D., & Zhang, X. (2016). Identification of potential sources of mercury (Hg) in farmlands soils using a decision tree method in China. *International Journal of Environmental Research and Public Health*, *13*(11), 1111. doi:10.3390/ijerph13111111 PMID:27834884

Zhou, C., Huang, C., Wang, J., Huang, H., Li, J., Xie, Q., Liu, Y., Zhu, J., Li, Y., Zhang, D., Zhu, Q., & Huang, C. (2017). LncRNA MEG3 downregulation mediated by DNMT3b contributes to nickel malignant transformation of human bronchial epithelial cells via modulating PHLPP1 transcription and HIF-1a translation. *Oncogene*, *36*(27), 3878–3889. doi:10.1038/onc.2017.14 PMID:28263966

Zhou, L., Chen, N., Chen, Z., & Xing, C. (2016). ROSCC: An efficient remote sensing observation-sharing method based on cloud computing for soil moisture mapping in precision agriculture. *IEEE Journal of Selected Topics in Applied Earth Observations and Remote Sensing*, *9*(12), 5588–5598. doi:10.1109/JSTARS.2016.2574810

Zhu, C., Tian, H., Cheng, K., Liu, K., Wang, K., Hua, S. et al. (2016). Potentials of whole process control of heavy metals emissions from coal-fired power plants in China. *Journal of Cleaner Production, 114*, 343–351. https://doi.org/. 05.008 doi:10.1016/j.jclepro.2015

Zwolak, A., Sarzynska, M., Szpyrka, E., & Stawarczyk, K. (2019). Sources of soil pollution by heavy metals and their accumulation in vegetables: A review. *Water, Air, and Soil Pollution*, *230*(7), 164. doi:10.100711270-019-4221-y

About the Authors

Joan Nyika is a Postdoctoral Researcher at the department of Civil Engineering Science of the University of Johannesburg. Her research is in the hydrobiogeochemistry field with a focus on tracking soil and water pollutants in different environs, understanding their causes, migration and effects in an era when pollution is a growing environmental concern globally. Dr. Nyika received her doctorate (in Science, Engineering and Technology) and master's (in Land and Water Management) degrees from the University of South Africa and University of Nairobi, Kenya in 2021 and 2017, respectively. She has been involved in various research activities where she has authored and co-authored more than 50 publications (article journals and book chapters) in internationally refereed journals. She has reviewed articles for several journals in her area of specialization. Additionally, she serves as a Lecturer at the Technical University of Kenya where she is involved in teaching, mentoring and guiding undergraduate students. Her area of expertise is mainly on land and water resources management. Some of the modules she has been and is teaching include Sustainable Water Management, Watershed Management, Water Quality Monitoring, Soil Science, Soil Management and Pollution Management.

Megersa Olumana Dinka is a graduate with PhD from University of Natural Resources and Applied Life Science (Vienna) in 2010. He also did postdoctoral research from 2012 to 2014 at Tshwane University of Technology, South Africa. He has expert knowledge in water resource engineering discipline specific to hydrology, hydraulics and water management aspects. His research interests are in hydrologic modelling, remote sensing and GIS irrigation, water treatment, climate change watershed (natural resources) management and irrigation engineering. He has about 15 years of experience as academician and 19 years of experience as a researcher. He has taught various courses and modules at undergraduate and postgraduate levels successfully. Most of the courses he taught are related to hydrology, hydraulics and watershed (water) management aspects. Currently he is teaching Hydrology, Hydraulics and Water Treatment Technology Modules at University of Johannesburg. He is also serving as the chair of the Department of Civil Engineering Science. Moreover, he also supervised a number of postgraduate students (26 MSc and 3 PhD) successfully and still supervising a number of undergraduate and postgraduate students at University of Johannesburg. He has published a number of peer reviewed scientific articles and book chapters in accredited international journals and book publishers. His research areas of focus include land-use and land-cover change detection, land and water resources management, spatial mapping in GIS, Irrigation performance assessment. He has reviewed a number of articles for various international journals.

Index

T

V

W

Z

www.ingramcontent.com/pod-product-compliance
Lightning Source LLC
LaVergne TN
LVHW080629011026
849160LV00014B/1160
* 9 7 8 1 6 6 8 4 7 1 1 7 3 *